Multifunctional Adaptive Microwave Circuits and Systems

Multifunctional Adaptive Microwave Circuits and Systems

Edited by

Michael B. Steer

North Carolina State University

and

W. Devereux Palmer

U.S. Army Research Office, Research Triangle Park, North Carolina

SciTech PUBLISHING

SciTech Publishing
Edison, NJ
www.scitechpub.com

SciTech Publishing
911 Paverstone Drive, Suite B
Raleigh, NC 27615
919) 847-2434, fax (919) 847-2568
scitechpublishing.com.

Published by SciTech Publishing, Edison, NJ

Editor: Dudley R. Kay
Production Director: Susan Manning
Production Coordinator: Robert Lawless
Cover Design: Kathy Gagne
Printed: Cadmus Communications

ISBN: 9781891121777

Library of Congress Cataloging-in-Publication Data:

Steer, Michael B.
Multifunctional adaptive microwave circuits and systems/ edited by Michael B. Steer and W. Devereux Palmer.
p. cm.
ISBN 978-1-891121-77-7 (hbk.: alk. paper) 1. Microwave devices. 2. Microwave circuits. 3. Radio circuits. I. Palmer, W.D. (William Devereux), 1957- II. Title.
TK7876.S735 2008
621.381'32--dc22

Printed in the United States of America

Michael dedicates this book to his oldest son
Cormac

and Devereux dedicates this book to his wife
Colleen

Contents

Preface

Adaptive, tunable and reconfigurable radio frequency (RF) circuits have the ability to adjust to new communication requirements and also to realize multi-band and broadband radar, radio and sensor systems using greatly reduced hardware. These have become known as MARRS systems for Multi-functional, Adaptive Radio Radar and Sensor systems. MARRS is also the name of a MURI (Multidisciplinary University Research Initiative) funded by the U.S. Department of Defense through the U.S. Army Research Office. MARRS systems can be adapted to new communication protocols and new uses.

In the most general terms a MARRS system is one that can receive and demodulate any electromagnetic signal from any direction, return a modulated signal in any specific direction, be reconfigurable to implement a variety of radio and sensor functions, and adapt to its local (ambient) environment to improve signal reception, reject interference, and compensate for multipath effects. While many commercial products are multi-featured, particularly modern cellular phones that may include wireless communications, digital music playback, digital cameras, and GPS navigation, each feature typically requires a separate hardware subsystem, and so the system is not truly multifunctional.

Perhaps the best commercial example of a true multifunctional system (albeit not an RF system) is the combination printer, fax, scanner, copier, and telephone commonly known as the multifunction printer. These devices consist of several functional blocks including scanner, print and fax engines, and telephone and data interfaces, all of which have both software and hardware components managed by a central controller. Different combinations of these building blocks are used depending on the desired function, see Table 1. Note that the functions cannot be performed simultaneously. The lack of simultaneity at worst will prevent the user from completing all desired tasks within the desired time frame. For multifunctional RF systems, particularly military systems, lack of simultaneity can create more serious problems.

The academic, government, and industrial research communities have been pursuing development of the technology required to realize multifunctional systems [1]. RF research in the commercial sector is focused on enabling development of multi-market, multi-carrier cellular phones, voice radios for first responders with inter-operability between systems used for fire, police, and emergency medical services, and automotive telematics, navigation, and collision avoidance radar [2–5]. For military systems, the goal is the capability of performing simultaneous communications, radar, and electronic warfare functions from the same, compact, hardware system. This goal has been pursued by the U.S. Department of Defense in a number of programs of record including the Defense Advanced Research Projects Agency (DARPA) Intelligent RF Front Ends (IRFFE) program [6], the Navy Advanced Multifunction RF Concept (AMRFC) [7, 8], the Air Force Sensorcraft concept [9, 10], and the Army Multifunction RF (MFRF) Advanced Technology Objective [11, 12]. In all cases, engineering challenges exist in component technologies (wide instantaneous bandwidth or tunability, wide dynamic range, high linearity, high efficiency), architectural approaches (optimal partitioning of analog and digital circuitry, optimization of multiple incompatible objectives, holistic system design), and concept of operations (time slicing between functions, prioritization). All of these challenges can be met with MARRS components and systems. Some background and examples follow.

Consumer electronics, including cellular phones, have relatively short product lifetimes and each model is manufactured in large volumes with cost as a major driver. Taking cellular phones

Table 1 Functions and components of a modern multifunctional printer

Function component	Print	Fax cend	Fax receive	Scan scan	Copy copy	Voice call
Engine		X	X			
Print Engine	X		X		X	
Scan Engine		X		X	X	
Phone Interface		X	X			X
Data Interface	X			X		

as an example, these units commonly support multiple communication protocols and sensors. A modern cellular phone handset typically will support the GSM standard around 900 MHz, an enhanced high-data capacity of GSM/Edge, around 1800 MHz, WCDMA at both frequencies, Bluetooth at 2.54 GHz, Wi-Fi at 5.6 GHz, as well as the GPS satellite navigation service at 1575.42 MHz (L1) and 1227.60 MHz (L2). Handsets targeted for less developed parts of the world also must support analog radio around 800 MHz to ensure connectivity. All of these protocols will operate using a common antenna or antennas but each has its own RF electronic subsystem including power amplifier and low noise receiver. Each subsystem is connected to the antenna(s) through a semiconductor or MEMS (micro electro-mechanical system) switch. The antenna itself is designed to be a good antenna at each of the operating frequencies, possibly using different feed points for the different frequency bands. Currently the use of different electronics for each application is the preferred approach, as each set of RF electronics can be optimized for the particular modulation format and operating power and each RF electronic subsystem can be obtained from the most competent or lowest cost provider. However, MARRS components are seeing increasing use in individual subsystems for applications such as tunable matching networks and in tuning antennas for optimum performance against a variety of situations such as different operating frequencies and different loading on the antenna.

The lifetime of a cellular phone handset varies from nine months to three years in different parts of the world. The turnover is driven by advances in features and new communication protocols, and the desire by consumers to have the new features. Producing a complex device with the ability to reconfigure and adapt to accommodate new and perhaps unforeseen features and communications protocols is impractical—better to manufacture and sell a new radio. Consumer RF electronics are produced in such large quantities, a hundred million for a particular model is not uncommon, that design of a fixed, integrated solution allows simultaneous optimization of both cost and size. The situation is quite different for non-consumer electronics, both commercial and military. Commercial non-consumer RF radio, radar and sensor systems include point-to-point microwave communications links, radio and television broadcast stations, and radar for commercial aviation. All military systems, even those manufactured in relatively large numbers supporting dismounted infantry, can be considered non-consumer. With these systems production volumes are relatively small, nothing like the millions of consumer electronics, they have long expected lives with thirty years being typical, and ideally can be reconfigured for multiple functions. This is particularly true of military radio, radar and sensor systems where the desired frequency coverage can be very broad, and missions evolve to address emerging threats. An example of the requirement for broad frequency coverage and evolving mission can be found in the operational requirements for the Joint Tactical Radio System, where future radios will be required to cover frequencies from 3 MHz to 50 GHz and provide inter-operability with over 30 legacy, civilian, and developmental waveforms and communications protocols [13].

In all current broadband systems filtering dominates size and cost. Fixed filters have excellent performance in terms of low insertion loss and frequency selectivity. A filter bandwidth of 100 to 500 MHz generally has sufficient frequency selectivity to manage dynamic range in most situations. But covering a bandwidth of 20 GHz with a bank of filters 500 MHz wide would require 40 filters even with no overlap. With overlap, which typically is needed for complete frequency coverage, nearly 80 filters would be required. Each of these filters requires manual tuning and this is most often the dominant cost in a system. Not surprisingly few such systems exist and are mostly very expensive electronic warfare systems that must monitor a large part of the spectrum. For long-lived systems changes in the regulatory environment often require that operating power levels and operating frequency bands change accordingly. Modifications in system performance also are necessary to address changes in the external electromagnetic environment including unintentional interference and, in the case of military systems, electronic attack. Currently the most common component used for this application is a tunable notch filter, which is difficult to design for high performance over a wide frequency range.

Some have suggested software-defined radio as a potential solution to the problem [14]. The ultimate software-defined radio is envisioned as an antenna connected to an analog to digital converter for the receive function and a digital to analog converter for the transmit function, all controlled by a baseband processor that can be reprogrammed for different applications. However this vision is viable neither now nor in the near future, because of the limited dynamic range, or number of bits, of available converters and the slow growth rate of the number of bits over time. What is viable now is the software-defined radio where software and algorithms control analog RF electronics. Recent advances in silicon RF electronics have led to the ability to fabricate analog, digital, and RF components on the same substrate. Advances in heterogeneous integration have led to the ability to mount MEMS and tunable electronic components on semiconductor chips. It is not technically or economically unreasonable to consider using one million or even 100 million silicon transistors to control and optimize a single RF circuit block. To reduce this vision to practice, MARRS components are essential.

The design of fixed RF front ends proceeds with considerable intuition, extensive computational modeling and simulation, and iterative design, build, test cycles. The design cycle is compressed by using optimizers available in most microwave computer-aided design tools. Top level system design decisions that determine what subsystems and components will be used and within what architecture, are much less amenable to optimization as many of the constraints result in discrete choices and so are harder to implement as a cost-function based assessment. Still, designing a relatively narrow band fixed technology is much easier than designing a reconfigurable and tunable broadband system. Designing a broadband system with fixed attributes over temperature, for various modulation formats, and for different power levels, would result in a system of lower efficiency and performance than a comparable set of discrete systems. For example, a fixed design that is required to operate over the entire 120°C of the mil-spec temperature range will by necessity require engineering tradeoffs resulting in reduced performance over parts of the range. A MARRS system, on the other hand, may be able to self-adjust over a range of 10°C or so, and reconfigure itself appropriately if the temperature changes beyond that. So in a MARRS system the controlling algorithms must know how to do design and must be aware of the behavioral models of the controlled circuitry in order to be able to design a system for the particular application being used at the time, and then rapidly switch to another system requirement. A systematic approach to the system level design required is presented in Chapter 1 on RF System Design.

Chapters 2 and 4 describe developments of the most viable tunable and reconfigurable components. These are Micro Electro-Mechanical System (MEMS) switches and varactors, described in Chapter 2. Chapter 4 describes electrically-tunable varactors made from ferroelectric materials of which Barium Strontium Titanate (BST) is the most common. Chapter 3 presents applications in tunable switches, tunable filters, tunable matching networks, and tunable antennas. These tunable circuits can be implemented using the MEMS and tunable ferroelectric components described in Chapters 2 and 4. Chapter 5 describes concepts and architectures for RF transmitters designed to optimize performance by adapting to variable loads, power levels, and linearity requirements. Chapter 7 describes reconfigurable and tunable RF filters using both semiconductor and ferroelectric varactors. One of the significant differences found with such tunable filters is that the RF voltage across the varactors must be controlled in order to minimize nonlinear effects. This imposes limits on the quality factor, Q, of the resonators. Thus the filter design must trade dynamic range for frequency selectivity, and the high selectivity required in most radio, radar and sensor systems is pushed from the limited frequency selectivity of the hardware filter to the baseband digital signal processing.

Chapter 8 describes retro-directive antenna arrays, which are antenna systems that automatically sense the direction of arrival of the received signal and re-transmit a signal along the same path without the need for banks of phase shifters or other complex beamforming electronics. If the received signal is subject to significant multipath propagation, as would be the case in an urban setting, the returned signal follows the same paths. Significant advantages of such an antenna include reduced transmit power requirements due to the high directivity, and improved operation in a cluttered environment. Chapter 9, the final chapter, describes a multifunction radar, communication and radiometry system that utilizes a retro-directive antenna array.

This book summarizes the findings of a five year project entitled MARRS: Multifunctional Adaptive Radio Radar and Sensors that ran from 2001 to early 2007. This work was funded by the United States Army Research Office under the Multidisciplinary University Research Initiative or MURI program. The work continues with the authors of the chapters still actively involved in the topics described and many other researches at universities are working on the topics. Nearly all of the concepts are now being used in industry and put into practice by engineers. The work has been a catalyst for what is possible and we hope that the book will continue to inspire.

Appreciation is extended to the many authors and former graduate students who contributed to the advances described in this book. Much of this book was organized and rendered consistent by Ms. Claire Sideri of North Carolina State University. Her devotion to getting the book completed are very mush appreciated.

Michael B. Steer and W. Devereux Palmer

June 2008

1

RF System Design

Michael B. Steer, Kevin G. Gard and Alan M. Victor

1.1 INTRODUCTION

System design is a creative and iterative activity combining choice of architecture, experience with the performance of subsystems, and an understanding of the performance that can be achieved from technology evolution. It is far more difficult to develop a new system design using new technologies for which experience has not been gained. Electronic system design is also driven by relating and recognizing the impact of technology and component capabilities on system performance. This hierarchical and cross-hierarchy feedback is captured in Figure 1-1.

One of the outcomes of system design is the transformation of system performance metrics to generally indirect specifications of individual subsystems. For example, the ultimate performance of a communication radar or sensor system can often be expressed as a bit error rate (BER) but the performance of an amplifier subsystem is usually specified in terms of noise figure and two-tone intermodulation performance. It can be shown that two-tone intermodulation performance and BER are roughly related if the actual signal statistics are taken into account and the interfering signals are assumed to be Gaussian [15–28]. However, the precise relationship depends on the details of the nonlinear characteristics of the subsystem and this is developed through experience. System-level simulation of BER from nonlinear characteristics is in its infancy and so again experience is a core requirement. With new technologies providing adaptive multifunctionality it is necessary to develop a greater understanding of distortion processes and their impact on system performance.

Optimization of system performance is overly constrained as not every subsystem can be freely optimized. Many times there are only a discrete number of subsystem choices (e.g., different filter technologies) and the cost of choosing an alternative subsystem may have deep ramifications—so experience is the best guide to subsystem selection. Some components can be continuously tunable providing much needed flexibility in system optimization. An example is a variable-gain amplifier for which the gain can be chosen over a broad range by choosing a terminal voltage.

1.2 RF SYSTEM CASCADE

An RF system is nearly always arranged as a cascade of subsystem components. This generally maximizes operating frequency and efficiency as feedback limits bandwidth and increases DC power consumption. Figure 1-2 shows an RF receiver system as a cascade of subsystems. This receiver architecture uses one stage of mixing or heterodyning. Signals are captured by the antenna where the signal quality is described by the signal-to-interference ratio (SIR) within the final

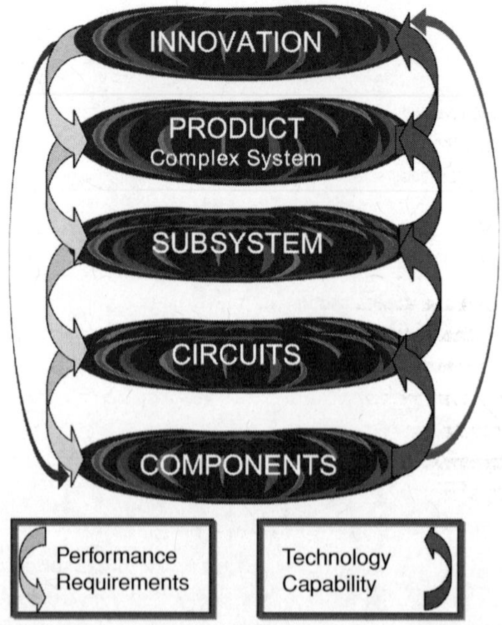

Figure 1-1 The electronic system design process is interactively driven by performance requirements coming down the hierarchy and capabilities moving up. The process results in innovative products and engineering solutions.

bandwidth specified by the receiver chain. That is, the bandwidth of the interference is determined by the bandwidth of the signal presented to the analog-to-digital converter (ADC). The converter signal presented to the ADC must have a specified SIR so that digital signal processing (DSP) can provide sufficient frequency selectivity (or filtering) in addition to that provided by the bandpass filters (BPFs). In addition, the ADC must have sufficient dynamic range so that the signal presented to the ADC can be resolved in the presence of interference. The goal of the RF hardware (to the left of the ADC), often called the RF front-end, is to provide the required spurious-free dynamic range (SFDR) which is a measure of both noise and of interference introduced by the subsystems. Higher SFDR than required generally comes at the price of higher DC power requirements. In the RF hardware the mixer is an intentional nonlinear element that provides frequency translation, or down-conversion, of the information-bearing modulation on the received RF signal. The signal that is down-converted is called the IF (for intermediate frequency). Usually this frequency is kept

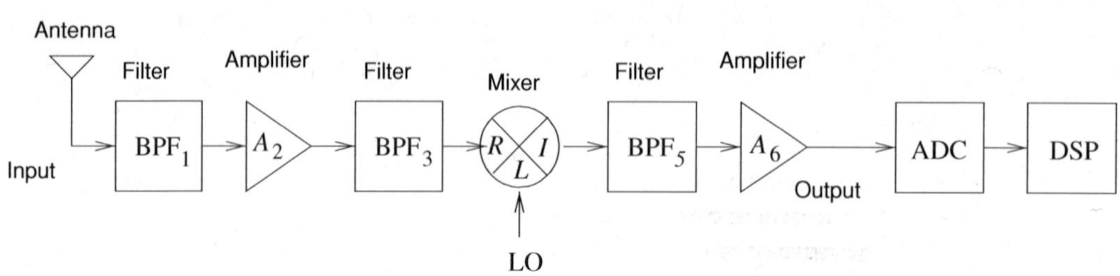

Figure 1-2 A receiver cascade system beginning with an antenna to receive a signal with interference and final processing in a digital signal processor (DSP).

as high as the ADC can handle as the IF filter, here just BPF_5, size is then kept small. The size of filters tends to dominate the size of the RF system. The impact on size also applies to the RF filters, here just BPF_1. The first RF filter can be quite large especially if the input BPF filter is channelized so that a number of discrete channels are separately presented. In an adaptive front-end this filter would be tunable so that the size of the filter is reduced compared to a fixed channelized filter. In many radio, radar and sensor systems the number of bits and bandwidth of the ADCs is at the point where the Q of the RF filter can be considerably relaxed. This has enabled adaptive and tunable filters with Qs of 100 or less to be used as the required additional frequency selectivity can be achieved in the DSP. The important point here is that it is only now that the adaptive, tunable and reconfigurable technologies described in this book are feasible. That is, advances in digital technology, and in particular the number of bits and sample rate of ADC's as well the speed of digital signal processors, has enabled relaxation of the performance parameters of the RF front-end. Conventionally RF filters have provided primarily frequency selectivity, now one of the major functions of the FR filter is management of dynamic range by eliminating out of bound signals. Thus the function of frequency selectivity is partitioned between the RF hardware and DSP.

In the receiver system of Figure 1-2 the mixer is intentionally nonlinear and unfortunately this results in spurious interference tones compromising the dynamic range of the receiver. In addition, the amplifiers are nonlinear and also contribute distortion. Every component—the linear filters, antenna, and active components—contribute noise. Sections 1.3, 1.4 and 1.5 discuss distortion in RF systems and present a strategy to reduce distortion in a cascade system by using linear filters. In Section 1.6 SFDR is defined and in Section 1.7 an improved strategy, called the contribution method, is presented for designing a system for required SFDR. In an adaptive multifunctional system the RF cascade is reconfigured and tuned so it is as though the RF system is continuously being redesigned. Thus it is important to have a systematic approach to system design rather than the budget method that is traditionally used.

In circuit design behavioral models describe the terminal characteristics of a circuit and can be used in efficient determination of system performance and as an aid in design. It has also been realized that measurement-derived behavioral models can be more accurate than the original circuit-level models used in developing a circuit design. In part this is because of uncertainties in the device-level models but also because of modifications made after finalization of the design. Section 1.9 describes behavioral modeling concluding with a discussion of the relationship of discrete-tone intermodulation distortion and distortion with the digitally modulated signals used in radio and radar systems.

The work described here is the basis of an intelligent controller for adaptive RF systems. The system controller must have a model for the performance of an adaptive system so that systems can be reconfigured and tuned without human intervention. Generally it is not permissible to adapt to an unstable or high-distortion state as this would result in loss of signal and transmission of spurious tones. The contribution approach presented in this chapter is part of the knowledge that must be embedded in the system controller.

1.3 CHARACTERISTICS OF NONLINEAR RF CIRCUITS

Nonlinear RF electronics have evolved to use constant voltage rails—often a single power rail and a ground rail, or positive and negative rails equally offset from ground. Between the rails one or more transistors are arranged such that power is transferred from DC power to an amplified form of the AC signal. A consequence of this design is that signal voltages in the circuit are generally confined to being between the voltages of the rails. As a result the signal has exponential-like compression at the extremes of its range. This can be seen with the FET amplifier of

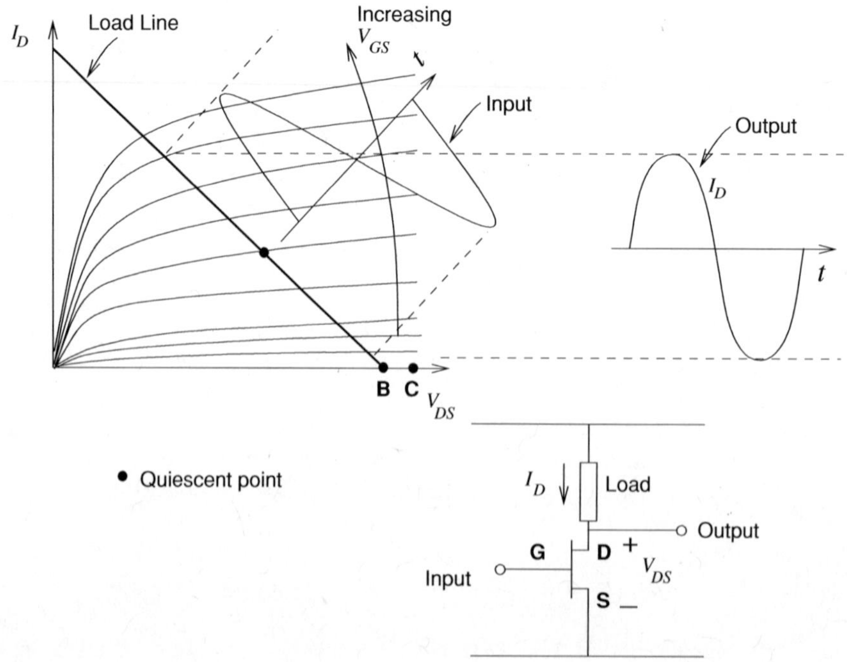

Figure 1-3 Current–voltage characteristic of a transistor used in an amplifier showing the quiescent point of a Class A amplifier.

Figure 1-3[1]. When the gate-source voltage V_{GS} is small, the transistor operates in the subthreshold region and the drain-source current I_{DS} is exponentially dependent on V_{GS}. At low V_{DS} there is a similar exponential dependence of I_{DS} on V_{DS}. The exponential dependence at the extremes occurs with all transistor types as it is an essential function of semiconductors wherein the available charges for conduction are exponentially dependent on field strength (just as they are on temperature). At the extremes the transistor reduces to a model consisting of two offset back-to-back diodes. Away from the extremes I_D is a scaled version of the input waveform. Mathematically this can modeled as offset back-to-back exponentials which is described mathematically by a tanh function relationship between the input and output quantities:

$$y = \tanh(x) = \frac{e^x - e^{-x}}{e^x + e^{-x}} \tag{1.1}$$

where the input is x and the output is y. This function is shown in Figure 1-4. The tanh function is the core of many RF behavioral models as the asymptotic response, here at large negative x and large positive x, is captured. Such an asymptotic response can only be approximated by a power series expansion of a device's input–output characteristic.

In actuality, and referring to Figure 1-5, compression of the output waveform at the extremes is modeled better by providing functional dependencies on the input voltage as follows:

$$y(t) = F_1(x(t)) \tanh\left[F_2\left(x(t)\right)\right]. \tag{1.2}$$

Commonly F_1 and F_2 are third-order power series.

1 Sometimes an inductor is used at the load allowing the drain voltage (at D) to reach twice the voltage rail.

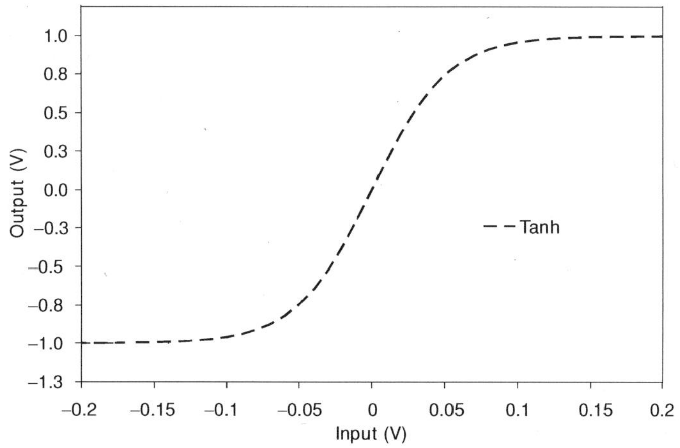

Figure 1-4 Tanh function response typical of analog electronic circuits.

1.3.1 Modeling an Amplifier Subsystem

One of the simplest amplifier subsystems is that of Figure 1-3. Here a GaAs MESFET is used as the FET. This transistor has the model of Figure 1-6 where the elements C_{GS}, C_{DS}, and C_{GD} are linear and the diodes and transconductance are nonlinear. The drain-source current $I_{DS}(t)$ is modeled as [29, 30]

$$I_{DS}(t) = i_{DS}(t) + i_{RDS}(t) \tag{1.3}$$

where

$$i_{DS}(t) = \left(A_0 + A_1 v_1(t) + A_2 v_1^2(t) + A_3 v_1^3(t)\right) \tanh\left(\gamma v_{DS}(t)\right) \tag{1.4}$$

is in the asymptotic form described by Equation (1.2). The other parameters are

$$i_{RDS}(t) = \frac{v_{DS}(t) - V_{DS}^{DC}}{R_{DS}} \tag{1.5}$$

and

$$v_1(t) = v_{GS}(t - \tau(t))\left[1 + \beta\left(V_{DS}^0 - v_{DS}(t)\right)\right], \quad \tau(t) = A_5 v_{DS}(t). \tag{1.6}$$

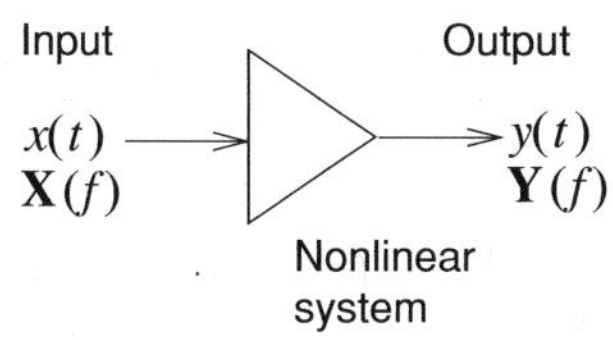

Figure 1-5 A unilateral bandpass nonlinear system.

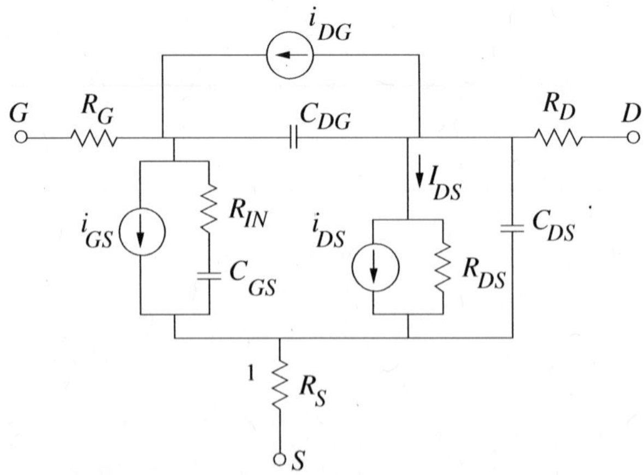

Figure 1-6 Equivalent circuit of the Curtice MESFET model [29, 30].

The forward-biased gate current $i_{GS}(t)$ and the drain-gate avalanche current $i_{DG}(t)$ are modeled by the Shockley diode equation

$$i_{GS}(t) = I_0\left(e^{v_{GS}(t)/V_T} - 1\right) \tag{1.7}$$

and the exponential function

$$i_{DG}(t) = I_0\left(e^{(v_{DG}(t)-V_R)/V_T} + 1\right) \tag{1.8}$$

where V_R is the drain-gate reverse breakdown voltage. The values of the linear components and the coefficients in Equations (1.4)–(1.8) are from References [31, 32] and are listed in Table 1-1.

The simulation results for both single-tone and two-tone inputs of the MESFET amplifier are compared with the experimental data in Figures 1-7 and 1-8 for single-tone and two-tone excitations.

Table 1-1 The element and parameter values of the MESFET equivalent circuit in Figure 1-6.

Element	Value	Parameter	Value
C_{GS}	0.52785 pF	A_0	0.016542
C_{DS}	0.25137 pF	A_1	0.0500214
C_{DG}	0.087 pF	A_2	0.02012
R_G	2.9 Ω	A_3	−0.00806592
R_{IN}	10 Ω	γ	2.16505
R_S	2.4 Ω	A_5	1.0×10^{-12}
R_D	5.3 Ω	β	−0.0394707
		V_{DS}^0	1.0 V
		R_{DS}	218.5 Ω
		I_0	1.0×10^{-9} A
		V_R	15 V

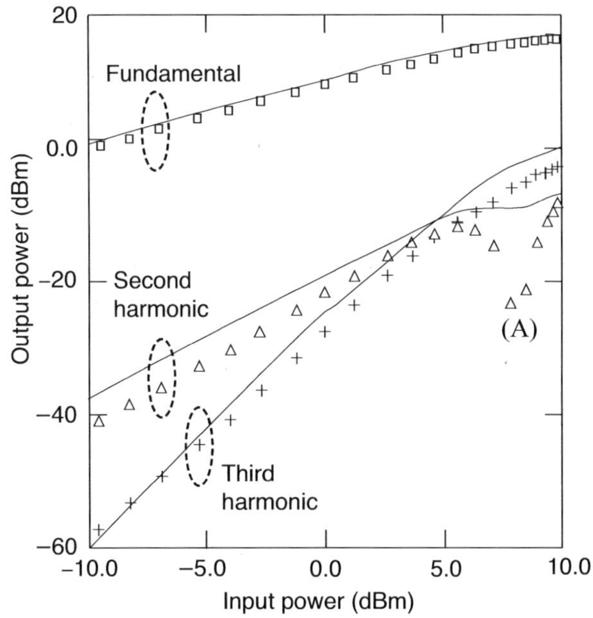

Figure 1-7 The results of a single-tone test at 2.3 GHz. Shown are the measured values (points) and simulated results (lines) [31].

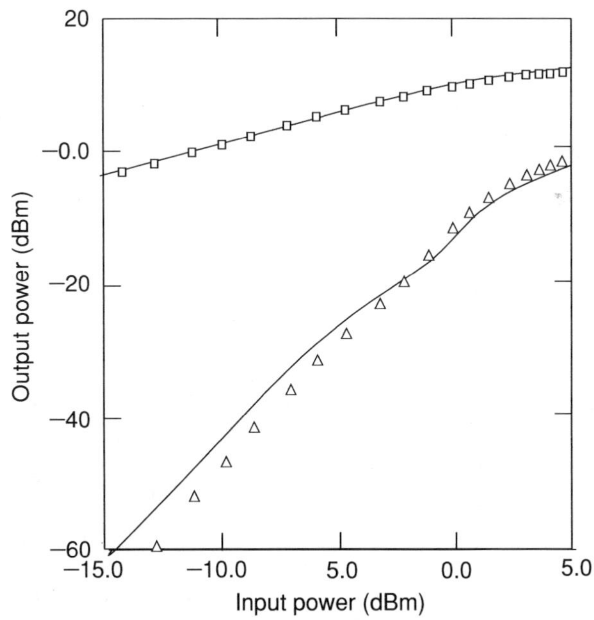

Figure 1-8 The results of the two-tone test. Shown are the measured values (points) and the simulated results as a function of input power for the power output at 2.35 GHz, (top curve) and the third-order intermodulation product at 2.3 GHz (bottom curve) [31].

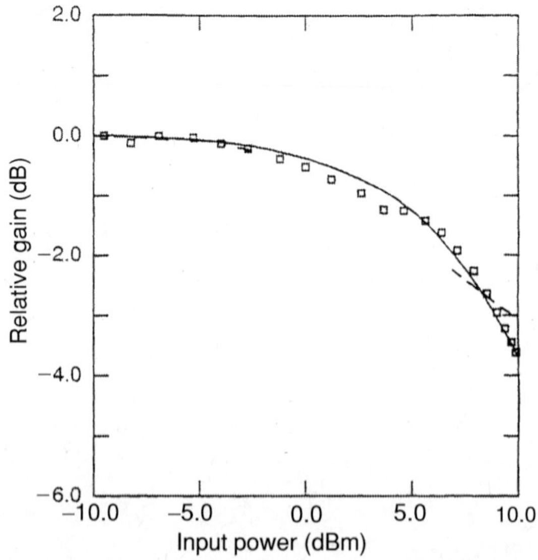

Figure 1-9 The relative gain of the common source amplifier as a function of input power. The curve is the simulated result and the points are measured values. The transistor is biased at $V_{GS} = -0.1$ V and $V_{DS} = 3.0$ V.

For single-tone excitation, a 3 GHz input signal was used, and, for two-tone excitation, two equal-amplitude signals input at 2.35 GHz and 2.4 GHz are used. The output power at the fundamental and the second and third harmonics as a function of input power are shown in Figure 1-7 for a single-tone test, and the output power at 2.35 GHz and the IP3 product at 2.3 GHz are shown in Figure 1-8 for a two-tone test. Greater detail of the saturated single-tone response is given in Figure 1-9. In the next section the nonlinear processes that lead to these characteristics are explored with the aim of developing an intuitive understanding of nonlinear circuit effects.

1.4 NONLINEAR SYSTEM RESPONSE TO SINGLE-FREQUENCY EXCITATION

With one-tone excitation (i.e., a single sine wave), the spectra of the input and output of the nonlinear system of Figure 1-5 are as shown in Figure 1-10. Here the system nonlinearity converts power at f_1 to power at DC (this is commonly referred to as rectification), and to power at the harmonics ($2f_1$, $3f_1$, ...), as well as to power at f_1. At least this is the conventional view of the nonlinear response to a single-frequency sinusoidal excitation. However, if the nonlinearity is reactive, subharmonics can also be generated.

The response of a GaAs amplifier to a one-tone input signal is shown in Figure 1-7 where both measured and simulated responses are plotted. As the input at f_1 is increased, the fundamental response (also at f_1) increases linearly but eventually compresses and reaches saturation. Second and third harmonics are also produced with the simulated responses having 2:1 and 3:1 slopes respectively. Rarely are these exact response slopes observed experimentally but generally the slopes are very close. The dip in the second-harmonic response at high powers is commonly observed and is the result of different mixing products destructively interfering. This process is best explained in terms of intermodulation response in the next section.

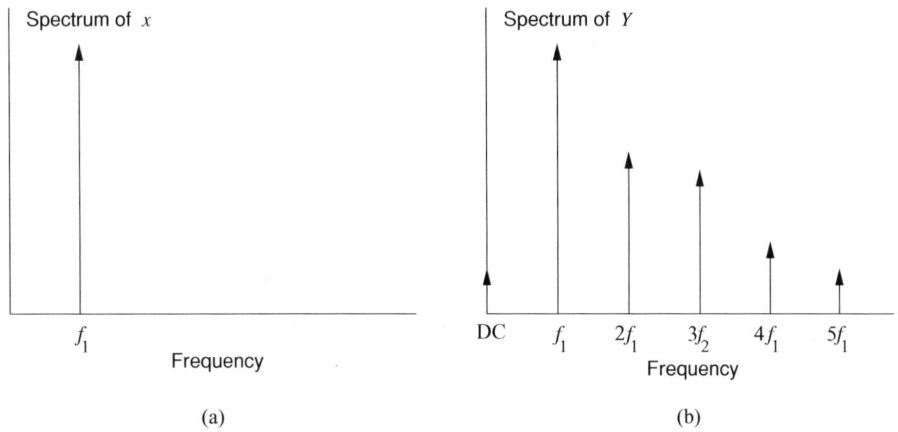

Figure 1-10 Spectra in a general nonlinearity with one tone excitation: (a) spectrum of the input x; and (b) spectrum of the output y.

1.5 NONLINEAR SYSTEM RESPONSE TO MULTIFREQUENCY EXCITATION

The nonlinear response to single-frequency sine wave excitation of a nonlinear circuit is simple to describe. It is much more complicated to describe the nonlinear response to multifrequency sinusoidal excitation. If the excitation of an analog circuit is sinusoidal or narrowband then specifications of distortion are generally in terms of frequency-domain phenomena, e.g., intermodulation levels, gain, and the 1-dB-gain compression point. In the time domain, the nonlinear behavior is evident as saturation or clipping of a waveform so that a sinusoidal waveform becomes a perturbed periodic waveform. However, with multifrequency excitation by signals that are not harmonically related, the waveforms in the circuit are not periodic.

Consider the nonlinear response of the system of Figure 1-5 to the two-tone excitation shown in Figure 1-11(a). The frequencies f_1 and f_2 are, in general, nonharmonically related and components at all sum and difference frequencies ($mf_1 + nf_2$, $m, n = -\infty, \ldots, -1, 0, 1, \ldots, \infty$) of f_1 and f_2 will appear at the output of the system. If the nonlinear system has a quadratic nonlinearity then the spectrum of the output of the system is that of Figure 1-11(b). With a general nonlinearity, so that k can be large, the spectrum of the output will contain a very large number of components. An approximate output spectrum is given in Figure 1-11(c). Also shown is a truncated spectrum which will be used in the following discussion. Most of the frequency components in the truncated spectrum of Figure 1-11(c) have been named: DC results from rectification; f_3, f_4, f_5, f_8, f_9, f_{10} are called intermodulation frequency components; f_4, f_5 are commonly called image frequencies as well; f_1, f_2 are the input frequencies; and f_6, f_7 are harmonics.

All of the frequencies in the steady-state output of the nonlinear system result from intermodulation—the process of frequency mixing. In other words, each frequency component is the summation of intermodulation products. However, it is usual to refer those frequencies additional to the input and output frequencies, and their harmonics, as intermodulation frequencies. It is unfortunate that the term intermodulation is used in two related but slightly different contexts. The terms "intermodulation frequency" or "intermodulation component" refer to the undesired frequencies generated in the mixing process as described here. The term "intermodulation product" (IP) refers to the entire nonlinear process of sinusoids (or even a single-frequency component) mixing to produce components at any frequency.

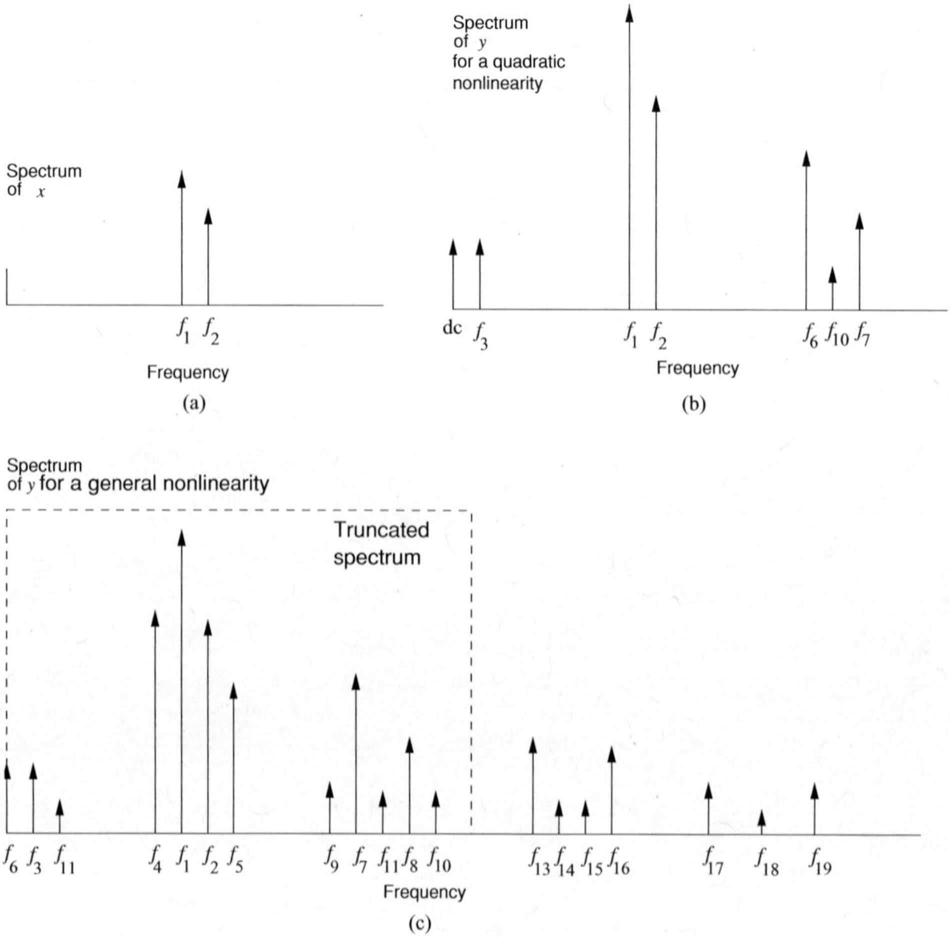

Figure 1-11 Spectra of the unilateral system of Figure 1-5: (a) spectrum of the input x; (b) complete spectrum of the output y for a quadratic nonlinearity; and (c) approximate spectrum of y for a general nonlinearity.

1.5.1 Intermodulation

Intermodulation is the generation of spurious frequency components at the sum and difference frequencies of the input frequencies. In the truncated spectrum f_3, f_4, f_5, f_8, f_9, f_{10}, and f_{11} are intermodulation frequencies. Focusing on just one of these, f_4, it is seen from Table 1-2 that there are a large number of low-order IPs contributing to this effect. The f_4 and f_5 tones are the most troublesome distortion tones as once generated they cannot be removed by filtering. In a nonlinear system the first-order response is usually large but this does not describe the origins of intermodulation. The lowest-order mixing that gives rise to intermodulation is a second-order process and it is seen that for the truncated spectrum there are four second-order IPs. These can destructively and constructively interfere so that the power of an intermodulation frequency component can vary erratically at high input powers as the phases of the IPs vary.

The measured and simulated response of a microwave amplifier excited by two tones is shown in Figure 1-8. Both the fundamental (top) and intermodulation response (bottom) increase with

Table 1-2 Descriptions of intermodulation products up to third order contributing to intermodulation at f_4.

Order	Output	Input
first order	f_4	f_4
second order	f_4	$f_1 - f_3$
second order	f_4	$-f_1 + f_8$
second order	f_4	$-f_2 + f_6$
second order	f_4	$-f_5 + f_{10}$
third order	f_4	$2f_1 - f_2$
third order	f_4	$f_1 + f_2 - f_5$
third order	f_4	$-f_1 - f_3 + f_6$
third order	f_4	$f_1 - f_6 + f_8$
third order	f_4	$f_1 + f_6 - f_{10}$
third order	f_4	$f_1 + f_7 - f_9$
third order	f_4	$f_1 - f_7 + f_{10}$
third order	f_4	$f_2 - 2f_3$
third order	f_4	$-f_2 + f_3 + f_8$
third order	f_4	$-f_2 - f_3 + f_{10}$
third order	f_4	$f_2 + f_6 - f_7$
third order	f_4	$f_2 + f_8 - f_{10}$
third order	f_4	$f_2 - f_9 + f_{10}$
third order	f_4	$-f_3 - f_4 + f_8$
third order	f_4	$f_3 - f_5 + f_6$
third order	f_4	$-f_3 - f_5 + f_7$
third order	f_4	$f_5 + f_6 - f_9$
third order	f_4	$f_5 - f_7 + f_8$

increasing input power with slopes of approximately 1:1 and 3:1. The 3:1 slope is the reason why this intermodulation product is called the third-order intermodulation (IP3). The simulated response does follow the ideal slopes exactly but the measured response do not[2].

1.5.2 Suppression of Intermodulation Distortion

The purpose of this subsection is to demonstrate that knowledge of nonlinear processes can be used to reduce distortion. It is shown how changes in architecture, here insertion of a bandpass filter, BPFs can be used to suppress intermodulation distortion in an RF system cascade. Consider the two-stage amplifier system of Figure 1-12 with subsystem A described by

$$y = a_0 + a_1 x + a_2 x^2 \tag{1.9}$$

subsystem B described by

$$z = b_0 + b_1 y + b_2 y^2 \tag{1.10}$$

2 It is beyond the scope of this chapter to explain the measured deviation from the 3:1 slope but the underlying reason is that integer calculus (behind conventional circuit simulation technology) is not able to fully capture drift-diffusion processes which are best described by fractional calculus.

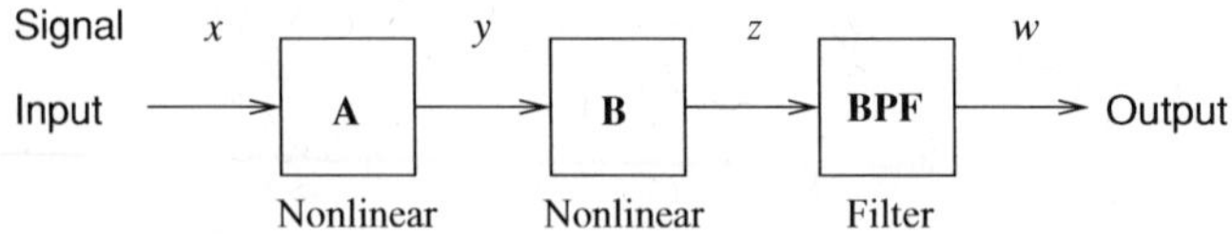

Figure 1-12 Cascaded nonlinear system.

and BPF is an ideal bandpass filter allowing only signals in the amplifier band to pass. The intermediate system response is described by

$$z = \left(b_0 + a_0 b_1 + a_0^2 b_2\right) + \left(a_1 b_1 + 2a_0 a_1 b_2\right) x$$
$$+ \left(a_2 b_1 + a_1^2 b_2 + 2a_0 a_2 b_2\right) x^2 + 2a_1 a_2 b_2 x^3 + a_2^2 b_2 x^4. \tag{1.11}$$

If x is composed of two sinusoids at frequencies f_1 and f_2 with amplitudes c_1 and c_2 respectively,

$$x = c_1 \cos(2\pi f_1 t) + c_2 \cos(2\pi f_2 t) \tag{1.12}$$

then y will have 7 frequency components and z will have 21 components each of which is the sum of intermodulation components. The number of frequency components can be determined using trigonometric expansions of Equations (1.9) and (1.11). This is a laborious process. An easier way is to realize that the output will consist of all possible sum and differences of f_1 and f_2 up to fourth order. That is, the frequency of a component of the output is $n_1 f_1 + n_2 f_2$ where $|n_1| + |n_2| \leq 4$. Thus the frequencies of the components of the output z are given in Table 1-3. Note that positive and negative frequencies are treated as referring to the same component.

Table 1-3 Frequency components and intermodulation paths of a cascade system with interstage filtering.

Order	n_1	n_2	**Frequency**
Zeroth	0	0	0 (dc)
$\|n_1\| + \|n_2\| = 0$			
First	1	0	f_1
$\|n_1\| + \|n_2\| = 1$	0	1	f_2
Second	2	0	$2f_1$
$\|n_1\| + \|n_2\| = 2$	0	2	$2f_2$
	1	1	$f_1 + f_2$
	1	−1	$f_1 - f_2$
Third	3	0	$3f_1$
$\|n_1\| + \|n_2\| = 3$	0	3	$3f_2$
	2	1	$2f_1 + f_2$
	1	2	$f_1 + 2f_2$
	2	−1	$2f_1 - f_2$
	1	−2	$f_1 - 2f_2$
Fourth	4	0	$4f_1$
$\|n_1\| + \|n_2\| = 4$	0	4	$4f_2$
	3	1	$3f_1 + f_2$
	1	3	$f_2 + 3f_1$
	3	−1	$3f_1 - f_2$
	1	−3	$f_2 - 3f_1$
	2	2	$2f_2 + 2f_1$
	2	−2	$2f_2 - 2f_1$

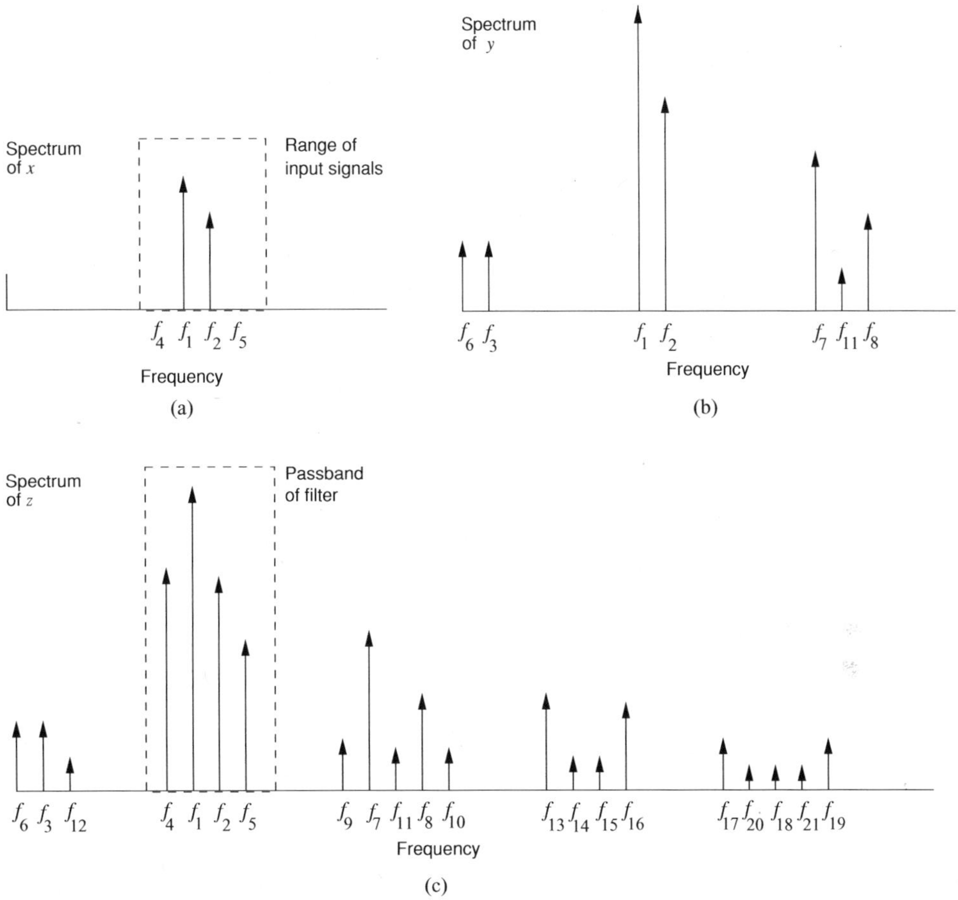

Figure 1-13 Spectrum of x, y, and z for the system of Figure 1-12

Bandpass filtering of z yields w, see Figure 1-14, which has 4 frequency components. The frequency components of z can be evaluated directly from substitution of Equation (1.12) into Equation (1.11), or in stages, first determining the spectrum of y and then the spectrum of z in terms of the spectral components of y. In the following this second approach is used as it provides greater insight.

The spectra of x, y, z and w are given in Figures 1-13 and 1-14 for the system of Figure 1-12. The intermodulation products giving rise to the spectrum of y are defined in Table 1-4 in terms of the frequency components of x. Similarly the IPs giving rise to the spectrum of z are defined in Table 1.5 in terms of the frequency components of y.

The magnitudes of the frequency components of y are given in Table 1.6. These magnitudes are evaluated by calculating the individual intermodulation products from a trigonometric expansion and summing those of the same frequency.

Returning to the main thrust of this discussion, to reduce the distortion presented at the output it is not sufficient to simply reduce the bandwidth of BPF as then allowable signal frequencies may be excluded from the output. However, by inserting the BPF between A and B the distortion frequencies produced by A can be eliminated before the signal is input to B, Figure 1-16. The spectra of z and w are then considerably simplified, see Figures 1-17 and 1-18.

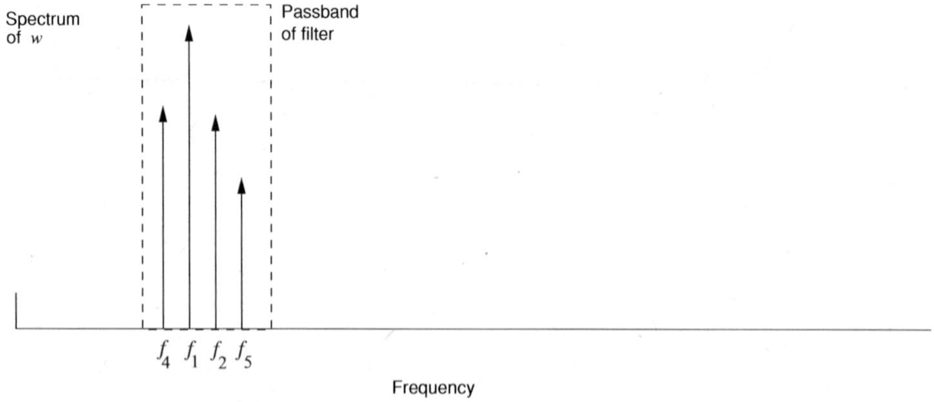

Figure 1-14 Spectrum of w.

1.6 DYNAMIC RANGE

The previous section provided an understanding of nonlinear distortion processes encountered with amplifiers and many other RF-active components. Using a nonlinear characterization in system design requires a higher abstraction and dynamic range is one metric that can be used. While system-level simulations are sometimes possible, relatively simple algebraic metrics are essential in developing the system architecture and are the starting point in system design.

While modern communication and radar systems use digitally modulated signals, two-tone signals are used to both approximately characterize nonlinearity and in manual calculations. In the previous section it was seem that for low powers the fundamental response has a 1:1 slope with respect to the input while the IP3 response varies as the cube of the level of input tones when both tones vary by the same amount as is common in a two-tone test. Thus IP3 has a 3:1 logarithmic slope with respect to the input. Since the relations are linear in a log-log sense it is possible to describe the nonlinear performance of an amplifier by a single quantity called the dynamic range (DR) or by the spurious free dynamic range (SFDR). SFDR and DR also capture noise properties.

Table 1-4 Intermodulation components of y (and of z) with bandpass filtering between subsystems A and B.

Order	Output	Input
first order	f_1	f_1
first order	f_2	f_2
second order	f_3	$f_2 - f_1$
zero order	dc, f_6	$0f_1 + 0f_2$
second order	f_7	$2f_1$
second order	f_8	$2f_2$
second order	f_{11}	$f_1 + f_2$

Table 1-5 Intermodulation components of z (no filtering).

Order	Output	Input
first order	f_1	f_1
second order	f_1	$f_2 - f_3$
second order	f_1	$f_{11} - f_2$
second order	f_1	$f_7 - f_1$
first order	f_2	f_2
second order	f_2	$f_1 + f_3$
second order	f_2	$f_{11} - f_1$
second order	f_2	$f_8 - f_2$
first order	f_3	f_3
second order	f_3	$f_2 - f_1$
second order	f_3	$f_{11} - f_7$
second order	f_3	$f_8 - f_{11}$
second order	f_4	$f_1 - f_3$
second order	f_4	$f_7 - f_2$
second order	f_5	$f_2 + f_3$
second order	f_5	$f_8 - f_1$
zero order, f_6	dc	$0f_1 + 0f_2 + 0f_3 + 0f_6 + 0f_7 + 0f_8 + 0f_{11}$
first order	f_7	f_7
second order	f_7	$2f_1$
second order	f_7	$f_{11} - f_3$
first order	f_8	f_8
second order	f_8	$2f_2$
second order	f_8	$f_{11} + f_3$
second order	f_9	$f_7 - f_3$
second order	f_{11}	$f_1 + f_2$
second order	f_{11}	$f_7 + f_3$
second order	f_{11}	$f_8 - f_3$
second order	f_{12}	$2f_3$
second order	f_{12}	$f_8 - f_7$
second order	f_{14}	$f_{11} + f_1$
second order	f_{15}	$f_{11} + f_2$
second order	f_{15}	$f_8 + f_1$
second order	f_{16}	$f_2 + f_8$
second order	f_{17}	$2f_7$
second order	f_{18}	$f_7 + f_8$
second order	f_{19}	$2f_8$
second order	f_{20}	$f_7 + f_{11}$
second order	f_{21}	$f_8 + f_{11}$

Table 1-6 Magnitudes of the frequency components of y.

Frequency	Amplitude
f_1	a_1c_1
f_2	a_1c_2
f_3	$a_2c_1c_2$
dc, f_6	$a_0+\frac{1}{2}a_2c_1^2+\frac{1}{2}a_2c_2^2$
f_7	$\frac{1}{2}a_2c_1^2$
f_8	$\frac{1}{2}a_2c_2^2$
f_{11}	$a_2c_1c_2$

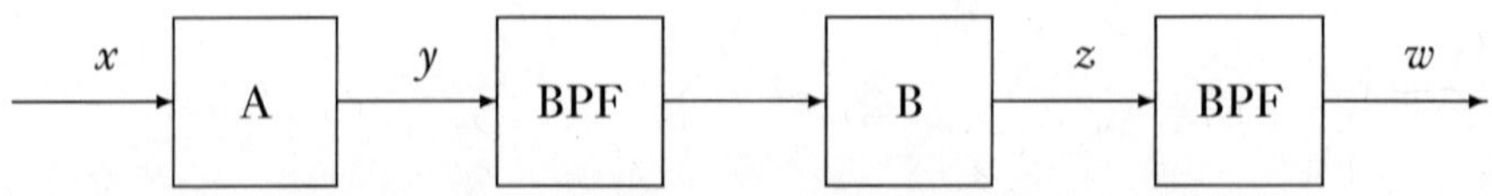

Figure 1-15 Nonlinear system with bandpass filters (BPF).

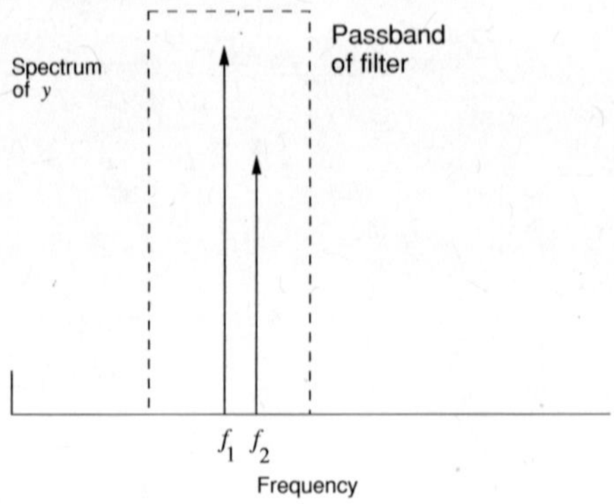

Figure 1-16 Spectrum of y after filtering.

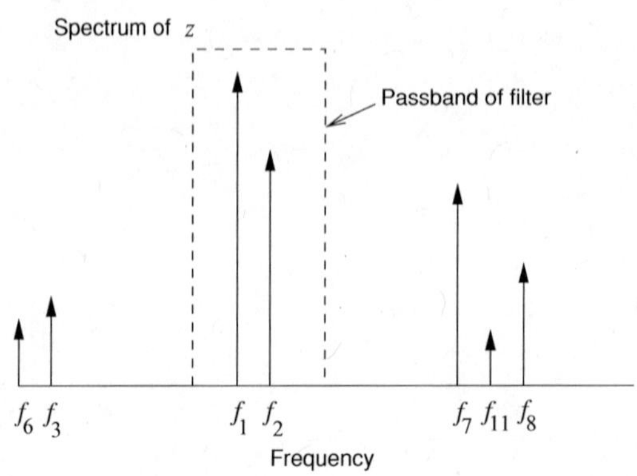

Figure 1-17 Spectrum of z with bandpass filter between subsystems A and B.

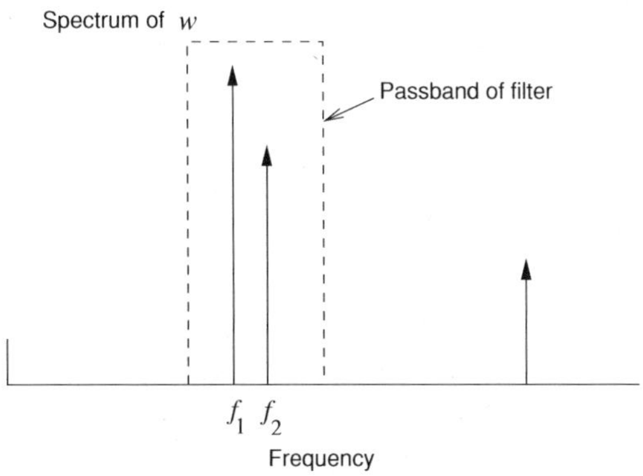

Figure 1-18 Spectrum of w with bandpass filtering between subsystems A and B.

In the following, an expression for SFDR is developed in terms of input-referenced quantities ($SFDR_i$). A similarly referenced dynamic range DR (DR_i) is also developed. SFDR describes the difference between a signal and the noise floor whereas DR incorporated the minimum detectable signal (MDS) [33] which is the noise level plus a minimum signal-to-noise ratio (SNR), SNR_{MIN}, expressed here in decibels. Figure 1-19 illustrates the input–output relationship of a one-tone signal to the IP3 response of a subsystem and also graphically defines the dynamic ranges. The

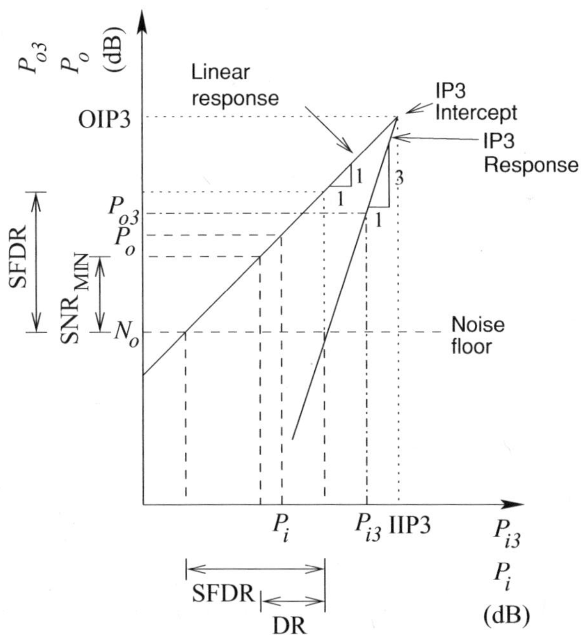

Figure 1-19 Output power versus input power of a stage or system plotted as output power in decibels versus input power in decibels. The IP3 response is a result of two-tone intermodulation and the input power is the combined power of the two signals that have equal amplitude. Extrapolations of the 1:1 linear response and the 3:1 third-order intermodulation response intersect at the IP3 point.

Table 1-7 Magnitudes of the frequency components of z without filtering.

Frequency	Amplitude
f_1	$a_1b_1c_1 + 2a_1a_2b_2c_1c_2^2 + \frac{1}{2}a_1a_2b_2c_1^3$
f_2	$a_1b_1c_2 + 2a_1a_2b_2c_1^2c_2 + \frac{1}{2}a_1a_2b_2c_2^3$
f_3	$a_2b_1c_1c_2 + a_1^2b_2c_1c_2 + \frac{1}{2}a_2^2b_2c_1^3c_2 + \frac{1}{2}a_2^2b_2c_1c_2^3$
f_4	$\frac{3}{2}a_1a_2b_2c_1^2c_2$
f_5	$\frac{3}{2}a_1a_2b_2c_1c_2^2$
dc, f_6	$b_0 + \frac{1}{2}b_2\left(a_1^2c_1^2 + a_1^2c_2^2 + 2a_2^2c_1^2c_2^2 + \frac{1}{2}a_2^2c_1^4 + \frac{1}{2}a_2^2c_2^4\right)$
f_7	$\frac{1}{2}a_2b_1c_1^2 + \frac{1}{2}a_1^2b_2c_1^2 + a_2^2b_2c_1^2c_2^2$
f_8	$\frac{1}{2}a_2b_1c_2^2 + \frac{1}{2}a_1^2b_2c_2^2 + a_2^2b_2c_1^2c_2^2$
f_9	$\frac{1}{2}a_2^2b_2c_1^3c_2$
f_{10}	$\frac{1}{2}a_2^2b_2c_1c_2^3$
f_{11}	$a_2b_1c_1c_2 + a_1^2b_2c_1c_2 + \frac{1}{2}a_2^2b_2c_1^3c_2 + \frac{1}{2}a_2^2b_2c_1c_2^3$
f_{12}	$\frac{3}{4}a_2^2b_2c_1^2c_2^2$
f_{13}	$\frac{1}{2}a_1a_2b_2c_1^3$
f_{14}	$a_1a_2b_2c_1^2c_2$
f_{15}	$\frac{3}{2}a_1a_2b_2c_1c_2^2$
f_{16}	$\frac{1}{2}a_1a_2b_2c_2^3$
f_{17}	$\frac{1}{8}a_2^2c_1^4$
f_{18}	$\frac{1}{4}a_2^2b_2c_1^2c_2^2$
f_{19}	$\frac{1}{8}a_2^2b_2c_2^4$
f_{20}	$\frac{1}{2}a_2^2b_2c_1^3c_2$
f_{21}	$\frac{1}{2}a_2^2b_2c_1c_2^3$

point of intersection of the extrapolated linear output (of power P_o) and third-order (IP3) output (of power P_{ip3}) is called the third-order intercept point (IP3 intercept). The point is identified by the output-referred intercept power (OIP3) or by the input -referred IP3 intercept power (IIP3) and these are key parameters in describing the linearity of nonlinear subsystems.

In the linear-gain region P_o versus P_i has a slope of 1:1 and so that

$$P_{\text{dBm},i} = P_{\text{dBm},o} - G_{\text{dB}} \tag{1.13}$$

where G_{dB} is the gain in decibels. P_o is used here as the output power with $P_{\text{dBm},o}$ indicating the output power in dBm. P_i and $P_{\text{dBm},i}$ are similarly defined. Using input quantities

$$\text{IIP3}_{\text{dBm}} = \text{OIP3}_{\text{dBm}} - G_{\text{dB}} \tag{1.14}$$

where again the dBm subscript indicates that the quantity is expressed in decibels referred to 1 milliwatt. The nonlinearity of RF-active components result in harmonics and intermodulation components. With the narrow-band amplifiers of communication and radar systems output filters conveniently filter out harmonics. However, intermodulation distortion cannot be filtered out as

Table 1-8 Magnitudes of the frequency components of z with bandpass filtering between the nonlinear subsystems.

Frequency	Amplitude
f_1	$a_1 b_1 c_1$
f_2	$a_1 b_1 c_2$
f_3	$a_1^2 b_2 c_1 c_2$
dc, f_6	$b_0 + \frac{1}{2} b_2 a_1^2 c_1^2 + \frac{1}{2} a_1^2 b_2 c_2^2$
f_7	$\frac{1}{2} a_1^2 b_2 c_1^2$
f_8	$\frac{1}{2} a_1^2 b_2 c_2^2$
f_{11}	$a_1^2 b_2 c_1 c_2$

these components are within the main passband. The intermodulation components are therefore spurious tones. Generally just one of these defines the maximum spurious tone and nearly always it is the third-order intermodulation tone with a two-tone input. Consideration of the maximum spurious tone and the noise floor defines SFDR.

Examining Figure 1-19 leads to the following inequality describing the linear gain of the amplifier:

$$\frac{\mathrm{OIP3}_{\mathrm{dBm}} - P_{\mathrm{dBm},o}}{\mathrm{IIP3}_{\mathrm{dBm}} - P_{\mathrm{dBm},i}} = \frac{\mathrm{OIP3}_{\mathrm{dBm}} - P_{\mathrm{dBm},o}}{(\mathrm{OIP3}_{\mathrm{dBm}} - G_{\mathrm{dB}}) - P_{\mathrm{dBm},i}} = 1. \tag{1.15}$$

Here P_o, P_i, OIP3 and IIP3 are the output power, the input power, the input-referred IP3 intercept, and the output-referred IP3 intercept. The third-order response is characterized by first introducing an equivalent input power $P_{\mathrm{dBm},i3}$ (P_{i3} expressed in dBm) defined as the average power of the two-tone signal that generates an IP3 of power $P_{\mathrm{dBm},o3}$. Noting that $P_{\mathrm{dBm},o3}$ varies with a 3:1 logarithmic slope with respect to $P_{\mathrm{dBm},i3}$ then

$$\frac{\mathrm{OIP3}_{\mathrm{dBm}} - P_{\mathrm{dBm,o3}}}{(\mathrm{OIP3}_{\mathrm{dBm}} - G_{\mathrm{dB}}) - P_{\mathrm{dBm,i3}}} = 3 \tag{1.16}$$

or

$$P_{\mathrm{dBm,i3}} = \tfrac{1}{3}\left(2 \times \mathrm{OIP3}_{\mathrm{dBm}} + P_{\mathrm{dBm,o3}} - 3G_{\mathrm{dB}}\right). \tag{1.17}$$

The SFDR can now be defined when the third-order intermodulation product of two-tone excitation is the dominant spurious tone. SFDR is defined as the difference between P_{i3} and P_i when they produce IP3 and linear output respectively that are both equal to the output noise power N_o, see Figure 1-19. That is when $P_o = P_{o3} = N_o$. So replacing $P_{\mathrm{dBm},o}$ in Equation 1.17 by N_o then

$$P_{\mathrm{dBm,i3}} = \tfrac{1}{3}\left(2 \times \mathrm{OIP3}_{\mathrm{dBm}} + N_{\mathrm{dBm},o} - 3G_{\mathrm{dB}}\right) \tag{1.18}$$

and

$$P_{\mathrm{dBm},i} = N_{\mathrm{dBm},o} - G_{\mathrm{dB}}. \tag{1.19}$$

Note that the difference between the linear output and the third-order intermodulation only reduces as the input power increases above P_{i3}. Thus the output-referred SFDR is

$$\begin{aligned} \mathrm{SFDR}_{\mathrm{dB},o} &= P_{\mathrm{dBm,i3}} - P_{\mathrm{dBm},i} \\ &= \tfrac{2}{3}\mathrm{OIP3}_{\mathrm{dBm}} + \tfrac{1}{3}N_{\mathrm{dBm},o} - G_{\mathrm{dB}} - N_{\mathrm{dBm},o} + G_{\mathrm{dB}} \\ &= \tfrac{2}{3}\left(\mathrm{OIP3}_{\mathrm{dBm}} - N_{\mathrm{dBm},o}\right). \end{aligned} \tag{1.20}$$

A similar development can be used to define the input-referred SFDR:

$$\mathrm{SFDR}_{\mathrm{dB},i} = \tfrac{2}{3}\left(\mathrm{IIP3}_{\mathrm{dBm}} - N_{\mathrm{dBm},i}\right). \tag{1.21}$$

Note that N_i is the input-referred noise and includes noise applied to the subsystem as well as the noise produced internally in the subsystem and referred to the input. SFDR provides a combined measure of distortion and noise. However, for usable dynamic range, DR, the minimum acceptable SNR must be considered. The minimum SNR, $\mathrm{SNR_{MIN}}$ is determined by the communication or radar modulation format, error coding and acceptable BER. So in defining DR the input power of the desired signal must increase sufficiently to produce an SNR of at least $\mathrm{SNR_{MIN}}$. Since the desired spurious level is still at the noise floor, this implies a direct subtraction in dB of the desired SNR.

Therefore, the input-referred third-order dynamic range, preferred for use with receivers, is

$$\mathrm{DR}_i = \frac{2}{3}\left(\mathrm{IIP3} - N_{\mathrm{dBm},i}\right) - \mathrm{SNR_{MIN}} \tag{1.22}$$

and the output-referred dynamic range, preferred for use with transmitters, is

$$\mathrm{DR}_o = \frac{2}{3}\left(\mathrm{OIP3} - N_{\mathrm{dBm},o}\right) - \mathrm{SNR_{MIN}}. \tag{1.23}$$

1.7 CASCADE SYSTEM DESIGN

RF systems are typically architected as a cascade of subsystems as shown in Figure 1-20. Feedback across subsystems is rarely used as this adversely impacts noise, bandwidth and efficiency. Each subsystem is characterized by gain (or loss) and for active devices their SFDR. System design involves both architectural choices and the selection of gain of adjustable gain stages. Design is not straightforward with many discrete choices and intuition is an important guide. This section presents what is called the contribution method that provides a good initial assignment of Noise Figure (NF), gain and linearity requirement to individual stages and enables informed assessment of trade-offs during system design. Application of the approach to the design of a receiver preselector

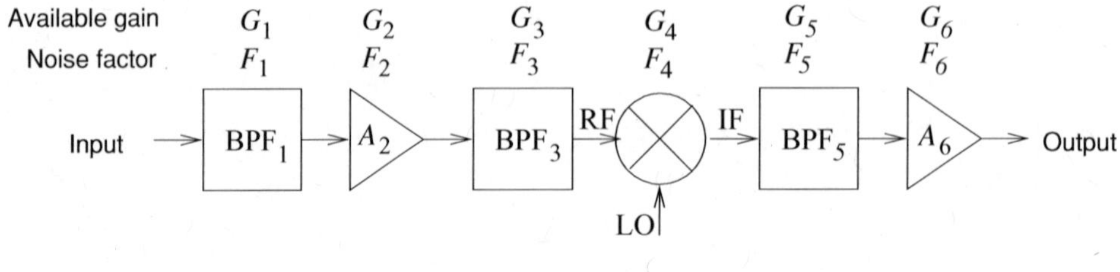

Figure 1-20 Cascade system for receive down converter or transmit up converter.

cascade for maximum SFDR is used to illustrate the concept. The contribution method provides a systematic approach to the control of adaptive and reconfigurable circuits as the control algorithm itself must embody knowledge of system design.

Traditionally the design of receiver and transmitter cascade systems for specified gain, noise, and distortion is based on budget analysis in which initial assignments are made to each stage in a cascade. This approach is well documented and based on the work of Friis [34] and others [35] that provides calculation of the total noise figure, SNR, and gain of a system from the parameters of individual subsystem stages. This work was extended [36] to address cascaded nonlinear distortion by using the intercept method to obtain the system SFDR. The system optimization goal is, in general, maximizing overall SFDR subject to the constraints of cost and availability of subsystems with the required attributes.

In realizing a cascade system typically one or more offending stages either dominates the SNR or significantly degrades overall distortion. In some cases a redistribution of stage parameters (such as gain) reduces the degradation. In other cases a change in architecture is required. Thus design is aided by having first a good initial assignment of noise and intermodulation performance of each stage, and second a clear indication of the contribution of each stage as design progresses. In budget analysis noise factor, F, gain, G, and IP3 levels are initially assigned individually to each stage in the cascade as in Figure 1-20. Generally maximum gain is assigned to the first-active stage and as a result this stage essentially establishes the system noise level. As a consequence active devices in latter stages nearly always require significantly higher linearity performance than necessary to meet the system SFDR objective. The cost of a suboptimum design is often the use of more expensive stages than necessary, or higher linearity requirements of some stages leading to increased supply power. In addition, with budget analysis system performance, i.e., SFDR, is difficult to assess as the system design evolves.

The stage contribution method is a good scheme for developing initial stage assignments as well as providing an understanding of trade-offs as system design progresses. In simple terms, the contribution method assigns to each active stage a percentage contribution to the overall SFDR. Stage contribution is balanced as no single stage dominates noise or distortion and all contributions to SFDR are equally weighted.

While the preceding discussion specifically relates to receiver design, the procedure applies to transmitter design with minor modification. Linearity is a significant issue with transmitters operating with complex modulation. Improper distribution of stage gain and loss impacts targeted goals such as modulation occupied bandwidth, SNR, and efficiency.

1.7.1 Contribution Framework

The contribution approach is developed in this subsection and is based on an extension of noise and distortion analysis of cascaded stages [37]. The noise performance of a stage is characterized by its noise factor $F = N_o/(GN_i)$ where G is the available power gain of the stage, N_o is the output noise power, and N_i is the input noise power of a resistor held at standard ambient temperature (290 K). Distortion is characterized by the amount of intermodulation distortion (IP3) produced in a two-tone test. The intersection of the linear-stage response and the projected third-order IMD response is called the third-order Intercept Point (IP3), see Figure 1-19. Receiver overall distortion performance is best captured by the Input-referred power of the IP3 intercept (IIP3) which is often expressed in decibels as IIP3_{dBm}. (With transmitter systems it is more appropriate to use the power of the output-referred IP3 point (OIP3).) Several subsystems in an RF chain functionally constrain and fix their contribution to noise, gain and distortion. For example, passive elements such as filters have characteristics that are specified in terms of their passband and stopband performance and this establishes their noise contribution for a specific implementation.

Frequency translation subsystems such as mixers also have characteristics that are governed by their frequency-domain response such as image rejection and noise and distortion performance is not independently selectable. For clarity of the discussion that follows, passive stages such as filters and attenuators are combined with the preceding active stage.

1.7.2 Noise Contribution

First consider the contribution of individual stages to system noise and assume that the stages are matched so that available and delivered input and output noise powers are equivalent. The n th stage, with all ports having equal bandwidth B, has output noise power

$$N_{o,n} = F_n G_n k T_o B \tag{1.24}$$

where k is Boltzmann's constant, and F_n and G_n are the noise factor and available power gain of the n-th stage respectively. The excess noise power of successive stages is additional to that of the first. So while the output noise power contribution of the first stage is

$$N_{o,1} = F_1 G_1 k T_o B \tag{1.25}$$

the excess output noise power of the second stage is

$$N_{o,2} = (F_2 - 1) k T_o B G_2. \tag{1.26}$$

Then the total noise power at the output of a two-stage cascade is

$$N_{o,2}^T = (F_2 - 1) k T_o B G_2 + N_{o,1} G_2. \tag{1.27}$$

So the total noise power at the output of the m th stage is

$$N_{o,m}^T = \sum_{n=2}^{m} \left[(F_n - 1) k T_o B \prod_{i=2}^{n} G_i \right] + F_1 k T_o B \prod_{n=1}^{m} G_n. \tag{1.28}$$

Thus an m-stage cascade has total cascaded system noise factor $F^T = N_{o,m}^T / (G^T N_{i,1})$ with G^T being the total cascaded available gain and $N_{i,1}$ the noise power input to the first stage. In terms of the parameters of individual stages

$$F^T = F_1 + \sum_{n=2}^{m} \frac{F_n - 1}{\prod_{i=n}^{m} G_{i-1}}. \tag{1.29}$$

The link between Equations (1.28) and (1.29) enables the noise contribution of each stage to be determined. Treated separately, Equation (1.28) provides the output noise power of a cascade such as a receiver. Although the extension of Equation (1.28) to Equation (1.29) is normally associated with the derivation of the receiver noise factor, it can also be used in transmitter noise analysis.

Now the noise contribution of a stage can be defined. The gain accumulated at the j th stage (the total cascade gain up to and including the j th stage) is

$$G_j^A = \sum_{n=1}^{j} G_n \tag{1.30}$$

(and so $G^T = G^A_m$ for an m-stage cascade). The fractional noise contribution of the j th stage to the total output noise is then defined as

$$C_j^N = \text{stage noise contribution} \equiv \left(\frac{F_j - 1}{\left(\frac{G_j^A}{G_j}\right) F^T} \right). \tag{1.31}$$

This is one component of the SFDR contribution of a stage.

1.7.3 Intermodulation Contribution

The two-tone intermodulation distortion of a cascade is assessed using the cascade intercept method and noting that for narrow channel bandwidth relative to the operating frequency, a worst-case additive distortion scenario is justified [38]. For receiver systems the contribution to nonlinear distortion is captured by IIP3 and for all stages this will be referred to the input of the cascaded system. Keeping with the nomenclature of the previous section, the accumulated IP3 of the j th stage referred to the system input is

$$\text{IIP3}_j^A = \frac{\text{IIP3}_j}{G_{j-1}^A} \tag{1.32}$$

and is $\text{IIP3}^A_{\text{dBm},j}$ when expressed in dBm. Note that all prior gain and loss up to the j th stage modifies the intercept when it is referred to the input of the cascaded stages. Combining the contribution of individual stages yields the total system IIP3:

$$\text{IIP3}^T = \left(\sum_{n=1}^{m} \frac{1}{\text{IIP3}_n^A} \right)^{-1}. \tag{1.33}$$

This leads to the definition of the fractional contribution of the j th stage to the system IIP3:

$$C_j^{\text{IP3}} = \text{IIP3 contribution} \equiv \frac{\text{IIP3}^T}{\text{IIP3}_j^A}. \tag{1.34}$$

1.7.4 Complete Stage Contribution

Intermodulation distortion is the upper bound on the total system undistorted response while noise sets the lower bound. So C_j^{IP3}, Equation (1.34), and C_j^N, Equation (1.31), denote the contributions of each stage to the dynamic range. In the above development the contributions of individual stages to noise and intermodulation levels were both referred to the input of a cascade. From these a stage contribution graph can be produced providing a clear picture of the impact of various stages on overall performance and identifies the stage(s) that dominates noise, distortion, and dynamic range performance of the system. Redistribution of the gain, noise or distortion contributions of the stages alters the overall SFDR and the correlation of these changes is clearly seen.

The cascade shown in Figure 1-20 represents an RF pre-selector (or down-converter) module and will be used as an example of receiver design using the contribution method. Stage assignment progresses from the filter input and terminates at the IF mixer output. Note that the same system cascade diagram (Figure 1-20) is applicable in reverse in transmit mode but with up-conversion.

Inspection of Equation (1.31) demonstrates the possibility of providing equal noise contribution by each stage to the total cascaded noise figure. If each of the stage parameters are appropriately chosen, a balanced noise contribution can be achieved where C_j^N has the same value for all j. In so doing, the required gain and noise factor is distributed and not necessarily bundled in the first few stages. The intent here is to couple the cascaded noise factor with the distortion of the cascade. The result of this procedure is that the minimum required gain is found for a given noise factor. As a result the intermodulation target is met with the lowest possible individual intercepts assigned to each stage. A further consequence is that it generally leads to minimized power consumption—a particularly important attribute of a transmitter cascade. In general it is possible to change the relative contributions of stages through choice of gain but sometimes changes in architecture are necessary. For the preselector case to be discussed in Section 1.7.6, the initial contribution assignment of stages is straightforward since the relation between gain, noise figure, and losses is clearly defined in the cascade formula. However, further optimization and trade-off analysis is possible. For example, the usual approach of establishing SNR in the early stages of the cascade results in stages further along the cascade needing to have higher intercept values. Consequently these stages will generally have additional power consumption. Trade-offs of the contributions of individual stages to distortion and noise will lead to constrained total power consumption while still achieving the required system SFDR.

1.7.5 Design Methodology for Maximizing Dynamic Range

In this section a design methodology is developed for creating a trade-off assessment to maximize SFDR of an RF receiver or transmitter front end consisting of cascaded stages. The distortion intercept of the cascade is either referred to the input for a receiver or to the output for a transmitter. The noise factor (or noise power) of the cascade uses the same reference. As mentioned earlier, for a relatively narrowband channel, a worst-case additive distortion scenario is justified [38] and represents an upper bound on the total system distortion. In particular this means that the distortion intercepts of the various stages are directly combined to yield the overall intercept. In the following discussion a receiver system is specifically addressed but with minor changes the discussion also relates to a transmitter system.

A central component of the methodology is the use of a stage contribution graph that provides a clear picture of stage performance identifying which stage or stages in the cascade dominate performance. This is not readily apparent with the budget spreadsheet approach. An example of this graph, which is used in a case study that follows, is shown in Table 1-9. Calculations of noise and distortion contributions provide an indication of which stage tends to raise the total noise factor or distortion compared to all others in the cascade. Redistribution of gain, noise, or distortion will alter this relationship and lead to changes in system SFDR. Passive stages generally do not restrict nonlinearity but they do contribute noise. So, to simplify the following discussion, passive stages immediately preceding an active stage and the active stage itself will be considered a single stage.

Table 1-9 Cascade stage assignments based on balanced contribution. In some cases there is no contribution or it is not required.

NF_i(dB)	$Gain_i$ (dB)	$IIP3_i$ (dBm)	Stage i	$Gain_i$	F_i
3.979	−3.97	–	1 filter	0.4	2.5
3.01	1.761	−0.969	2 LNA	1.5	2
3.979	−3.97	–	3 filter	0.4	2.5
2.041	–	−3.18	4 Mixer	x	1.6

First of all consider the conventional system design approach where the noise figure of the receiver cascade is established on the premise of maximizing the gain of the first stage so that the SNR of the system is also set by this stage. Thus a target system SFDR determines the overall IP3 performance. Also in the conventional design approach the first stage has negligible impact on IIP3. Then a reasonable choice in the design process is to select stages following the first as contributing equally to the reduction in IP3. Thus the minimum acceptable IIP3 (in dBm) of the j th stage (i.e., $\mathrm{IIP3}_j$) of an m-stage cascade that is required to meet the target total IIP3 in dBm (denoted as $\overline{\mathrm{IIP3}}^T_{\mathrm{dBm}}$) is obtained from Equations (1.32) and (1.33) as

$$\mathrm{IIP3}_{\mathrm{dBm},j} = \mathrm{IIP3}^A_{\mathrm{dBm},j} + G^A_{\mathrm{dB},(j-1)} \tag{1.35}$$

where

$$\mathrm{IIP3}^A_{\mathrm{dBm},j} = \overline{\mathrm{IIP3}}^T_{\mathrm{dBm}} + 10 \log m. \tag{1.36}$$

In addition, the gain or loss of preceding stages will modify this IIP3 as indicated in Equation (1.32). This modification includes filter insertion to meet selectivity requirements such as image rejection while contributing a predictable loss based on resonator unloaded Q and bandwidth. In addition, G_j will reflect additional losses to the adjacent and alternate channels if a suitable filter is used, thus further modifying the contribution factor of latter stages. For the preselector study, this arrangement provides for equal IMD contribution of each stage (that is each stage has the same IP3 referred to the system input). For example, with a cascade of three stages each stage would need to meet a minimum system IIP3 of 4.8 dB in excess of the target system IIP3. If the noise cascade were based on balanced-stage noise contributions (C^N_j being the same for all stages) and associated gains (G_j being the same for all stages), then the required individual IIP3 values tend to be the minimum.

An important point is that this approach permits the establishment of a baseline for further optimization of the cascade. Also the contribution diagram (as opposed to the budget-level diagram) permits immediate notification as to which stage(s) is the limiting contributor. Most budget diagrams are based on ad hock assumptions and prior experience with little mathematical basis from which to start optimization. For example, in transmitter cascade design the intent might be to provide as high an intercept as possible in the lower frequency up-conversion stages. This approach is based on devices operating at lower frequencies having very good linearity and most cost effective in achieving high IP3. Alternatively, it is clear from a contribution viewpoint that the power amplifier IP3 will have a major impact on the reduction of linearity in this case. Thus assuming an equal contribution assignment permits a clear assessment of required trade-offs.

Maximizing system dynamic range becomes an exercise in maintaining the lowest noise power and highest IIP3 value throughout the cascade. The link between these parameter values is the distribution of gain and loss [39]. Referring to Figure 1-19, for a linear cascade the gain slope defined by the logarithmic relation of P_1 (the fundamental input signal level) to P_o (the fundamental output signal level) is 1:1. A change in P_1 to P_{1x} leads to a linear change in the output power to P_{ox}. For a well-behaved nonlinearity third-order distortion is dominant and the logarithmic relation of the power of the intermodulation frequency component P_3 to the power of the two-tone input level is 3:1. Consequently the rate of change between the distortion terms and the linear terms is 2:1. Establishing the point at which the distortion power created by third-order distortion is just equal to the noise power at any point in the cascade permits the dynamic range to be assessed at that point in the cascade. The relationships required were developed in Section 1.6 with reference to

Figure 1-19 and the cascade performance described by the SFDR referred to the receiver input is, in decibels,

$$\text{SFDR}_{\text{dB},i} = \tfrac{2}{3}\,(\text{IIP3} - N_i)\,. \tag{1.37}$$

For a transmitter the SFDR is most commonly referred to the output:

$$\text{SFDR}_{\text{dB},o} = \frac{2}{3}\,(\text{OIP3} - N_o) \tag{1.38}$$

and $\text{SFDR}_i = \text{SFDR}_o$. In the above N_i and N_o are the total input and output noise powers and are assigned a noise-floor value dependent on the cascade noise factor target and linear system gain. In one final step it should be noted that the RF system must be designed for a minimum SNR so that the effective dynamic range is as described in Section 1.6 on Page 14.

1.7.6 Design of a PreSelector Down Converter

A preselector down-converter with the architecture of Figure 1-20 is designed here considering the first four stages only and using the balanced contribution method for the initial stage assignment. The system operates at 1.5 GHz with a first IF of 250 MHz and an IIP3 (input-referred third-order intercept point) design target of 0 dBm with a noise figure target of 10 dB. The initial stage assignments for balanced noise and IMD contributions are shown in Table 1-9. The stage assignments shown in Table 1-9 meet the target specifications exactly with no single stage contributing any more than the required noise factor or nonlinearity (distortion) to meet the target. The low noise amplifier (LNA) gain is low permitting a significant reduction of third-order intermodulation distortion (IP3) in the mixer (or mixer IP3). However, if the mixer noise figure could potentially be high. With this in mind the process of system cascade trade-off and selection of actual element parameters continues.

Design proceeds with a selection of available modules for stages and continuous but limited variation of stage parameters (through bias control for example). The following stages were chosen for the stages: a MMIC amplifier for Stage 2 (NEC part UPC2745) and a MMIC mixer for Stage 4 (MCL part SYM-2500). Bandpass dielectric resonator filters were chosen for Stages 1 and 3. In particular, a tuned lumped-element Chebychev bandpass filter with 0.1 dB ripple at 1.5 GHz was used to provide selectable loss in Stage 3 and this proved to be important in establishing balanced stage contribution and improved dynamic range. Continuous control parameters include controlling the mixer IMD contribution by changing the LNA gain (through bias control), changing the mixer IIP3 (by varying the LO drive level), and re-tuning the Stage 3 filter. Table 1-9 provides data based on the initial contribution and a progression of designs. The noise and distortion contributions of stages with the initial stage assignment, initial MMIC performance, and optimized design are shown in Figure 1-21 and 1-22. Figure 1-23 depicts the calculated dynamic range. Based on the balanced contribution method, the SFDR is 109.5 dB normalized to a 1 Hz bandwidth. The initial value provided by the devices selected is 108 dB. After appropriate tradeoff and adjustable loss in Stage 3, the DR achieved is 111 dB. Design proceeds by monitoring SFDR continuously throughout with contribution at the initial assignment, nominal operating conditions of the component devices (recommended by the vendor), and contributions of individual stages are indicated by the (a), (b), and (c) sequence in Figures 1-21–1-23. The same final system design could be achieved using budget assignment but with much greater design effort.

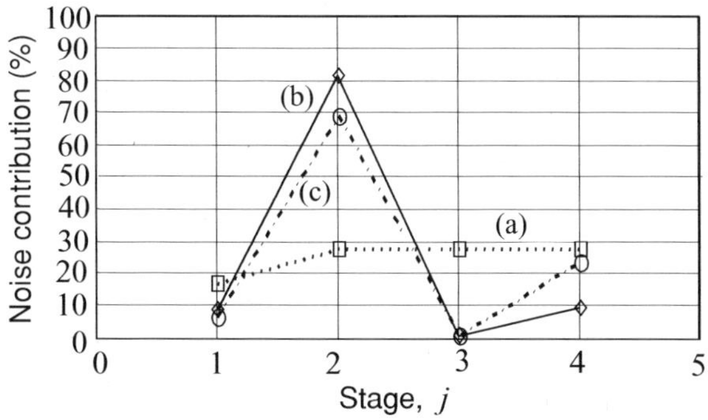

Figure 1-21 Noise figure contribution: (a) initial calculation of balanced noise figure contribution; (b) measured MMIC contribution under nominal operating contributions; and (c) and final settings after altering bias, filter loss, and mixer LO drive levels for improved dynamic range.

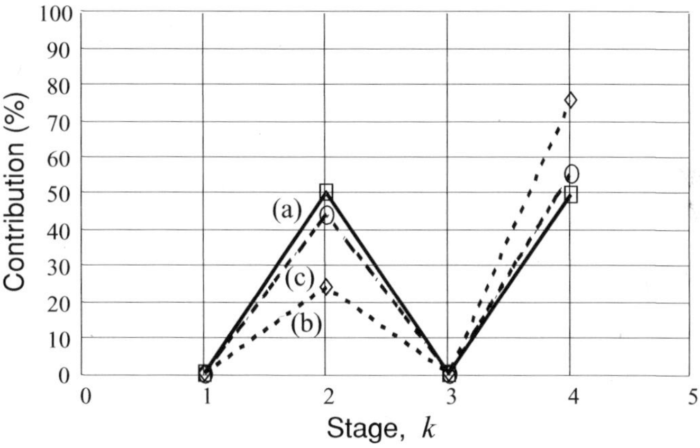

Figure 1-22 Input intercept (IP3) contribution: (a) initial calculation; (b) original measured MMIC devices; and (c) final performance after alteration of bias, LO drive, and filter loss adjustment in Stage 3. Final settings in (c) led to the highest dynamic range for the MMIC set chosen.

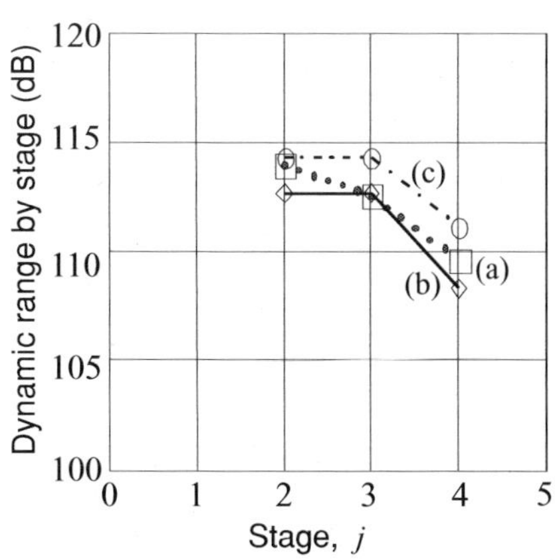

Figure 1-23 SFDR by stage: (a) initial assignment; (b) initial system performance after selection of modules but before optimization; and (c) final performance after setting bias, LO drive, and filter loss in Stage 3.

1.7.7 Summary

A method based on stage contribution for assessing the performance requirements in receiver and transmitter design was described in this section. Compared to the budget analysis method, the contribution technique permits better initial insight into appropriate distribution of stage gain, loss, noise, and distortion to meet a particular target specification and also indicates the trade-off's involved in design. The balanced contribution assignment defines a system that meets the noise figure target value exactly albeit with no system margin. Specifically, the gain assignments are exactly the minimum required to meet the system goal and no excess gain—that would tax linearity requirements and increase power consumption—is necessary. Thus the contribution approach highlights the minimum required stage intermodulation performance and serves as a good baseline for trade-off analysis.

1.8 WIRELESS DIGITAL COMMUNICATIONS

The system design procedure described in the previous section was based on signals comprising discrete tones. Actual signals, however, are more sophisticated. Modern wireless communication systems use digital modulation in which the RF phasor is transitioned from one state to another in precise movements with either the state or the transition representing one or more bits. The scheme of transitions defines a modulation format which can be classified as being either a constant envelope or non-constant envelope modulation. The envelope is the amplitude modulation imparted on a radio frequency carrier. Constant envelope modulation includes frequency shift keying (FSK), frequency modulation (FM), Gaussian mean shift keying (GMSK), phase shift keying (PSK), and other techniques involving detection of symbols by frequency or phase shifts of the carrier envelope. Nonconstant envelope modulation includes amplitude modulation (AM), quadrature phase shift keying (QPSK), offset QPSK (OQPSK), differential QPSK (DQPSK), quadrature amplitude modulation (QAM), and other techniques involving detection of variations in envelope amplitude and phase.

Interestingly, amplitude modulation is not inherently required for information transmission for phase shift keying (PSK) modulation schemes. For instance, a QPSK signal consists of two digital data streams, equal in amplitude, modulated in quadrature onto a carrier signal. The resulting signal would have a constant envelope; however, the occupied bandwidth is quite large and the first sidelobe of the $\sin(x)/x$, or sinc, spectrum will only be 13 dB down from the carrier in the middle of the adjacent channel. Typically, a lowpass filter is applied to each digital data stream to minimize or limit the out of band spectrum of the signal. The filters impart some finite memory on the data stream which results in amplitude variations as the ringing energy from a previous data pulse add with the current data pulse.

Amplitude variations of the modulated signal are characterized by measured waveform statistics. The peak-to-average power ratio (PAR), reported in decibels, is a popular statistic for describing signals with amplitude variation. A signal with high PAR requires that the RF system cascade have amplifiers with high linearity to handle the average power requirements and the peak amplitude excursions without generating excessive out of band distortion. However, it is possible for a signal with a higher PAR to exhibit less nonlinear distortion than a signal with lower PAR [40]. The reason for this apparent inconsistency is because the signal peak is a singular point measurement with, typically, a low probability of occurrence. Thus PAR is an incomplete statistic for determining the linearity requirements for a transmitter to carry a signal.

The amplitude probability density function (APDF) is a more complete statistical description of amplitude variations of a modulated signal. The APDF defines the maximum and minimum variation along with the relative probability of occurrence of amplitudes within the variation. The

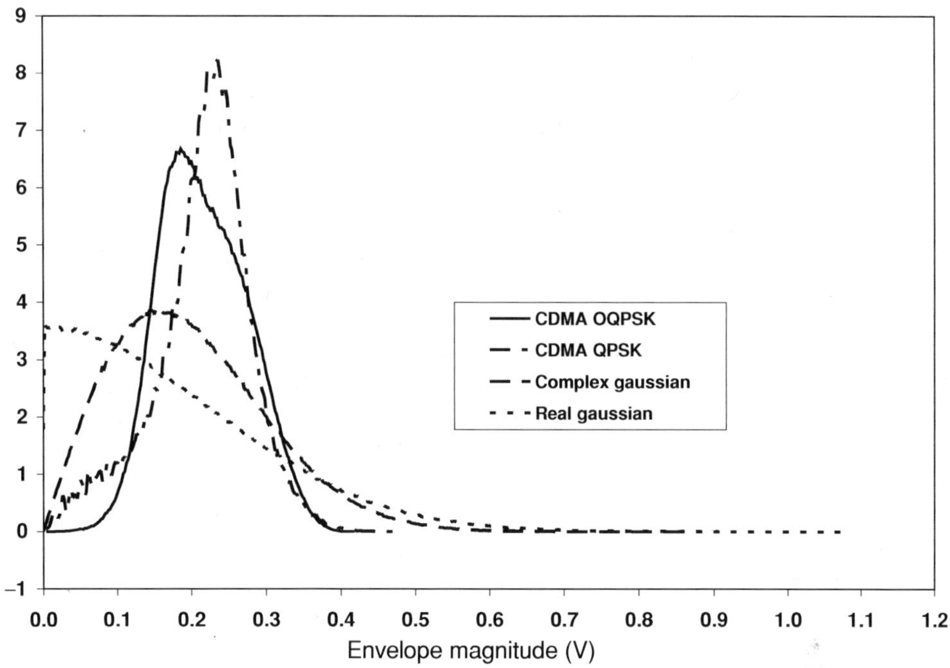

Figure 1-24 Amplitude PDF for CDMA and Gaussian modulation signals.

APDF is typically estimated from a histogram of amplitudes, with a uniform bin size, by

$$f(A) = \frac{N}{\Delta A \times N_c} \tag{1.39}$$

where N is the number of counts per bin, ΔA is the bin amplitude width, and N_c is the total number of samples. The shape of the amplitude density between the mean and peak amplitude influences the sensitivity of a particular signal to spectral regrowth due to nonlinear gain compression or expansion. For example, Figure 1-24 shows the amplitude probability density function (APDF) for a CDMA mobile transmitter using OQPSK modulation, the same signal using QPSK modulation, a real Gaussian signal, and a complex Gaussian QPSK signal where the average power of each signal is set to 0 dBm. Gaussian signals are of particular interest as their simple statistics lends them, and their interaction with nonlinearities, to quasi-analytic treatment. The OQPSK signal has the quadrature data stream offset in time by half the symbol rate while the in-phase and quadrature data for the QPSK signal are clocked together. The real Gaussian signal comprises carrier-modulated samples of a Gaussian process passed though the IS-95 reverse link baseband transmitter filter while the complex Gaussian signal is the quadrature sum of two independent samples of a filtered Gaussian process. The PAR for each signal is shown in Table 1-10. The shape of the amplitude density function after the mean differs for both signals where a significant portion of the QPSK amplitude above the mean resides close to the mean while the OQPSK amplitude density function is more linear after the mean. Thus it is difficult to determine, a priori, which signal will be more sensitive to nonlinear gain compression or expansion even though the QPSK has a higher PAR than OQPSK.

The Gaussian signals are particularly interesting because of the difference between the real and complex Gaussian signals. The complex Gaussian signal is the quadrature sum of two real

Table 1-10 Decibel peak to average ratio (PAR) for CDMA and Gaussian signals.

Signal modulation	PAR
OQPSK CDMA	5.4
QPSK CDMA	6.6
Real Gaussian	13.5
Complex Gaussian	11.8

Gaussian processes, so an intuitive guess would suggest that the amplitude distribution should be wider for the complex signal. Exactly the opposite is true; the PAR of the complex Gaussian signal is 1.7 dB less than that of the real Gaussian signal. An explanation for this is that the average power of each real Gaussian input signal is scaled down by 3 dB to yield the correct complex Gaussian power level; however, it is an unlikely event that two peaks from each of the real Gaussian signals will occur at the same time leading to a peak signal that is less than 3 dB plus the peak Gaussian amplitude. Thus the PAR is reduced since the peak distributions do not add in power. Again, from just the PAR, it is difficult to determine which signal will yield the least amount of distortion for the same output power level.

1.8.1 Bandpass Nonlinearities

Bandpass analysis simplifies the formulation of the nonlinear response by separating the envelope of the modulation from its carrier signal. However, this analysis assumes that the modulation bandwidth is narrow in comparison to the carrier frequency such that distortion terms from other tones and harmonics related to the carrier do not overlap in the resulting output spectrum. A carrier signal with amplitude and phase modulation is expressed as

$$w(t) = A(t)\cos\left[\omega_c t + \theta(t)\right] \tag{1.40}$$

where $A(t)$ and $\theta(t)$ are the respective amplitude and phase components of the modulation. The carrier modulation is often referred to as the complex envelope and is expressed either in polar form

$$\tilde{z}(t) = A(t)e^{j\theta(t)} \tag{1.41}$$

or in rectangular form

$$\tilde{z}(t) = i(t) + jq(t) \tag{1.42}$$

where $i(t)$ and $q(t)$ represent the in-phase and quadrature components of the baseband input signal. The modulated carrier expressed in terms of the complex envelope is

$$w(t) = \frac{1}{2}\tilde{z}(t)e^{j\omega_c t} + \frac{1}{2}\tilde{z}^*(t)e^{-j\omega_c t}. \tag{1.43}$$

The concept of bandpass nonlinearities was developed in the 1950s by information theorists who needed a simple model of a nonlinear channel to analyze the degradation in SNR of a modulated RF carrier passed through a nonlinear circuit followed by a bandpass filter centered at the carrier frequency [41]. The output bandpass filter eliminated all other distortion components except those centered about the carrier frequency. The resulting analysis is simplified by eliminating the need to consider other nonlinear terms harmonically related to the carrier frequency. Various behavioral models for the nonlinearities are used including power series [42], Chebyshev transformations [43,44], and Bessel function expansions [45].

The simplest characterization of a bandpass nonlinear circuit is by its instantaneous, or memoryless, large signal input-output response. For a modulated carrier this requires sweeping the input amplitude of an unmodulated carrier and measuring the corresponding change in gain and phase at the output port of the circuit. The amplitude gain response is known as the amplitude modulation to amplitude modulation transfer (AM-AM) characteristic and the amplitude phase response is the amplitude modulation to phase modulation (AM-PM) characteristic. The bandpass nonlinear response in terms of the complex envelope is

$$\tilde{G}\left[\tilde{z}(t)\right] = F\left[A(t)\right] e^{j\{\theta(t)+\Phi[A(t)]\}} \tag{1.44}$$

where $F[A(t)]$ and $\Phi[A(t)]$ are the AM-AM and AM-PM response functions respectively. Once the AM-AM and AM-PM response is obtained either by measurement or simulation, the data can be used directly with a numerical interpolation routine or used to fit model parameters to best fit the model to the data.

The power of the AM-AM and AM-PM representation is the ease of extraction of the characteristic and simplification of nonlinear analysis. The input-output characteristic is measured by sweeping the input power or signal level and measuring the resulting gain and phase at the output using a vector network analyzer or vector voltmeter relative to a fixed reference signal. Likewise, the characteristic can be simulated using a SPICE transient steady-state shooting method, or harmonic-balance circuit simulation engine to simulate the steady-state gain and phase transfer function as the input signal is swept in power. The negative tradeoff in using a memoryless nonlinear model is that frequency dependencies in the linear and nonlinear response are not modeled resulting in modeling errors for devices with strong frequency variation over the modulation bandwidth or over the bandwidth of distortion terms which contribute significantly to the output spectrum.

1.8.2 Power Spectrum Estimation

Engineers routinely use a fast Fourier transform (FFT) or a discrete Fourier transform (DFT) to estimate the power spectrum of signal. A Fourier transform is performed on the resulting output complex envelope signal to obtain the output power spectrum [46]

$$S(f) = \left| \int_{-\infty}^{\infty} \tilde{z}(t) e^{j2\pi f t} dt \right|^2 \tag{1.45}$$

where f is frequency. This approach is known as the *direct method* of spectral estimation of the output power spectrum. While this method is straightforward in its application, it provides little insight into how the nonlinear model and signal interact together to produce the output spectrum. At best, different parameters can be altered, a new model extracted, the signal run through the new model, the resulting spectrum observed, and the sensitivities deduced.

The *indirect method* is an alternative approach to calculating the power spectrum by first calculating the autocorrelation function of the signal at the output of the model then computing the Fourier transform of the autocorrelation function. The autocorrelation function is the convolution of a signal with its complex conjugate

$$\tilde{R}(\tau) = \int_{-\infty}^{\infty} \tilde{z}(t)\tilde{z}^*(t+\tau)dt \ . \tag{1.46}$$

Once the autocorrelation function is calculated the power spectrum is estimated from the Weiner–Khinchine theorem [47]

$$\tilde{S}(f) = \int_{-\infty}^{\infty} \tilde{R}(\tau)e^{j\omega\tau}d\tau \tag{1.47}$$

where $\omega = 2\pi f$. If $\tilde{z}(t)$ is a series expansion, then the autocorrelation function is a summation of products of all combinations of terms in the series, and the resulting power spectrum is a summation of individual spectra from the Fourier transform of each term in the series. This provides insight into how parameters in the nonlinear model interact with the signal to affect changes in the resulting power spectrum.

1.9 BEHAVIORAL MODELS

SFDR and DR are crude measures of nonlinear system performance but circuit simulation using digitally-modulated signals is prohibitive. One approach to modeling the effects of nonlinear distortion in a reasonable time is to use behavioral models. Determination of BER and spectral regrowth for actual signals requires simulation with long bit streams at as low a level as possible. With behavioral models this simulation can be done with the actual RF waveform for millions of RF cycles. For this reason equation-based models called behavioral models are used often using the statistics of the modulated RF signal and rarely with the actual waveform. The most widely accepted RF behavioral models are based on the models of Blackman [48,49]. Blackman introduced a model using one tone AM-AM and AM-PM conversion characterization.

In developing this model, a single sine wave (a one-tone test) is applied at the input of a subsystem and the relative amplitude and phase of the fundamental output measured. The harmonic response can rarely be measured as RF components are band limited. Referring to Figure 1-26, the response to a one-tone test is described by

$$x(t) = \text{Re}\{\boldsymbol{X}e^{j\omega t}\} = |\boldsymbol{X}|\cos(\omega t + \angle\boldsymbol{X}) = \boldsymbol{X}\cos(\omega t) \tag{1.48}$$

since the phase of $\boldsymbol{X}$ is zero and

$$y(t) = \text{Re}\{\boldsymbol{Y}e^{j\omega t}\} = |\boldsymbol{Y}|\cos(\omega t + \angle\boldsymbol{Y}) = |\boldsymbol{Y}|\cos(\omega t + \tau(|\boldsymbol{X}|)). \tag{1.49}$$

Typical results of a one-tone test are shown in Figure 1-25. This is a complex envelope model and has been successfully used for predicting nonlinear distortion with constant envelop signals such as FM signals. This model does not consider long-term memory effects arising from thermal effects and the delayed onset of RF breakdown. A more fundamental problem is that low frequency effects, such as at the difference frequency with a two-tone signal are not captured at all. The circuit impedance at the low frequency has been shown to have a significant effect on spectral regrowth.

Behavioral models developed using discrete-tone characterization can be used to capture the response of an RF front-end to stochastic signals. It is important that the model architecture capture all nonlinear effects and that the proper translation of the derived envelope model be transformed

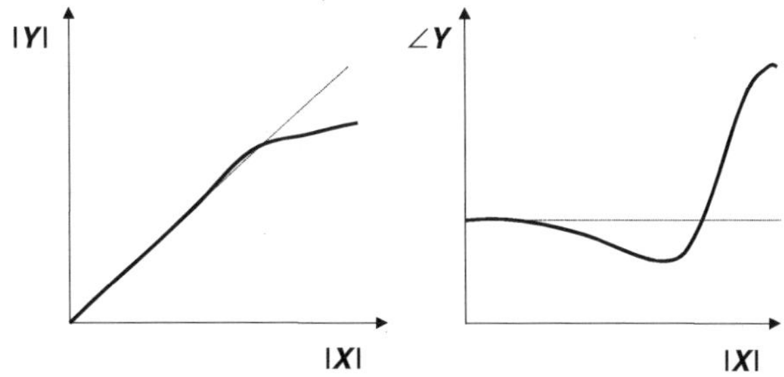

Figure 1-25 AM-AM and AM-PM characterizations from a one-tone test: (a) AM-AM data; and (b) AM-PM data.

into an instantaneous model. Here the requirements for the transformation are presented, a multislice model architecture that can capture complex behavior from narrow-band measurements is presented together with an extraction procedure.

Behavioral models are used extensively for characterizing RF front-ends. The underlying rationale for employing these models is that relatively simple measurements can be used to develop the model and that these can be used with more complex signals. Traditionally single-tone characterization is performed to develop AM-AM and AM-PM transfer functions which can be represented as the RF output phasor as a complex function of the input phasor. The AM-AM and AM-PM models can be conveniently represented in a variety of simulations environments including system, transient (spice), and harmonic-balance simulators. A natural extension is to use two-tone characterizations. Several concerns arise in extending this modeling approach to richer input signals including digitally modulated signals that can be described as stochastic signals with prescribed characteristics:

1. The need for an instantaneous model to treat general signals while one-tone and two-tone measurements yield envelope characterizations.
2. The need for a behavioral modeling architecture and extraction procedure for two-tone characterizations that can still be implemented in system, transient, and harmonic-balance simulators.
3. Verification that discrete-tone characterizations can be used to model the performance of RF front-ends with stochastic signals.
4. A general understanding of the relationships and differences of discrete and stochastic signals.

1.9.1 Instantaneous and Envelope Models

In general RF front-ends can only be partially observed because of bandpass input and output filtering, see Figure 1-26. The one-tone characterization of this type of system involves expressing the phasor of a tone at the output while the amplitude of an RF phasor is varied at the input. The AM-AM and AM-PM results are thus in terms of amplitudes and so provide an envelope

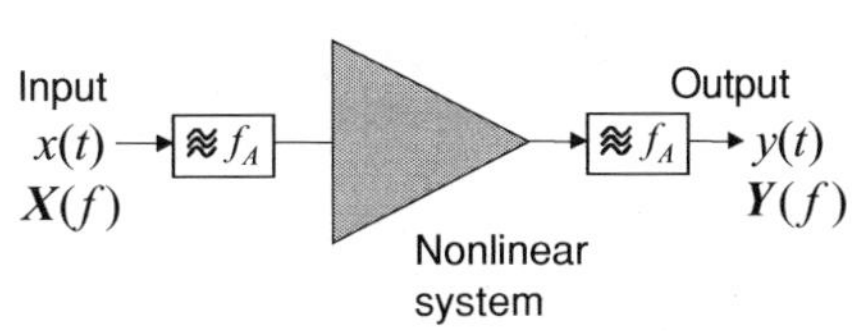

Figure 1-26 A unilateral bandpass nonlinear system.

characterization. Fitting a function to these characterizations results in an envelope RF behavioral model. This type of model can be used in system simulators as long as phasors are used. It is not appropriate to use these models with more complex signals as the instantaneous forms of the models are required to uncover the true nonlinearity of the model. To illustrate this point ignore phase variations and consider the envelope forms of a limiter response modeled using a tanh model:

$$v_o = L \tanh\left(\frac{g v_{\text{in}}}{L}\right) \tag{1.50}$$

or as a Cann/Rapp model:

$$v_o = \frac{g v_{\text{in}}}{1 + \left(\frac{g v_{\text{in}}}{L}\right)^{1/s}}\ . \tag{1.51}$$

These are standard limiter models where v_o is the amplitude of the output phasor and v_{in} is the amplitude of the output phasor. Varying g, L and s changes the linearity of the curves. These models have the envelope characteristics shown in Figure 1-27. The underlying instantaneous RF response is shown in Figure 1-28 and is fundamentally different. The clearest difference is that the instantaneous response is sharper than the envelope response. To understand this the instantaneous RF response, Figure 1-28, should be used. When a sine wave is applied to the nonlinear device the RF signal is swept over a large portion of the response as the amplitude increases. This softens the apparent AM-AM response so that the envelope characteristic, Figure 1-27, is softer. It is critically important to use the instantaneous response in translating from discrete-tone characterizations to models that can be used with stochastic signals.

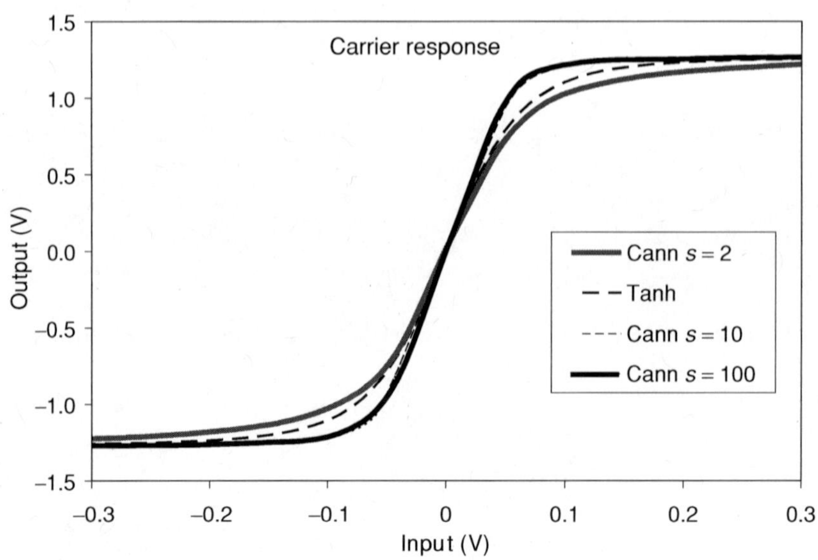

Figure 1-27 Carrier response of various limiter models.

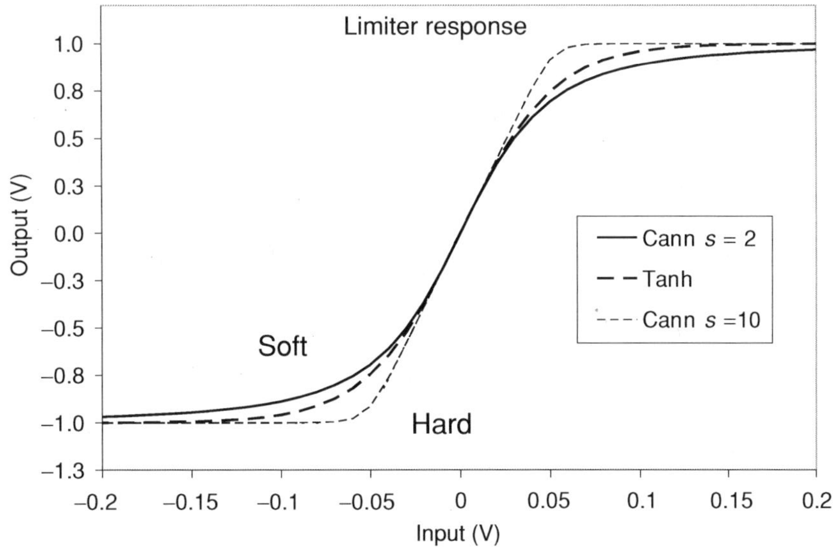

Figure 1-28 Instantaneous response of various limiter models.

1.9.2 Narrowband Behavioral Model

In this section behavioral modeling using partially-observed (bandlimited) characterizations is addressed. One of the significant difficulties with behavioral modeling is that intermediate frequencies cannot be directly modeled. In particular, narrowband communication signals passing through a nonlinear system result in the generation of signals at the envelope frequency as well as signals centered around the harmonics. To capture these effects the behavioral model should be partitioned so that the different frequency regimes can be addressed individually. Thus the behavioral model architecture is composed of a number of individual behavioral models, one for each frequency band, the output of each being frequency shifted to model the composite response. The individual behavioral models could be of any form.

An RF module can be described as a bandpass nonlinear system as shown in Figure 1-26. This model has bandpass filters at the input and the output. Even if the system does not have ideal bandpass filters, the internal spectral components will in general not be observable from external tests. The development of a useful behavioral model requires that a testing scheme be used that completely tests the internal response of the nonlinear system at the nonobservable frequencies. The input to the bandpass nonlinearity is represented by $x(t)$, which has a frequency domain representation $\boldsymbol{X}(f)$. Similarly the output is represented by $y(t)$ which has a frequency domain representation $\boldsymbol{Y}(f)$.

A two-tone test, with tones of frequency f_1 and f_2, generates signals internal to the nonlinear system that can be described by the spectrum in Figure 1-29. The frequency components generated fall in a number of bands: f_C denotes the difference frequency band resulting from the second-order mixing of tones within the main channel band f_A; f_B denotes the second harmonic band; and f_D denotes the third harmonic band. All of the components in band f_A see the same impedance levels. Each component is band f_B also see the same impedance as other components in band f_B. However, the components in band f_C see impedance levels that are frequency dependent across the band.

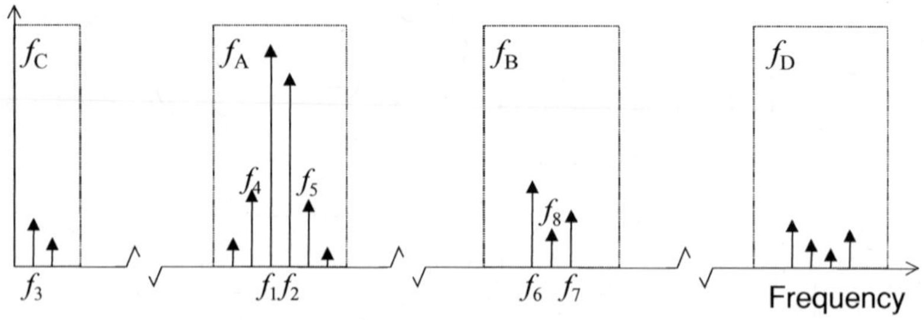

Figure 1-29 Spectrum of the internal signals of the nonlinear system.

1.9.3 Multislice Behavioral Model

The behavioral model currently being developed here is shown in Figure 1-30 where a quantity $z(t)$ with frequency representation $\boldsymbol{Z}(f)$ is introduced to described components internal to the bandpass system. The view implicit to this model is termed the intermodulation view and the various horizontal slices are termed intermodulation levels. There can be many levels dependent on the level of sophistication desired. $\boldsymbol{N}_A$, $\boldsymbol{N}_B$, and $\boldsymbol{N}_C$, represent frequency independent nonlinear transformations that could be described by complex power series or any other describing functions. The nonlinear transformations are frequency independent. $\boldsymbol{L}_A$, $\boldsymbol{L}_B$, and $\boldsymbol{L}_C$, represent linear transformations, which in general could be frequency dependent. In the case of frequency dependent one-tone AM-AM or AM-PM data, frequency dependence would be captured by $\boldsymbol{L}_A$. With a narrowband signal the frequency dependence in the f_A and f_B bands is negligible. In this case, all of the linear transformations are frequency independent except for $\boldsymbol{L}_C$, which captures the frequency dependence of the internal impedances at the envelope (difference) frequency. Of the linear transformations only $\boldsymbol{L}_C$ will be indistinguishable from the nonlinear transformations. The final multiplication stage translates components of $z(t)$ back to the f_A band.

Extraction of the multislice behavioral model is described in [50] and a very brief summary follows. So referring to Figure 1-30. The one-tone response is captured by the first slice of the model (with $\boldsymbol{N}_A$,$\boldsymbol{L}_A$,f_A) evaluated using a tone-tone AM-AM and AM-PM test for example. The

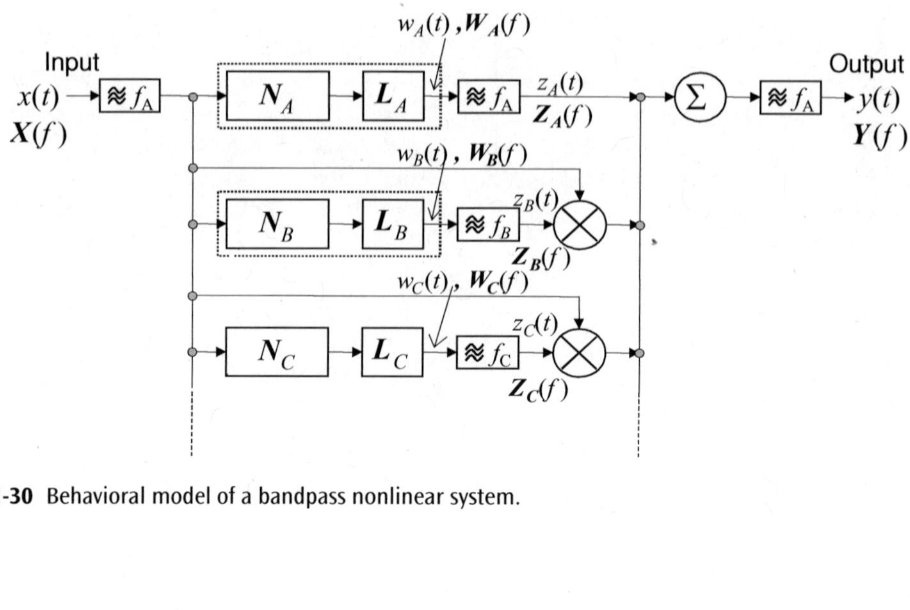

Figure 1-30 Behavioral model of a bandpass nonlinear system.

second slice of the model (with $\boldsymbol{N}_B, \boldsymbol{L}_B, f_B$) is evaluated by taking the measured response, subtracting the response of the ($\boldsymbol{N}_A$, $\boldsymbol{L}_A, f_A$) slice, and then dividing the result by the original spectrum. Then the components in frequency band f_B are used to fit $\boldsymbol{N}_B$ and $\boldsymbol{L}_B$.

1.10 CONCLUSION

This chapter presented a high-level view of RF systems and pointed out that system design is, among many parameters, an optimization of dynamic range, frequency selectivity and complexity. System cost as well as that of design are also important parameters with varying weightings depending on volume and commercial or military applications. All modern radar, radio, and sensor systems use digital modulation with RF signals having stochastic characteristics. However, system design, and even the design of subsystems, progresses using discrete-tone characterizations of nonlinear performance. With digitally-modulated signals one of the performance metrics is the proportion of RF power that is generated in the channels adjacent to the main channel. This is captured by the adjacent channel power ratio (ACPR). Bit error rate (BER) is also another important metric and this is increased by intermodulation and noise. It is difficult to work with ACPR and BER with digitally modulated signals. In both measurements and simulation reliance is on the two-tone test leading to the third-order intermodulation distortion ratio, IMD3, or the IP3 intercept point as preferred nonlinear performance metrics. The two-tone signal is referred to as a deterministic signal. As was seen, IMD3 (or the derived dynamic range) is the basis of system design. The inherent assumption is that there is a simple relationship between nonlinear response to deterministic and stochastic signals. There is an approximate relationship but this is a function of the technology used and of signal statistics [15]. So experience in working with particular signals is important and while the concept presented here systemizes system design there is still much that depends on intuition.

2

RF MEMS Components: Switches and Varactors

Linda Katehi and Dimitrios Peroulis

2.1 INTRODUCTION

This chapter focuses on fundamental principles of RF MEMS switches and varactors. In particular, the discussion here is on capacitive switches, metal-to-metal contact switches, and analog variable capacitors. Particular attention is paid to basic design procedures and solutions to common issues related to nonideal fabrication conditions, such as residual stress. Power handling is discussed in detail for both switches and varactors.

2.2 CAPACITIVE MEMS SWITCHES

Micromachining and MEMS are among the most promising enabling technologies for developing low-power, low-cost, miniaturized RF components for high-frequency applications. Several universities and companies have developed RF MEMS switches in the last decade that can be primarily classified as a) series or shunt, b) fixed-fixed membranes or cantilever beams, and c) capacitive or metal-to-metal contact type switches [51]. The main driving force behind this major research effort is the outstanding demonstrated RF performance of the MEMS switches from DC to 100 GHz compared to PIN diodes or FET transistors. Furthermore, electrostatically driven switches require only a few microwatts of DC power compared to several milliwatts that their solid-state counterparts dissipate. It is for this reason, as well as for the simplicity of their biasing networks that most of the developed switches are electrostatic in nature.

Most of these studies, however, have only limited their focus on the RF performance of MEMS switches and have provided little information on several important phenomena directly related to the inherent electromechanical characteristics of these structures. Their sheer interdisciplinary nature imposes a very tight coupling between the electrical and mechanical domains. For instance, thin-film residual stress and viscous damping may have a far greater influence on the performance of the device than intuitively anticipated. Moreover, the vast majority of the MEMS switches in the literature typically require a pull-in and hold-down voltage of 40–100 V and 15–30 V respectively. Whereas no difficulty exists in achieving these ranges in a typical laboratory environment, they may be quite challenging for handheld mobile phones, automotive vehicles and similar wireless devices that rely on low-voltage power supplies. In addition, Goldsmith *et al.* [52] have shown that the lifetime of capacitive switches strongly depends on the applied actuation voltage. In particular, for capacitive switches they experimentally observed a lifetime improvement of a decade for every

5–7 V drop on the switch pull-in voltage. Consequently, reducing the actuation voltage of MEMS switches may not only broaden the range of their possible applications, but also significantly enhance their performance. It would seem, therefore, that further investigations are needed in order to provide the MEMS engineer with complete and accurate information on the design and operation of these devices.

It is the purpose of this section to discuss important results on these issues. First, in Section 2.2.1 the design of the low-spring-constant beams that support the main switch structure is introduced. A number of designs that resulted in switches with pull-in voltages of as low as 6 V is also demonstrated. Section 2.2.2 discusses the effects of residual axial and gradient stresses on the device performance and shows how minor design and fabrication details may have a significant impact on the final structure. Section 2.2.3 presents experimental and theoretical results on the dynamic behavior of the low-voltage MEMS switch.

2.2.1 Spring Constant and Actuation Voltage

2.2.1.1 Design The mechanical design of most electrostatically based switches starts with the consideration of the required DC actuation voltage. Equation (2.1) presents a widely cited formula (e.g. [53]) for calculating the pull-in voltage of fixed-fixed beams, or air-bridges:

$$V_p = \sqrt{\frac{8K_z g_0^3}{27\epsilon_0 A}}\,. \tag{2.1}$$

K_z is the equivalent spring constant of the moving structure in the direction of desired motion (typically the z-direction), g_0 is the gap between the switch and the actuation electrode, ϵ_0 is the free space permittivity and A is the switch area where the electrostatic force is applied. Formula (2.1) implies that there are several ways that may decrease the required actuation voltage [54]. For instance, reducing g_0 can significantly lower the pull-in voltage. Although this solution can be partly applied to low-frequency applications (< 10 GHz), it will adversely affect the high-frequency off-state switch performance by compromising the switch isolation (for a series switch), or insertion loss (for a shunt switch). A second approach in lowering the pull-in voltage would be to increase the actuation area A. This area, however, has to stay within reasonable limits, primarily imposed by the desire for miniaturized circuits. The third alternative, which offers the maximum design flexibility for a low to moderate actuation voltage, is to lower the switch spring constant, hence designing a compliant switch. However, particular attention is needed here because this also reduces the restoring force of the switch, which increases its probability of failure due to stiction.

Figure 2-1 shows an SEM picture of a low- to moderate-voltage switch in a coplanar waveguide (CPW) configuration [55]. The switch consists of three movable metallic plates, one over each conductor of the CPW line. These plates are connected together with three short beams (connecting beams) and the whole structure is connected to the substrate at four points (anchors) through four beams. Because of their shape, these beams will be called serpentine springs or folded-suspension beams. The switch is typically suspended 4–5 μm above the CPW line and is electrostatically actuated when a DC voltage is applied between the switch and the CPW ground planes.

2.2.1.2 Fabrication Fabrication of the CPW switch of Figure 2-1 requires four masks and is described in detail in Reference [55]. Figure 2-2 summarizes the required steps, which were first introduced by Pacheco and Katehi [56]. The CPW line is typically made of Ti/Au (500/9000 Å) and is defined first through a lift-off process. A plasma-enhanced chemical vapor deposition (PECVD) of approximately 1200–2000 Å of Si_3N_4 follows. Since the switch is made of metal (typically Ni) this dielectric layer is primarily needed during the actuation stage to prevent a direct DC

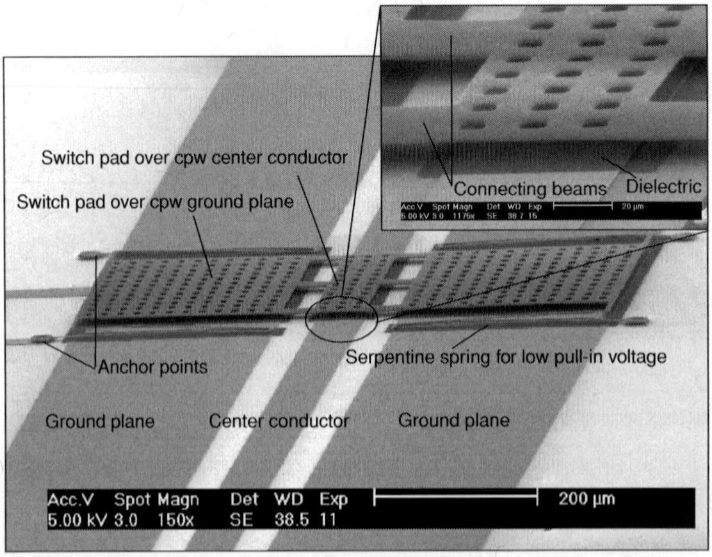

Figure 2-1 SEM picture of a low-voltage capacitive shunt switch over a CPW line (after [55]).

contact between the switch and the CPW line. Therefore, a positive photoresist intended to protect the Si_3N_4 underneath the switch is deposited and patterned, while the remaining dielectric layer is etched through an RIE process. After the photoresist removal, the sacrificial layer (polyimide or photoresist) is deposited and the switch anchor points are photolithographically defined. Afterward, a seed layer (typically Ti/Ni 2000/500 Å), is deposited, patterned, and

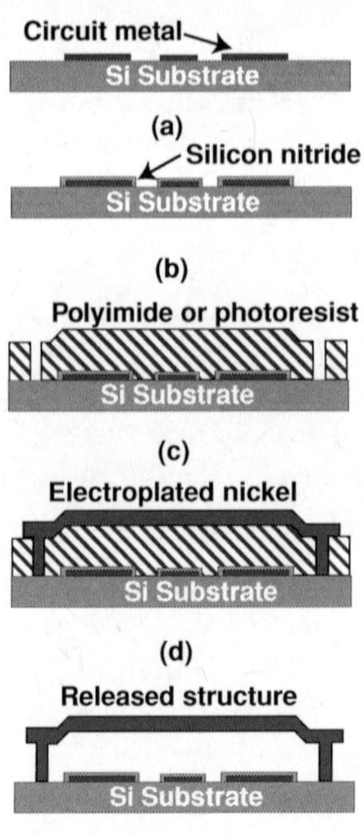

Figure 2-2 Summarized switch fabrication process.

electroplated. The last step is the removal of the sacrificial layer and the supercritical CO_2 drying of the structure.

2.2.1.3 Spring Design Since the mathematical details of the electrostatic actuation (including Equation (2.1)) have been extensively analyzed in the past [57, 58], the basic principle is briefly described here. When no DC bias is applied, the switch presents a very small shunt capacitance (typically in the order of 30–50 fF) between the center conductor and the ground planes. This is called the up or off state and the RF signal can propagate with minimal loss (typically with less than 0.1 dB at X-band). On the other hand, if the applied bias exceeds the actuation voltage, the switch collapses on the dielectric layer underneath, resulting in a significant shunt capacitance, which is equivalent to an RF short circuit. This is called the down or on state and virtually all the incident RF power is reflected back to the source.

As was previously mentioned, the switch of Figure 2-1 is connected to the substrate through four serpentine springs that are used to lower the switch spring constant and release the in-plane residual stress. If k_z is the z-directed spring constant of each of the springs, the total switch spring constant, K_z, is given by

$$K_z = 4k_z\,. \tag{2.2}$$

Compared to simple cantilever beams of equal total length, these springs have the additional advantage of occupying considerably less space, but they also have a higher spring constant. As will be shown, however, adding more meanders can significantly lower the spring constant without excessively increasing the required space. In the following the spring constant of an N-section meander, see Figure 2-3(a), will be calculated when a virtual force F_z is applied at its free end. An

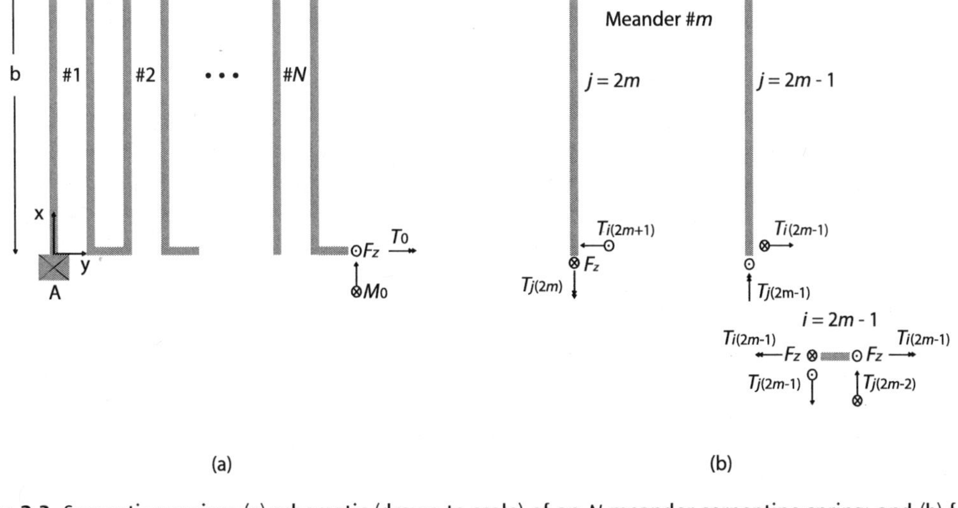

Figure 2-3 Serpentine spring: (a) schematic (drawn to scale) of an N-meander serpentine spring; and (b) forces, torques, and moments in the m th meander.

Table 2-1 Physical dimensions and material constants for the low-voltage switch.

Parameter	Dimension
Primary meander length (a)	20 μm
Secondary meander length (b)	240 μm
Switch thickness (t)	2 μm
Beam width (both beams) (w)	5 μm
Ni Young's modulus (E)	207 GPa
Ni Poisson's ratio (ν)	0.31
Shear modulus (G)	$E/(2(1+\nu))$
x-axis moment of inertia (I_x)	$wt^3/12$
z-axis moment of inertia (I_z)	$tw^3/12$
polar moment of inertia (I_p)	$I_x + I_z$
Torsion constant (J)	$0.413 I_p$ (see [60])

analytical solution for a similar folded meander has been obtained by Fedder [59] and the analysis here is based on his work.

Each meander of the whole spring is defined as the set of four beams: two primary beams of length a and two secondary beams of length b. Therefore, an N-meander spring has $2N$ primary beams and $2N$ secondary beams. The switch shown in Figure 2-1 for instance, has a single-section meander ($N = 1$) with $a = 20$ μm and $b = 240$ μm. All the necessary dimensions and material constants for the switch are given in Table 2-1. For the analytical calculation it is assumed that all six degrees of freedom of the anchor point (point A in Figure 2-3(a)) are fixed. Moreover, the guided-end boundary conditions are applied for the free-end point of the spring, since this point is attached to the main switch body. Consequently, a moment M_0 and a torsion T_0 are applied to this point to constrain the rotation angles around the x and y axes. The torsion and moment of each beam are then given by [59], Figure 2-3(b):

$$
\begin{aligned}
M_{a,i} &= M_0 - F_z\left[x + (i-1)a\right] \\
T_{a,i} &= T_0 + \left[\frac{1+(-1)^i}{2}\right] F_z b \\
M_{b,j} &= (-1)^j T_0 - F_z x + \left[\frac{1+(-1)^j}{2}\right] F_z b \\
T_{b,j} &= (-1)^j \left(j F_z a - M_0\right)
\end{aligned}
\tag{2.3}
$$

where $M_{a,i}$ and $T_{a,i}$ ($M_{b,j}$ and $T_{b,j}$) are the moment and torsion of the i th primary beam (j th secondary beam) with $i = 1$ to $2N$ ($j = 1$ to $2N$). In these equations, x is the longitudinal dimension along each of the beams.

Following the virtual work method, the total elastic strain energy of the meander is given by

$$
U = \sum_{i=1}^{2N} \int_0^a \left(\frac{M_{a,i}^2}{2EI_x} + \frac{T_{a,i}^2}{2GJ} \right) dx + \sum_{j=1}^{2N} \int_0^b \left(\frac{M_{b,j}^2}{2EI_x} + \frac{T_{b,j}^2}{2GJ} \right) dx \tag{2.4}
$$

where I_x, I_z, G, and J are defined in Table 2-1. Finally, the spring constant is given by

$$k_z = \frac{F_z}{\delta_z} = \frac{F_z}{\partial U / \partial F_z} \tag{2.5}$$

along with the boundary conditions

$$\phi_0 \equiv \frac{\partial U}{\partial M_0} = 0 \quad \text{and} \quad \psi_0 \equiv \frac{\partial U}{\partial T_0} = 0\,. \tag{2.6}$$

These equations lead to the following expressions for the reactions M_0 and T_0

$$M_0 = \frac{\frac{2Na}{EI_x} + \frac{(2N+1)b}{GJ}}{2\left(\frac{a}{EI_x} + \frac{b}{GJ}\right)} a F_z \tag{2.7}$$

$$T_0 = -\frac{F_z b}{2} \tag{2.8}$$

and for the spring constant k_z

$$k_z = \left[\frac{8N^3a^3 + 2Nb^3}{3EI_x} + \frac{abN[3b + (2N+1)(4N+1)a]}{3GJ} - \frac{Na^2\left[\frac{2Na}{EI_x} + \frac{(2N+1)b}{GJ}\right]^2}{2\left(\frac{a}{EI_x} + \frac{b}{GJ}\right)} - \frac{Nb^2}{2}\left(\frac{a}{GJ} + \frac{b}{EI_x}\right)\right]^{-1}. \tag{2.9}$$

Although Equation (2.9) is lengthy, it is written in an intuitive way that facilitates the design of these meanders or similar beams. The first two terms of the denominator represent the percentage of the spring constant that is due to beam bending (first term) and twisting (second term). In other words, these terms depend solely on the meander geometry and the ability of the beam material to bend and twist. The last two terms of the denominator are due to the boundary conditions of the meander moving end and correspond to its inability to rotate around the x-(third term) and y-axis (fourth term). These two terms may have comparable magnitudes to the first two and considerably increase the switch spring constant.

Equation (2.9) has been verified by a commercial finite element (FEM) code [61]. The dimensions of Table 2-1 were input in the code and several linear simulations were performed for springs with one to five meanders. For each simulation a concentrated z-directed force of $F_z^{FEM} = 1\ \mu$N was applied at the tip of the spring along with the necessary guided-end boundary conditions. The resulting deflection Δ_z was then computed and the FEM spring constant was extracted as $k_{mz}^{FEM} = F_z^{FEM}/\Delta_z$. Excellent agreement between the analytically and numerically computed spring constants is observed in Figure 2-4, which graphically presents the two spring constants as a function of the number of the meanders. This graph also illustrates that the serpentine spring constant is not significantly reduced after including four or five meander sections. Hence, three or four meanders would be a good compromise between low spring constant requirements and space limitations.

2.2.1.4 Actuation Voltage Measurements Five switch designs with one, two, three, four, and five meanders in their folded-suspensions were fabricated and measured [55]. Except for the serpentine

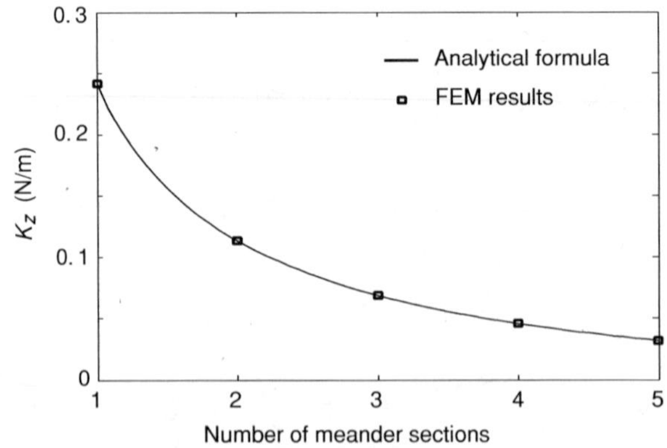

Figure 2-4 Analytically computed and FEM-simulated results of the *z*-directed spring constant of an *N*-section meander.

springs, all designs were identical and were fabricated on the same wafer by the same fabrication process. For each design the pull-in voltage was measured using an HP 4275A multifrequency LCR meter with an internal bias option. These measurements are presented in Table 2-2, which also compares the extracted switch spring constant from the measured pull-in voltage (based on Equation (2.1)) with the corresponding theoretical results. The theoretical values have been calculated for a switch thickness of $t = 2.5\ \mu\text{m}$ (because of overplating in the fabricated switches) and an initial gap of $g_0 = 5\ \mu\text{m}$. The reason for this higher gap is that, although the sacrificial layer thickness was 4 μm, the induced residual stress across the structure caused a slight out-of-plane deflection, which on average increased the total distance from the substrate to 5 μm [55]. Stress issues will be discussed in more detail in Section 2.2.2. Figure 2-5 also shows the measured DC-capacitance of the switches as a function of the applied bias voltage and the number of meanders.

This data reveals several discrepancies between the simulated and measured results. The first dissimilarity is the fact that the measured pull-in voltages are five to ten times higher than the theoretically calculated ones. The second, and the most interesting one, is related to the percentage of the spring constant reduction as the number of meanders is increased. For example, when the number of meanders was increased from one to two, the experimentally extracted spring constant was decreased by 46%, while the theoretical calculations predicted 53%. Although these results

Table 2-2 Actuation voltage measurements for several MEMS switches.

	Experimental		Theoretical	
Meanders	V_p **(V)**	K_z **(N/m)**	V_p **(V)**	K_z **(N/m)**
1	30	26.9	8.3	2.08
2	22	14.5	5.7	0.97
3	17	8.6	4.4	0.58
4	11	3.6	3.6	0.38
5	6	1.1	3.0	0.27

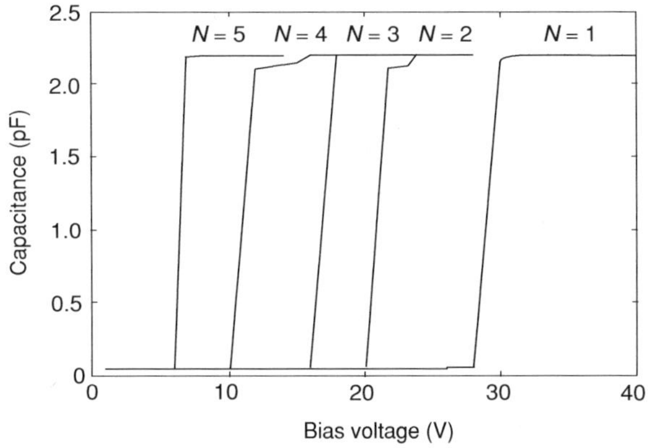

Figure 2-5 Measured DC switch capacitance as a function of the applied bias voltage and the number of meanders.

are in fair agreement, this is not the case for switches with more meanders. The switches with five meanders, for instance, had 70% lower spring constant than the ones with four. According to Equation (2.9), however, this number should be close to 30%. Figure 2-6 graphically illustrates these observations for all cases. All these issues are due to the high intrinsic axial stress built into the Ni layer during the fabrication and are discussed in Section 2.2.2.

2.2.1.5 Stiction and Top Electrode Design Although a low spring constant is essential in obtaining low-voltage switches, preventing down-state stiction is equally important. A low-voltage switch experiences a relatively weak restoring force while in the down state, which may not be sufficiently high to pull the switch up, particularly in humid or contaminated environments. This drawback of this family of switches can be at least partially overcome by including top electrodes. Of course this is done at the expense of more complicated fabrication. Section 2.2.4 includes a detailed discussion of the design, fabrication and measurements of a top electrode for the low-voltage switch. Additionally it discusses the advantages of the top electrode switch with regard to its power handling capabilities.

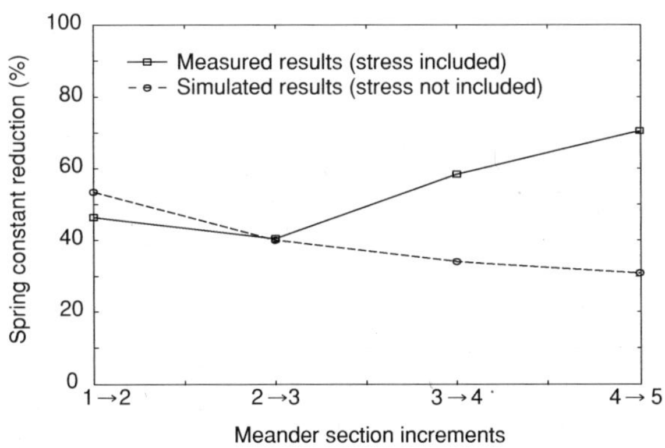

Figure 2-6 Experimental and theoretical percentage change of the switch spring constant as the number of meanders is increased.

2.2.2 Intrinsic Residual Stress Issues

Although the previous analysis allows for a first estimate of the switch spring constant and pull-in voltage, it does not account for any intrinsic residual stress on the structure. Residual stress, however, is developed during the fabrication of most microstructures and typically presents most of the major challenges in developing these devices. Under this stress, thin-film structures can experience undesirable deformations, which may be significant, particularly for high-aspect ratio structures. Additionally, many MEMS switches must satisfy very stringent requirements for reliable performance, including being planar over the circuit underneath it. Any undesirable buckling or curling may easily deteriorate the performance of the switch, or lead to the complete failure of the device. A lot of attention, therefore, has to be paid to residual stress and its effects on compliant structures before any successful devices can be developed.

When a thin film is deposited on a sacrificial layer at a temperature lower than its flow temperature, intrinsic stresses develop in the film [62]. A number of studies have been performed to theoretically explain the mechanisms of these stresses [63,64] and to experimentally measure their effects [65,66]. Nonetheless, in general, thin-film stress is complicated and heavily depends on the specifics of the fabrication process. There is also very little information for metallic microstructures built by thin films depositions and effective ways that can control its stress and/or its effects. This section illuminates the most important stress-related challenges for developing low-voltage switches.

2.2.2.1 Gradient Residual Stress A general uniaxial residual stress field in a thin film can be represented as [65]

$$\sigma_{\text{total}} = \sum_{k=0}^{\infty} \sigma_k \left(\frac{y}{h/2} \right)^k \tag{2.10}$$

where h is the film thickness and $y \in (-h/2, h/2)$ is the coordinate across the film thickness, with its origin at the midplane of the film. For a first order approximation, the total stress can be calculated as

$$\sigma_{\text{total}} \simeq \sigma_0 + \sigma_1 \left(\frac{2y}{h} \right) . \tag{2.11}$$

This equation implies that the total stress can be expressed as a superposition of the constant mean stress σ_0 (positive or negative depending on whether the film is in tension or compression) and a gradient stress σ_1 about the mid-plane (see Figure 2-7). The effects of the gradient stress are analyzed in this subsection and those of the mean stress in the following one.

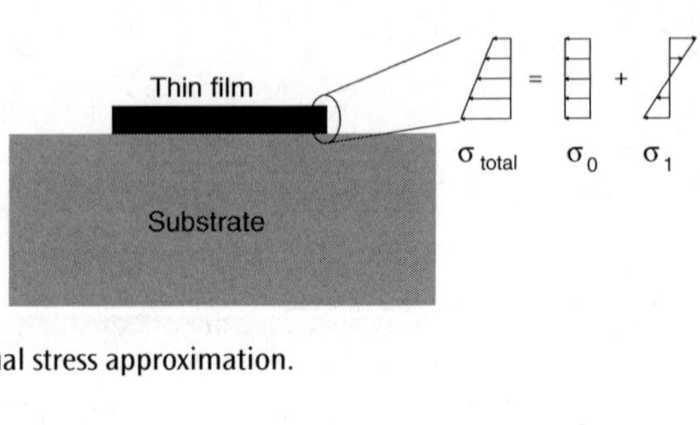

Figure 2-7 Thin film residual stress approximation.

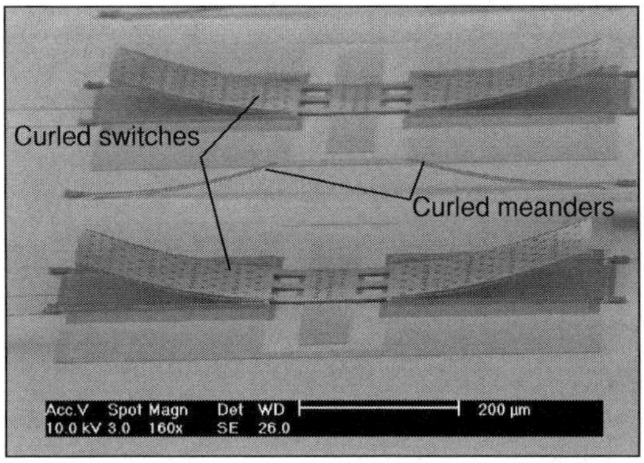

Figure 2-8 Switches with considerable deflection as a result of a poorly designed fabrication process.

It is widely known that residual gradient stress causes undesirable out-of-plane deformation. Figure 2-8 shows two examples of extremely warped switches. These switches were 4-μm thick and the maximum deformation, defined as the distance between the higher and lower switch points, was on average 22–24 μm. This deformation was recorded for switches 640-μm long (not counting the length of the meanders) but increased to 70–80 μm for switches close to 1 mm long. This substantial deformation renders both structures unusable for two reasons: a) the required actuation voltage is much higher than the design value (> 80 V), and b) the up-state switch RF capacitance is considerably higher than anticipated (C_{up} > 150 fF, instead of 50 fF). The sacrificial layer and the seed layer that resulted in such a stress were the polyimide DuPont PI2545 and an evaporated Ti/Ni (1500/500 Å) layer respectively. The switches were then electroplated in a Ni solution (nickel sulfamate, Barrett SN by Mac Dermid) with a steady current density of 4 mA/cm^2 for approximately 30 minutes. The induced stress with this fabrication process was repeatable over a period longer than six months.

This switch shape under residual gradient stress was also theoretically validated using SUGAR [67]. Figure 2-9 shows the simulated switch shape, which agrees very well with the fabricated switches. The maximum gradient stress value in the software was varied until the measured maximum deflection was obtained.

One way to alleviate this problem without increasing the actuation voltage is to selectively increase the switch thickness [68]. In this technique, the main switch body thickness is increased to 6–8 μm, but the springs remain 2-μm thick (Figure 2-10(a)). This process utilizes two electroplating steps: first the switch and the springs are plated, and subsequently the switch main body is

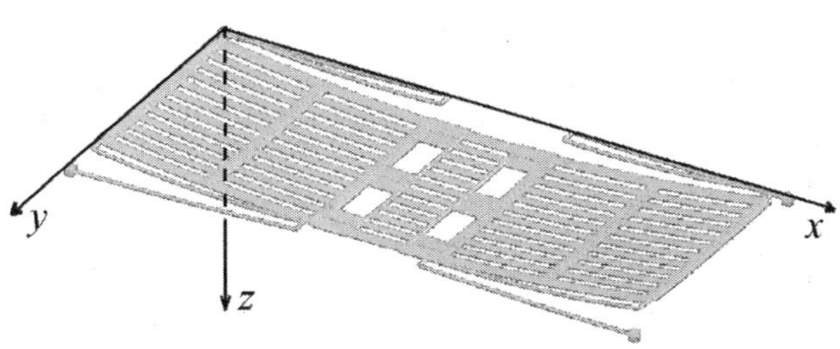

Figure 2-9 Simulated warped switch structure (by SUGAR). The maximum switch deflection is approximately 23 μm.

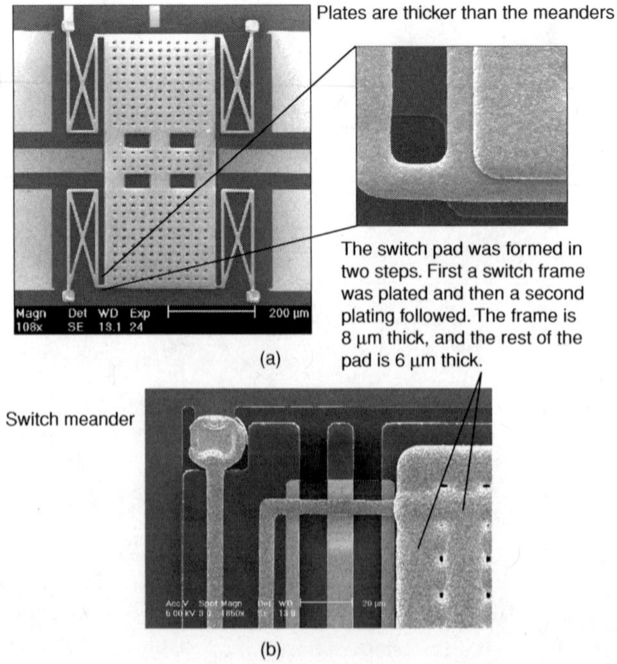

Figure 2-10 Fabricated switches with the selective electroplating process. The adhesion problems of the first switch were solved by increasing the seed layer area that was exposed to the second plating.

plated again until it reaches a thickness of 6–8 μm. Because of some adhesion difficulties between the two plated structures, the process was slightly changed by plating only a switch frame during the first step instead of the whole switch (Figure 2-10(b)). This improvement resulted in a 98% yield. The drawback of this technique, however, is that although it limits the switch warping to 1–3 μm and may prove useful for other types of MEMS devices, it also results in less conformal switches with lower down-state capacitance than the original structures.

It was experimentally found that a better solution is to sputter deposit the Ti layer (instead of evaporating) above the sacrificial layer. The deposition usually results in films of 1000–2500 Å of Ti. After the sputtering process is completed, the sample is immediately (in order to minimize Ti oxidation) taken to the Ni e-beam evaporator where 500 Å of Ni are deposited. It is also very important to point out that nothing else is changed in the process, including the Ni electroplating solution, current density and sacrificial layer etching. Furthermore, this process can be followed with either polyimide or photoresist with negligible differences.

Figure 2-1 is one example of the plane switches fabricated with this process and Figure 2-11 shows some details of three more examples. The switches shown in Figures 2-1 and 2-11(a) are 2-μm thick and suspended over a 40/60/40 μm CPW line. Although the latter switch was fabricated with three meanders instead of one, no appreciable difference in the warping profile was observed. The maximum measured deflection (as defined in Figure 2-8) for both of them is approximately 0.5 μm. This corresponds to less than 0.1% of the main switch length (640 μm) and does not affect the RF switch performance, since it is strongly localized at the end of the DC actuation pad.

The improved fabrication process has also been tested with switches only 1-μm thick, 680-μm long, and 240-μm wide. Two of these switches are shown in Figure 2-11(b) and (c). Both switches

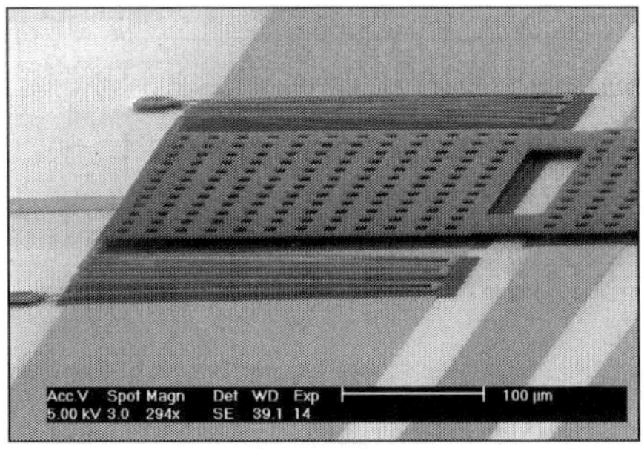

(a)

(b)

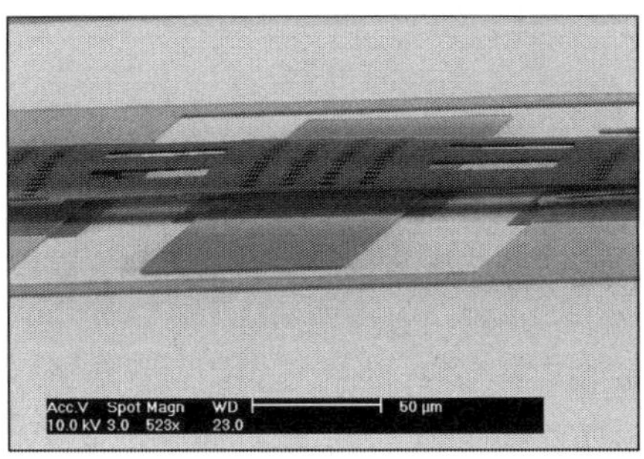

(c)

Figure 2-11 Planar switches fabricated with sputtered Ti layer instead of evaporated. (a) This three-meander switch is 640 μm long, 2 μm thick, its maximum out-of-plane deformation is less approximately 0.5 μm and the used sacrificial layer was polyimide; (b) and (c) these switches are 680 μm long, 1 μm thick and photoresist was their sacrificial layer. Their warping level is about 3 μm and is mostly along their short dimension.

include one meander, but in the first one the straight connecting beams have been replaced with a second meandering beam. Evidently both structures are fairly plane and their residual gradient stress is nearly negligible. The first one, however, exhibits somewhat straighter profile, which is related to the axial residual stress, analyzed in the next section. This meandering connecting beam was originally employed to increase the shunt switch inductance [69], but it also proved useful in releasing the in-plane switch stress. As for the three short connecting beams, the slightly

less-plane switch may not appear very appealing at first, but it is very useful in limiting the high-frequency up-state insertion loss, because it substantially reduces the up-state capacitance (by 50–70%). Furthermore, the pull-in voltage increase is negligible (both structures actuate with 15–20 V) because most of the electrostatic force is concentrated on the ground plane switch pads. Finally, both structures experienced some warping along their width, but this did not prove to be a problem. The switches were thin enough that were able to conform on the dielectric surface in the down-state.

2.2.2.2 In-Plane Residual Stress In-plane residual stress primarily increases the switch spring constant and it is therefore essential to control it within reasonable limits. It was experimentally found [66] that the induced tensile stress during the fabrication process was in the order of 150 MPa. A first-order approximation that shows the impact of this stress on the spring constant is analyzed in [70] where the deflection of a guided-end cantilever is calculated under simultaneous axial tension and concentrated transverse loading. The maximum deflection at the tip of the beam is given by

$$z_{\max} = \frac{W}{\gamma P}\left[\frac{(\cosh(\gamma l) - 1)^2}{\sinh(\gamma l)} - (\sinh(\gamma l) - \gamma l)\right] \tag{2.12}$$

where P is the axial tensile load, W is the transverse concentrated load at the tip of the beam, l is the length of the beam and γ is defined by

$$\gamma = \sqrt{\frac{P}{EI_x}}\,. \tag{2.13}$$

This leads to the following expression for the spring constant:

$$K_z^{\text{cant}} = \frac{\gamma P \sinh(\gamma l)}{2(1 - \cosh(\gamma l)) + \gamma l \sinh(\gamma l)}\,. \tag{2.14}$$

Figure 2-12 shows the variation of the normalized spring constant and the associated actuation voltage (with respect to the spring constant and actuation voltage of zero axial stress) for an axial tensile stress of 0–300 MPa. This figure clearly demonstrates the considerable impact of the axial

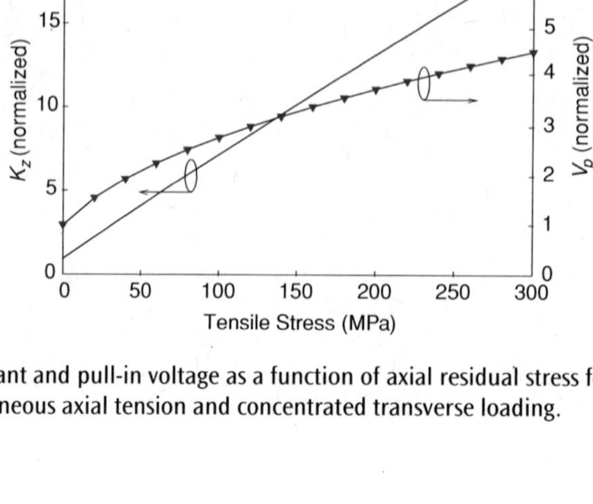

Figure 2-12 Spring constant and pull-in voltage as a function of axial residual stress for a guided-end cantilever with simultaneous axial tension and concentrated transverse loading.

stress on the switch actuation voltage. For instance, a tensile stress of 150 MPa would increase the pull-in voltage of a switch suspended by four cantilever beams by more than three times.

The serpentine beams shown in Figure 2-3 exhibit higher flexibility in handling the in-plane mean stress than the simple cantilever beam. In other words, as the number of meanders is increased, not only is the z-directed spring constant reduced, but also the lateral ones. To show this effect, the x- and y-directed spring constants were calculated with a similar process to that of Section 2.2.1. The y-directed spring constant can be expressed as

$$k_y = \frac{\frac{2S_{1y}}{Nb^2} EI_z}{2\left(a + \frac{2b}{3}\right) S_{1y} - (a+b)S_{2y} - 3(2N+1)a(a+b)^2} \tag{2.15}$$

where S_{1y} and S_{2y} are calculated by

$$S_{1y} = 4N^2a^2 + \left(4N^2 - 1\right) b^2 + 2\left(4N^2 + 1\right) ab \tag{2.16}$$

$$S_{2y} = 2N(2N-3)a^2 + \left(4N^2 - 1\right) b^2 + \left(8N^2 - 6N - 1\right) ab\ . \tag{2.17}$$

Similarly, the x-directed spring constant is given by

$$k_x = \frac{EI_z}{Na^2\left[S_{1x} + S_{2x} + \frac{8N^2a + (2N+1)(4N+1)b}{3}\right]} \tag{2.18}$$

where

$$S_{1x} = (2Na + (2N+1)b)\left[-\frac{2N+1}{2} + \frac{a(a+2\frac{b}{3})}{(a+b)(a+\frac{b}{3})}\right] \tag{2.19}$$

$$S_{2x} = -(2N+1)(a+b)\frac{a}{2(a+\frac{b}{3})}\ . \tag{2.20}$$

Linear FEM simulations verified the previous formulas and the results are given in Figure 2-13. From this figure it is clearly observed that the spring constant of the serpentine spring is greatly reduced as the number of meanders is increased. For instance, a spring with three meanders is

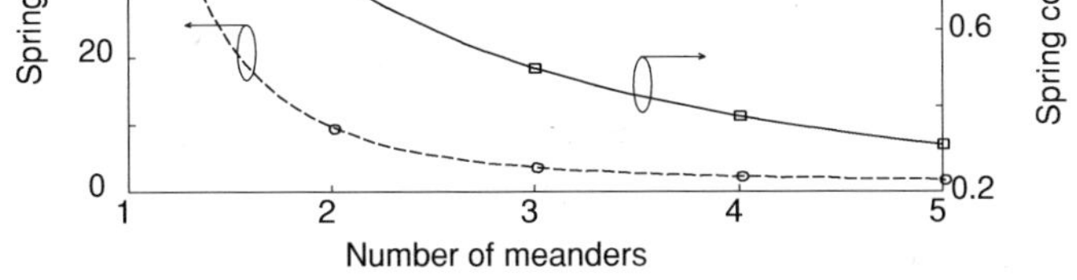

Figure 2-13 In-plane spring constants of the serpentine spring for various number of meanders.

177 times more flexible along the x-dimension than a spring with one meander. As a result, such a spring can help release the axial switch stress along its long dimension much more effectively than a spring with one meander. In other words, springs with many meanders are much more effective as stress buffers than springs with only one meander. A similar tendency exists for k_y, which is decreased by a factor of 5.2 when the meanders are increased from one to five. These results provide a qualitative explanation for the trends observed in Figure 2-6.

2.2.3 MEMS Switch Dynamic Behavior

Switching speed is one of the few disadvantages of MEMS components compared to PIN diodes and FET transistors. While their mass is typically very small (in the order of 10^{-12} to 10^{-9} kilograms), inertia due to mechanical movement still limits their response time—typically in the order of a few microseconds. The fastest switch so far has been developed by researchers at MIT Lincoln Laboratory [71]. It is a compact cantilever switch (less than 50 μm long) with a response time of approximately 1 μs. This low speed is primarily due to its small dimensions, mass and limited squeeze-film damping. It does require, however, a high pull-in voltage of 50–60 V, while it is typically actuated with 70–80 V. Low-voltage switches are generally expected to be slower since they typically have to move a relatively large actuation area. This is particularly true if the switch is expected to operate in air or other gas environment, such as N_2 (for limited humidity).

The switching speed is measured by recording the change in the power transmitted through the switch when a step voltage is applied at the bias of the device, see Figure 2-14 [72]. The RF input signal at 40 GHz is provided by an RF synthesizer, while the RF output signal is recorded by a high-frequency diode detector. The biasing signal is provided by a suitable combination of two DC power supplies and an inverter. Figure 2-15 presents two typical measurements taken with this setup for three-meander switches. The measured switches were suspended a mean distance of 5 μm above the CPW line and the applied bias voltage was only 20–30% higher than their actuation voltage. Both pull-in and release times were measured for these switches. Pull-in time is the time it takes the switch to touch the dielectric underneath it. On the other hand, the time that is required for the switch to move from the down-state to its original height (or within 5% from this value) is defined as the release time. Figure 2-15 shows that the actuation and release times are approximately 52 and 213 μs respectively. However, for the release time measurement, it takes

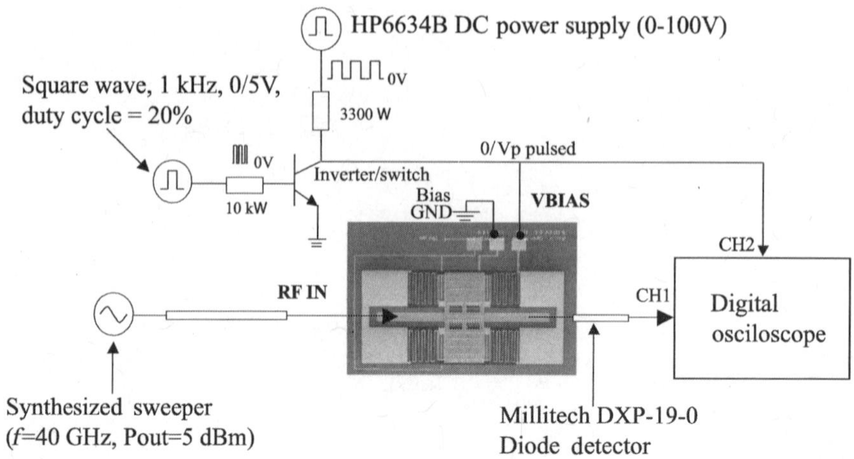

Figure 2-14 Switching speed measurement setup (Courtesy of Prof. Rebeiz, University of California, San Diego).

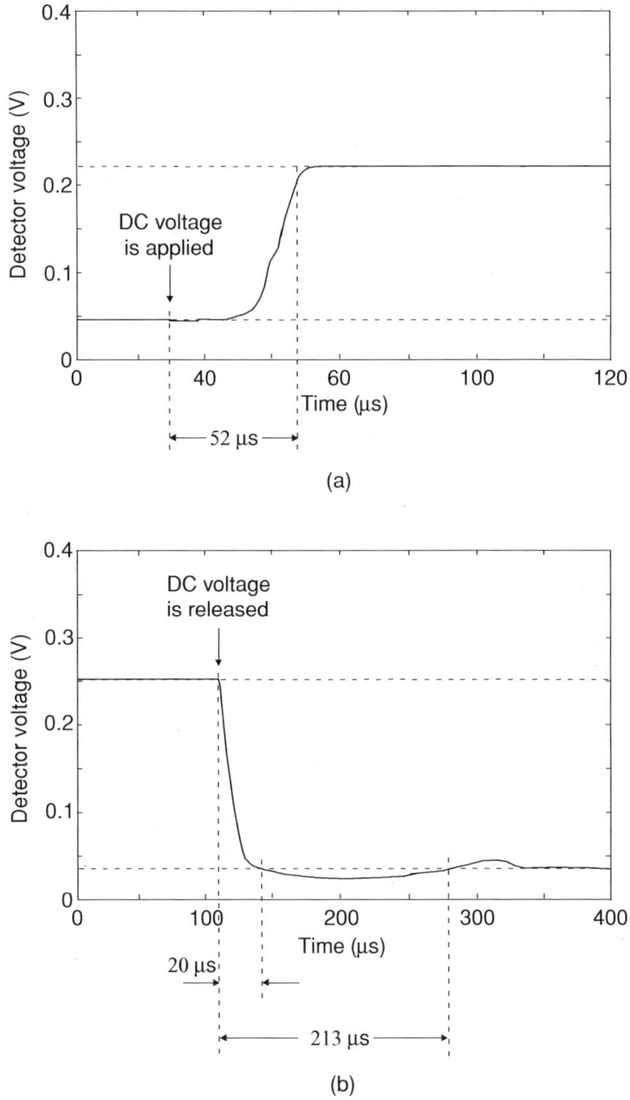

Figure 2-15 Measurements of the switching time for the (a) up-down; and (b) down-up movements. Because of the diode detector, high voltage level corresponds to low RF power and vice versa.

the switch only 25 μs to reach its normal height, but another 190 μs are needed to settle within 5% of its original height. It is also interesting to note that during the first 30 μs of the actuation stage the switch does not move significantly. These effects will be discussed in the remaining of this section.

To explain the experimental results a simple 1-D nonlinear model adopted by several researchers [57, 73, 74] is being employed here. This model treats the switch as a single lumped mass and applies classical Newtonian mechanics to predict its behavior under the applied electrostatic force. A model that would accurately predict the dynamic behavior of the MEMS structure should integrate a good understanding of several different phenomena including electrostatics, mechanics, residual stress, contact forces, compressible squeeze-film damping and impact effects on a microscale. A lot of these areas are currently under investigation and there is not a complete model that accounts for all of these effects. Furthermore, our switch is a relatively large structure that cannot be accurately approximated as a lumped mass. Nevertheless, the 1-D model can be used for qualitative explanation of the measured switch behavior and provide reasonable approximations for the switching times.

The following equation of motion is the basic formula for the 1-D model

$$mz'' + bz' + K_z z = F_e + F_c \tag{2.21}$$

where m is the switch mass, b is the damping coefficient, K_z is the switch spring constant in the direction of motion, A is the switch actuation area, g_0 is the initial gap, ϵ_r and t_d are the dielectric layer constant and thickness respectively, V is the applied DC voltage, F_e is the electrostatic force and F_c is the contact force when the switch touches the dielectric. Several approximations may be adopted for calculating the parameters in this model. For instance, although the viscous damping can be considered constant for small displacements, this is not the case when the switch is moving completely towards the substrate. The model presented here for the switching speed calculations is based on the discussion presented in [73] and [75] where these effects are taken into account. These equations can be summarized as follows

$$F_e = \frac{\epsilon_0 A V^2}{2\left(\frac{g_0 + t_d}{\epsilon_r - z}\right)^2} \tag{2.22}$$

$$b = \frac{K_z}{\omega_0 Q} \quad \text{where} \quad \omega_0 = \sqrt{\frac{K_z}{m}} \tag{2.23}$$

$$Q = Q_0 \left(1 - \left(\frac{z}{g_0}\right)^2\right)^{3/2} \left(1 + 9.638\left(\frac{\lambda}{g_0 - z}\right)^{1.159}\right) \tag{2.24}$$

$$F_c = \frac{m_1 A}{(g_0 - z)^3} - \frac{m_2 A}{(g_0 - z)^{10}}\ . \tag{2.25}$$

Equation (2.24) calculates the switch quality factor and takes into account the damping dependence on the switch height. If this is ignored, the second term of the right hand side should be replaced by one. Furthermore, the third term reduces the gas-flow resistance underneath the switch because of the slip effect, were particles can have fewer interactions before escaping [76]. The variable λ of this term is called the mean free path and is approximately 0.1 μm at STP. The damping coefficient, which is related to Q_0 by Equation (2.23), was derived in [77] for a square plate with area A as

$$b = \frac{3}{2\pi}\frac{\mu A^2}{{g_0}^3} \tag{2.26}$$

where μ is the air viscosity (at STP $\mu = 1.845 \cdot 10^{-5}$ Pa·s). For the switch dimensions and for a gap of 5 μm, $b \simeq 2.5 \cdot 10^{-4}$ Pa·s and $Q_0 = 0.64$ (for $K_z = 8.6$ N/m). However, the holes included in the switch allow the air underneath to escape more easily, thus reducing the damping coefficient and increasing the Q of the structure. Therefore, this value can be considered as a lower bound for the switch quality factor. In fact, the experimental results suggest a quality factor of about two. The final equation of the model, Equation (2.25), was used to provide a stable solution to the simulation when the switch contacts the dielectric layer.

Figure 2-16(a) shows the simulated results for the pull-in and release times. A spring constant of 8.6 N/m (Table 2-2) and an actuation voltage 25% higher than the pull-in voltage have been used for these simulations. These simulations provide a valuable insight into the measured dynamic behavior of the switch. The pull-in time, for instance, is approximately 50 μs from which about 30 μs are needed for the switch to move from 5–3.5 μm. The RF capacitance, however, does not change appreciably over this distance and this explains the relatively long period that is required to note any difference between the measured output power levels, see Figure 2-15. On the other hand, during the release stage the switch reaches its original height within 35 μs, but 140 additional μs are required for stabilization within 5% of its original height. In the same figure the simulated release time has also been plotted assuming a constant quality factor $Q = Q_0$. Evidently, taking into account the quality factor variation with height is of vital importance in obtaining meaningful simulations.

The problem of the long stabilization time can be easily corrected by decreasing the quality factor to about one. Figure 2-16(b) shows the simulated release time for quality factors of 2, 1.5, and 1. This figure clearly demonstrates that the release time can be decreased by more than a factor of three if a lower quality factor is achieved. This can be done by optimizing the hole orientation on the main switch structure and by reducing the number of holes per unit area. This change will not appreciably increase the pull-in time as Figure 2-16(c) shows. For example, with $Q = 1.5$, the pull-in time will be increased by only 10%, yet the release time will be reduced by 300%. Consequently, optimization of the switch holes can lead to a design with switching times in the order of 50–60 μs.

2.2.4 RF Power Handling

The previous section focused on the mechanical design, residual stress and switching speed of a low voltage RF MEMS switch. The maximum RF power that MEMS switches can successfully handle, is an equally important issue for a number of different applications, such as transmitters in satellite and/or earth-based communications stations. Research in this area could initiate a significant effort toward reducing the weight, cost, and power required for these stations.

The main purpose of this section is to provide a theoretical and experimental study that addresses the aforementioned issues of the switch. In particular, Section 2.2.4.1 briefly describes the RF properties used in the experiments reported in Sections 2.2.4.2–2.2.4.4. Sections 2.2.4.2 and 2.2.4.3 discuss the problems of self-actuation and stiction in the down state due to high RF power. These findings suggest a few guidelines for designing switches specifically suited for high-power applications. It will be shown for example that de-coupling the RF and DC actuation pads can substantially increase the power handling capabilities of the MEMS switch. Finally, Section 2.2.4.4 presents a novel switch architecture with a top electrode above the switch to enhance its behavior during hot switching.

2.2.4.1 RF Switch Performance and Equivalent Circuits Figure 2-17 shows a lumped-element equivalent circuit for the low-voltage switch structure presented earlier in this section that is accurate up to 80 GHz [78]. Typical up and down-state measurements are presented in Figure 2-18 along with the simulated results from this model. The switch presents an up-state insertion loss of 0.04 and 0.17 dB at 20 and 40 GHz respectively. The return loss is less than −12 dB up to 40 GHz. Based on

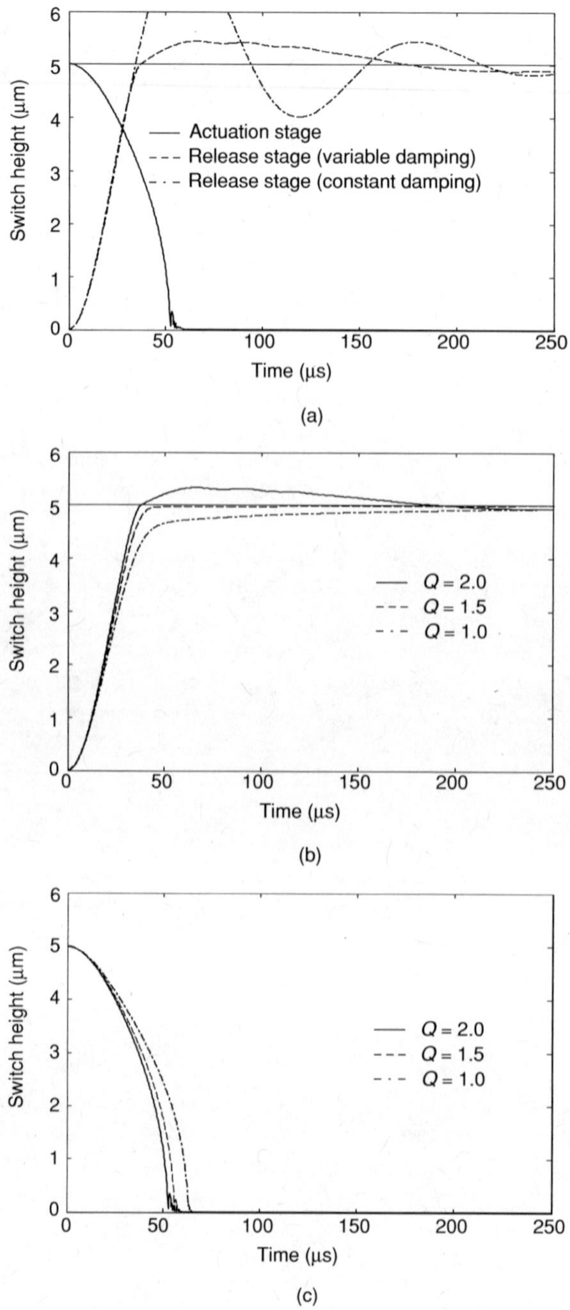

Figure 2-16 Pull-in characterization: (a) Simulated switch pull-in and release times; (b) comparison of simulated release times for different values of the quality factor; and (c) comparison of simulated pull-in times for different values of the quality factor.

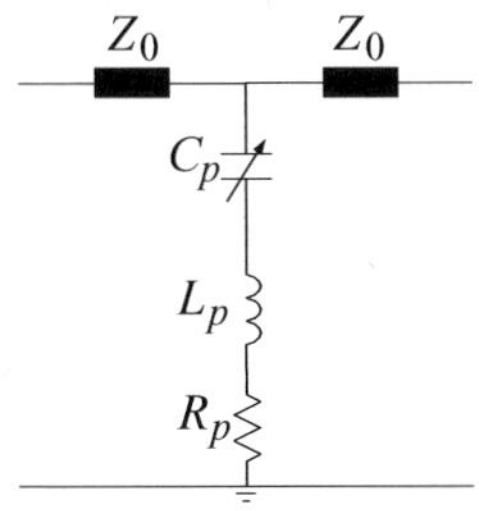

Figure 2-17 Lumped element equivalent model for the low voltage switch of Figure 2-1.

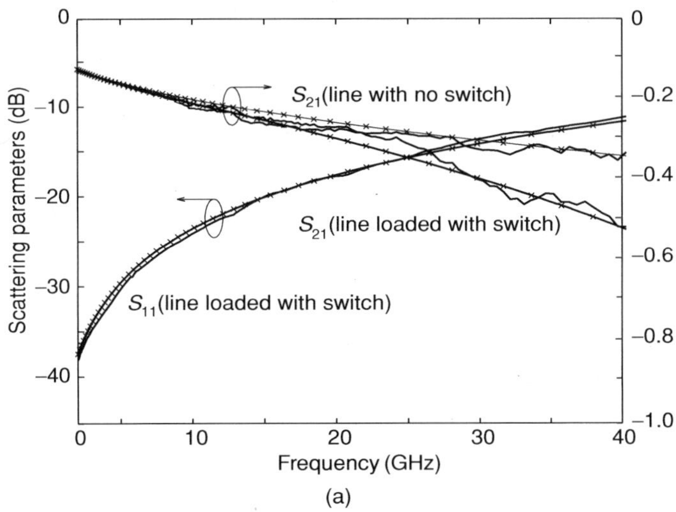

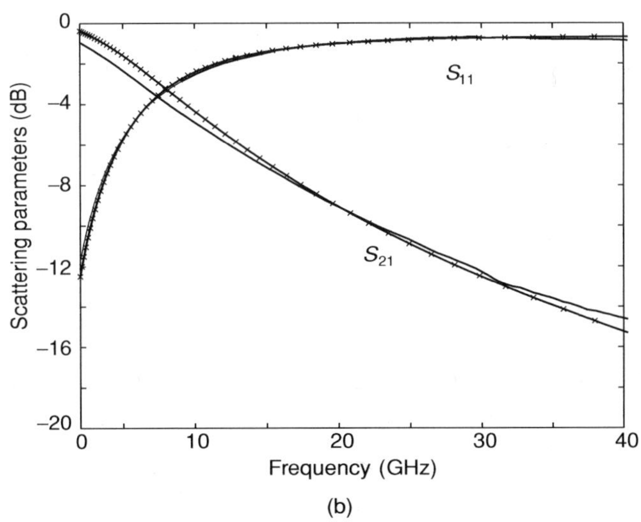

Figure 2-18 Measured and simulated RF performance of the switches used for the high-power measurements in the (a) up and (b) down states.

the measured return loss, an up-state capacitance of approximately $C_U = 40$ fF can be extracted. The down-state isolation depends strongly on the switch geometry, the dielectric thickness and roughness [55]. The maximum isolation can be also tuned to the desired frequency by changing the inductance L_p, which depends on the connecting beams of the switch [69]. The switches measured in this study have been fabricated with 2000 Å of Si_3N_4, and have the connected beams shown in Figure 2-1 that result in $L_p = 2 - 3$ pH. These switches exhibit a maximum measured isolation of approximately 15.5 dB at 40 GHz (Figure 2-18), which corresponds to a down-state capacitance of approximately $C_D = 800$ fF.

2.2.4.2 Case 1: Self-Actuation Due To High RF Power

2.2.4.2.1 Design and Modeling Although a DC voltage is applied to bias electrostatic switches, these structures respond to any voltage with frequency lower than the switch self-resonance, which is typically in the order of a few kHz. For example, a harmonic RF voltage $V(t) = V_0 \cos(2\pi f t)$ applied between two parallel metallic plates in air induces an electrostatic force [75]

$$F_{RF} = -\frac{1}{2}\frac{\epsilon_0 A \left(\frac{V_0}{\sqrt{2}}\right)^2}{g^2}(1 + \cos(4\pi f t)) \tag{2.27}$$

where A is the area of the plates and g the distance between them (fringing capacitance has been ignored). Although the high-frequency component of this force will not cause any significant movement, the low-frequency force is equal to the force that would have been experienced by the metallic plates, if an equivalent DC voltage

$$V_{eq} = \frac{V_0}{\sqrt{2}} = V_{RF}^{\text{rms}} \tag{2.28}$$

had been applied to them. Equation (2.27) implies that if the applied RF power is sufficiently high, the induced force may become sufficiently strong to pull the switch down without any DC bias voltage. Equivalently, the switch will be actuated if the voltage V_{eq} is higher or equal to the pull-in voltage given by

$$V_p = \sqrt{\frac{8K_z {g_0}^3}{27\epsilon_0 A}} \tag{2.29}$$

where K_z is the equivalent spring constant of the moving structure in the direction of desired motion (typically z-direction), g_0 is the gap between the switch and the actuation electrode, ϵ_0 is the free space permittivity and A is the switch area where the electrostatic force is applied. This phenomenon is called self-actuation and is characteristic of high RF power.

To model the self-actuation mechanism, the ideas presented in [75] have been followed. Consider an input power

$$P_{\text{in}} = \frac{V_0^2}{2Z_0} \tag{2.30}$$

to the CPW line underneath the switch (see Figure 2-19). Since the switch is in the up state, almost zero power will be reflected. Equations (2.28), (2.29) and (2.30) readily show that the

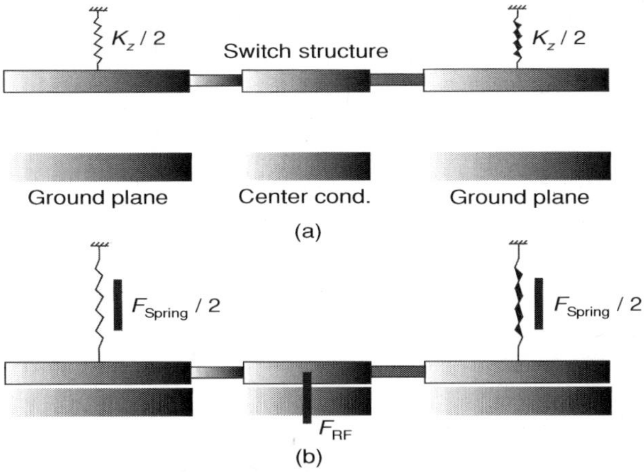

Figure 2-19 Schematic representation of the self-actuation mechanism: (a) the switch is initially in its up state over a CPW line; and (b) the incident power on the CPW line is higher than the pull-down power and the switch is actuated with no bias voltage present.

required power to actuate the switch is

$$P_{\text{act}} = \frac{8K_z g_0^3}{27\epsilon_0 A_{RF} Z_0} \tag{2.31}$$

where A_{RF} is the switch area that interacts with the applied RF power. This power will be referred to as the minimum pull-down RF power.

Although for typical air-bridge-type switches [53,79] A_{RF} can be approximated as the switch area over the center conductor A_{cc} of the CPW line (fringing fields are ignored in this approximation), it is not straightforward that this approximation holds for the switch of Figure 2-1. Since this switch is not directly connected to the CPW ground planes, part of the total attractive force will be distributed on the switch pads that are suspended over the CPW ground planes. Whereas this may initially suggest that the value of A_{RF} used in Equation (2.31) has to be larger than A_{cc}, this is not necessarily true. The reason lies in the fact that Equation (2.31) assumes a potential difference of $V_{eq} = V_0/\sqrt{2}$ between the switch and the center conductor, which is not the case for our switch. Our switch is not anchored to the ground planes and, therefore, its potential is higher than zero. Consequently, the potential difference between the switch and the center conductor will be lower than V_{eq}.

These arguments suggest that an equivalent value of the area A_{RF}—suitable for Equation (2.31)—can be calculated by comparing the actual attractive force F_{actual} with the well-known parallel-plate approximation

$$F_{pp} = \frac{1}{2}\frac{\epsilon_0 w V_{eq}^2}{g_0^2} \text{ N/m} \tag{2.32}$$

where w is the width of the CPW center conductor. The actual attractive force can be calculated by using a quasi-static approximation as sketched in Figure 2-20. The switch beam in this simulation is placed 4 μm above the CPW line and is also treated as a floating conductor with zero total charge. This represents the fact that the switch is not in direct contact with the CPW line. This

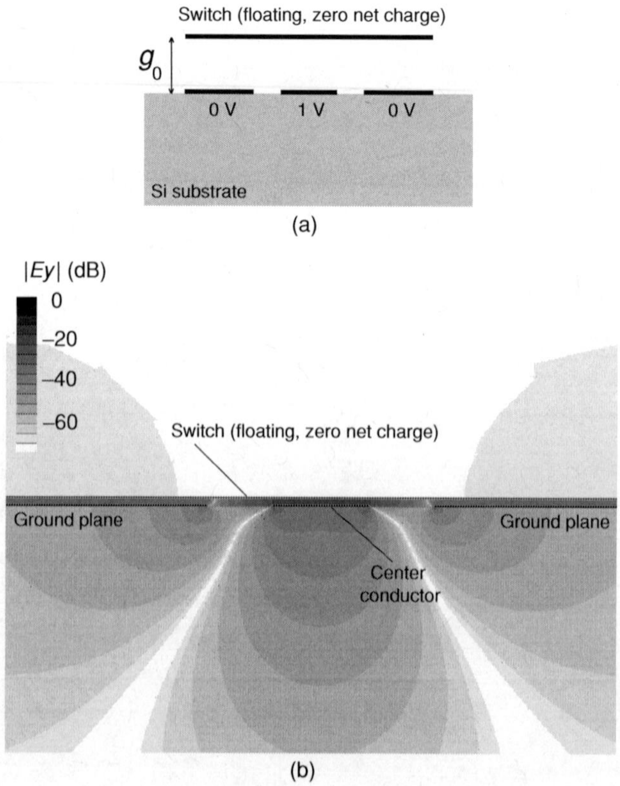

Figure 2-20 Quasi-static model: (a) 2-D quasi-static simulated model (not drawn to scale for clarity) of the switch over the CPW line for field and force calculations; and (b) detail of the simulated magnitude of the electric field vertical component. The results of this simulation were used to calculate the total attractive force between the switch and the CPW line.

electrostatics problem was numerically solved by Ansoft's Maxwell code [80]. The CPW line was excited with a total voltage difference of 1 V (see Figure 2-20) and the resulting attractive force was calculated to be

$$F_{\text{actual}} = 16.47\ \mu\text{N/m}\ . \tag{2.33}$$

Using the same geometrical parameters and with $V_{eq} = 1$ V Equation (2.32) results in

$$F_{pp} = 16.60\ \mu\text{N/m} \tag{2.34}$$

Comparing Equations (2.33) and (2.34) it can be seen that the following equation

$$A_{RF} \simeq A_{cc} \tag{2.35}$$

is a very good approximation to be used in Equation (2.31).

It is also interesting to note that in the design shown in Figure 2-1 the bias voltage is primarily applied between the CPW ground planes and the switch pads above them. As a result, the switch area responsible for the DC actuation A_{DC} is significantly larger than A_{RF}. In particular, as shown by the geometry of Figure 2-1,

$$A_{\text{DC}} = 2A_{gp} \tag{2.36}$$

where A_{gp} is the area of the switch pad above each of the ground planes. Additionally,

$$A_{gp} = 4A_{\text{cc}} \,. \tag{2.37}$$

Consequently, Equations (2.35), (2.36), and (2.37) lead to

$$A_{\text{DC}} \simeq 8A_{RF} \tag{2.38}$$

This is a significant advantage of this design compared to traditional CPW fixed-fixed beam switches [53,79] where

$$A_{\text{DC}} \simeq A_{RF} \qquad \text{(fixed-fixed beam cpw switches)} \tag{2.39}$$

Since A_{DC} and A_{RF} control the actuation voltage and pull-down power respectively (see Equations (2.29) and (2.31)), de-coupling of these areas significantly improves the design flexibility for this switch. Compared to the fixed-fixed beam geometries, for instance, a much higher RF pull-down power can be achieved for a given actuation voltage. Equivalently, a much lower actuation voltage can be obtained for a given spring constant and pull-down power. Typical values for A_{DC} and A_{RF} as well as for other switch parameters are given in Table 2-3. On the other hand, the drawback of this approach is that the achieved down-state capacitance is slightly reduced by the series combination of the capacitances C_1 and C_2, where C_1 (C_2) is the capacitance between the center conductor (ground plane) and the switch pad above it. The achieved capacitance is

$$C = \frac{C_1(2C_2)}{C_1 + 2C_2} = 0.89C_1 \tag{2.40}$$

Table 2-3 Typical parameters for the switches analyzed in this chapter.

Parameter	Dimension
Center conductor pad (A_{cc})	$250 \times 60\ \mu\text{m}^2$
Ground plane pad A_{gp}	$250 \times 250\ \mu\text{m}^2$
Spring constant[a] – 1 meander (K_{z1})	26.9 N/m
Spring constant – 2 meanders (K_{z2})	14.5 N/m
Spring constant – 3 meanders (K_{z3})	8.6 N/m
Spring constant – 4 meanders (K_{z4})	3.6 N/m
Spring constant – 5 meanders (K_{z5})	1.1 N/m
Mean switch height (g_0)	5 μm
CPW Line impedance (Z_0)	50 Ω
Down-state capacitance	0.8 pF
Dielectric layer thickness (t_d)	2000 Å
Effective dielectric constant (ϵ_r)	7.5
Extracted air-gap[b] (d_{gap})	600 Å

[a] All the spring constants are extracted from measurements in [55].
[b] Extracted by comparing the down-state capacitance and the capacitance that would have been achieved if the switch had perfect contact with the dielectric layer underneath it.

since $C_2 = 4C_1$ for the geometry of Figure 2-1. This shows that for a 10% reduction in the down-state capacitance, the actuation voltage can be reduced by a factor of $\sqrt{8} = 2.83$ for a given spring constant and maximum sustained RF power.

One more point that needs to be addressed with regard to Equation (2.31) is the treatment of the 8-μm holes in the switch structure. These holes facilitate the switch release process, partially relieving the residual stress in the structure and allowing the air underneath the switch to escape during actuation [55]. However, they also reduce the capacitance between the switch and the CPW line. Their effect is only of minor importance in the up state, because of the fringing fields that "fill" the area of the holes. Their effect on the down-state capacitance, however, is significant since the dielectric layer thickness is very small and the fringing field negligible. Consequently, the following formulas are used throughout this work for A_{RF}

$$A_{RF} = \begin{cases} 250 \times 60 = 15,000\ \mu\text{m}^2 & \text{up state} \\ 8,800\ \mu\text{m}^2 & \text{down state} \end{cases}$$

Figure 2-21 shows the minimum calculated RF power required to cause self-actuation for switches with one to five meanders and for various gaps. The parameters used for these calculations are given in Table 2-3. This plot reveals the significance of allowing adequate distance between the switch and the RF circuit, particularly for low spring constant switches.

2.2.4.2.2 Experimental Results and Discussion The experimental characterization of the switches was performed by the measurement set-up shown in Figure 2-22. The RF signal in this setup is produced by an X-band-synthesized sweeper. After passing through a variable 0–50 dB attenuator, it is amplified by an X-band traveling wave tube (TWT) amplifier that produces a maximum power level of about 7 W. The amplified signal is then guided through an isolator to the switch under test. A portion of the output signal is then routed to a power meter through a 10 dB coupler and a 20 dB attenuator. The measurements were performed at 10 GHz because of the limitations imposed by the TWT amplifier. Although the switch provides an isolation of only 3–4 dB at this frequency, this isolation is significantly higher than the measurement error and therefore it is still possible to determine the presence of any self-actuation issues.

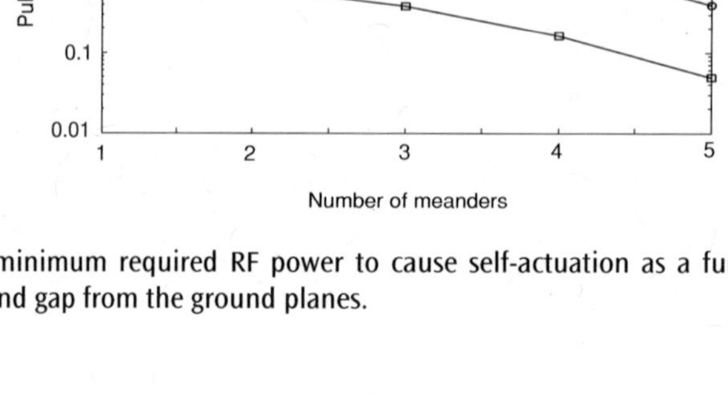

Figure 2-21 Calculated minimum required RF power to cause self-actuation as a function of the number of meanders and gap from the ground planes.

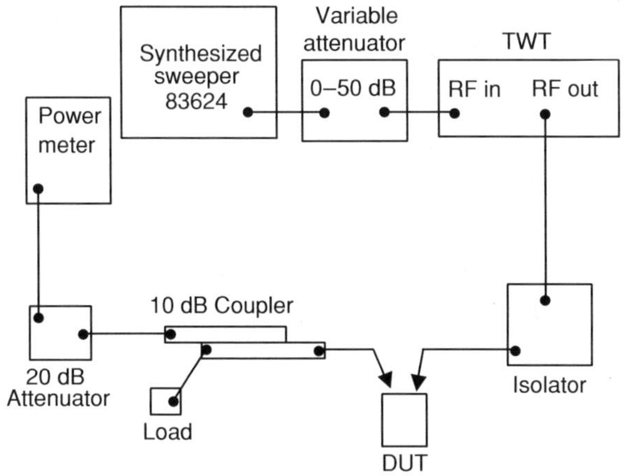

Figure 2-22 X-band power measurement setup.

Switches with one to four meanders were tested with this setup and were actuated with a bias voltage of 10–40 V. The mean gap g_0 was approximately 4.8–5.2 μm for all the switches. All the switches were initially tested in the up-state with an input power of approximately 0.7 mW. Then the RF power was slowly increased up to a maximum of about 5.5 W by decreasing the attenuation of the variable attenuator. No self-actuation was observed for any of the switches at any stage. The fact that the maximum recorded power was less than 7 W is explained by the losses of the connections and the cables introduced between the TWT and the DUT. The maximum output power of the TWT amplifier (7 W) was another limitation in this setup and therefore no self-actuation could be experimentally observed [81]. The same procedure was repeated with the switch in the down state. No catastrophic failure of the switch or of the dielectric underneath it was observed in these measurements. Only for high power levels (higher than 3 W) was slight bending of the switch springs observed. The springs returned to their original positions when the power was turned off. This is due to the thermal effects associated with the heat generated due to the input power. Figure 2-23 shows the recorded results for a typical switch with three meanders versus the reading of our variable attenuator. The 3–3.5 dB difference between the power levels of the switch up and down states is explained by the switch isolation at 10 GHz (see Figure 2-18). These results agree well with the theoretical expectations, since according to Equation (2.31) an input power of 48 W would be required for self-actuation.

2.2.4.3 Case 2: Switch Stiction Due To High RF Power

2.2.4.3.1 Modeling Although the minimum pull-down power may be in the order of several watts, this is not the case for the hold-down power, i.e. the minimum RF power required to hold a switch in the down position without any bias voltage. A simple schematic for this scenario is shown in Figure 2-24. Initially it is assumed that the switch is in the up state and an incident power P_{in} is applied to the CPW line underneath the switch. It is also assumed that this power is not sufficiently high to cause self-actuation, i.e. $P_{\text{in}} < P_{\text{act}}$. Then the switch is actuated by a bias voltage higher than the pull-in voltage (Figure 2-24(b)). An effective air gap of d_{gap} has been introduced to model the lack of perfect contact between the switch and the dielectric layer underneath it. This air gap is due to the surface roughness between the contact areas and the inability of the switch to conform perfectly on the dielectric layer underneath it. For example, the switches in this study

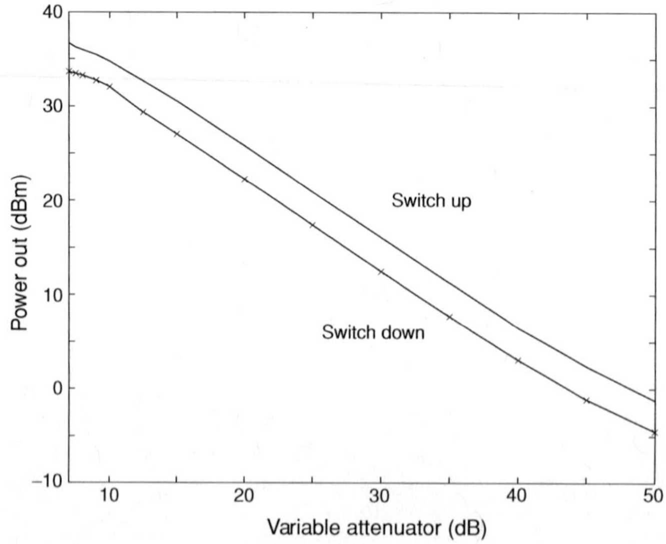

Figure 2-23 Measured power levels for the up and down switch states.

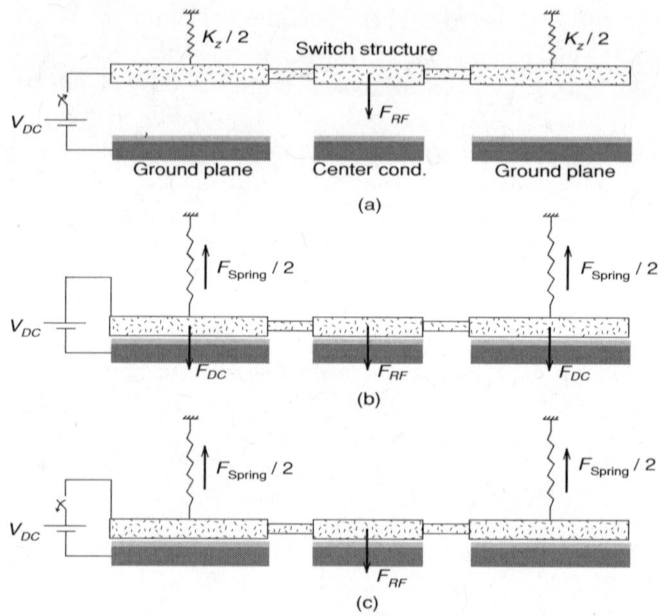

Figure 2-24 Simplified scenario for hold-down power calculations: (a) initially the switch is in the up state and an incident RF power (less than the pull-down power) is applied to the line underneath it; (b) at some point a bias voltage (higher than the actuation voltage) is applied and the switch is actuated; and (c) the bias voltage is removed afterwards, but the switch remains in the down state since the RF power exerts an RF force on the switch higher than the restoring force.

were approximately 3-μm thick and perfect conforming could not be achieved. The relatively low measured isolation (Figure 2-18(b)) is mostly due to this factor. By comparing the switch down-state capacitance (0.8 pF) and the capacitance that could have been achieved if the switch had perfect contact with the dielectric layer underneath it (2.6 pF), an effective air gap of about 600 Å can be deduced. The forces applied to the switch when in its down state are given by [75]

$$F_{RF} = -\frac{1}{2}\frac{\epsilon_0 A_{RF} |V_{\text{sw}}|^2}{\left(d_{\text{gap}} + \frac{t_d}{\epsilon_r}\right)^2}$$

$$F_{\text{Spring}} = K_z g_0 \tag{2.41}$$

and

$$F_{\text{DC}} = \frac{1}{2}\frac{\epsilon_0 (A_{\text{DC}}) V_{\text{DC}}^2}{\left(d_{\text{gap}} + \frac{t_d}{\epsilon_r}\right)^2}$$

where V_{sw} is the switch RF voltage. When the DC voltage vanishes, the switch will return to its initial state unless $|F_{RF}| \geq |F_{\text{Spring}}|$, or equivalently,

$$\frac{1}{2}\frac{\epsilon_0 A_{RF} |V_{sw}|^2}{\left(d_{\text{gap}} + \frac{t_d}{\epsilon_r}\right)^2} \geq K_z g_0. \tag{2.42}$$

Equations (2.28) and (2.30) cannot be applied directly in this case to estimate V_{sw} since they are associated only with the incident power on the switch. The switch being in the down state, however, generates a reflected voltage, which greatly reduces the total RF voltage across it. Consequently the total voltage can be calculated as $V_{\text{sw}} = V_{RF}(1 + \Gamma)$, where V_{RF} is the amplitude of the incident wave and Γ is the reflection coefficient due to the shunt switch capacitance on the line. Standard transmission line theory yields [75]

$$|V_{\text{sw}}| = \frac{|V_{RF}|}{\sqrt{1 + (\pi f C Z_0)^2}}. \tag{2.43}$$

For all frequencies below 80 GHz this voltage can be considered almost constant along the switch width, since the switch is at least an order of magnitude smaller than the wavelength. Combining Equations (2.28), (2.30), (2.42), and (2.43) the following formula is obtained for the RF power required to hold the switch down

$$P_h = \frac{2K_z g_0 (d_{\text{air}} + t_d/\epsilon_r)^2 \left[1 + (\pi f C_D Z_0)^2\right]}{\epsilon_0 A_{RF} Z_0}. \tag{2.44}$$

2.2.4.3.2 Experimental Results and Discussion The experimental setup of Figure 2-22 was also used to measure the hold-down RF power. In this case, however, the maximum incident RF power was limited to 0.8 W. (These measurements will be compared to the experiments presented in Section 2.2.4.4, where the maximum applied power is limited to 0.8 W). Switches with one, three, and five meanders were tested with this set-up and the results are shown in Table 2-4 [81]. No power-stiction problems were observed with one-meander switches, which is in agreement with our theoretical expectations. Furthermore, the measured and calculated hold-down powers for

Table 2-4 Measured and calculated hold-down RF power for switches with one, three, and five meanders.

Meanders	P_b (mW) (Calc.)	P_b (mW) (Meas.)
1	1335	—
3	426	500
5	54.6	63

three- and five-meander switches are also in reasonable agreement with the calculated powers from Equation (2.44). The minor discrepancies (in the order of 15%) are due to the variations in height and the down-state capacitance of the measured switches as compared to the nominal values used in the theoretical calculations.

These results indicate that the 1-D model is physically meaningful and can be used in predicting the performance of the same switches at different frequency bands. For instance, Equation (2.44) shows that the hold-down RF power is proportional to the square of the operating frequency. Consequently, if the same measurements had been performed at 40 GHz, the hold-down power for switches with one and five meanders would have been 13.6 and 0.55 W respectively. At this frequency the switch presents a much improved short for the RF signal, hence limiting the RF voltage across it to a much smaller value than before. Figure 2-25 graphically illustrates these results. One has to be careful, however, in interpreting these simulations. Thermal issues, for example, [82,83] may cause different problems well before high power levels are reached. Additionally, the dielectric layer underneath the switch is typically a temperature-sensitive material and it may lead to additional stiction problems at sufficiently high temperatures. This particularly applies to the case of non-heat-conducive substrates (e.g., quartz). These thermal effects have not been taken into account in our analysis and their study requires a completely different treatment.

One more interesting phenomenon is related to the hold-down power as a function of the switch down-state capacitance. Typically the down-state capacitance is limited by the ability of the switch to conform to the dielectric underneath it and the surface roughness of the dielectric and metals. Both of these factors influence the maximum capacitance by limiting the contact between the switch and the dielectric layer. It has already been mentioned that these effects can be modeled by an equivalent air gap d_{gap} that typically varies between 50 and 1000 Å. Higher down-state capacitance,

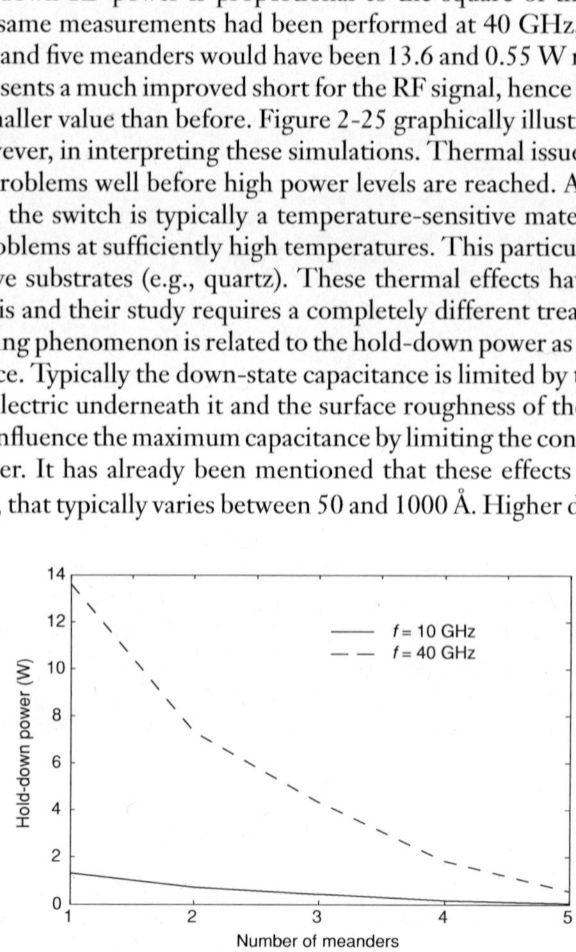

Figure 2-25 Simulated hold-down power for the same switch at two different frequencies. The sustained hold-down power is higher at 40 GHz because the switch presents a better short circuit at that frequency.

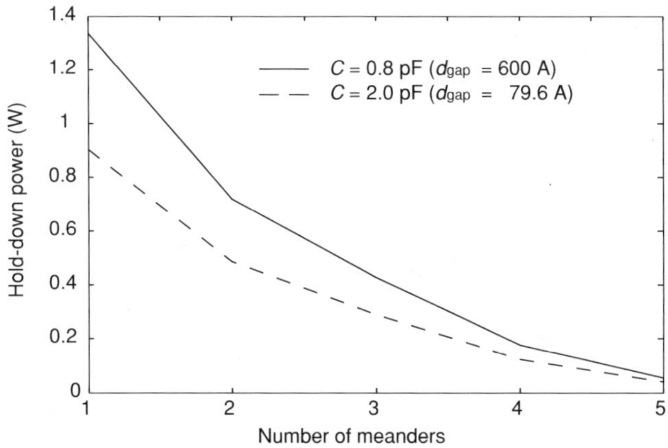

Figure 2-26 Simulated hold-down for different air gaps. A smaller air gap leads to a higher down-state capacitance but also to a lower hold-down RF power.

therefore, can be achieved by either increasing the switch area (which will also increase the up-state insertion loss) or by decreasing the equivalent air gap. If the switch area is increased, but the air gap remains essentially the same, higher hold-down power can be achieved for reasons described in the previous paragraph. On the other hand, if the switch capacitance is increased by limiting the effective air gap, a lower hold-down power may actually be obtained. The reason is that although a lower RF voltage exists across the switch, due to its higher capacitance, the equivalent RF force is also higher, see Equation (2.41). Figure 2-26 shows two examples using the values of Table 2-5 for capacitances of 0.8 and 2 pF. These correspond to air gaps of 600 and 79.6 Å respectively. This figure clearly shows the trade-offs between the maximum down-state isolation and power-stiction problems, particularly for switches with spring constant higher than 10 N/m.

2.2.4.4 Case 3: Switch with Top Electrode

2.2.4.4.1 Design and Fabrication The spring-restoring force is the only force that is usually considered for returning the switch to its initial state after it has been actuated. This force can be relatively weak for low-voltage switches, especially if the moving distance is limited (less than 2 μm). As a result, it may be insufficient for reliable switching in high-power applications. A possibility of having a top electrode above the switch is introduced and discussed here.

The top electrode is electrically isolated from the switch and the CPW line underneath it and it can be used to pull the switch from the down state to the up position. Figure 2-27 shows the fabrication process followed for this electrode after the switch is completed, but before it is released. In particular, in the first stages (the initial process can be found in [56]) the circuit underneath the switch is deposited, followed by the deposition and patterning of the dielectric layer (Si_3N_4). Spinning and patterning of the first polyimide layer (first sacrificial layer) follows and then a seed layer of Ti/Ni is deposited. This seed layer is electroplated to define the switch structure. After the seed layer is removed, a SiO_2 (approximately 5000 Å thick) layer is sputtered and patterned with an RIE process on top of the switch. This dielectric layer is introduced to prevent any DC short circuit between the switch and the top electrode. A new sacrificial layer of polyimide PI2545 is then spun at 3–4 krpm for 30 s. and cured at 150°C for 15 min. This results in a thickness of approximately 2–2.5 μm. It was experimentally found that layers with smaller thicknesses were particularly difficult to remove afterwards. The thickness of the first sacrificial layer was also adjusted to 1.4–1.6 μm

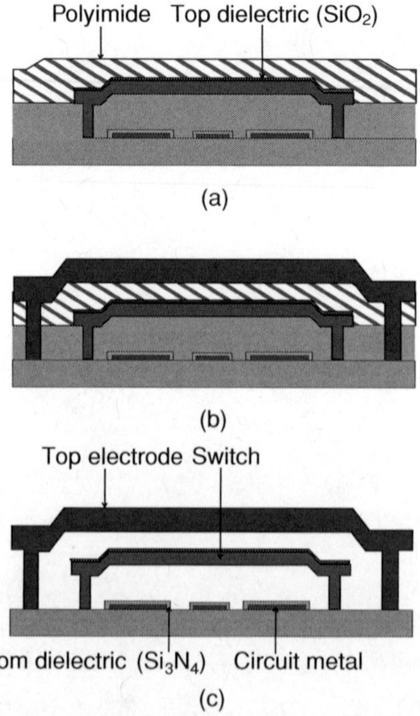

Figure 2-27 Fabrication process of the switch top electrode. (a) A SiO_2 layer is sputter deposited and defined on top of the switch. Then a polyimide layer is spun deposited and serves as a second sacrificial layer; (b) the anchor points of the top electrode are defined. A seed layer of Ti/Au/Ti is then deposited and electroplated to define the top electrode; and (c) the polyimide layers are etched and the structure is dried with a CO_2 supercritical process.

in order to limit the maximum distance between the top-electrode and the CPW line to about 4 μm. The anchor points of the top electrode are then patterned with a process very similar to the one followed for the switch anchor points. The next step is to deposit a seed layer of Ti/Au/Ti (500/2000/500 Å). The seed layer is electroplated with Au typically up to 4–5 μm to shape the top electrode. Eventually the two sacrificial layers are etched in hot PRS2000 and the whole structure dried using a standard supercritical CO_2 process.

Figure 2-28 shows two examples of fabricated switches (with three and four meanders) with top electrodes. Due to the switch symmetry, one electrode above each of the switch actuation pads is fabricated. However, no top electrode is placed above the center conductor pad, because such an electrode would deteriorate the RF performance of the switch due to additional parasitic capacitance. This does not significantly reduce the electrostatic force between the top electrodes and the switch, since the center conductor pad is significantly smaller that the other two pads. An additional advantage of the top electrodes is that they are extremely stiff, hence enhancing switch stabilization. Even without accounting for any axial stresses, the calculated spring constant for each of these electrodes is in the order of 2300 N/m. A DC voltage in excess of 230 V would be, therefore, required for any significant movement.

2.2.4.4.2 Modeling Besides switch stabilization, the main purpose for introducing the top electrodes is to enhance the power handling capabilities of the low-voltage switches. The most useful (and at the same time worst-case) scenario that was theoretically and experimentally

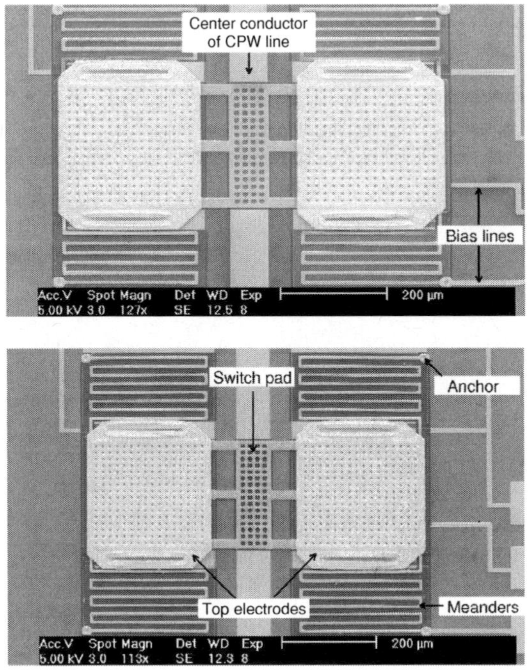

Figure 2-28 Two examples of switches with fabricated top electrodes.

investigated [81] is illustrated in Figure 2-29. It is assumed that the switch is initially in the down-state position, either because of self-actuation, or because of DC biasing. In both cases, the RF power is considered sufficient to counteract the restoring force, forcing the switch to remain down. To move the switch to the up state, a DC voltage is applied between the switch and the top electrode. If this voltage is sufficiently high the switch will move upwards and eventually touch the top electrode. In obtaining the general force expressions it is assumed that the switch has moved by a distance z from the dielectric layer (Figure 2-29). In the following equations $g_0 = 1.5\ \mu$m is the distance between the circuit and the initial switch state ($F_{\text{spring}} = 0$), and $g_{\text{up}} = 4\ \mu$m is the distance between the circuit and the top electrode. The forces acting on the switch are then given by

$$F_{RF} = -\frac{1}{2}\frac{\epsilon_0 A_{RF}|V_{\text{sw}}|^2}{\left(\frac{d_{\text{gap}}+t_d}{\epsilon_r+z}\right)^2}$$

$$F_{\text{spring}} = K_z\,(g_0 - z) \tag{2.45}$$

and

$$F_{UP} = \frac{1}{2}\frac{\epsilon_0 A_{\text{DC}} V_{\text{up}}^2}{\left(\frac{g_{\text{up}}+t_{d2}}{\epsilon_{r2}-z}\right)^2}\ .$$

The meaning of the symbols is the same as in Formula (2.41), with the exeptions a) V_{up} is the applied DC voltage between the top electrodes and the switch, and b) t_{d2} and ϵ_{r2}, are the thickness

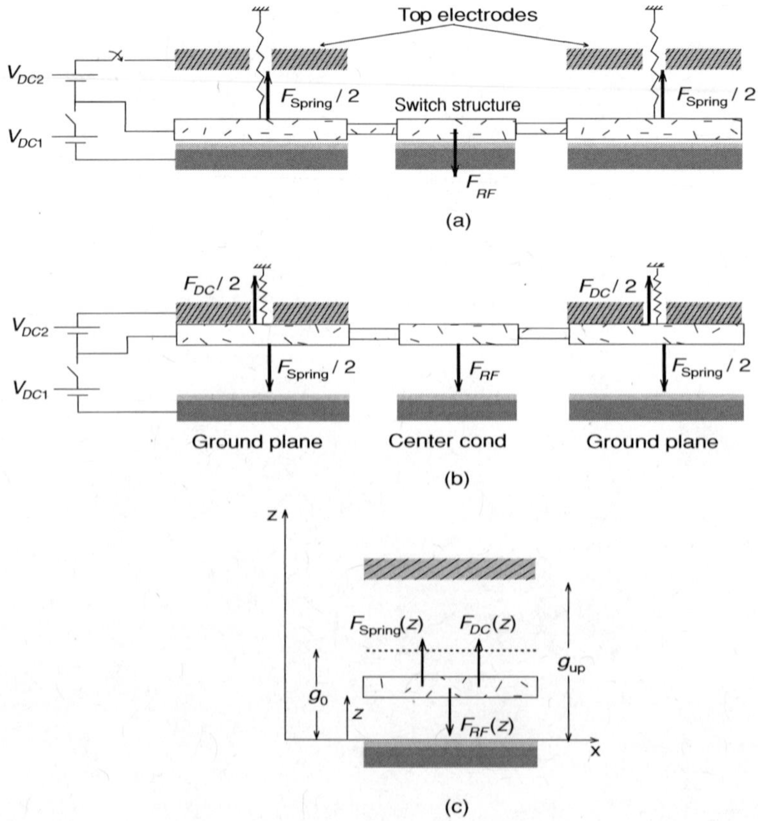

Figure 2-29 Simplified scenario for modeling the top-electrode effect in power-stiction problems: (a) the switch is in the down state either by self-actuation or by DC biasing; (b) a DC voltage is applied between the switch and the top electrode and the switch is moved upwards and stops to the top electrode; and (c) simple 1-D schematic that demonstrates the geometrical details of the switch design.

and dielectric constant of the SiO_2 film underneath the top electrodes. It has also have to be noted that A_{DC} is smaller than $2A_{\mathrm{gp}}$ because of the holes that exist in both the switch and the top electrode. Consequently, the fringing fields are weak in the areas of the holes and can be ignored. Hence, A_{DC} includes only the overlapping area between the top electrodes and the switch and is given by $A_{DC} = 2 \times 38,0000\ \mu\mathrm{m}^2 = 76,000\ \mu\mathrm{m}^2$.

Based on these equations, a simple 1-D dynamic simulation can be performed, similar to the one presented in [55] for switching speed calculations. However, for low spring-constant switches, the restoring force can be ignored for high RF power levels. Consequently, an analytical expression for the required top-electrode voltage as a function of the incident RF power and the switch geometry can be obtained. As the following results will show, this is a valid assumption in our case. The minimum required top-electrode DC voltage to pull the switch up is given by the (approximate) requirement $F_{\mathrm{UP}} \geq F_{RF}$, or equivalently,

$$V_{\mathrm{up}} \geq \frac{g_{\mathrm{up}} + \frac{t_{d2}}{\epsilon_{r2}}}{d_{\mathrm{gap}} + \frac{t_d}{\epsilon_r}} \sqrt{\frac{A_{RF}}{A_{\mathrm{DC}}} \frac{Z_0 P_{\mathrm{in}}}{1 + (\pi f C Z_0)^2}}\,. \tag{2.46}$$

This equation underlines the significance of separating the DC from the RF pads. Additionally, it underscores the importance of obtaining a good RF short circuit in minimizing the required pull-up voltage. For example, if

$$\pi f C Z_0 \gg 1 \tag{2.47}$$

and

$$C = \frac{C_1(2C_2)}{C_2 + 2C_2} \simeq C_1 = \frac{\epsilon_0 A_{RF}}{d_{\text{gap}} + \frac{t_d}{\epsilon_r}} \tag{2.48}$$

(where C_1 and C_2 are defined in Equation (2.40), equation (2.46) yields

$$V_{up}^{\min} \simeq \frac{1}{\pi f C_{\text{up}}} \sqrt{\frac{A_{RF} P_{\text{in}}}{A_{\text{DC}} Z_0}} \tag{2.49}$$

where

$$C_{\text{up}} = \frac{\epsilon_0 A_{RF}}{g_{\text{up}} + \frac{t_{d2}}{\epsilon_{r2}}} \tag{2.50}$$

is the fictitious capacitance between the switch center conductor pad and an imaginary electrode at a distance g_{up} from it.

2.2.4.4.3 Experimental Results and Discussion The theoretical results have been validated by fabricating and testing several switches with top electrodes using the experimental setup of Figure 2-22 [81]. Although at 10 GHz the switch provides a typical isolation of only 3–4 dB, the advantages of the top electrode can still be exposed. These measurements were preformed with the RF power continuously on (hot switching) by alternating the bias voltage between the switch/the CPW ground plane, and the switch/top electrode. Mostly switches with five meanders were tested, since these presented the lowest spring constant. Figure 2-30 shows typical results of the experimentally measured top-electrode voltage that was needed to pull up a five-meander switch with input power

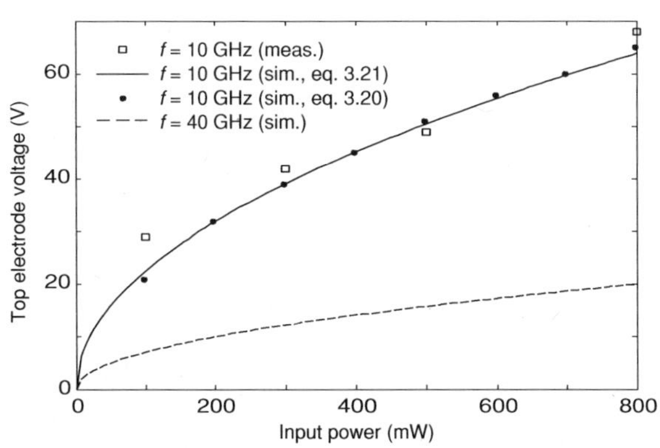

Figure 2-30 Measured and simulated pull-up voltage for switches with five meanders and fabricated top electrodes. The required pull-up voltage is significantly lower at 40 GHz than at 10 GHz, because the RF voltage across the switch is substantially reduced at 40 GHz.

of up to 0.8 W. At this power, a voltage of 68 V was required to pull the switch up, which was slightly lower than the break-down voltage of the SiO_2 film underneath the top electrode. This was the limiting factor in applying more power to the system. Figure 2-30 shows the theoretically calculated pull-up voltage based on Equation (2.46) and on dynamic simulations that included the spring-constant force given in Equation (2.45). In both simulations the dielectric layers underneath (2000 Å of Si_3N_4) and above the switch (5000 Å of SiO_2) were also taken into account. This figure clearly shows that the theoretical results agree very well with the measured values of the pull-up voltage. Additionally, it validates the assumption of ignoring the switch spring constant in these calculations as Equation (2.46) is obviously an excellent approximation of the pull-up voltage.

It is worth noting that the measured and simulated pull-up voltages are substantially higher than the measured pull-in voltage of 6 V [55]. For instance, for an input power of 0.8 W, the pull-up voltage was measured to be almost 70 V. Although this may initially seem to be a contradiction of the low-voltage properties of the switch, it is important to remember that, as Equation (2.46) shows, the pull-up voltage depends heavily on the frequency of operation and down-state switch capacitance. The low-voltage switch considered here is a high-frequency device and is not well suited to X-band operation. The switch isolation is only 2–5 dB in this band. However, the measurements were taken at 10 GHz because of the frequency of operation of the TWT. Despite the low switch isolation in this band, these measurements prove that the 1-D model is physically correct and can be successfully used for predicting the switch performance in different frequency bands. If the same measurements had been performed at 40 GHz, where the switch is more useful with a down-state $S_{21} = -15.5$ dB, the pull-up voltage would have been substantially lower. Figure 2-30 presents the simulated 40-GHz pull-up voltage for the same switch and under the same conditions. The pull-up voltage is now 10 and 20 V for input powers of 0.2 and 0.8 W respectively. This analysis shows that a number of power handling issues need to be considered during the design process of an RF MEMS switch.

2.3 SPRING-LOADED DC-CONTACT SWITCH

Metal-to-metal contact MEMS switches and micro-relays have received significant attention over the past few years [84] and offer an attractive alternative to capacitive switches. Several different designs with outstanding RF performance have been developed by Rockwell Scientific [85], RADANT and Northeastern University [86], the University of Michigan [72,87], MIT Lincoln Laboratory [71], and others. Figure 2-31 shows some of these designs.

The aforementioned switches typically rely on electrostatic or (less frequently) on thermal actuation mechanisms. In the typical switch structure (see Figure 2-32) a movable beam or plate is attracted toward an actuation electrode because of an appropriately applied electric field. In other words, the electrostatically generated force serves as the actuation force for this structure. Moreover, an electric contact is achieved when the movable beam contacts the appropriate metal pad underneath it. The electrostatically generated force, therefore, serves also as the necessary contact force between the two metals. In summary, during electrostatic actuation, the applied force is responsible for 1) actuating the switch structure and 2) forming a low-resistance metal-to-metal contact. These two functions, however, do not necessarily need the same force. For instance, although the Lincoln Laboratory switch actuates with a bias voltage of 36 V, 80 V are needed for a contact resistance of less than 1 Ω [71]. However, such a high voltage may severely impact the switch lifetime by increasing the risk of a mechanical or dielectric failure [52]. Furthermore it is suspected that most of the contact damage is due to the ultra high acceleration that the beam hits the contact during electrostatic actuation [84]. In addition, metal-to-metal contact switches are typically very sensitive to several fabrication parameters, such as intrinsic stress. Minor inaccuracies

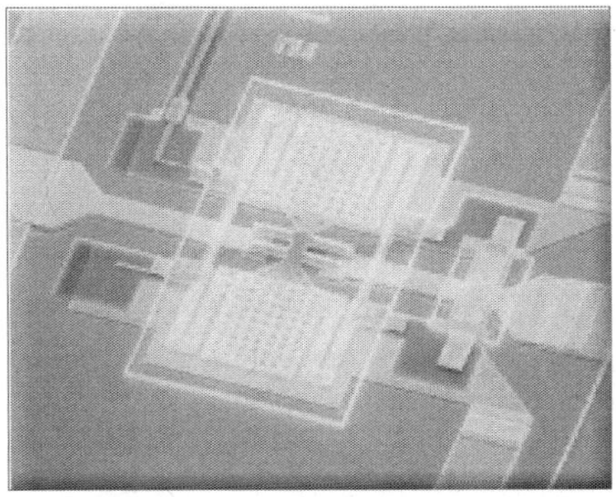

(a)

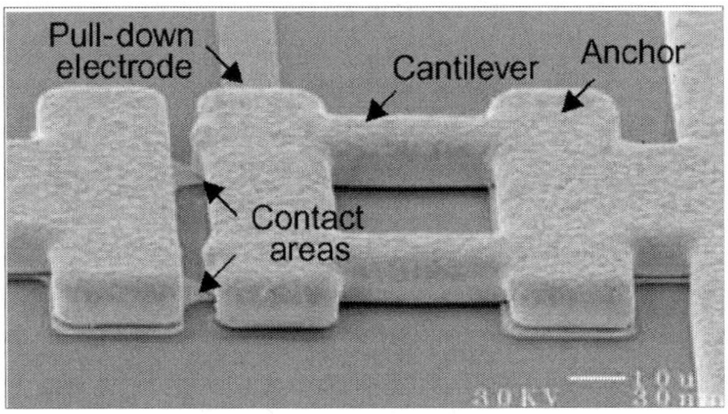

(b)

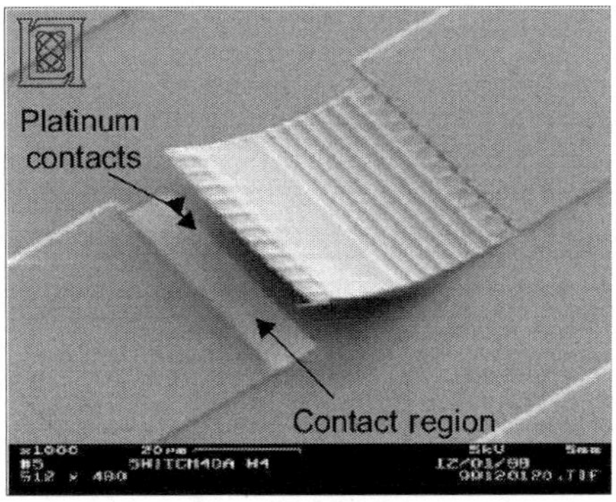

(c)

Figure 2-31 DC-contact (metal-to-metal contact) switches developed by (a) Rockwell Scientific, (b) Analog Devices & Northeastern University, and (c) MIT Lincoln Laboratory.

during the fabrication process may significantly compromise their performance. Residual gradient stress, for example, may substantially deteriorate the metallic contact and substantially increase the required actuation voltage as have already been discussed in this chapter.

A different approach of designing DC-contact switches that alleviates the aforementioned problems is discussed in this chapter. This method relies on 1) de-coupling the actuation and contact forces, and 2) constructively employing gradient stress effects [88]. Electrostatic actuation is still

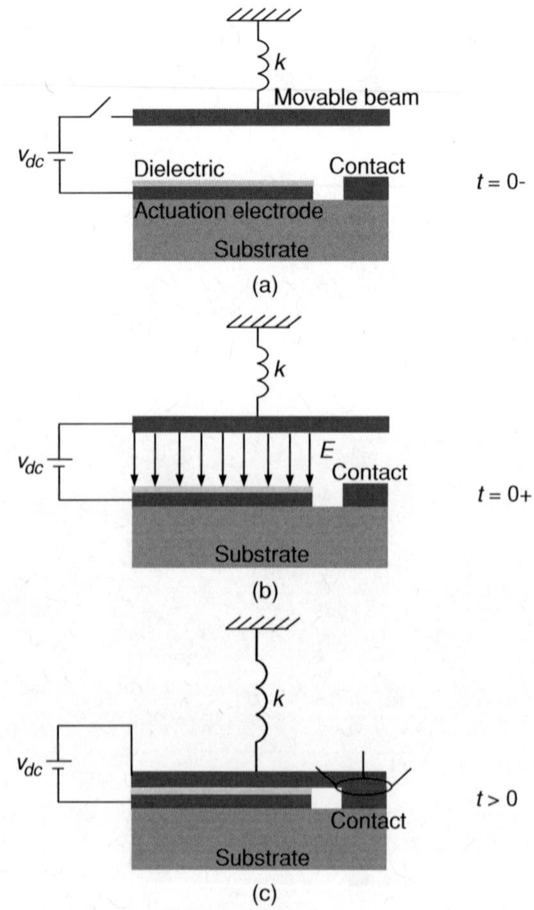

Figure 2-32 Traditional architecture of DC-contact switches: (a) a schematic of the switch structure before the bias voltage is applied; (b) an electric field is generated between the movable beam and the actuation electrode immediately after the bias voltage is applied to the structure. This field results in the electrostatic actuation force; and (c) after sufficient time (typically in the order of a few μs) the movable beam contacts the designated metal and a short circuit is created. The electrostatically generated force is used as the contact force between the two metals.

maintained, since it is relatively fast (switching speed in the order of a few μsec) and has very low power requirements. The metal contact, however, is achieved through a stress-originated force.

2.3.1 Initial Switch Design

2.3.1.1 Design Concept The basic metal-to-metal switch concept is illustrated in Figure 2-33. The moving structure is a cantilever beam, which it will be referred to as the movable cantilever beam. This beam is fabricated with a high-gradient-stress process that causes a significant upward deflection when the switch is released, Figure 2-33(a). A second and much stiffer cantilever beam (static beam) is fabricated using a low-stress process above the movable cantilever. Although the two beams are initially separated by a distance determined by the fabrication process, the two cantilevers come into direct contact immediately after the structure is released because of the

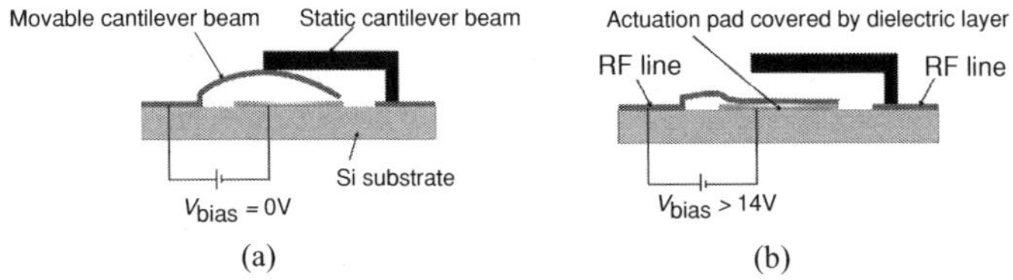

Figure 2-33 Basic switch concept and operation: (a) when no bias is applied, the movable cantilever touches the stiff static beam leading to a metal-to-metal contact; and (b) if the DC bias is higher than the actuation voltage of the movable beam, this beam deflects downwards leading to an open circuit.

movable cantilever deflection. The movable and static beams are connected to two RF lines as shown in Figure 2-33(a) and thus a closed RF contact is achieved. This is the up or on state of the switch. An actuation pad covered by a dielectric layer is also fabricated underneath the cantilevers. The down or off state of the switch is realized when a DC potential is applied between the movable cantilever and the actuation pad. If this potential is higher than the movable beam's actuation voltage, this beam deflects and the contact points are separated (Figure 2-33(b)). Although the static cantilever is a very stiff beam, as a precaution it is always maintained at the same potential as the actuation pad. Therefore, this beam does not deflect under the influence of the applied DC bias.

Typically a low on-state insertion loss, high off-state isolation, and low actuation voltage are the desired properties of such a switch. The on-state insertion loss is dominated by the contact resistance, which depends on the movable beam geometry, stiffness, and distance from the static beam. It also depends on the fabrication procedure and especially on the types of metals used and their surface smoothness. The following sections discuss this issue in more detail. On the other hand, the off-state isolation is primarily determined by the distance between the two beams and their overlap area. Finally, because of the movable beam shape and stress, it is difficult to analytically calculate the switch actuation voltage. Nonetheless, a lower bound may be obtained assuming that the beam is at a mean distance of D from the actuation pad. Then the following formula [89] for the actuation voltage may be used:

$$V_{\min}^{th} = \frac{8}{3}\left(\frac{EID^3}{3\epsilon_0 AL^3}\right)^{\frac{1}{2}} \tag{2.51}$$

where E = Young's modulus of the movable beam material, I = same beam moment of inertia, ϵ_0 = free space dielectric permittivity, A = effective area between the movable beam and the actuation electrode, and L = movable beam length. Although this formula may be useful for a first-hand calculation, it will only yield a low bound since no stress effects are introduced in it.

2.3.1.2 Fabrication Process A seven-mask process, summarized in Figure 2-34, is required for the fabrication of the proposed switch.

- The process starts with the preparation of a 400 μm silicon substrate with 2000 Å of SiO_2. A lift-off process is first performed to define the bias lines and the actuation pad. These lines are made of a thin SiCr layer (1200 Å) that is sputtered on top of the negative photoresist [87]. These lines are essential for biasing the switch without affecting the RF signal [72].
- A second lift-off process defines the RF lines, which are made of evaporated Ti and Au layers of 500 and 9000 Å respectively.

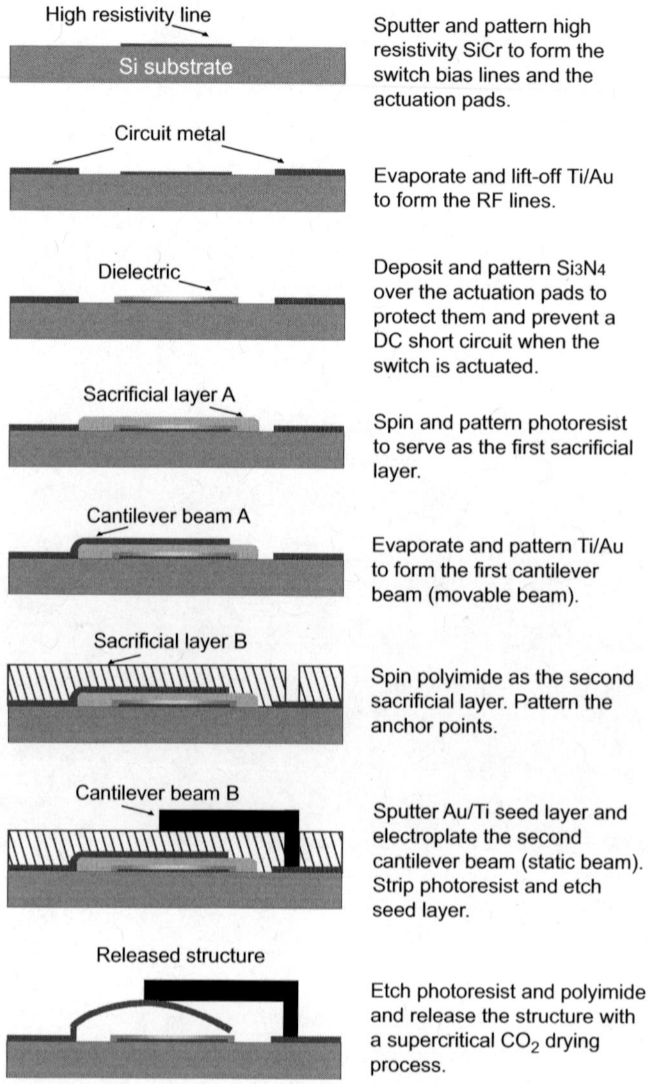

Figure 2-34 Process flow for the fabrication of a metal-to-metal MEMS switch.

- The actuation pad and bias lines are covered by 3000 Å of Si_3N_4 deposited by a PECVD process and patterned with a dry RIE etch.
- A positive photoresist is then spun and patterned to form the first sacrificial layer. However, it is important to note that an additional hard-baking of 160°C for 3.5 min follows the traditional lithography. This is essential in preventing the photoresist sacrificial layer from bubbling and destroying the switch when hard-baked in subsequent steps. Also, for low-actuation voltage switches the thickness of this layer is between 0.8–1.2 μm.
- The fifth mask is used to define the movable cantilever beam—2000 Å of Ti followed by 7000 Å of Au are evaporated on the whole wafer for this purpose. These layers are subsequently etched except for the locations of the first beam. This is an important step since the induced stress on

the beam will determine the deflection magnitude and slope. If the residual gradient stress is not sufficiently high, the movable beam will deflect only a few thousand angstroms, resulting in a poor contact. On the other hand, if excessive stress is introduced, adhesion problems between the movable beam and the sacrificial layer will render the rest of the process impossible.

- After this beam is formed, the second sacrificial layer is defined by spinning and soft-baking a 2–2.5 μm layer of polyimide DuPont PI2545. The anchor points for the second beam are then defined similarly to the process reported in [56].
- The seed layer (2000 Å of Au) necessary for electroplating the second beam is sputter deposited and patterned. It was experimentally found that there is no need for any adhesion layer between Au and polyimide, especially since this layer is only used as a seed layer. Furthermore, an adhesion layer of Ti or Cr underneath the Au layer would deteriorate the desired metal-to-metal contact.
- The seed layer is then patterned and electroplated to about 6 μm to form a stiff gold beam (the static beam).
- Finally, the two sacrificial layers are etched and the structure is dried and released with a standard supercritical CO_2 process.

SEM pictures of three series-coplanar waveguide (CPW) switches fabricated in this process are shown in Figure 2-35. The first structure, Figure 2-35(a), presents a switch with no static beam fabricated. This picture clearly shows the shape of the movable beam after the switch is released. The fabrication process of this beam induces such a stress that its center contacts the static cantilever with a zero slope. A completed switch with the static beam included is shown in Figure 2-35(b). The static beam is a low stress and thick cantilever connected to the left part of the CPW center conductor. A large rectangular hole has been also introduced in this beam in order to minimize the overlap area of the two beams and thus increase the off-state isolation. The movable cantilever is connected to the right part of the center conductor in both structures. Therefore, the two beams become part of the RF circuit in both the up and down states. Finally, the third structure (Figure 2-35(c)) is similar to the second one, except for the movable beam geometry. This beam splits into two parts (secondary movable beams) after the contact point with the static beam. This geometry has the advantage of having a short and thus stiffer static beam. Furthermore, the distance that the RF signal is traveling out of the CPW line is now half of that for the structure of Figure 2-35(b), which results in less mismatch loss.

Some of the bias details are also visible in these pictures, which also illustrate the need for high resistance bias lines. In other words, since the bias circuitry comes into direct contact with the RF lines, a strong coupling between the RF and DC signals will occur unless these high resistance lines are introduced.

2.3.1.3 Results and Discussion The RF performance of these switches was measured by an 8510C vector network analyzer with the help of an Alessi probe station and GGB picoprobe 150 μm pitch coplanar probes [88]. A standard wideband TRL calibration [90, 91] was used. Both geometries of Figure 2-35(b) and Figure 2-35(c) gave similar results overall, but the second one gave a better on-state performance. Figure 2-36 presents the measured and simulated RF performance of the switch in Figure 2-35(c).

Figure 2-36(a) shows that the switch presents an up-state insertion loss of 0.14 and 0.38 dB at 2 and 20 GHz respectively. Comparing the low-frequency loss values with a CPW line of the same total length (500 μm) and metallization (0.9 μm of Au), a contact resistance of 0.7 Ω is extracted. Moreover, the measured return loss of the switch is less than -18 dB from 2 to 40 GHz. This slight mismatch is due to the fact that the switch is not at the same level as the CPW line. In other words, although the movable beam is only 3 μm away from the Si substrate at the contact area, it

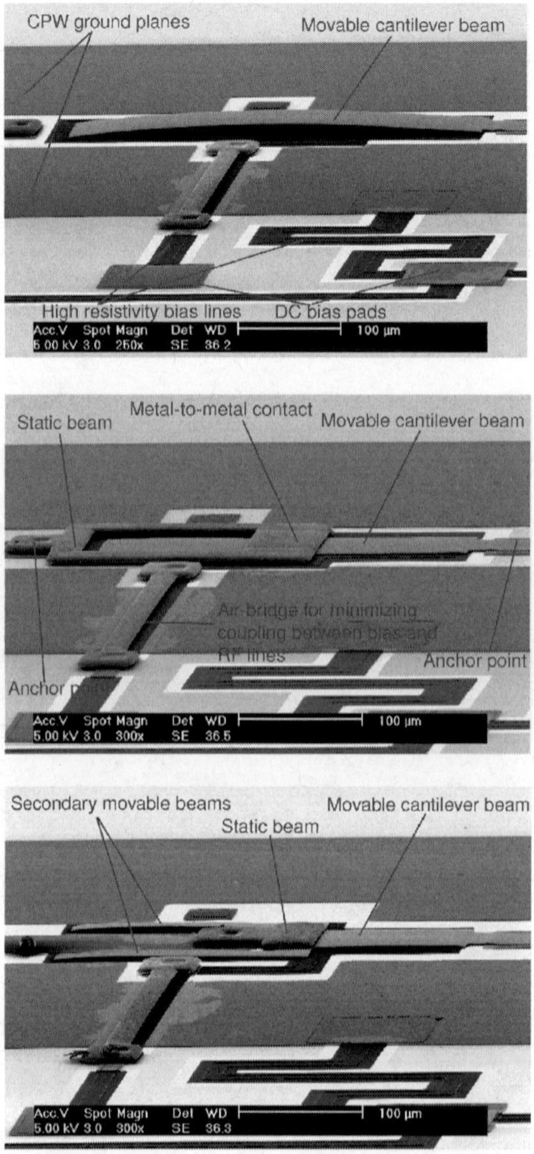

Figure 2-35 SEM pictures of fabricated metal-to-metal switches.

still has a parabolic shape with a mean distance of about 6–7 μm from the substrate. This is the distance that was used in the full wave simulation [92].

Although outstanding up-state results were observed, the down-state measurements reveal an inferior performance. The measured switch isolation is 28 dB at 2 GHz and remains higher than 10 dB up to about 20 GHz. This corresponds to a down-state capacitance of approximately 30 fF. However, the photoresist and polyimide thicknesses were 0.8 and 2 μm respectively and, as a result, the theoretical down-state distance between the movable and the static beams is about $d_{tb} = 2.8$ μm. Considering that the overlap area between the two beams is $70 \times 30 = 2100$ μm^2, a

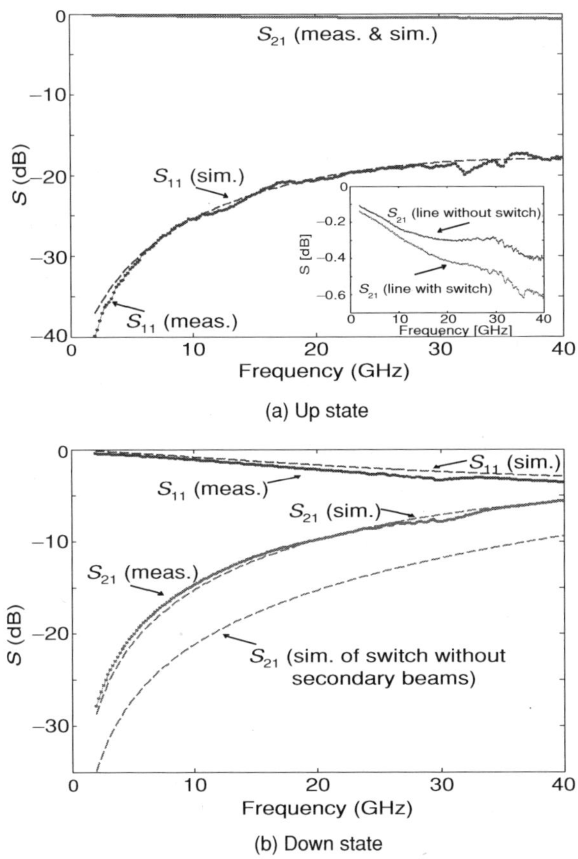

Figure 2-36 Measured and simulated results of the metal-to-metal switch shown in Figure 2-35(c) for the (a) up; and (b) down state.

down-state capacitance of 6.6 fF is calculated using the well-known quasi-static formula. Even if fringing field capacitance is considered, the down-state switch capacitance is still three times smaller than 30 fF. On the other hand, the simulated results agree very well with the experimental values presented in Figure 2-36(b). The results reveal that the dominant coupling mechanism of the switch with secondary beams, see Figure 2-35(c), is not the down-state capacitance, but rather the two secondary beams themselves. The simulation was repeated for a hypothetical switch without these two beams and the results show a 7 dB improvement in the switch isolation, Figure 2-36(b). However such a switch could only be possible if the stress distribution on the movable beam could be altered, thus ensuring that this beam would deflect the same way without the secondary beams. Such a switch is presented in the next section.

2.3.2 Optimized Switch

The switch design illustrated in Figure 2-37 does not suffer from the high coupling between the secondary beams and the CPW line of the switch in Figure 2-35(c). This design can be realized provided that the required deflection of the movable beam can be achieved. This deflection, which depends of course on the stress distribution of this beam, is typical of metals with mismatched thermal expansion coefficients [93].

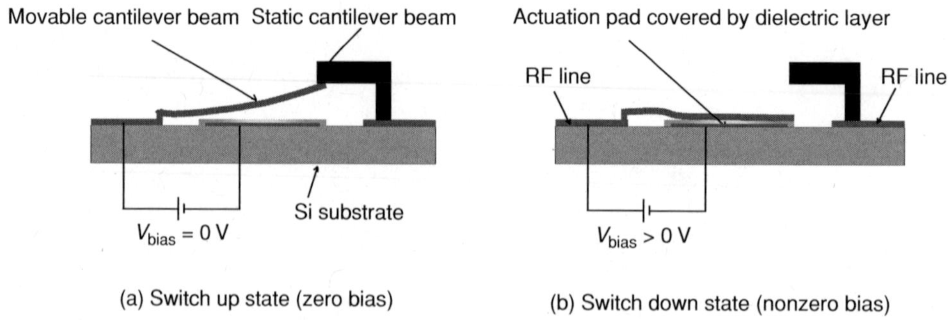

Figure 2-37 Optimized switch design. The coupling between the input and output RF lines is minimized with this design.

This stress distribution was successfully achieved by sputtering a layer of Ti/Au 2500/7500 Å at 7mT Ar pressure [88]. This step replaces the evaporation of the Ti/Au used in the initial switch implementation. The difference between the two stress distributions is caused by the different deposition processes. Figure 2-38(a) shows a switch without the top static beam. The movable beam is 230-μm long and its tip deflection is approximately 75 μm. This is the average tip deflection among 5 measured beams. Furthermore, its deflection profile follows closely the one of Figure 2-37.

One more improvement of this design has been the fabrication of beam stoppers underneath the movable beam as shown in Figure 2-38(b). These stoppers are fabricated by introducing an additional lift-off process (Ti/Au 500/500 Å) immediately after the formation of the high-resistivity lines and actuation pads. They are later protected by the PECVD Si_3N_4. The major advantage of these stoppers appears in the switch down state as they considerably limit the contact area (and therefore adhesion force) between the movable beam and the Si_3N_4 layer. This improvement significantly alleviated several stiction problems observed in the previous design.

As previously mentioned, a low contact resistance requires a high contact force. This force is typically in the order of tens of μN [94, 95] but it depends heavily on the surface profile and cleanliness. A first-order calculation of this force in the pre-stressed cantilever beam with the top beam is necessary in estimating the required residual stress. Two methods—one analytical and one experimental—were employed to compute this force. The first method follows the analysis found in [96–98] and is based on the theory of bimorph cantilevers. The second one, is based on test structures specifically fabricated to measure the contact force. As shown in the next two sections, the combination of these methods provides a good estimate of the contact force.

2.3.2.1 Bimorph Cantilever Model Figure 2-39 shows a schematic of the Ti/Au bimorph cantilever and the forces and moments acting on a cross section along the length of the cantilever. The variables used as well as their values can be found in Table 2-5. Because of the common interface between the two different materials, the forces and moments must balance:

$$P_1 = P_2 = P \tag{2.52}$$

$$P\frac{t_1 + t_2}{2} = M_1 + M_2 = \frac{E_1 I_1}{r} + \frac{E_2 I_2}{r} \tag{2.53}$$

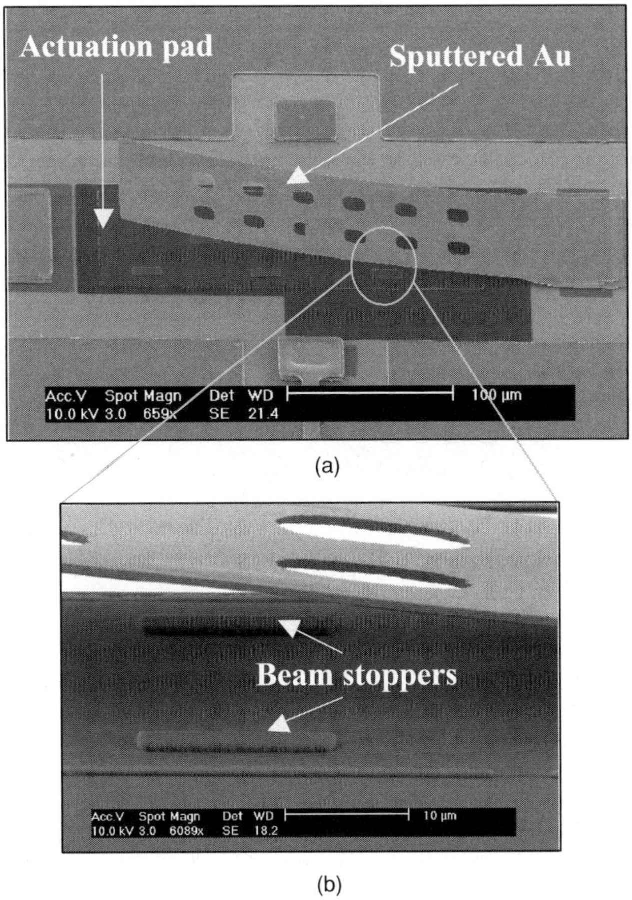

Figure 2-38 Fabricated switch: (a) SEM picture of the fabricated switch without the top static beam using the optimized process. (b) This detail shows two of the fabricated beam stoppers underneath the movable beam. These stoppers are important in the down state because they limit the contact area (and therefore adhesion force) between the movable beam and the Si_3N_4 layer underneath it.

since beam theory yields

$$M_1 = \frac{E_1 I_1}{r}, \quad M_2 = \frac{E_2 I_2}{r} \tag{2.54}$$

where r is the radius of curvature of the cantilever beam and I_1, I_2 are the moments of inertia for the Ti and Au layer respectively. It is important to underline that the bi-axial modulus of elasticity has to be used in all analytical formulas about the bimorph beam. The thermal mismatch between the two layers of the cantilever beam is uniform across all directions, causing the beam to deform along its length as well as along its width. This plate-like behavior increases the beam's stiffness and its characterized by the biaxial modulus of elasticity defined as [99]:

$$E_b = \frac{E}{1 - \nu} . \tag{2.55}$$

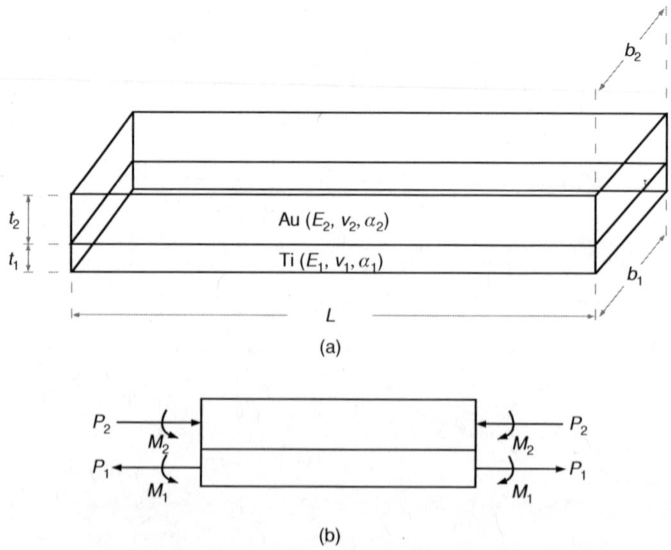

Figure 2-39 Bimorph movable cantilever: (a) schematic; and (b) force and moments acting on the cross section of a segment along the length of the bimorph cantilever.

Although this will not be made explicit in the following, the biaxial modulus of elasticity will be used whenever the symbol of modulus of elasticity E is written.

The normal strain of Ti and Au also has to be the same at their interface. Hence:

$$\alpha_1 \Delta T + \frac{P}{E_1 t_1 b_1} + \frac{t_1}{2r} = \alpha_2 \Delta T - \frac{P}{E_2 t_2 b_2} + \frac{t_2}{2r} \tag{2.56}$$

where ΔT is the equivalent temperature difference that would result in the radius of curvature r.

Table 2-5 Variables used in the bimorph cantilever model and their respective values.

Variable	Description	Value	Dimension
E_1	Ti Young's modulus of elasticity	116	GPa
E_2	Au Young's modulus of elasticity	80	GPa
ν_1	Ti Poisson's ratio	0.34	none
ν_2	Au Poisson's ratio	0.42	none
$E_{1,b}$	Ti bi-axial modulus of elasticity	$E_1/(1-\nu_1)$	GPa
$E_{2,b}$	Au bi-axial modulus of elasticity	$E_2/(1-\nu_2)$	GPa
α_1	Ti coefficient of thermal expansion (20°C)	8.9	μm/m-°C
α_2	Au coefficient of thermal expansion (20°C)	14.4	μm/m-°C
L	Bimorph cantilever length	230	μm
b_1	Ti layer width	80	μm
b_2	Au layer width	80	μm
t_1	Ti layer thickness	0.25	μm
t_2	Au layer thickness	0.75	μm

From Equations (2.53) and (2.56) the inverse of the radius of curvature can be found:

$$k = \frac{1}{r} = \frac{6b_1b_2E_1E_2t_1t_2(t_1+t_2)(\alpha_2-\alpha_1)\Delta T}{(b_1E_1t_1^2)^2+(b_2E_2t_2^2)^2+2b_1b_2E_1E_2t_1t_2(2t_1^2+3t_1t_2+2t_2^2)} . \tag{2.57}$$

The radius of curvature is related to the beam's maximum deflection d_{max} (at the tip) with the following formula [96]

$$d_{\mathrm{max}} = 2r\left(\sin\frac{L}{2r}\right)^2 . \tag{2.58}$$

The force of a bimorph cantilever is distributed linearly along its length. An equivalent force F_{eq} at the tip can be defined, however, as the necessary force to counteract the thermally induced force. This force will be given by

$$F_{\mathrm{eq}} = K_{\mathrm{eq}}d_{\mathrm{max}} \tag{2.59}$$

where K_{eq} is the equivalent spring constant of the cantilever. In order to calculate K_{eq} the equivalent flexural rigidity $(EI)_{\mathrm{eq}}$ of the beam needs to be known. This quantity is [98, 100]

$$(EI)_{\mathrm{eq}} = \frac{(b_1E_1t_1^2)^2+(b_2E_2t_2^2)^2+2b_1b_2E_1E_2t_1t_2(2t_1^2+3t_1t_2+2t_2^2)}{12(b_1E_1t_1+b_2E_2t_2)} . \tag{2.60}$$

Thus the equivalent force at the tip of the cantilever can be calculated:

$$F_{\mathrm{eq}} = \frac{3(EI)_{\mathrm{eq}}d_{\mathrm{max}}}{L^3} . \tag{2.61}$$

Based on the values given in Table 2-5 the equivalent contact force is calculated to be

$$F_{\mathrm{eq}} = 16.4\ \mu\mathrm{N} . \tag{2.62}$$

The next section presents a number of test structures fabricated to experimentally validate this model.

2.3.2.2 Test Structures for Contact Force Estimation Figure 2-40 shows a schematic diagram of a test structure suitable for measuring stress-induced contact forces. The structure consists of two cantilever beams C_b and C_s. C_b is the Ti/Au cantilever beam used in the DC-contact switch as the movable beam. C_s is a stress-free Au cantilever beam that deflects upwards under the force applied by C_b. The contact force between the two beams may be extracted by measuring the final deflection δ of the stress-free beam and using the following equation (from Equation (2.61)):

$$F_s = \frac{3E_sI_s\delta}{L_s^3} = \frac{E_sw_st_s^3\delta}{4L_s^3} \tag{2.63}$$

where E_s, $I_s = w_st_s^3/12$, and L_s are the Young's modulus (not bi-axial), moment of inertia and length of the stress-free cantilever C_s. It should be noted that since C_s has a finite stiffness and deflection, F_s will always be smaller than F_{eq} calculated in the previous section.

The accuracy of this method depends on 1) the ability to fabricate stress-free cantilever beams with zero out-plane deflection and 2) the ability to accurately measure δ. Low-stress cantilever

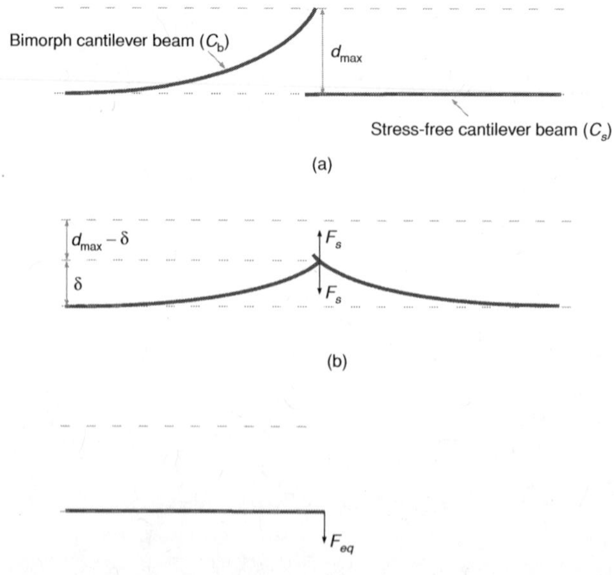

Figure 2-40 Bimorph cantilevers: (a) schematic of the bimorph cantilever beam (C_b) and the stress-free cantilever beam (C_s) (under no external forces the maximum deflections for C_b and C_s are d_{max} and 0 respectively); (b) the stress-free beam C_s deflects upwards by δ under the force applied by C_b (the contact force of the two beams is F_s); and (c) the bimorph cantilever C_b experiences zero deflection under the influence of an external force F_{eq} applied to its tip (this is the maximum possible contact force).

beams with zero out-of-plane deflection were fabricated by sputtering 2 μm of Au with a low-stress process (10 mT Ar sputtering pressure) [88]. The out-of-plane deflection of these beams was measured with an optical microscope with a measurement accuracy of ± 1 μm.

Figure 2-41 shows a microphotograph of a test structure fabricated for measuring the contact force F_s. Three different types of these test structures were fabricated. Each of these structures

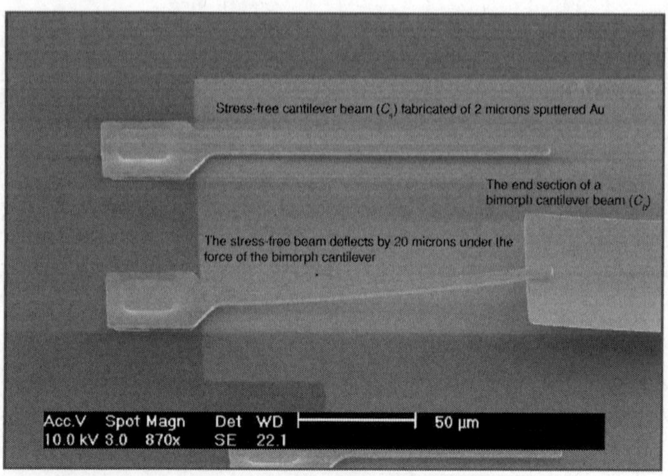

Figure 2-41 Microphotograph of a fabricated test structure for measuring the contact force between a bimorph cantilever and a stress-free cantilever beam.

Table 2-6 Measured results of the fabricated structures for measuring the contact force F_S.

W^a (μm)	δ^b (μm)	F_S^c (μN)	$F_{eq}^{\delta d}$ (μN)
40	7.5	17.5	14.5
20	11	12.8	13.6
10	20	11.7	11.3

[a] Width of cantilever C_s.
[b] Measured deflection of cantilever C_s caused by the force F_s.
[c] Contact force between cantilevers C_b and C_s (see Figure 2-41).
[d] Force applied to the tip of the bimorph cantilever necessary to reduce its tip deflection from $d_{\max}$ to δ. This force can be computed from Equation (2.61) if $d_{\max}$ is replaced by $d_{\max} - \delta$.

included an $L = 230$-μm long Ti/Au (0.25/0.75 μm) bimorph cantilever beam fabricated underneath an $L_s = 140$-μm long and $t_s = 2$-μm thick stress-free cantilever. The width w of the stress-free cantilever was varied between 40 and 10 μm. Table 2-6 summarizes the measured results of these test structures. Each deflection on this table is the average deflection obtained from three identical beams.

Based on the measured deflections and Equation (2.63) the calculated contact force varies between 11.7 (for $w_s = 10$ μm) and 17.5 μN (for $w_s = 10$ μm). These extracted force values compare well with the analysis of the previous section. This can be verified if Equation (2.61) is used to calculate F_{eq}. In this case $d_{\max}$ is replaced by $d_{\max} - \delta$ in order to account for the finite stiffness of the stress-free cantilever beam. Based on the results of these sections and the stiffness of the top beam of the switch, the contact force between the two beams in the actual switch structure is estimated to be approximately 19–20 μN. This is a reasonable region for Au-Au contact, but higher forces would be required for other types of contact (e.g. hardened Au or refractory-metal contacts).

2.3.2.3 Measured Results and Discussion Figure 2-42 shows a microphotograph of a fabricated switch. The movable and stiff cantilever beams can be clearly seen in this figure. An optical

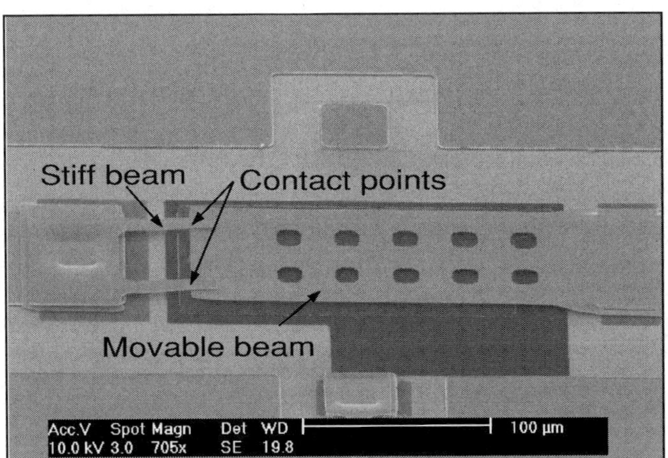

Figure 2-42 Microphotograph of a fabricated optimized spring-loaded DC-contact switch.

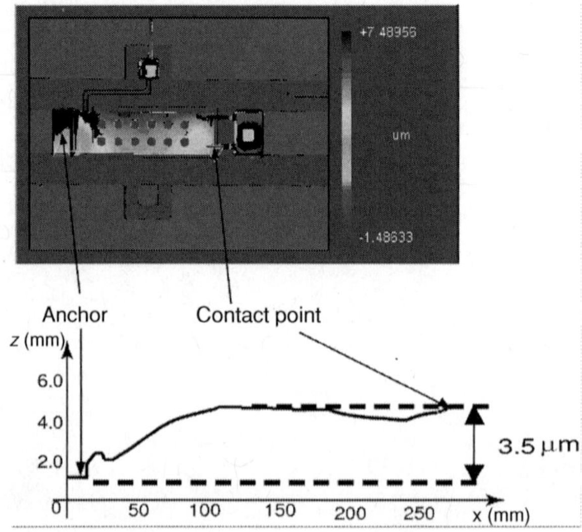

Figure 2-43 Interferometric image of the switch shown in Figure 2-42.

interferometric image of this switch was also acquired and is presented in Figure 2-43. This image shows that the movable beam, despite the pressure, remains reasonably flat and close to the actuation pad underneath it.

Six identical switches were measured and their average actuation voltage was 50–60 V. Figure 2-44 shows the de-embedded on-state measured S-parameters from these switches. The term "de-embedded" refers to calibrated measurements that do not include the loss of any transmission lines connecting the switch to the measurements set up. The on-state switch loss is on average 0.05 and 0.14 dB at 2 and 40 GHz respectively. The observed slope in the measurements is due to the fact that the switch is a 1-μm thick, 80-μm wide and 230-μm long Au beam that introduces an additional small loss versus frequency. The switch return loss is below -19 dB up to 40 GHz. This slight mismatch is due to the fact that the movable beam is approximately 3.5 μm higher than the CPW line. Figure 2-45(a) shows the switch equivalent circuit whose component values can be fit to the measured data. This fitting results in an on-state contact resistance of $R_s = 0.5\ \Omega$. This contact resistance is among the lowest reported in the literature and shows that stress-induced forces are sufficiently high to result in very low contact resistance. It is also interesting to note that the impedance of the line in the model is approximately 62 Ω to match the measured return loss.

Figure 2-46 presents the off-state switch isolation versus frequency. On average, the measured isolation is 38 and 13 dB at 2 and 40 GHz respectively. This is a substantially improved performance compared to the initially obtained results. Based on the measured isolation, the switch off-state series capacitance is $C_s = 10$ fF on average. The simulated switch performance based on this fitted capacitance is also plotted in Figure 2-46.

2.4 ELECTROSTATICALLY-TUNABLE ANALOG RF MEMS VARACTORS WITH HIGH CAPACITANCE RANGE AND POWER HANDLING CAPABILITIES

A number of different high-frequency switches have been presented in the previous sections. These devices and their associated circuit designs rely on electrostatic actuation because of their simplicity and their very low biasing power requirements. This actuation mechanism, however, results in the well-known pull-in effect that forces the microswitch to collapse when it moves (approximately) one

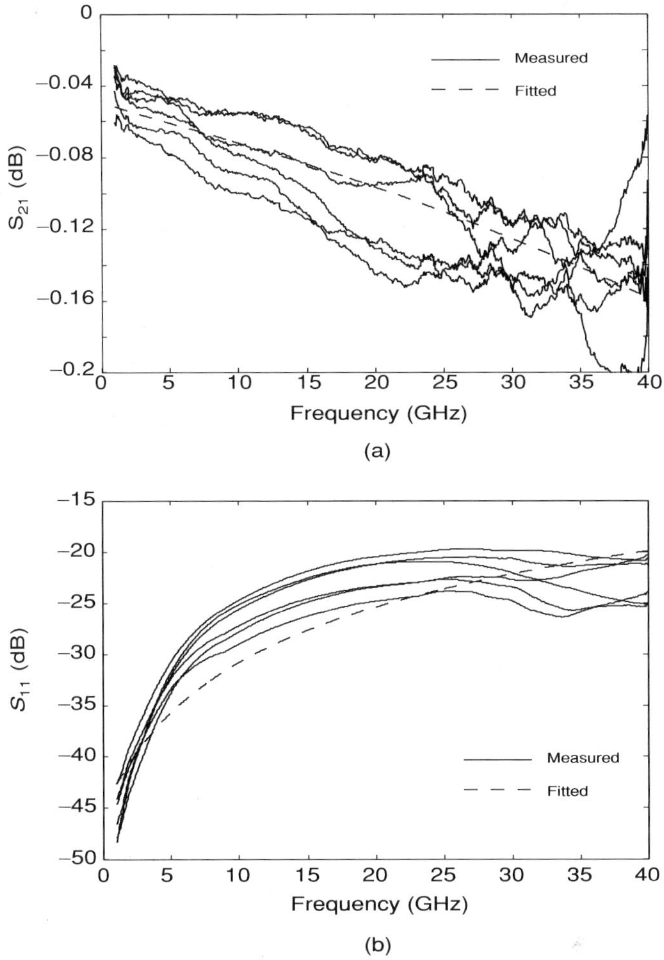

Figure 2-44 On-state measured (a) insertion loss and (b) return loss of six identical spring-loaded switches. The fitted S-parameters that are based on the equivalent circuit of Figure 2-45(a) are also presented.

third of its initial height. Although this effect does not significantly impair the switching operation, it can be a serious limitation for MEMS variable capacitors and applications that require analog rather than digital tuning. Voltage control oscillators, tunable filters, and matching networks are a few examples where analog tuning can be particularly beneficial.

Two main approaches have been proposed in the literature to overcome this issue. The first one relies on digital MEMS variable varactors that are typically implemented by a bank of MEMS switches [101]. In the second approach the value of a MEMS capacitor is changed in a continuous way by tuning the gap between two parallel plates [102–104]. The main advantage of the digital tuning technique is the ability to achieve very high tuning ranges, similar to the on/off capacitance ratios of MEMS switches. Furthermore, the digital approach leads to components with relatively low sensitivity to fabrication process variabilities and vibration. Nevertheless, the required switch bank can occupy a large space and the interconnecting lines typically result in high losses and low self-resonant frequencies. On the other hand, the analog MEMS varactors have low losses, occupy

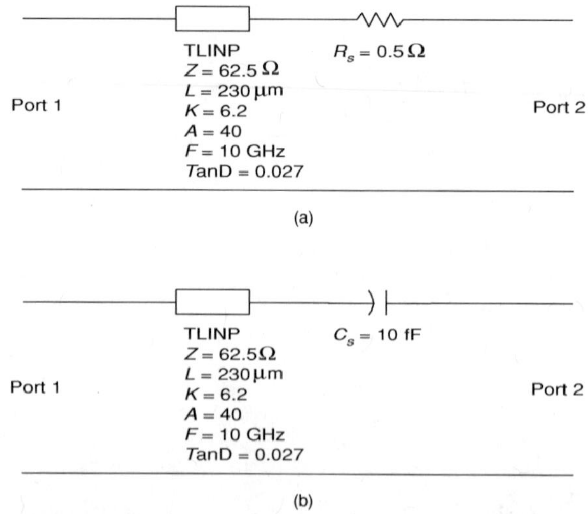

Figure 2-45 Equivalent circuits: (a) On- and (b) off-state extracted ADS equivalent circuit of the spring-loaded DC-contact switch.

less space, and are relatively easy to implement. Their tuning range, however, is relatively limited and can be sensitive to vibration, residual stress, and temperature. A further advantage of the analog MEMS varactors is that they are typically non-contact MEMS devices and are, therefore, inherently more reliable than contact type of MEMS (digital varactors or switches).

This section focuses on presenting analog MEMS varactor with very high-tuning range and power handling capabilities. This technique is based on the idea of an extended tuning range varactor proposed by Zou *et al.* in 2001 [102]. Figure 2-47 shows the traditional analog MEMS varactor as well as the extended tuning range varactor presented by Zou. In the conventional design the RF capacitance is sensed between two parallel plates with an overlap area A and initial spacing x_0. When a DC voltage is applied between them the spacing is decreased up to a minimum (theoretical) value of $2x_0/3$. If the bias voltage is increased beyond this limit, the top plate will snap on the bottom one and the ability to smoothly tune the gap x will be lost. This is the well-known

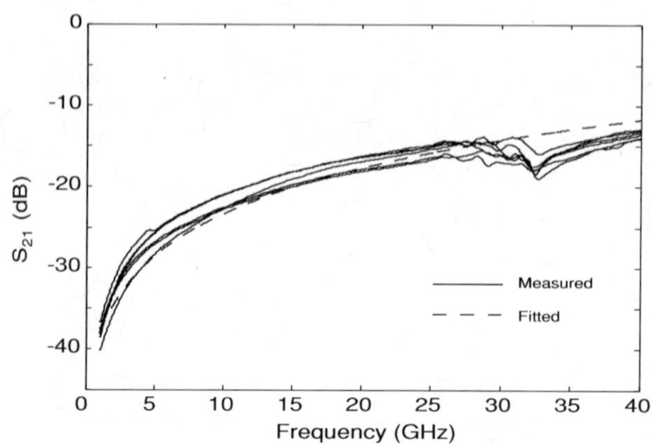

Figure 2-46 Off-state measured and fitted isolation of six identical spring-loaded DC-contact switches.

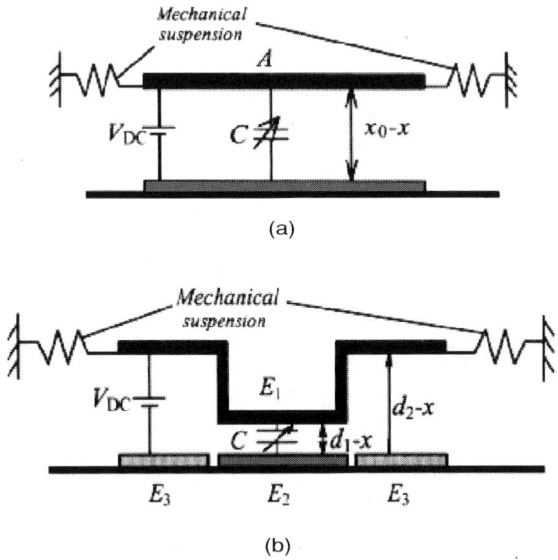

Figure 2-47 MEMS varactor: (a) schematic of the traditional analog MEMS varactor based on tuning the gap between two parallel plates or beams in a continuous way; and (b) extended tuning range analog MEMS variable capacitor as presented by Zou [102].

pull-in effect and limits the tuning range of this varactor to a maximum value of 50%. This value is a theoretical limit and is rarely achieved in practice because of residual stress in the beam. Practical circuits based on this approach typically achieve a tuning range of approximately 20–30% [105].

One way to increase the tuning range of the parallel-plate varactor is to allow the gap x to be varied beyond the pull-in point. This is achieved by the extended tuning-range varactor, schematically illustrated in Figure 2-47(b). The key idea in this design is to separate the DC biasing plates (E_1 and E_3) from the RF capacitive plate (E_2). The initial spacing d_2 between the movable structure and plates E_1 and E_3, in particular, is designed to be larger than the gap d_1 that controls the RF capacitance. By taking into account that when a bias voltage V_{DC} is applied, all plates are pulled down by a distance x, the tuning range TR of the structure can be calculated as

$$\mathrm{TR} = \frac{C - C_0}{C_0} = \frac{\frac{\epsilon A}{d_1 - x} - \frac{\epsilon A}{d_1}}{\frac{\epsilon A}{d_1}} = \frac{x}{d_1 - x} \tag{2.64}$$

As long as $x < d_2/3$ the structure will not suffer any pull-in effect and the movable beam can be pulled infinitely close to the plate E_2, resulting in a (theoretically) infinite tuning range.

Despite this encouraging theoretical result, practical circuits are still limited by the residual stress in the structure. Zou *et al.* achieved a measured tuning range of 70%, while Dussopt and Rebeiz measured 46% out of a very similar device [103]. The main reason behind this is the nonuniform movement of the movable beam or plate because of the residual stress in it. Other researchers have focused on a three-plate structure to balance the pull-in effect [106]. The maximum-measured tuning range in this case was 86%. Of course the pull-in effect can be completely eliminated by considering other actuation mechanisms. Feng *et al.*, for example, designed a variable capacitor based on thermal actuation with a measured capacitance range of 270% (3.7 : 1). This extended range, however, comes at a price of considerably higher power consumption (in the order of milliwatts) and slower speed (response time is in the order of milliseconds for thermally driven devices).

The varactor discussed here also follows the extended tuning range approach, but it is based on a two-metal, two-sacrificial-layer process and results in a measured tuning range of nearly 4 : 1 (300%). Furthermore, it still uses electrostatic actuation, thus preserving the advantage of low DC power consumption. Section 2.4.1 discusses the design of this varactor. The fabrication process follows in Section 2.4.2, while RF experimental results are reported in Section 2.4.3. Reliability and power handling results from this device are also presented in Sections 2.4.4 and 2.4.5.

2.4.1 Analog Varactor Design

Figure 2-48(a) shows a schematic of the varactor [107, 108]. This device is composed of two Au beams. The first beam (lower beam) is a 0.7-μm thick beam that is suspended $g_1 = 2\ \mu$m above a 50/80/50-μm CPW line, while the second one (upper beam) is 13-μm thick and is suspended $g_2 = 6\ \mu$m above the lower beam. Figures 2-48(b) and (c) illustrate two possible implementations of the lower beam. The first implementation, Figure 2-48(b), shows a folded-suspension lower beam design (FSLB design), while the second one, Figure 2-48(c), shows a fixed-fixed lower beam design (FFLB design). The second difference between the two implementations is the fact that in the FSLB design a dielectric layer covering the CPW line prevents direct metal contact between

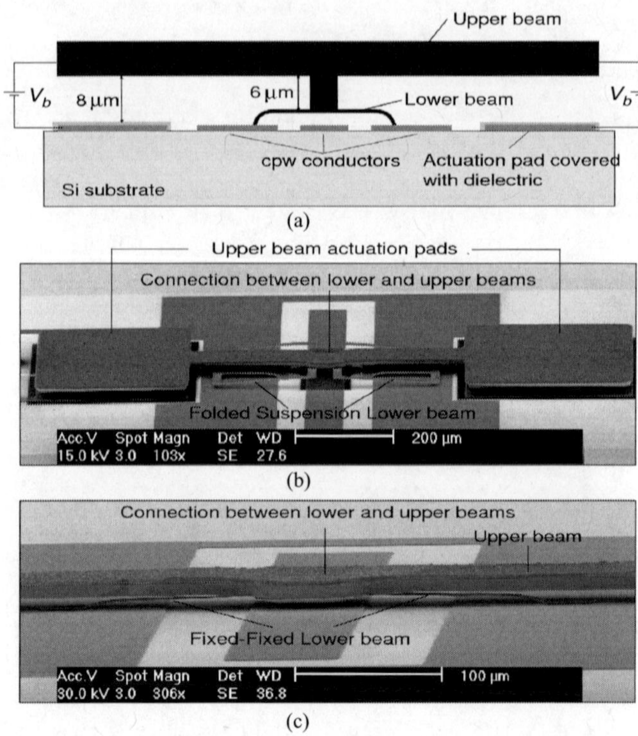

Figure 2-48 Analog MEMS varactor: (a) schematic diagram; (b) first implementation of the analog MEMS varactor. This varactor uses a folded-suspension lower beam (FSLB) and a dielectric layer underneath the lower beam; and (c) second implementation of the analog MEMS varactor This varactor uses a fixed-fixed lower beam (FFLB) and it does not have any dielectric layer that could prevent a DC-contact between the lower beam and the center conductor of the CPW.

the lower beam and the center conductor of the CPW, while this is not the case in the FFLB design. Both implementations are based on the same design principles but they yielded slightly different experimental results.

In both designs the anchor points of the lower beam are connected to the ground planes of the CPW line. Furthermore, in both designs the upper beam (which is suspended a distance $g_2 = 6\ \mu m$ above the lower beam) has no anchor points, but is connected to the middle of the lower beam. Although the upper beam is almost 1 mm long, it is very stiff, because it is made of a 13-μm thick low-stress electroplated Au. The upper beam also forms two 360×300-μm^2 pads (upper pads) that are suspended a total distance of $g_{tot} = g_1 + g_2 = 8\ \mu m$ above the two electrostatic actuation pads covered with a dielectric layer as shown in Figure 2-48(a).

When an electrostatic potential V_b is applied between the upper beam and the two actuation pads, the lower beam deflects, because its spring constant is three orders of magnitude lower than that of the upper beam. Additionally, because of the height difference between the two beams, the pull-in instability of the upper beam does not occur until it moves by approximately 2.5 μm downwards. This means that the upper beam does not collapse before the lower beam touches the center conductor of the CPW line (in the FFLB case) or the dielectric layer covering it (in the FSLB case). When this happens (at $V_b = V_{ct}$), both beams have moved by about 2 μm, which is the maximum allowable distance the lower beam can travel. Consequently, in both cases the tuning range of the varactor can be designed to be very high and is limited only by fabrication issues, such as residual stress, thickness and roughness of the dielectric layer.

2.4.2 Fabrication Process

Figure 2-49 summarizes the six-mask process that is necessary for the fabrication of the varactor. The fabrication is only briefly presented here since it is similar to the one of the metal-to-metal contact switch. The varactors are fabricated on a high-resistivity silicon substrate (approximately 2000 Ω-cm) with 2000 Å of thermally grown SiO_2. The fabrication starts with a lift-off process of Ti/Au 250/9000 Å that is performed to define the 50/80/50 CPW lines and the actuation pads. PECVD deposition of 3000 Å Si_3N_4 follows, which is patterned with a dry RIE process. This dielectric layer is mainly deposited to protect the actuation pads from potential direct contact with the upper beam. The first sacrificial layer (photoresist SC1827 by Shipley) is subsequently spun at 3.5 krpm and patterned. This sacrificial layer is post baked at a 180°C hotplate for 3.5 min to avoid any out-gasing in the remainder of the process. The lower beam deposition and patterning follows. This beam is made of low-stress 0.7-μm thick Au and is anchored at two or four points, depending on the design (see Figure 2-48). A second sacrificial layer (photoresist AZ 9260 by Clariant) is spun at 4 krpm and patterned. This photoresist results in a 6-μm layer after being post baked in a 150°C oven for 1 h. A thin layer (1000 Å) of low-stress Au is subsequently sputter deposited on top of the second sacrificial layer. This layer is electroplated to 12–13 μm to create the very stiff upper beam. The final steps are the etching of the sacrificial layer and drying of the varactors with a conventional supercritical CO_2 process. It is worth noting that no adhesion layer is required between the sputtered Au and the photoresist in both depositions described in this process. Such an adhesion layer (e.g., Ti or Cr) could easily create a significant mismatch between the two metals (Ti/Au or Cr/Au) and result in a significant warping of the structure.

2.4.3 Initial Results and Discussion

2.4.3.1 FSLB Tuning Range The RF performance of these varactors was measured from 2–40 GHz using an 8510C vector network analyzer, an Alessi probe station, and GGB picoprobe 150 μm

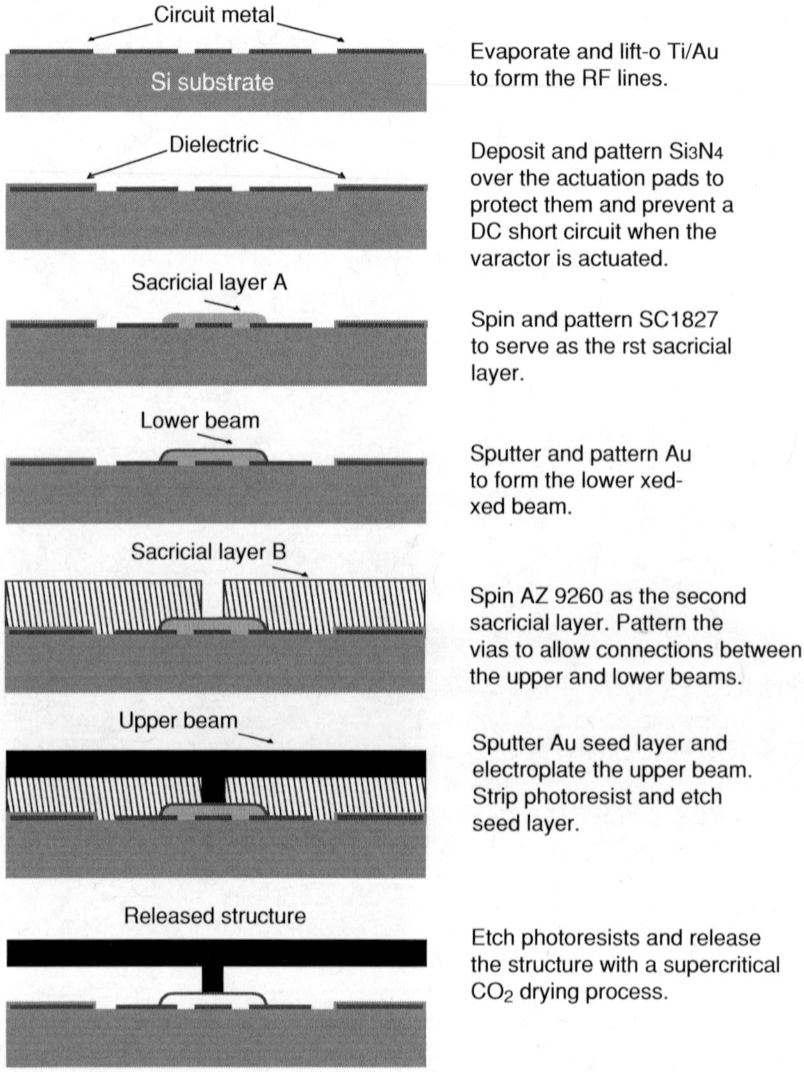

Figure 2-49 Process flow for the fabrication of the MEMS varactor.

pitch coplanar probes. Standard wideband TRL calibration [90, 91] procedure was used. The RF shunt capacitance of each of the varactors was extracted by fitting the RF results to the model shown in Figure 2-51. Figure 2-50(a) presents the measured reflection coefficient of a typical varactor following the geometry of Figure 2-48(b).

These measurements indicate an almost ideal capacitive response of the varactor for the various bias voltages. The capacitance, however, does not vary smoothly with voltage, and shows a deviation from the theoretically expected curve at bias voltages of 22–23 V. This intermediate discontinuity becomes more obvious in Figure 2-50(b) where the extracted capacitance is plotted as a function of bias voltage. Varactor C in this plot is the varactor whose results are shown in Figure 2-50(a), while the three other curves represent the extracted capacitances from three other varactors with similar geometries.

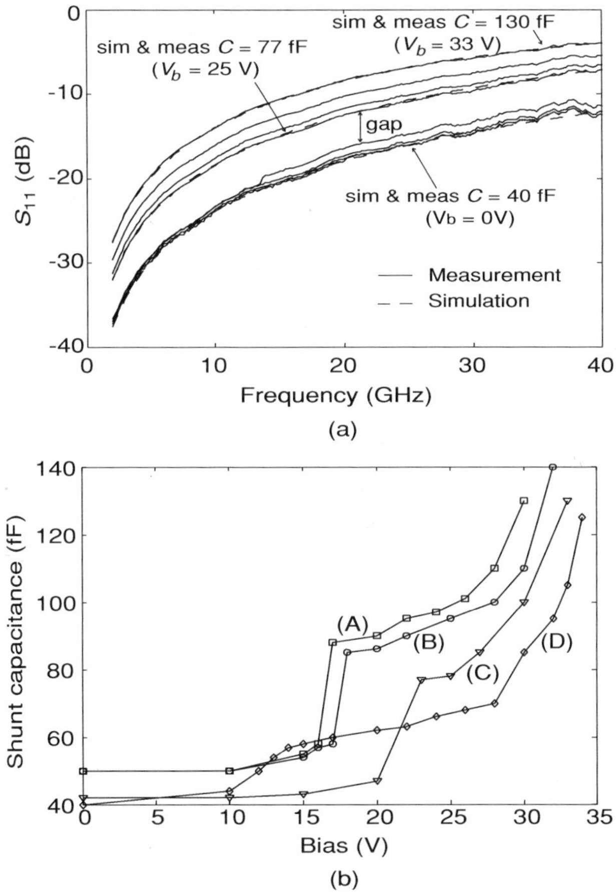

Figure 2-50 Performance of the varactor shown in Figure 2-48(b): (a) typical measured and simulated reflection coefficient; and (b) shunt capacitance presented by several varactors as a function of the applied bias voltage. Varactor C is the varactor whose reflection measurements are shown in part (a) of this figure. It is interesting to note the discontinuity for each varactor that occurs at different voltage levels.

Figure 2-50 demonstrates that the tuning range for all varactors is between 160–213% and the maximum and minimum capacitance values do not vary by more than 12 and 20% respectively. Table 2-7 summarizes these results and also provides the measured pull-in voltage and respective capacitance for each one of these varactors. This table shows that the maximum possible capacitance could not be achieved in an analog way. This was due to the residual stress in the lower beam that forced it to warp upward and thus increase g_1, g_2, or equivalently decrease $(g_2 - g_1)/g_1$. It is

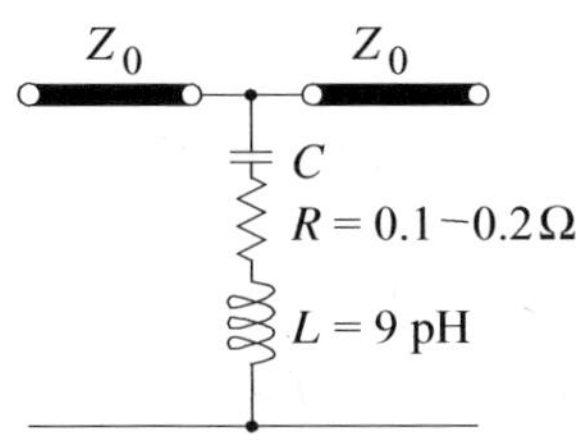

Figure 2-51 Equivalent circuit used to extract the capacitance values vs. bias voltage.

Table 2-7 Summary of the results for the varactors presented in Figure 2-50.

Varactor	A	B	C	D
Min. measured cap. value (fF)	50	51	42	40
Max. measured cap. value (fF)	130	140	130	125
Pull-in voltage (V) (± 1 V)	31	33	34	34
Pull-in capacitance (fF)	270	265	230	240
Measured tuning range (%)	160	175	210	213

worth noting, however, that although the maximum tuning range could not be achieved because of residual stress, the device still demonstrated a significantly improved tuning range compared to previous designs.

It is also worth noting that the intermediate discontinuity was observed for all varactors but at different voltage levels. Figure 2-50(b) shows that the voltage level where this discontinuity is observed varies by about 50%. Due to this effect it will be very difficult to realize capacitance values close to the discontinuity region. Varactor C, for instance, cannot easily provide the values from 47 to 77 fF. This is due to residual compressive stress and the upward initial curling of the lower beam. As the biasing voltage is slowly increased, the device starts moving downwards, which increases the compressive stress in the structure. As the compressive stress is increased, buckling of the structure occurs at some point. This abrupt movement of the device is observed as the discontinuity shown in Figure 2-50. After buckling, the device's movement continues uninterrupted until pull-in occurs. Devices with tensile residual stress would not experience this problem. For more details please see Section 2.4.5

The measurements presented in Figure 2-50 were performed by slowly increasing the biasing voltage from 0 V to a value slightly less than the pull-in voltage. The reverse measurements were performed by decreasing the voltage from its maximum value to 0 V. Figure 2-52 shows an example of such a measurement. This figure shows that a hysteresis loop exists in the operation of the varactor and that the observed intermediate discontinuity appears at different voltage levels in

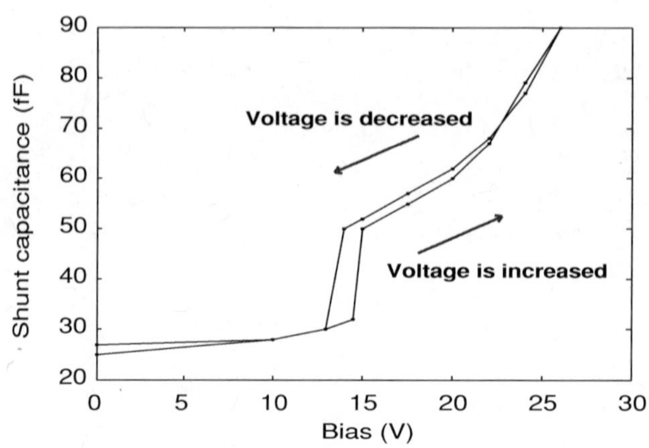

Figure 2-52 Hysteresis loop for a varactor when the bias voltage is increased from zero to a value close to the pull-in voltage and is subsequently decreased back to zero.

the forward and backward biasing schemes. This hysteresis loop is repeatable and very similar for all the varactors of Figure 2-50. Section 2.4.4 discusses this in more detail. The hysteresis loop is a direct result of the existence of the intermediate discontinuity.

2.4.3.2 FFLB Tuning Range The measurements presented in the previous subsection were repeated for the FFLB varactors [107]. The results were in general similar, but the achieved tuning range was higher and reached nearly 300%. This improved tuning range was due to the fact that the fixed-fixed lower beam did not significantly deflect due to the process residual stress as it was the case for the FSLB designs. As a result, the lower beam could move almost the total distance of $g_1 = 2\ \mu$m without collapsing due to the pull-in effect of the upper beam. In order to verify this, the FFLB varactor has fabricated without the dielectric layer between the lower beam and the center conductor of the CPW for the line. As a result, if the voltage was increased past the pull-in voltage, a direct metal contact would be observed and a low contact resistance would be recorded.

Figure 2-53(a) presents the reflection coefficient measurements for a typical varactor of the geometry shown in Figure 2-48(c). The measurements demonstrate that the lower beam does not

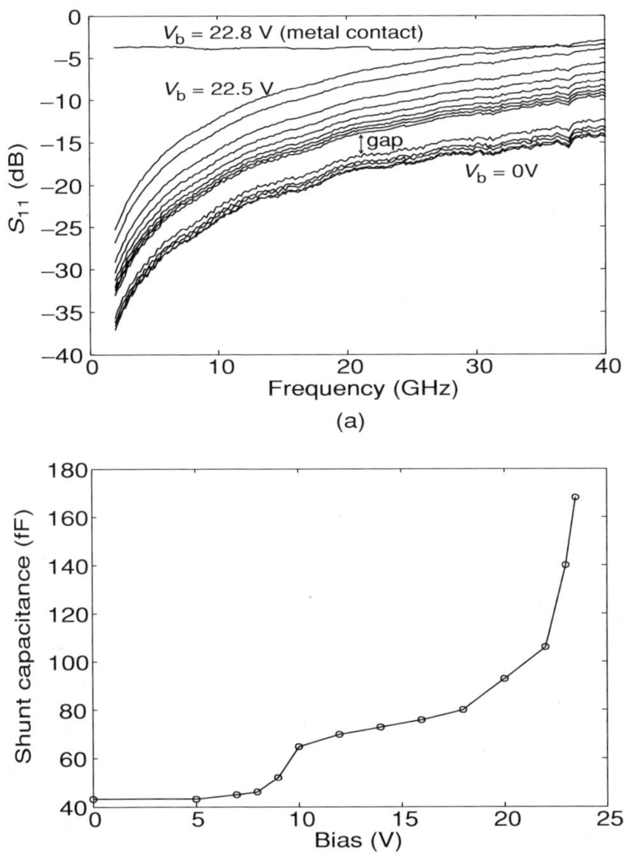

(b)

Figure 2-53 Characteristics of the varactor showed in Figure 2-48(c): (a) typical measured reflection coefficient and (b) extracted shunt capacitance of the MEMS varactor as a function of the applied bias voltage. The tuning range of this capacitor is nearly 300%. Although the intermediate discontinuity is present in this varactor too, its effect is in general milder than the ones under the FSLB design.

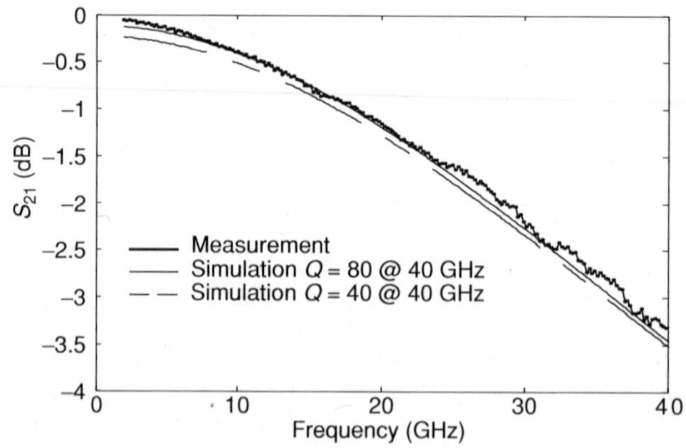

Figure 2-54 Measured and simulated insertion loss of the varactor showed in Figure 2-48(c) for $C = 168$ fF. The simulations were performed for various quality factors and two examples are presented in this Figure. These simulations indicate that the quality factor is at least 80 at 40 GHz.

contact the center conductor of the CPW line even with a voltage of 22.5 V. However, when the voltage is increased to 22.8 V, a weak metal contact occurs, as the measured reflection coefficient at this biasing voltage indicates. The extracted capacitance values (except for $V_b = 22.8$ V) are presented in Figure 2-53(b). Based on the parallel plate model approximation, and these capacitance values, the gap between the lower beam and the center conductor at 22.5 V is approximately 0.36 μm.

The fact that the upper beam does not collapse before the lower beam touches the CPW center conductor is also proven by the fact that the contact resistance can be precisely controlled by the bias voltage as shown in [109].

2.4.3.3 Resonant Frequency and Quality Factor The resonant frequency of both FSLB and FFLB designs are high and are dominated by the highest capacitance value and the parasitic inductance of the designs. The parasitic inductance depends on the springs of the lower beam and is in the order of 10 pH for both designs [109]. Consequently the resonant frequency of these varactors is very high (> 100 GHz) and could not be determined by the measurement techniques presented in this work.

The quality factor of these devices is also high and is dominated by the resistive losses of the CPW line and the lower beam. Figure 2-54 shows the insertion loss of a FFLB varactor with 168 fF capacitance and the simulated insertion loss for various quality factors at 40 GHz. This figure demonstrates that the quality factor is higher than 80 at 40 GHz.

2.4.4 Hysteresis Experimental Characterization

It is interesting to see how the hysteresis loop shown in Figure 2-52 varies over time. In this characterization, the varactor was driven by a 100 Hz bipolar square wave. The measurements were performed under low-RF input power (10 mW) and the varactor capacitance value was extracted from the recorded return loss. The test was performed in a regular laboratory environment.

During the first few hundreds of cycles, the varactor under test (VUT) was following exactly the same hysteresis loop. However, after a few thousand cycles, it was observed a tendency of the hysteresis loop to shift toward lower voltages. This trend has been recorded for both branches of the hysteresis loop. Figures 2-55 and 2-56 present the measured results for the forward (varactor

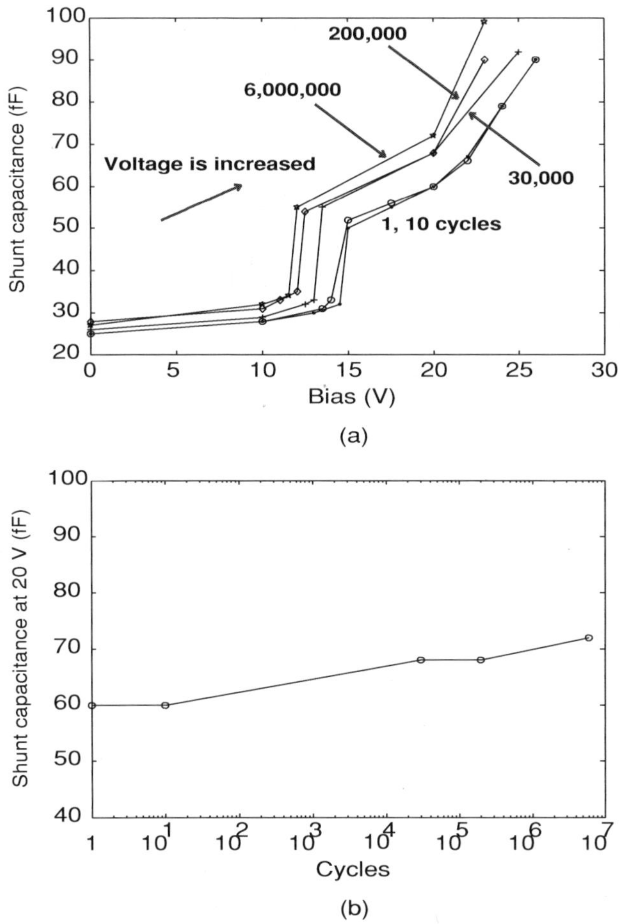

Figure 2-55 Typical varactor: (a) measured forward (varactor moves downwards) branch of a typical varactor hysteresis loop up to 6 million cycles; and (b) measured varactor capacitance at 20 V recorded in the downward movement.

moves downward) and return (varactor moves upward) branches respectively. Both branches show similar trends. In particular, after 30,000 cycles, the intermediate discontinuity recorded in the forward branch shifted from 15 to 13.5 V, while this shift was also measured to be 1.5 V (from 14 to 12.5 V) for the return branch. After cycling the varactor for an additional 170,000 cycles, a further shift of 1 and 0.7 V was recorded for the forward and return branches respectively. The test was continued until a total of 6 million cycles. As Figures 2-55 and 2-56 show, the hysteresis loop again shifted slightly to lower voltages. Based on these measured performance, it can be also seen that the varactor spring constant is reduced with cycling. For example, Figures 2-55b and 2-56b show that the varactor's capacitance values at a specific voltage (20 V for instance) is increased with cycling. In other words, a higher deflection is achieved for the same applied voltage. The varactor was still operational after this cycling. The results presented are typical of several tested varactors.

Although more studies are necessary for a quantitative characterization of these results, a qualitative explanation can be based on the phenomenon of creep. Whereas creep is often observed, it is not widely reported in the literature. However, it is known as memory effect in micromirrors [110]

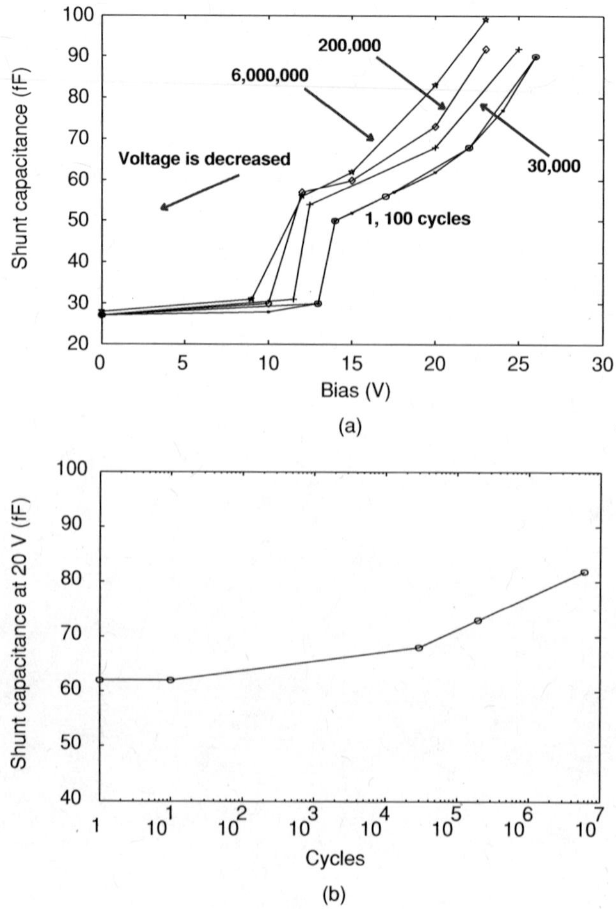

Figure 2-56 Typical varactor: (a) measured return (varactor moves upwards) branch of a typical varactor hysteresis loop up to 6 million cycles; and (b) measured varactor capacitance at 20 V recorded in the upward movement.

and it has been also observed in thin metal plated cantilevers [111]. More specifically, Vickers-Kirby *et al.* [111] developed an apparatus to experimentally characterize over time the voltage required for a constant deflection of MEMS cantilevers. A number of different materials were studied and it was found that Au was the softest and most sensitive material to creep. Figure 2-57 presents the results obtained for 100-μm long cantilevers [111]. The observed creep behavior can be attributed to defect hoping, dislocation motion and grain boundary sliding that are caused by mechanical stress. Since creep is of mechanical origin, harder materials are less sensitive to creep. Si was the hardest material in this study and Si devices showed the least voltage drop over time (< 0.4 V). On the other hand, Au was the softest material and exhibited the greatest voltage drop (almost 5 V).

The behavior of the structure considered here can be also explained on the basis of creep of a different kind. Creep is not necessarily associated only with constant deflections, but may be also developed in a structure during cycling even at room temperature [112]. As it has been observed in other structures, residual stress can be partially relieved by creep or plastic strain [112]. Since the spring constant of the varactor here is dominated by stress, any relief of the stress in the structure

(a)

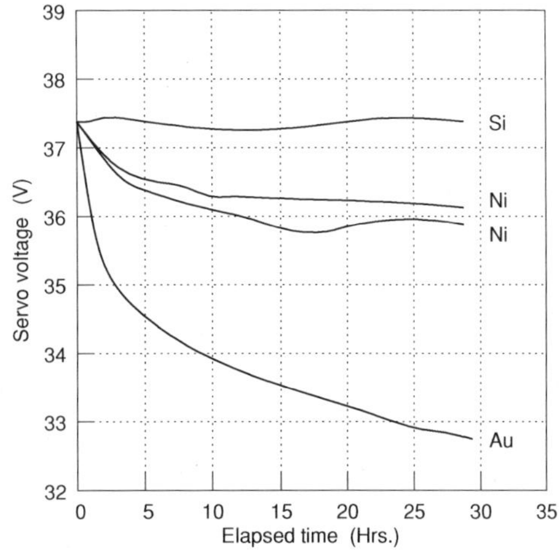

(b)

Figure 2-57 Accelerometer: (a) SEM picture of an electroplated nickel MEMS tunneling accelerometer used in [111] to characterize the required bias voltage to sustain a certain beam deflection over time; and (b) long term bias stability versus time for different cantilever materials in air (after [111]).

will reduce the spring constant of the varactor and equivalently the required pull-in voltage. This is indeed observed in the varactors made of Au, see Figures 2-55 and 2-56. This issue can be alleviated by using harder metals (e.g. Ni) for the lower beam of the varactor and by reducing the residual stress.

2.4.5 Improved Design and Power Handling Results

After significant design and process optimization an improved varactor has been implemented [113]. Accurate stress control and a new design of the lower beam (crab legs) are the main improvements.

Figure 2-58 shows the improved MEMS device. The new varactor is also composed of two beams, namely the lower and upper beams, which are 1 and 10 μm thick respectively. The lower beam is suspended approximately 4 μm above the center conductor of a 70/120/70 μm CPW and is anchored at its ground planes. The upper beam is approximately 8 μm above the substrate and it extends over two square pads that are used for actuating the device. These square pads are completely isolated from the CPW line. Figure 2-59 shows a simplified schematic of these ideas.

As already mentioned the upper beam cannot deflect substantially due to its large stiffness. As a result, it pushes the lower beam downward when an electrostatic potential is applied between the varactor and the actuation pads. Because of the height difference between the two beams, the pull-in instability for the upper beam does not occur until it moves by approximately 2.5 μm downwards. Consequently, the tuning range of the varactor can be very high. As shown in this section though, it is limited by the desired power handling.

2.4.5.1 Low-Power Measurements Small signal measurement was first performed using an 8510C vector network analyzer to measure the S parameters of a typical varactor from 2 to 40 GHz at various bias voltages. The varactor's capacitance was then extracted from the measured S parameters. Figure 2-60 shows typical measured and simulated reflection coefficients and the extracted C-V curve. The simulated S-parameter data was calculated using the Sonnet electromagnetic software

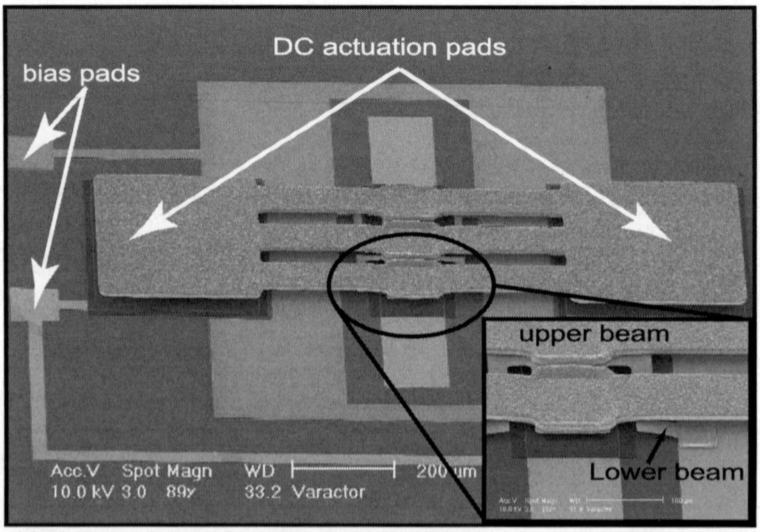

Figure 2-58 Improved contact-less analog RF MEMS varactor [108].

package [114]. The C-V curve and the equivalent circuit parameters were extracted using the Agilent Advanced Design System circuit simulator (ADS [115]). The varactor has a low capacitance of 0.095 pF at no bias and a maximum capacitance of 0.3 F at a bias voltage of 55 V. This corresponds to a tuning range of approximately 3.15 : 1. As it can be observed, no discontinuity appears in the improved design.

2.4.5.2 Power Handling Modeling Lack of any contacts renders the device at hand immune to welding-induced stiction. As a result, self-actuation will be the dominant power-induced failure mechanism. Following the analysis in [81, 113] and by referring to Figure 2-59 the self-actuation power can be approximated to be

$$P_{max} = \frac{\left(8kg_0^3\right)\left(1 + (\pi f C Z_0)^2\right)}{27 A_1 \epsilon_0 Z_0} \tag{2.65}$$

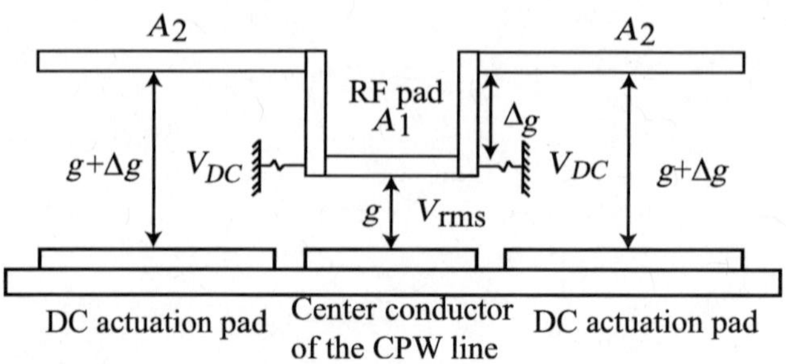

Figure 2-59 Simple schematic diagram of the RF MEMS analog varactor. A1 and A2 are the areas of the RF plate and the DC actuation pads respectively. The varactor is connected to the ground plane of the CPW line with four crab-leg springs (not shown in this figure).

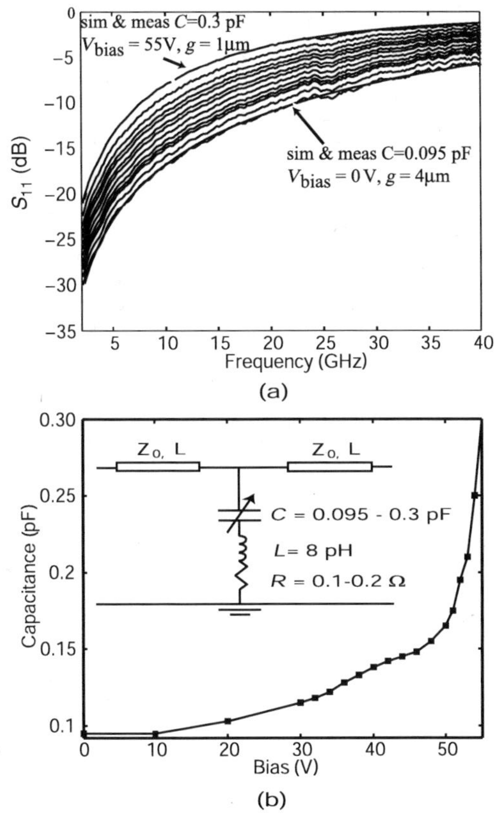

Figure 2-60 Varactor characteristic: (a) measured and simulated return of the MEMS varactor for several bias voltages; and (b) extracted C-V curve and equivalent circuit of the varactor.

where $k = 19.5$ N/m [108] is the z-directed spring constant of the varactor, $g_0 = 4\ \mu$m is the initial gap of the lower beam, f is the operating frequency, C is the varactor's capacitance, $Z_0 = 50\ \Omega$ is the system impedance, and $A_1 = 3.4 \times 10^{-8}$ m^2 is the RF actuation area.

All the varactor's parameters are summarized in Table 2-8. The fringing field effect has been taken into account in the above by the following linear approximation:

$$C = \frac{\epsilon_0 A_1}{g}\left(0.89 + 0.11\frac{g}{1\ \mu\text{m}}\right) \tag{2.66}$$

where g is the lower beam gap and varies between 1 to 4 μm. This equation results in a fringing field capacitance of 34 and 1% when $g = 4$ and $g = 1\ \mu$m respectively. This is consistent with the measurements of Figure 2-60 since the parallel plate capacitances for $g = 4$ and $g = 1\ \mu$m are 0.075 and 0.301 pF respectively.

Based on this analysis, Figure 2-61 shows the calculated minimum allowable gap between the lower beam and the CPW center conductor before self-actuation occurs. The calculation has been performed at $f = 10$ GHz. Notice that although the minimum gap increases as a function of the incident power, the device remains functional but with a reduced tuning range. Figure 2-61 also shows the calculated and measured (please see next subsection) tuning ranges as a function of

Table 2-8 Basic varactor parameters.

Parameter	Symbol and Value
RF area between lower plate and CPW center conductor	$A_1 = 3.4 \times 10^{-8}$ m^2
DC actuation area	$A_2 = 1.1 \times 10^{-7}$ m^2
Effective spring constant	$k = 19.5$ N/m
Lower beam gap	$g_0 = 4\ \mu$m
Upper beam gap	$g_0 + \Delta g = 8\ \mu$m
Actuation voltage	$V_{DC} = 55$ V

the incident power. This graph shows that the 1- and 10-W tuning ranges are 2.3 : 1 and 1.25 : 1 respectively. It is worth noting that the device provides a reasonable tuning range even at powers as high as 10 W. The experimental data associated with this graph are discussed in the next subsection.

2.4.5.3 Experimental Apparatus for High-Power Measurements The high-power experimental characterization of the device is performed using the apparatus shown in Figure 2-62. An X-band Traveling Wave Tube (TWT) amplifier is used to get high input power and a directional coupler (HP11692D) is employed to monitor the incident and reflected powers. The magnitude of the reflection of the device under test (DUT) can then be calculated from P_{incident} and $P_{\text{reflected}}$. However, when such setup is used to measure devices with very low reflections such as the MEMS varactors, special attention needs to be paid to the measurement errors. In particular, there are two major contributors in the test setup that introduce errors in the measurement: 1) the coupler directivity and 2) the reflections from the connecting cable and probes.

The HP11692D is a 2–18 GHz broadband 22 dB coupler with directivity higher than 26 dB from 8 GHz to 18 GHz. The high directivity of the coupler greatly reduces the effect of the input signal (Port 1) on the reflection port (Port 2). However, a typical picoprobe 40 A microwave probe from GGB Industries Inc. has a better than 18 dB return loss from DC to 40 GHz. The signal reflected by the calbes/probes then combines with the reflected signal from the DUT and introduces inaccuracies. This error can be quite high if the DUT's return loss is close to that of cables and probes.

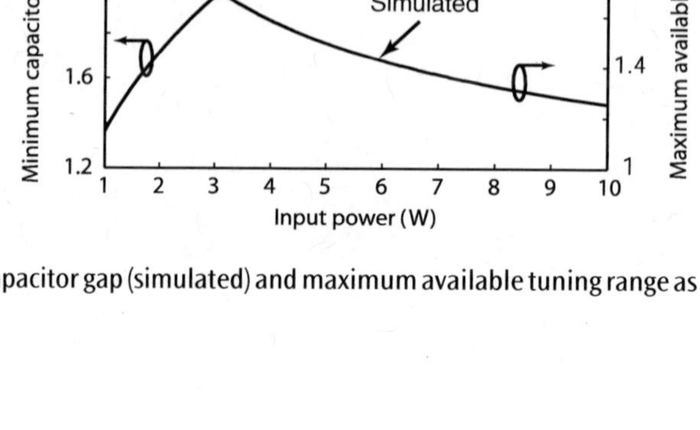

Figure 2-61 Minimum capacitor gap (simulated) and maximum available tuning range as a function of the incident power.

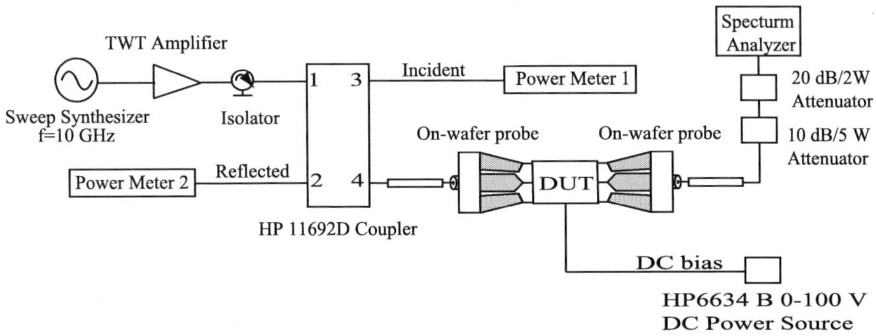

Figure 2-62 Experimental setup implemented for high-power characterization.

In order to calibrate these errors out the following procedure was followed [113]. The varactor was first tested with the setup shown in Figure 2-62 at very low power level (10 dBm). Power meter 1 was used to monitor the input power while power meter 2 was used to record the reflected power. This data is then related (it will be called *power reflection*) to the capacitance value extracted from the network analyzer measurements. For example, when the bias voltage was 30 V, a *power reflection* of −14.43 dB was measured . From the C-V curve in Figure 2-60, it is known that the capacitance is 0.115 pF at 30 V. Therefore, a *power reflection* of −14.43 dB is related to 0.115 pF. This one-to-one mapping between the *power reflection* with the capacitance is performed for all capacitances from 0.095 fF to 0.3 pF. Effectively, the *power reflection* data collected at a low input power level (10 dBm) serves as the calibration scheme.

2.4.5.4 High-Power Measurements, Hot-Tuning Range After this mapping is completed, the high power C-V curve can be extracted from the power measurements. Figure 2-63 shows these capacitance values at different input power levels ranging from 0.01 to 4 W. In these measurements the hot tuning range of the varactor was extracted for every incident power level. From these curves, it is readily observed that there are no power-related failures for powers as high as 4 W. The maximum output power of the TWT was limited to 6 W, which did not permit testing these devices at higher power levels (approximately 2 W are lost in the cables and transitions from the TWT to the varactor). It is interesting to note, however, that at high power levels, the hot tuning range

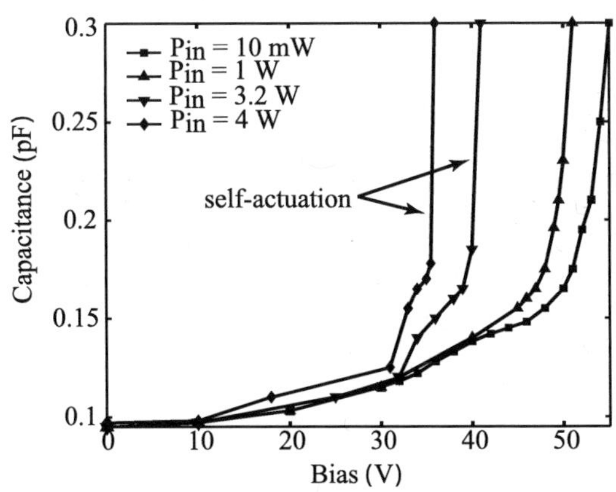

Figure 2-63 Measured varactor tuning range at different input power levels (hot tuning).

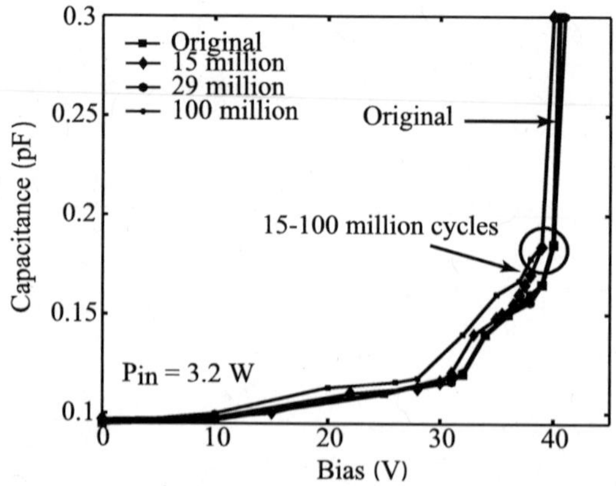

Figure 2-64 Measured varactor C-V curve before and after 15, 29 and 100 million hot cycles at input power of 3.2 W.

of the varactor is reduced due to self-actuation. For example, at 4 W input power, the maximum capacitance that the varactor can provide is about 0.18 pF at 35.5 V. This corresponds to about 40% reduction of the tuning range. Figure 2-61 summarizes these data. Good agreement is observed between the simulated and measured results. The 8–15% error is due to the approximations involved in the simple 1D model.

The same setup for hot tuning range measurement in Figure 2-62 is used to cycle the varactor at an input power of 3.2 W (hot tuning). The DC bias source is replaced with a square wave generator which generates a 0-37 V square wave at 2 kHz. A 3 V safe margin from the pull down voltage (40 V) is used to avoid self-actuation during this cycling test. The varactor is cycled up to 100 million cycles in an open laboratory environment with no nitrogen protection or hermetic packaging. Figure 2-64 shows the varactor C-V curves at $P_{in} = 3.2\ W$ after 15, 29, and 100 million cycles. The varactor's performance is quite consistent during the whole cycling process except that the C-V curve shifts slightly toward the low-voltage side.

After the cycling test is finished, the hot tuning rage of the varactor at various input power levels is measured again. Figure 2-65 shows the measured results after 100 million hot cycling tests as a

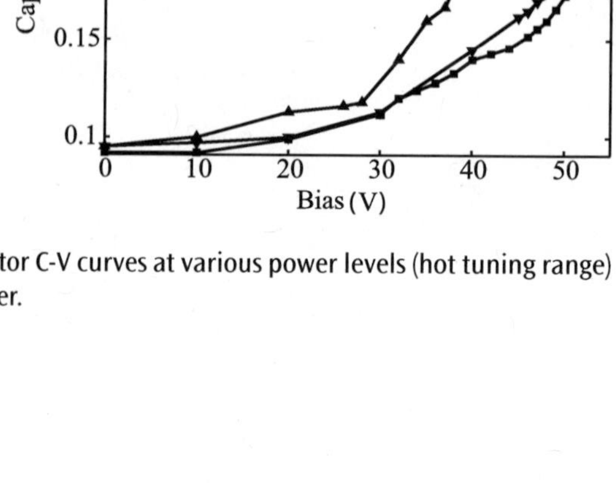

Figure 2-65 Measured varactor C-V curves at various power levels (hot tuning range) after 100 million cycling test with 3.2 W power.

function of bias voltage and input RF power. The measurement results agree very well with the results shown in Figure 2-63 that were taken before the cycling test. The consistency can be clearly seen from the maximum capacitance value that the varactor provides and the corresponding pull-down voltages at different power levels before and after cycling test. The measurements shown in Figure 2-65 were taken several days after the experiments had been started. During this period the TWT's maximum power level was dropped by 0.6 dB, which explains the difference between the maximum power reported in Figures 2-63 and 2-65. It is also worth noting that the devices were perfectly functional after the hot cycling tests were stopped at 100 million cycles.

2.5 CONCLUSION

A number of design techniques for MEMS devices have been presented in this chapter. We started by considering a wide variety of options for achieving low-voltage electrostatic switches and concluded that this is really challenging particularly for robust high-frequency high-power switches. Even if both pull-down and pull-up electrodes are implemented, it is difficult to achieve voltages below 20 V for 40 GHz switches capable of handling over 1W. While lower voltages are theoretically possible for lower powers, it is important to carefully consider the trade-offs in terms of stiction/reliability, bandwidth and/or switching speed. Based on the lessons learned from this study, we presented a metal-to-metal contact switch design based on stress-induced rather than electrostatic forces. This design offers an interesting possibility of decoupling the actuation voltage from the contact resistance. Consequently, high-contact forces that lead to low contact resistance ($< 1\Omega$) may be achieved with moderate voltages (50-60 V). Furthermore, this design eliminates the excessively high and potentially damaging acceleration and impact forces around the contact region that are often observed with conventional electrostatically actuated switches. Finally, K-band contactless analog variable capacitors with tuning ranges that approach 4:1 and power handling in the order of 4 W (hot tuning) are presented. This is achieved by successfully decoupling the DC and RF capacitances while effectively controlling the impact of residual stresses.

3

RF MEMS for Reconfigurable Circuits and Antennas

Linda Katehi and Dimitrios Peroulis

3.1 INTRODUCTION

This chapter focuses on tunable circuits and antennas based on the RF MEMS components described in Chapter 2 (An exception is the exploration of tunable antennas in Section 3.6 where conventional diodes are used for faster innovation). Emphasis is placed on design techniques directly applicable to achieving reconfigurability.

3.2 HIGH-ISOLATION SWITCH NETWORKS

As discussed in Chapter 2, single capacitive shunt switches typically provide a maximum isolation of 25–30 dB. If a higher total isolation is desired, multiple parallel switches can be used. Such a technique has been widely adopted in PIN diode reflective switches and phase shifters [116]. The main idea is to use a τ- or a π-match technique to increase the on-state isolation and minimize the off-state insertion loss. In this section we discuss similar techniques for high-isolation switch packets and in particular the various trade-offs including bandwidth, circuit compactness and maximum isolation that need to be considered in such a design.

3.2.1 The π-Match

The simplest high-isolation switch network is the π-match circuit shown in Figure 3-1. Each shunt element represents a MEMS switch with admittance

$$Y_{sw} = Z_{sw}^{-1} = \left[R_p + j\left(\omega L_p - \frac{1}{\omega C_p}\right)\right]^{-1} \equiv G_{sw} + jB_{sw} \tag{3.1}$$

where, as mentioned before, C_p, L_p, and R_p are the switch capacitance, inductance, and resistance respectively. The scattering parameters of the π-match network are given by [116]:

$$S_{11} = S_{22} = \frac{BY_0 - CZ_0}{2A + BY_0 + CZ_0} \tag{3.2}$$

$$S_{21} = S_{12} = \frac{2}{2A + BY_0 + CZ_0} \tag{3.3}$$

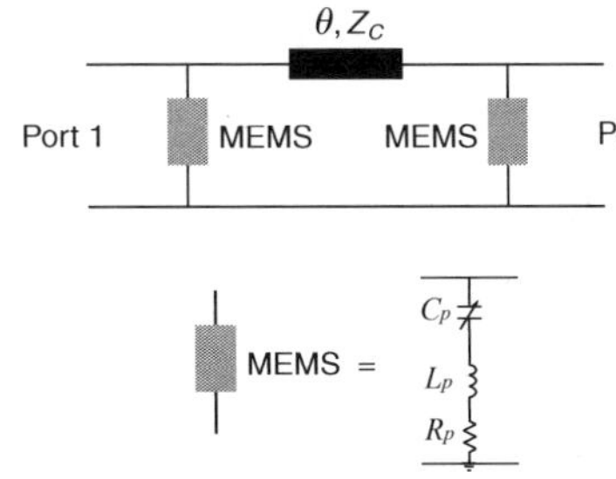

Figure 3-1 The π-match network.

where A, B, C, and D represent the $ABCD$ parameter matrix of the network given by

$$A = D = (\cos\theta - B_{sw} Z_C \sin\theta) + jG_{sw} Z_C \sin\theta \tag{3.4}$$

$$B = jZ_C \sin\theta \tag{3.5}$$

$$C = 2G_{sw}(\cos\theta - B_{sw} Z_C \sin\theta) + jZ_C\left[2B_{sw}Y_C\cos\theta + \left(Y_C^2 + G_{sw}^2 - B_{sw}^2\right)\sin\theta\right]. \tag{3.6}$$

Although these equations may be lengthy to be used as a design aid, they can be greatly simplified by considering the dominant switch components in the up and down states. In the up state, for instance, the dominant component is the small up-state capacitance. Hence

$$Z_{sw} \simeq -j\frac{1}{\omega C_U} \quad \text{(up state)}. \tag{3.7}$$

On the other hand, in the down state and at the switch resonant frequency, the complex impedance Z_{sw} is simply equal to R_p

$$Z_{sw} = R_p \quad \text{(down state at the switch resonant frequency)}. \tag{3.8}$$

The design goals for the π-match design can be summarized as follows

- Up state: $S_{11} = 0$. This goal can be exactly true at only one frequency in the design bandwidth. We will refer to this frequency as the cancellation frequency $\omega_{r,\text{up}}$ [117]
- Down state: S_{21} =min at the switch resonant frequency $\omega_{r,\text{dn}}$

Based on these two requirements, the optimal values for the transmission line electrical length θ and characteristic impedance Z_C can be determined. The optimal value θ_{opt} can be determined from the first goal and by Equation (3.2):

$$S_{11} = 0 \Leftrightarrow BY_0 = CZ_0. \tag{3.9}$$

This equation along with Equation (3.7) leads to

$$\theta_{opt} = \tan^{-1}\left(\frac{2C_U\omega_{r,\text{up}} Z_C Z_0^2}{Z_C^2 - Z_0^2 + \left(C_U\omega_{r,\text{up}} Z_C Z_0\right)^2}\right). \tag{3.10}$$

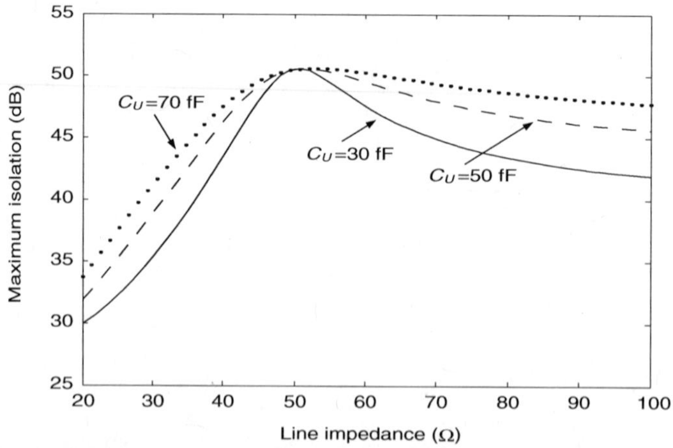

Figure 3-2 Maximum down-state isolation of the π-match circuit at the switch resonant frequency. The following values have been assumed for this graph: $C_U = 30$, 50, and 70 fF, $R_p = 2\ \Omega$, $Z_0 = 50\ \Omega$, and $f_{r,\mathrm{up}} = 15$ GHz.

It is interesting to note that any (practical) value of Z_C can be chosen. The most commonly chosen values are: 1) $Z_C = Z_0$ [53] which leads to the simplified equation

$$\theta_{\mathrm{opt}} = \tan^{-1} \frac{2}{C_U Z_0 \omega_{r,\mathrm{up}}} \tag{3.11}$$

and 2) $Z_C \gg Z_0$ which leads to minimal electrical length.

Nevertheless, these choices do not necessarily lead to the maximum possible isolation at the switch resonant frequency. The optimum value $Z_{C,\mathrm{opt}}$ may, however, be computed from the second goal, namely $S_{21} = 0$ at the switch resonant frequency $\omega_{r,\mathrm{dn}}$. Considering Equations (3.3), (3.4), (3.5), and (3.8) results in

$$\alpha = \frac{1}{S_{21}} = \frac{\left(1 + \frac{Z_0}{R_p}\right)^2 + Z_C^2 \tan^2 \theta_{\mathrm{opt}} \left[\frac{1}{R_p} + \frac{1}{2Z_0} + \frac{Z_0}{2}\left(\frac{1}{Z_C^2} + \frac{1}{R_p^2}\right)\right]^2}{1 + \tan^2 \theta_{\mathrm{opt}}}. \tag{3.12}$$

Figure 3-2 graphically illustrates this equation for $C_U = 30$, 50, and 70 fF, $R_p = 2\ \Omega$, $Z_0 = 50\ \Omega$, and $f_{r,\mathrm{up}} = 15$ GHz. We can clearly observe that the maximum isolation at the switch resonant frequency is achieved for $Z_C = Z_0$, while its sensitivity depends on the value of the up-state switch capacitance.

Equations (3.10) and (3.12) can be successfully employed to design a low-loss and high-isolation switch network with two shunt capacitive switches. These networks, however, are fairly narrowband since the second design goal has been the maximization of isolation around the switch-resonant frequency. If broader bandwidth is desired, the two switches need to have different resonant frequencies. The analysis can be completed in this case in a similar way, but it is more convenient to use a CAD package (e.g., [115]). Furthermore, a CAD package greatly simplifies the design of switch packets with more than two switches, which are particularly suited for applications that require higher isolation over broader bandwidth. The following section presents several examples of switch networks with two, three, and four shunt switches.

3.2.2 Switch Networks: Multiple Switches

As already seen, an optimal characteristic impedance $Z_{C,\text{opt}}$ exists for the transmission line section connecting the switches in a switch packet depending on the design goals. If circuit compactness is desired, Z_C has to be the highest impedance allowed by the fabrication technology. If, on the other hand, the maximum isolation is desired at the switch resonant frequency, Z_C has to be selected based on Equation (3.12). These design considerations are also true for switch packets with more than two switches. Furthermore, the down-state isolation bandwidth should be also added to the design considerations of the switch networks. Narrowband designs with maximum isolation are possible if switches with identical resonant frequencies are used, while higher bandwidths but with lower (theoretical) isolation are obtained with switches that have significantly spread resonant frequencies. Such designs are discussed in detail in this section.

3.2.2.1 Measured Results with Printed Switches Figure 3-3 shows two groups of high-isolation switch network designs. Group A shows switch networks with 2, 3, and 4 switches connected with $Z_C^A = Z_0 = 50\ \Omega$ transmission line sections. The length of these lines is $l_C^A = 1400\ \mu\text{m}$. The major goal of these designs is to achieve maximum isolation. Group B, on the other hand, includes the same designs but with $Z_C^B = 84\ \Omega$, $l_C^B = 250\ \mu\text{m}$. The high-impedance line is made of an 80/20/80 μm CPW. These designs are focused on achieving a high isolation with the minimum possible area.

Switches with two different resonant frequencies are considered to compare broadband and narrowband designs. The networks of Figure 3-3 were first fabricated with the switches printed in the down position so that only their RF performance would be characterized [118]. (The following section presents the results of the same networks with the switches fabricated 4–5 μm over the CPW lines.) The measured down-state RF performances of the printed single switches are presented in Figure 3-4. Based on the discussion of chapter 2, the equivalent circuits of these switches can be easily extracted from these measurements (see Table 3-1).

Figure 3-5(a) presents the measured results of the group A high-isolation switch networks. Only the SPS1 switch was used in all the designs in order to obtain the maximum isolation at the lowest possible frequency. Figure 3-5(b) shows the measured results from the group B switch packets using the same type of switch. As expected, in both cases, the higher the number of switches used, the higher the obtained isolation is. Per design goals, group A results in higher isolation in the 5–10 GHz band since it incorporates the optimum line impedance $Z_C = 50\ \Omega$. This advantage, nonetheless, becomes less significant as the number of switches is increased. This is due to the leaking modes in the Si substrate that limit the achieved isolation as the recorded "noise floor" shows. It is also interesting to note that, although both groups have the SPS1 switch, group B is inherently more broadband than group A.

The same networks were designed and fabricated [118] using both the SPS1 and SPS2 switches in an effort to obtain more broadband networks. Table 3-2 shows the exact configuration of the switches for the high-isolation networks. The measured results from these circuits are given in Figure 3-6 for both group A and group B designs. Both designs are significantly more broadband that the ones using only the SPS1 switch, but the improvement is more dramatic for the designs of group A. It can also be observed that group A results in approximately 10 dB higher isolation between 7–20 GHz, which is consistent with the analysis of Section 3.2.1. Group B, however, is still more broadband than group A and maintains an isolation of higher than 40 dB even up to 40 GHz for all designs.

In summary, it is seen that extremely high-isolation switch networks are possible. The design procedure for these packets depends on the number of switches in the packet and the isolation/bandwidth requirements. If the maximum possible isolation is desired, switch packets with only

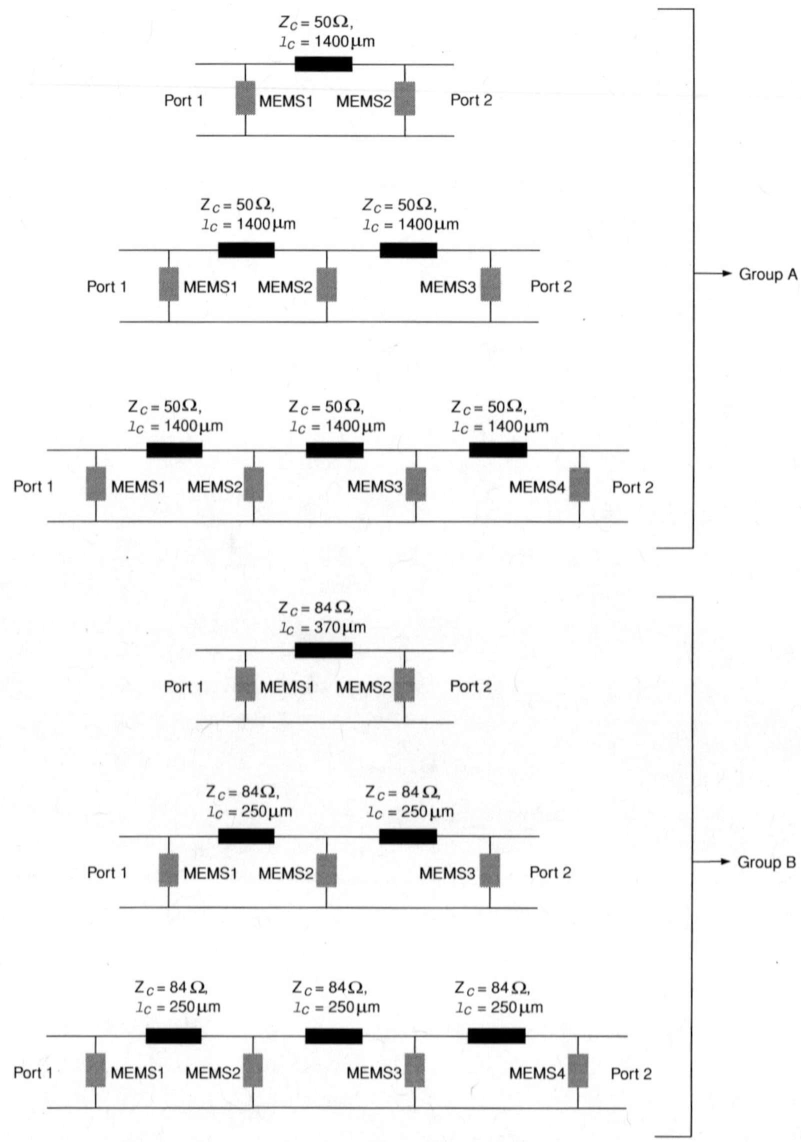

Figure 3-3 High-isolation switch networks. Group A focuses on obtaining the maximum possible isolation, while compactness is the primary goal of group B designs.

one switch type and with the optimum transmission line impedance should be implemented. The measurable isolation of networks with more than 2 switches, however, is typically limited by the surface wave modes in the Si substrate (this is not true if micromachining is employed to etch the substrate, see for example [119, 120]. Hence, the isolation requirements of the network can be relaxed. The benefits in this case are 1) minimization of circuit real estate by using high impedance transmission line sections and 2) increased bandwidth by using switches with different resonant frequencies.

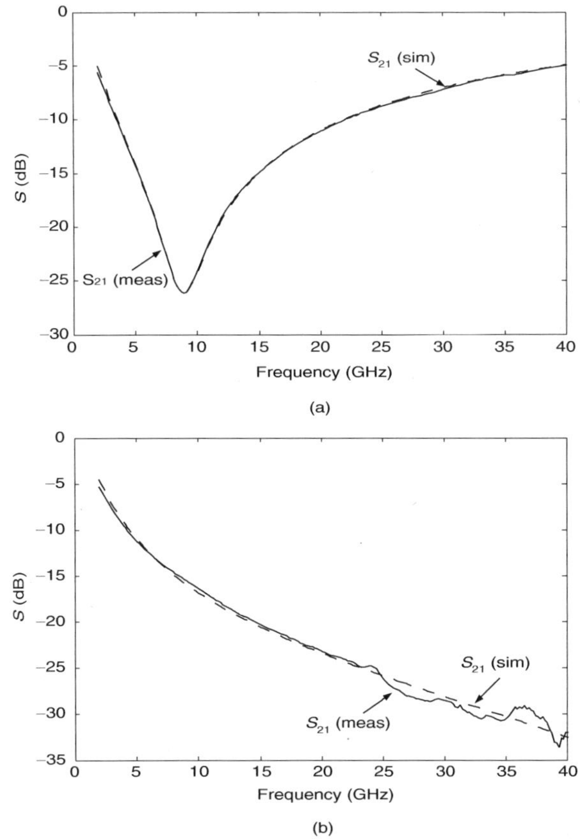

Figure 3-4 Measured down-state S-parameters of two printed switches: (a) single printed switch 1 (SPS1 in Table 3-1); and (b) single printed switch 2 (SPS2 in Table 3-1).

3.2.2.2 Measured Results with Suspended Switches In an effort to verify the results with the printed switches and characterize the behavior of the high-isolation networks with suspended switches, the most complex 4-switch networks of group A were built and measured [118]. Figure 3-7 shows the up- and down-state S-parameter of this network. The up-state insertion loss is 0.94 dB at 10 GHz and 1.31 dB at 25 GHz. Based on TRL calibration [90,91], the loss of the simple through lines is 1.55 dB/cm at 10 GHz and 1.89 dB/cm at 25 GHz. Consequently, the loss of a through line with the same length (5250 μm) as the high-isolation network would be 0.81 dB and 1 dB at 10 and 25 GHz respectively. This shows that the total up-state loss of the four switches is only

Table 3-1 RF equivalent down-state capacitance, inductance, and resistance for the two different switches used in the high-isolation networks. These values correspond to the switch characteristics when printed in the down position.

Switch type	C_D (pF)	L_p (pH)	R_p (Ω)
SPS1	4.2	76	1.3
SPS2	4.2	1.5	0.2

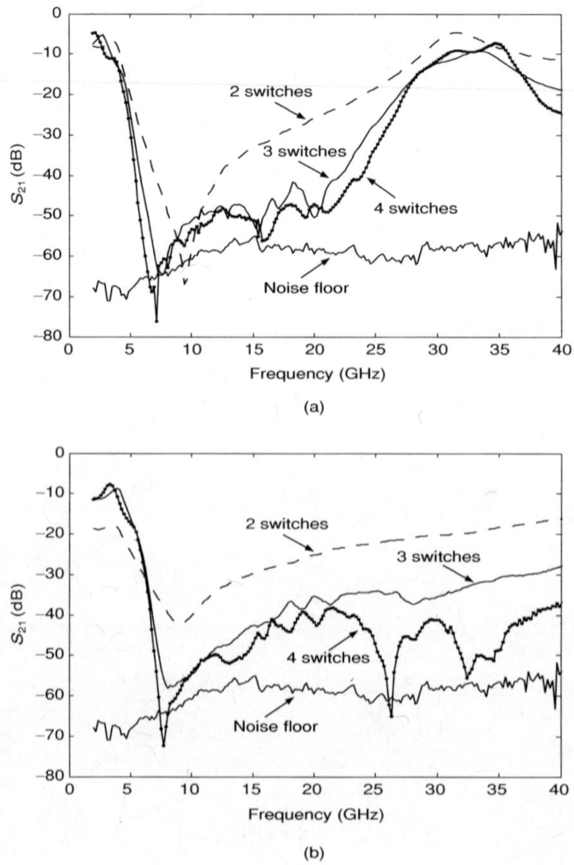

Figure 3-5 Measured down-state isolation of the (a) Group A and (b) Group B switch networks (printed switches). All networks use only the SPS1 switch (see Table 3-1).

0.13 dB and 0.31 dB at 10 and 25 GHz respectively. The return loss of the switch packet is less than −13.5 dB up to 30 GHz.

The down-state performance of the network shows an isolation greater than 40 dB from 10 to 26.5 GHz. The measured isolation follows closely the one obtained from the same network with the switches fabricated in the down position (Figure 3-7(b)). The only difference is that the

Table 3-2 Types of switches used in the broadband high-isolation networks (see Figure 3-3).

	Network			
Number of switches	**MEMS1**	**MEMS2**	**MEMS3**	**MEMS4**
2	SPS1	SPS2	–	–
3	SPS2	SPS1	SPS2	–
4	SPS1	SPS2	SPS2	SPS1

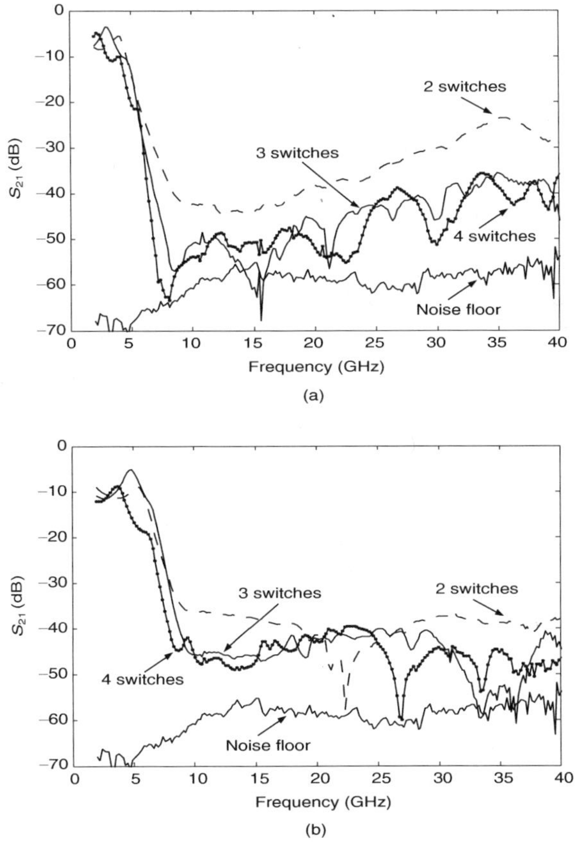

Figure 3-6 Measured down-state isolation of the (a) Group A and (b) Group B switch networks (printed switches). All networks use both switch types according to the configuration shown in Table 3-2.

achieved down state capacitance is approximately 1.2 pF. This is due to the residual stress in the fabricated switches and the roughness of the contacting surfaces that did not allow the switches to conform perfectly on the dielectric layer underneath them.

3.2.3 Summary

This section has presented several design techniques for achieving high isolation from capacitive MEMS switches. A variety of switch packets that address different design goals have been presented. Switch networks with up to four switches show an insertion loss of only 0.03 dB and 0.08 dB at 10 and 25 GHz respectively per switch. They also provide very high isolation that is only limited by the surface wave substrate modes. Theoretical and experimental results of these packets have shown that a compromise between the highest possible isolation, bandwidth, and circuit size is typically required for practical circuits.

3.3 SWITCHABLE FILTERS BASED ON LUMPED MEMS COMPONENTS

Capacitive RF MEMS switches are more general devices than relays, since they have the capability to locally change the RF impedance of a transmission line. Several examples that demonstrate the applicability of this property are introduced here. In particular, we discuss a unique technique of

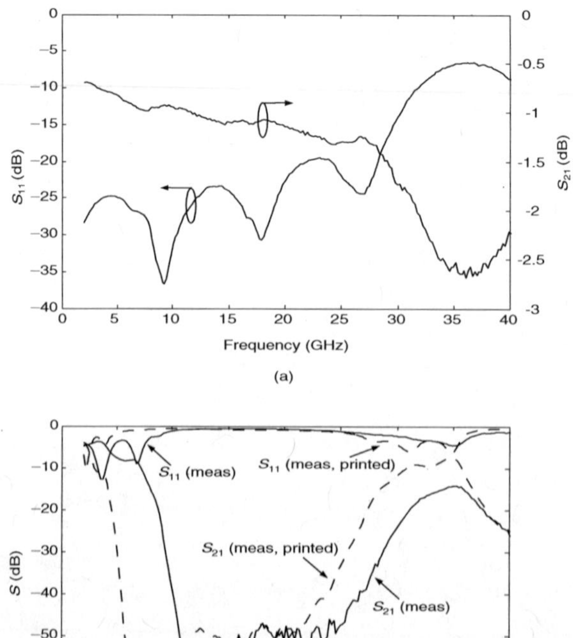

Figure 3-7 Measured (a) up- and (b) down-state *S*-parameters of the Group A 4-switch high-isolation network (normally suspended switches). The down-state isolation compares well with the measured isolation of the same network with the switches fabricated in the down position, except that the achieved down state capacitance is about 1.2 pF.

employing MEMS switches for digitally tuning CPW-based lumped elements and the potential of this technique in compact switchable filter design. Three lumped tunable components for microwave frequencies are first introduced (shunt capacitor, series inductor, and shunt inductor) and then used in lowpass and bandpass tunable filters. Due to the lumped elements employed in these filters, moderate quality factors (20-60) are to be expected.

3.3.1 Tunable Shunt Capacitor

The easiest way to implement a digitally tuned capacitor is to use a capacitive MEMS switch. For example, a metallic plate suspended over a CPW line that can be electrostatically actuated and short the line is a well-known technique. The switch in the up state presents a very small capacitance (typically in the order of 20 to 50 fF), while the down state capacitance is in the range of 1 to 4 pF. Therefore, a tuning ratio of 40 to 80 can be achieved. While this simplistic model of a capacitive MEMS switch may be adequate for switching-network design, it is not sufficient for filter design due to the parasitic components that cannot be neglected in the latter case.

The low-voltage switch already discussed in Chapter 2 is used in this chapter as a vehicle to demonstrate the previous concepts. Of course similar techniques can be readily applied to most of the capacitive switches with similar properties. Figure 3-8(a) illustrates the layout of this switch in

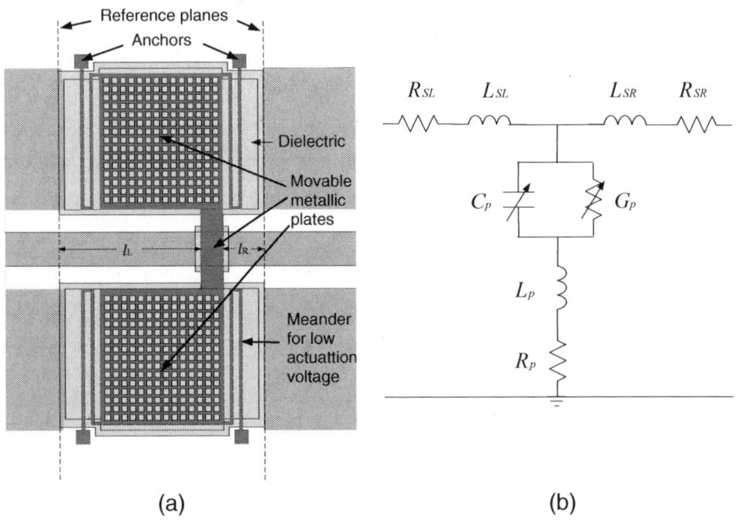

Figure 3-8 Capacitive MEMS switch: (a) used as tunable capacitor over a CPW line; and (b) its equivalent circuit.

a configuration that will be later used for the filter implementation. Figure 3-8(b) shows a lumped-element equivalent T-network that can be used to accurately simulate the RF performance of this switch.

The capacitance C_p is given by the series connection of the capacitance C_{ct} and C_{gnd}, where C_{ct} (C_{gnd}) is the capacitance formed between the center conductor (ground plane) of the CPW and the plate on top of it. Following [53], the up- and down-state capacitances can be expressed as

$$C_p = \begin{cases} \varepsilon_0 \dfrac{A}{d + t_d/\varepsilon_r} + C_{frng} & : \text{ up state} \\ \varepsilon_0 \varepsilon_r \dfrac{A}{t_d + d_{gap}} & : \text{ down state} \end{cases} \tag{3.13}$$

where A is the plate area on top of the center conductor, d is the distance between the switch and the CPW in the up-state ($d \simeq 4--5$ μm for the switches in this work), t_d is the thickness of the dielectric material on top of the CPW line ($t_d = 1200A$), ε_r is the dielectric constant of the material used ($\varepsilon_r \simeq 7.5$ for PECVD Si_3N_4), and C_{frng} is the additional capacitance due to the fringing field effect. Finally, d_{gap} is an effective thickness that models the inability of the switch to achieve a perfect contact with the dielectric layer underneath it. For the geometry shown in Figure 3-8(a), the extracted up and down capacitances from the measured S-parameters are $C_p^{up} = 17$ fF and $C_p^{down} = 920$ fF.

R_p and L_p represent the resistance and inductance of the switch beam and can be modeled with a full wave simulator. Typical values range from 0.1–2 Ω and 1–80 pH for the R_p and L_p respectively. They are primarily dominated by the beams connecting the switch plates (see Chapter 4). G_p depends on the quality of the dielectric material and can be readily modeled using the formula:

$$G_p = \begin{cases} 0 & : \text{ up state} \\ \omega C_p \tan\delta & : \text{ down state} \end{cases} \tag{3.14}$$

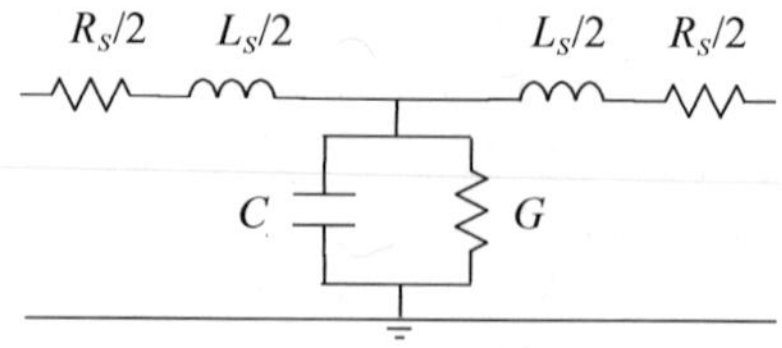

Figure 3-9 Equivalent circuit of a short length of transmission line.

where $\tan\delta$ is the loss factor of the dielectric material used as the insulator. The inductive sections L_{SL} and L_{SR} are used to model the short transmission line sections of lengths l_L and l_R respectively (see Figure 3-8). These inductances are important since their values can be comparable to those of the inductors used in a filter. L_{SL} can be approximated as:

$$L_{SL} \simeq \frac{2Z_0}{\omega}\tan\left(\frac{\beta l_L}{2}\right) \simeq \frac{Z_0\beta l_L}{\omega}. \tag{3.15}$$

A similar formula holds for L_{SR}. Alternatively short transmission line sections can be employed to model the finite (but small) switch electrical length. The resistances R_{SL} and R_{SR} represent the ohmic loss of the same transmission line sections and can be modeled as:

$$R_{SL} \simeq 2\alpha Z_0 l_L \quad \text{and} \quad R_{SR} \simeq 2\alpha Z_0 l_R \tag{3.16}$$

where α is the measured attenuation of the transmission line (approximately 1.2 dB/cm for the circuits measured in this work).

3.3.2 Tunable Series Inductor

It is a well-known fact that a relatively high-Q inductor can be implemented with a short section of a transmission line assuming that the length is less than approximately a tenth of a wavelength [121] (Figure 3-9). The component values of this equivalent circuit can be calculated by formulas similar to the ones given in Equations (3.14), (3.15), and (3.16). The quality factor of this inductor can be estimated if the small shunt capacitance and conductance are neglected, as follows :

$$Q_{S,\text{ind}} = \frac{\omega L_S}{R_S} = \frac{\tan\left(\frac{\beta l}{2}\right)}{\alpha l} \simeq \frac{\beta}{2\alpha}. \tag{3.17}$$

This equation results in a quality factor of 19 at 10 GHz with a measured attenuation of 1.2 dB/cm, which is typical for high-resistivity Si wafers (2000 Ωcm before the wafer oxidization). Higher Q values can be readily achieved by using higher resistivity Si wafers (e.g. 10000 Ωcm), high-quality quartz wafer (with a loss tangent of 0.0004), or by employing well-known buck micromachining techniques [122–124]. Independently of the available technology, the short transmission line section results in a (conventional) inductor with the highest possible quality factor.

The reconfigurability of this element can be achieved in association with the tunable capacitor. The primary goal is to load a short transmission line section with RF MEMS switches and then actuate the appropriate switches in order to get the desired equivalent series inductance between the actuated switches. An example that explains this idea is illustrated in Figure 3-10, which shows a short transmission line with four capacitive MEMS switches numbered from one to four. Assuming that 1) the up-state capacitance is negligible, and 2) all four switches are identical, the series inductor can be tuned by actuating different sets of switches. For instance, if S_1 and S_4 are

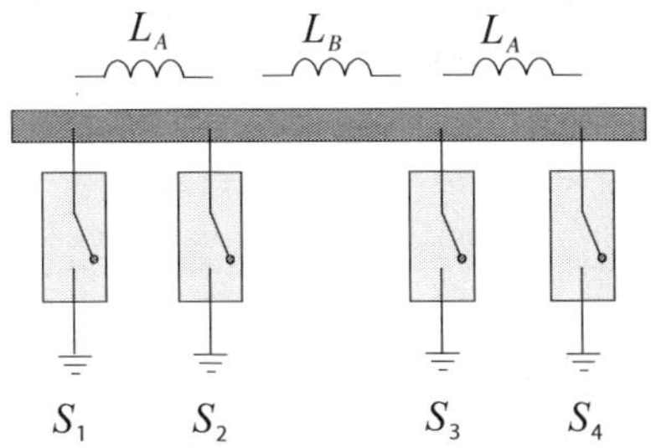

Figure 3-10 Tunable series inductor with four switches.

down, then the inductance value between these switches is $L_{\text{ind}}^{1,4} = L_B + 2L_A + 3(L_{SL} + L_{SR})$. On the other hand, when S_2 and S_3 are down, this value becomes $L_{\text{ind}}^{2,3} = L_B + L_{SL} + L_{SR}$ resulting in a tuning ratio of

$$\mathcal{R}_{\text{ind}} \equiv \frac{L_B + 2L_A + 3\,(L_{SL} + L_{SR})}{L_B + L_{SL} + L_{SR}} = 1 + 2\frac{L_A + L_{SL} + L_{SR}}{L_B + L_{SL} + L_{SR}}. \tag{3.18}$$

The major advantage of this technique is that the inductor tuning ratio can be quite high and is totally controlled by the placement of the switches on the transmission line. Moreover, although two switching pairs have been used in this example, more switching pairs can be easily integrated with the transmission line section and the inductor can take more discrete values. The minimum and maximum inductance values are given by

$$L_{\text{min}} = L_{SL} + L_{SR} \quad \text{and} \quad L_{\text{max}} = L_B + 2L_A + 3(L_{SL} + L_{SR}). \tag{3.19}$$

3.3.3 Tunable Shunt Inductor

An inductive CPW stub is usually employed to implement a shunt resonator. Although stubs as long as $\lambda/4$ have been used [125], only short stubs are necessary here. Such short inductive stubs exhibit similar properties (in phase shifting and quality factor) to the series inductors presented in the previous subsection. Therefore, their equivalent circuits are not substantially different from the ones already presented and will not be repeated.

A metallic plate suspended on top of the inductor and the ground plane as shown in Figure 3-11(a) can be utilized for tuning this component. The plate completely covers the inductive stub and a part of the ground plane. Furthermore, the holes have been removed from the plate just above the stub in order to achieve a higher capacitance (for clarity in the figure, the metallic plate is transparent). In the up state, the plate should not substantially affect the value of the inductance. In the down state, however, the stub is effectively shorted by the capacitively formed RF short and the inductance value changes to an effective inductance L_{eff}.

The 3-D electromagnetic simulator—IE3D [92] has been employed to model the inductance in the up and down position and demonstrate this concept. The simulated results are presented in Figure 3-11(b), which clearly indicate that a tuning range of approximately 2:1 can be realized through this technique. Two ideal inductance values are also mentioned in Figure 3-11, which correspond to the inductances that could be realized if it were possible to perfectly short the stub in the down state and leave it unaffected in the up state. Obviously, an even higher tuning range could be achieved if these minimum and maximum inductance values could be obtained. A metal-to-metal contact switch could, therefore, provide results closer to the ideal situation. Additionally, the tuning range can be further optimized by changing the geometrical features of the stub.

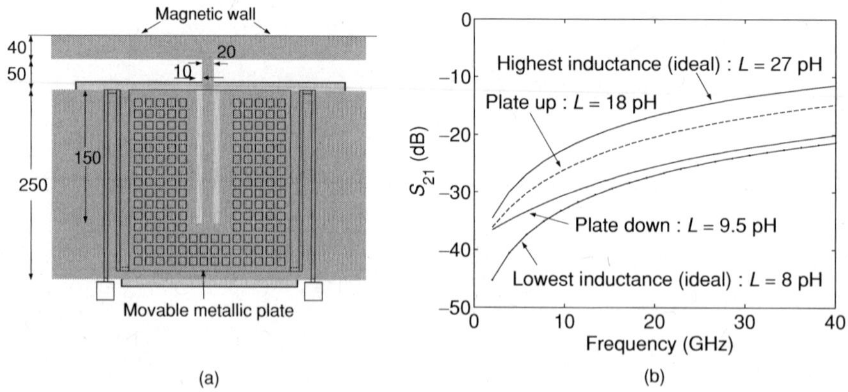

Figure 3-11 Tunable shunt inductor: (a) layout (dimensions are in μm); and (b) simulated response.

3.3.4 Tunable Lowpass Filter

The tunable components analyzed in the previous section can be combined to form potentially useful tunable circuits. To demonstrate this concept, two switchable filters are presented using these ideas. A 3:1 switchable lowpass filter is presented in this section where the basic concepts are explained. A more innovative and complex 2:1 bandpass filter design that employs all of the three tunable components in this chapter is discussed in the next section.

3.3.4.1 Design Approach Figure 3-12 shows the well-known LC ladder topology of a prototype lowpass filter. The tunable lumped elements presented in this chapter are adequate for reconfiguring this topology. The basic idea is to implement the shunt capacitors by shunt capacitive MEMS switches and the series inductors of the filters with high-impedance section transmission lines. The filter topology can then be readily altered by appropriately selecting the group of MEMS switches to be actuated.

Figure 3-13 illustrates an example of this technique. The initial filter topology represents a third-order lowpass filter when switches 1 and 5 are activated, while it is transformed to a fifth-order filter if switches 2, 3, and 4 become active. The desired bandwidth of the two filters can be achieved by appropriately selecting the capacitance values of the MEMS switches and the length of the transmission lines (equivalent inductors) connecting them. It is interesting to note that this technique allows the tuning of every single element of the filter. This is necessary for a true reconfiguration of the filter.

3.3.4.2 Implementation and Results These techniques were applied to implement a 3:1 Chebyshev tunable lowpass filter [78]. The basic design starts by considering separately the two desired filters. For this implementation two 0.2 dB Chebyshev filters are considered with 10 (third-order filter) and 30 GHz (fifth-order filter) bandwidth respectively. Figure 3-14 shows the two lowpass

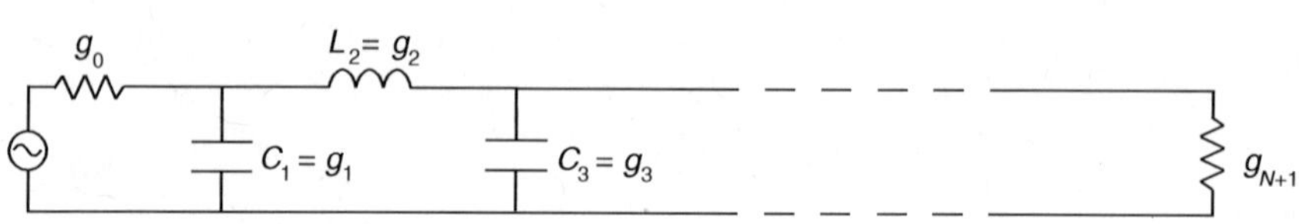

Figure 3-12 Lowpass filter topology.

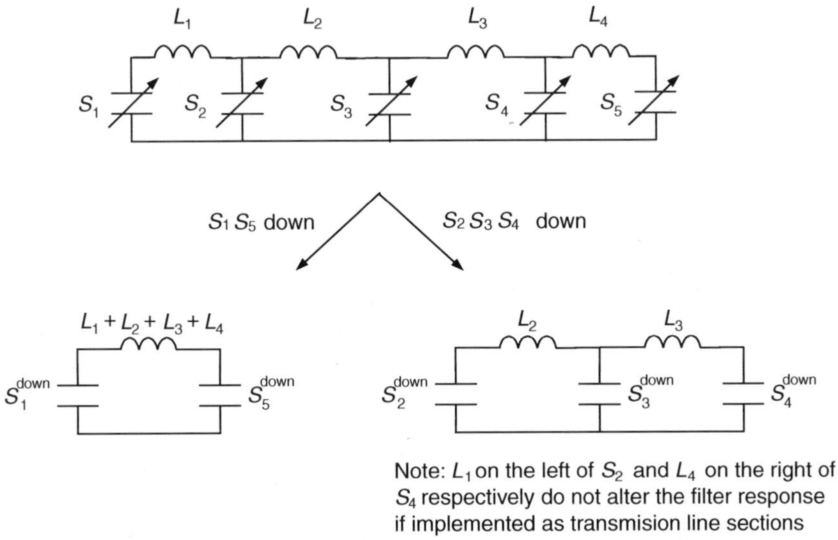

Figure 3-13 Basic concept for the lowpass filter reconfiguration.

prototypes as well as the lumped-element value components for the desired bandwidths assuming a 50 Ω system. These two filters are combined according to the idea illustrated in Figure 3-13. Each capacitor is then implemented by a shunt MEMS switch and each inductor with a high-impedance transmission line. Figure 3-15 shows the final fabricated filter, a simple extracted equivalent circuit for the two stages and the final measured and simulated results.

In general the measured results can be well explained by the extracted equivalent circuits. The low-frequency filter in particular matches very well with the model. This filter's insertion loss is approximately 1.1 dB at the 10 GHz passband bandwidth with a return loss of less than −15 dB. However, there is an increase of 0.8 dB in the insertion loss at the 30 GHz band. This is partly due to the increased return loss at the level of −8 dB around 20 GHz. This discrepancy can

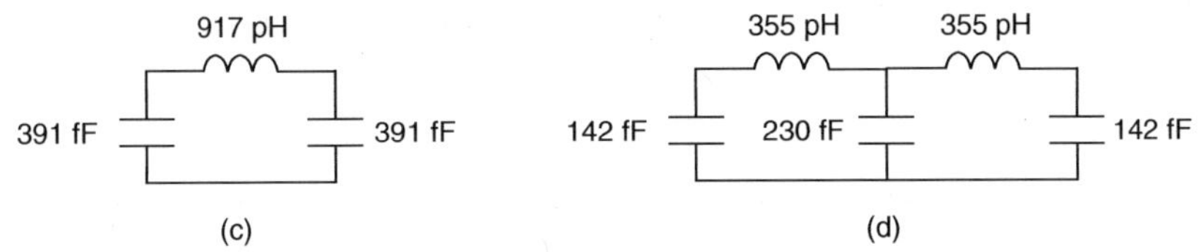

Figure 3-14 Filter prototypes: (a) lowpass prototype for 0.2 dB third-order Chebyshev filter; (b) lowpass prototype for 0.2 dB fifth order Chebyshev filter; (c) lumped-element implementation of (a) with 10 GHz bandwidth in a 50 Ω system; and (d) lumped-element implementation of (b) with 30 GHz bandwidth in a 50 Ω system.

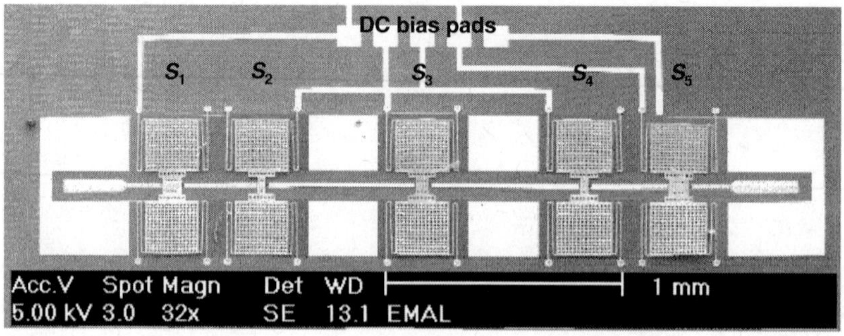

(a)

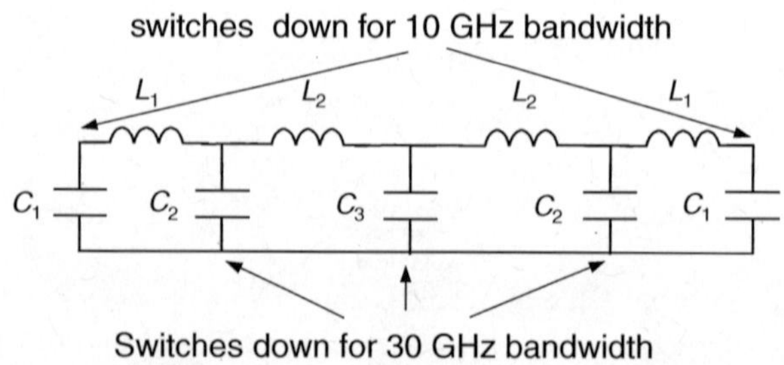

(b)

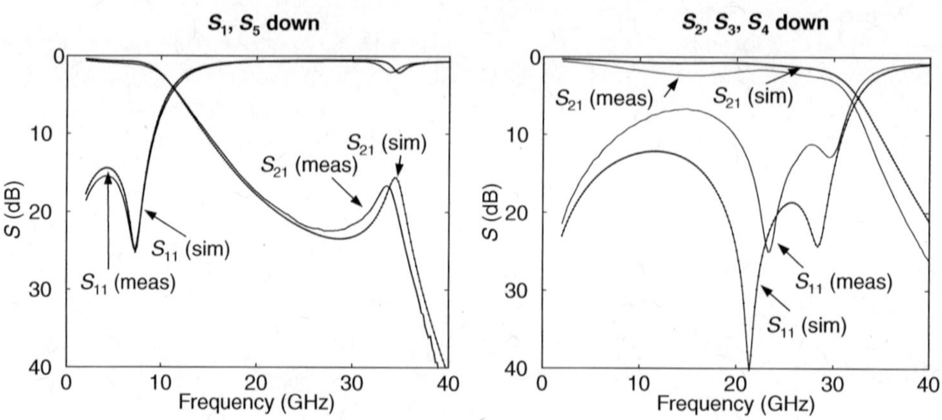

(c)

Figure 3-15 Lowpass filter: (a) layout; (b) simplified equivalent circuit; and (c) simulated and measured response.

be attributed to the somewhat lower capacitances that were experimentally obtained when the switches 2, 3, and 4 were down.

3.3.5 Tunable Bandpass Filter

3.3.5.1 Design Approach While a conventional Chebyshev bandpass implementation with an LC ladder is possible, the presence of the small inductors L_{SL} and L_{SR} (see Section 3.3.1) makes the accurate implementation of the shunt LC resonator difficult (Figure 3-16). This is particularly

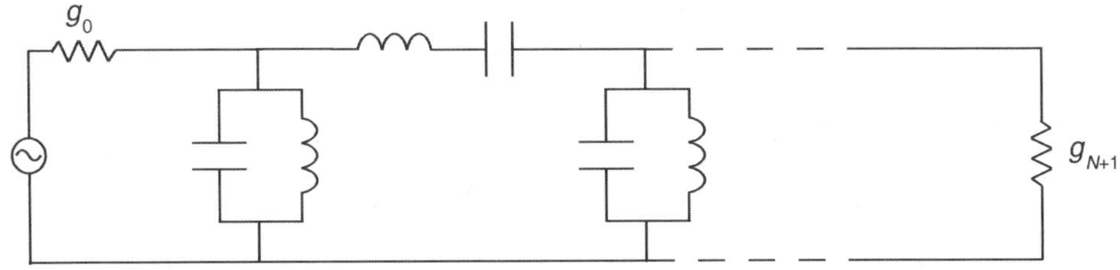

Figure 3-16 Traditional LC-ladder bandpass filter implementation. The switch (shunt capacitor) electrical length significantly complicates the realization of shunt resonators of this filter.

true when more than one switches are placed in parallel with the shunt inductor for a multiband operation. L_{SL} and L_{SR} in such a case can easily become comparable to the inductor of the shunt resonator, leading to a significantly distorted filter response. Consequently, the traditional LC ladder scheme is suboptimal for the implementation of lumped-element tunable bandpass filters.

Instead a new network particularly suited for MEMS switch reconfigurability has been investigated in [78]. The major advantages of this network are a) it results in very compact implementations, b) it requires only one mechanically moving device (the shunt capacitive switch) for its reconfiguration, and c) it preserves the absolute bandwidth of the filter as the center frequency is tuned, since every one of its components is tuned.

Figure 3-17 presents the conceptual steps followed to design this network. The first basic idea relies on the fact that the circuits of Figure 3-17(a) and Figure 3-17(b) are equivalent for short transmission-line lengths. In other words, for $\beta l < \pi/6$ they result in (almost) the same reflection coefficient. Consequently, by placing different shunt capacitors in various distances from the short circuit of Figure 3-17(b) we can obtain different shunt LC resonators. This technique can be used for frequencies that satisfy $\beta l < \pi/6$. Based on these one-port networks, a two-port network can

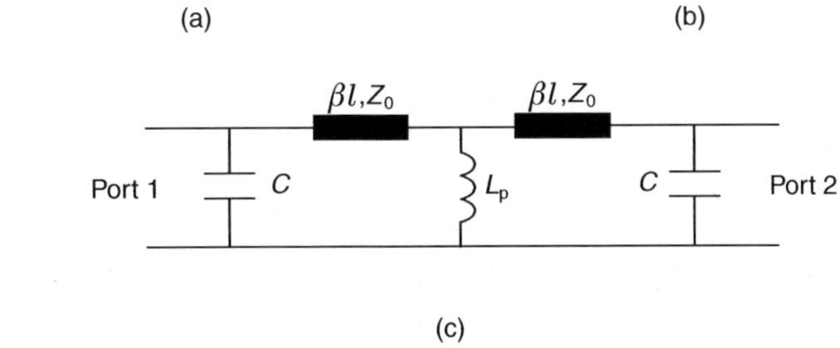

Figure 3-17 Design approach for the synthesis of the bandpass filter topology. The two one-port networks shown as circuits (a) and (b) are (almost) equivalent if $\beta l < \pi/6$. Network (c) is the finally proposed two-port filter.

be synthesized as shown in Figure 3-17(c). The shunt inductor L_p is necessary for the desired bandpass response.

The frequency response analysis of this filter can be readily obtained by following the well-known even–odd analysis method. Figures 3-18 and 3-19 graphically illustrate the detailed steps of the analysis.

3.3.5.2 Even- and Odd-Mode Analysis The even-mode excitation is shown in Figure 3-18(b). Because of the symmetry emphasized in Figure 3-18(d), the circuit is finally transformed to the topology shown in Figure 3-18(e). From this topology we have:

$$V_A^e = \frac{V_g}{2}\frac{C \parallel (L_s + 2L_p)}{Z_0 + C \parallel (L_s + 2L_p)} = \frac{V_g}{2}\frac{s\frac{\omega_{0e}}{Q_e}}{s^2 + s\frac{\omega_{0e}}{Q_e} + \omega_{0e}^2} \tag{3.20}$$

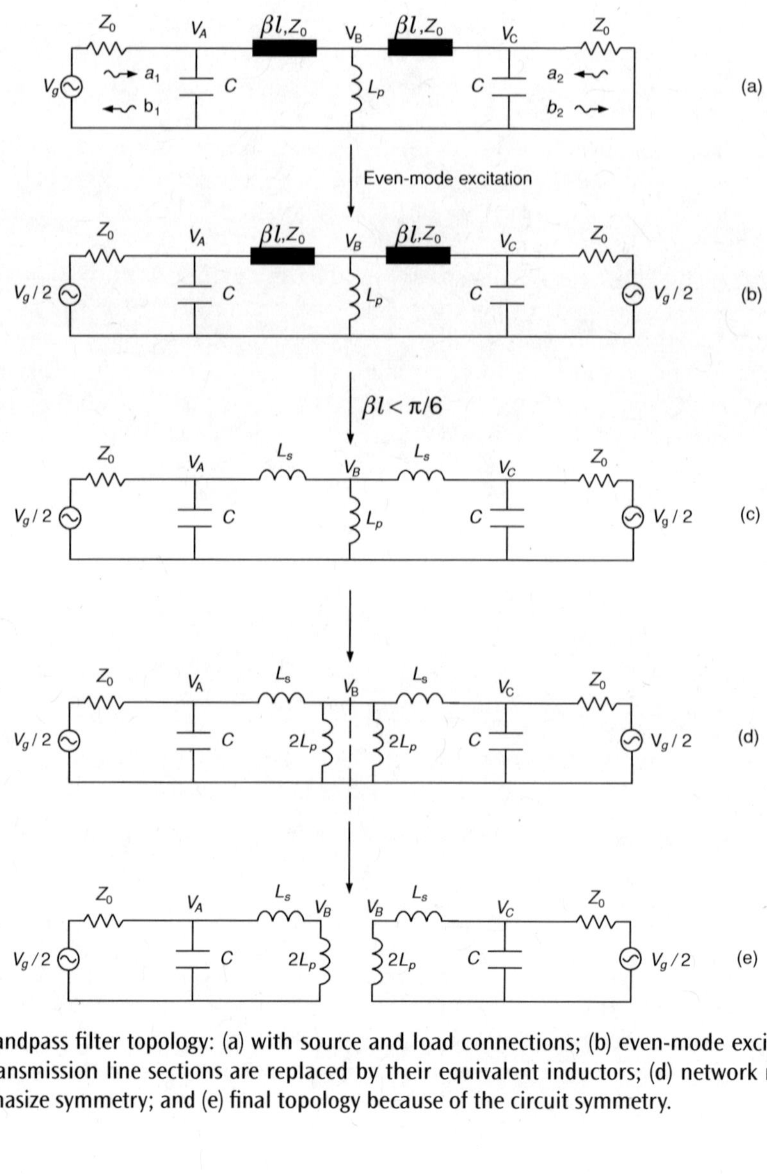

Figure 3-18 Bandpass filter topology: (a) with source and load connections; (b) even-mode excitation; (c) short transmission line sections are replaced by their equivalent inductors; (d) network redrawn to emphasize symmetry; and (e) final topology because of the circuit symmetry.

where

$$\omega_{0e} = \frac{1}{\sqrt{C(L_s + 2L_p)}} \quad \text{and} \quad Q_e = \frac{1}{Z_0 C \omega_{0e}} = Z_0 \sqrt{\frac{C}{L_s + 2L_p}}. \tag{3.21}$$

The voltages V_B and V_C are then readily given by

$$V_B^e = V_A^e \frac{2L_p}{L_s + 2L_p} \quad \text{and} \quad V_C^e = V_A^e \quad \text{(because of symmetry).} \tag{3.22}$$

The odd-mode excitation is shown in Figure 3-19. Due to anti-symmetry we have

$$V_A^o = -V_C^o \quad \text{and} \quad V_B^o = 0. \tag{3.23}$$

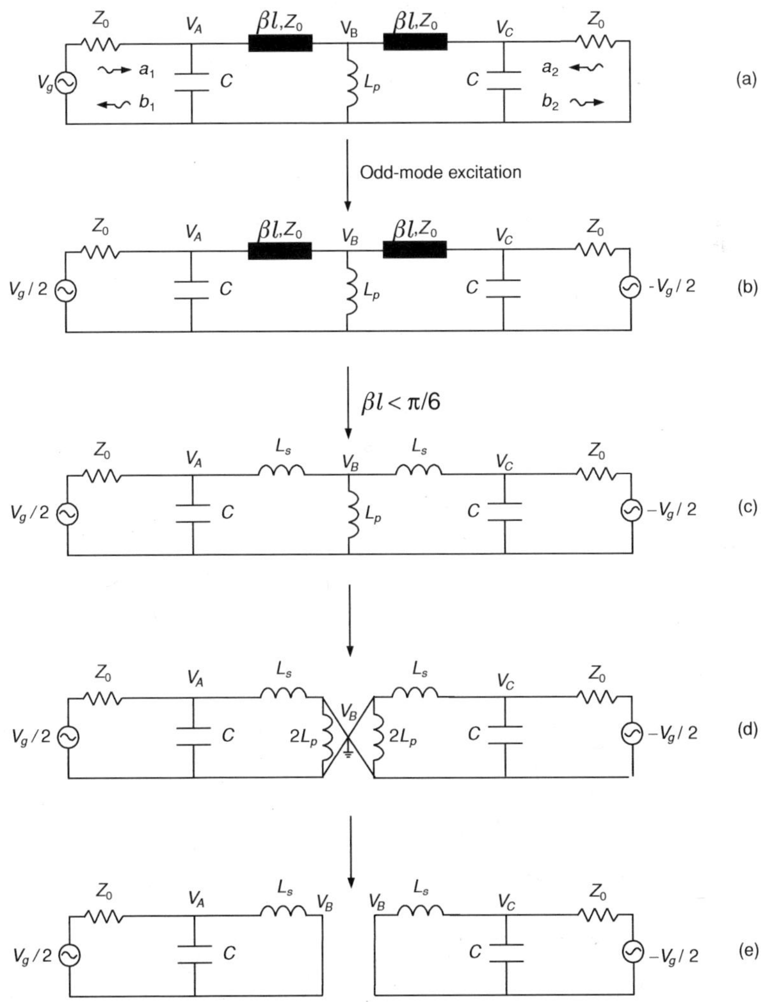

Figure 3-19 Bandpass filter topology: (a) with source and load connections; (b) odd-mode excitation; (c) short transmission-line sections are replaced by their equivalent inductors; (d) network redrawn to emphasize symmetry; and (e) final topology because of the circuit symmetry.

Additionally, using a similar procedure as for the even-mode analysis

$$V_A^o = \frac{V_g}{2} \frac{s \frac{\omega_{0o}}{Q_o}}{s^2 + s \frac{\omega_{0o}}{Q_o} + \omega_{0o}^2} \tag{3.24}$$

where

$$\omega_{0o} = \frac{1}{\sqrt{CL_s}} \quad \text{and} \quad Q_o = \frac{1}{Z_0 C \omega_{0o}} = Z_0 \sqrt{\frac{C}{L_s}}. \tag{3.25}$$

By applying superposition the total voltages on the filter are given by

$$V_A = V_A^e + V_A^o = \frac{V_g}{2} \left[\frac{s \frac{\omega_{0e}}{Q_e}}{s^2 + s \frac{\omega_{0e}}{Q_e} + \omega_{0e}^2} + \frac{s \frac{\omega_{0o}}{Q_o}}{s^2 + s \frac{\omega_{0o}}{Q_o} + \omega_{0o}^2} \right] \tag{3.26}$$

$$V_B = V_B^e + V_B^o = \frac{V_g}{2} \frac{2L_p}{L_s + 2L_p} \frac{s \frac{\omega_{0e}}{Q_e}}{s^2 + s \frac{\omega_{0e}}{Q_e} + \omega_{0e}^2} \tag{3.27}$$

$$V_C = V_C^e + V_C^o = \frac{V_g}{2} \left[\frac{s \frac{\omega_{0e}}{Q_e}}{s^2 + s \frac{\omega_{0e}}{Q_e} + \omega_{0e}^2} - \frac{s \frac{\omega_{0o}}{Q_o}}{s^2 + s \frac{\omega_{0o}}{Q_o} + \omega_{0o}^2} \right]. \tag{3.28}$$

The return loss of the filter is now calculated as

$$S_{11} = \frac{b_1}{a_1} |_{a_2=0} \tag{3.29}$$

by taking into account that

$$V_A = \sqrt{Z_0} \left(a_1 + b_1 \right). \tag{3.30}$$

The final expression is

$$S_{11} = \frac{s \frac{\omega_{0e}}{Q_e}}{s^2 + s \frac{\omega_{0e}}{Q_e} + \omega_{0e}^2} + \frac{s \frac{\omega_{0o}}{Q_o}}{s^2 + s \frac{\omega_{0o}}{Q_o} + \omega_{0o}^2} - 1. \tag{3.31}$$

Similarly it can be shown that the insertion loss of the filter is

$$S_{21} = \frac{s \frac{\omega_{0e}}{Q_e}}{s^2 + s \frac{\omega_{0e}}{Q_e} + \omega_{0e}^2} - \frac{s \frac{\omega_{0o}}{Q_o}}{s^2 + s \frac{\omega_{0o}}{Q_o} + \omega_{0o}^2} \tag{3.32}$$

or equivalently

$$S_{21} = \frac{s \frac{\omega_0}{Q} \left(\omega_{0o}^2 - \omega_{0e}^2 \right)}{\left(s^2 + s \frac{\omega_{0e}}{Q_e} + \omega_{0e}^2 \right) \left(s^2 + s \frac{\omega_{0o}}{Q_o} + \omega_{0o}^2 \right)} \tag{3.33}$$

where

$$\frac{\omega_0}{Q} = \frac{\omega_{0e}}{Q_e} = \frac{\omega_{0o}}{Q_o} = \frac{1}{CZ_0} \tag{3.34}$$

Evidently, this filter has two second order poles at

$$f_{pe} = \frac{1}{2\pi\sqrt{C(L_s + 2L_p)}} \quad \text{and} \quad f_{po} = \frac{1}{2\pi\sqrt{CL_s}} \tag{3.35}$$

one zero at $s = 0$ and three zeros at infinity.

3.3.5.3 Implementation and Results Based on the above equations, a bandpass filter with a tunable center frequency at 15 and 30 GHz has been designed and implemented [78]. This filter is illustrated in Figure 3-20 along with its equivalent circuit, measured and simulated results. As described before, the filter includes the switched inductor (shunt CPW stub) in its center, which is controlled by the center switch. This inductor is tuned between 16 pH (15 GHz operation) and 9 pH (30 GHz operation). This geometry clearly shows that all elements of the filters (shunt capacitors, series inductors, shunt inductors) are tunable. Because of this key feature, this filter is also capable of preserving its absolute bandwidth even when switched over an octave frequency range. It is worth noting that the total area of this fourth order filter is less than 1 mm^2.

The measured response at 15 GHz shows an insertion loss of 3.5 dB, which agrees very well with the theoretical expectations. Both the center frequency and the bandwidth, are also correctly predicted from the theory. At 30 GHz the measured results show a loss of 4 dB. This band also shows a discrepancy between theoretical and experimental results for frequencies below 30 GHz. The reason for this discrepancy lies in the facts that 1) the tuning range of the shunt inductor is smaller that the one predicted by the theory and 2) the capacitive shorting of the inductor is not sufficient for low frequencies (see Section 3.3.3). The results can be improved if a metal-to-metal contact switch is used to tune the shunt inductor, particularly for designs at lower frequencies.

3.4 HIGH-POWER TUNABLE MATCHING NETWORKS

Analog MEMS varactors are very well suited for high-power tunable circuits because of lack of any metal-to-metal contacts that are typically sensitive to welding and stiction. Consequently, they can be exploited in areas where conventional RF MEMS switches cannot easily provide viable solutions. High-power RF MEMS tuners is one of these areas and it is presented in this section.

A number of performance measures are generally applied to characterize an impedance tuner. These typically include: (1) impedance tuning range, (2) loss of the tuner, and (3) the tuner's power handling capability. The tuner's loss is an important characteristic because the maximum VSWR that the tuner can achieve is primarily limited by its loss. At the same time, high-power handling is equally important since tuners are primarily focused on load-pull applications where RF power higher than 0.5 W is often required.

Several MEMS impedance tuners have been reported [126–130]. Tuners based on double-stub networks are reported in [126] and [127]. Both designs utilize a double-stub topology and employ MEMS switches to vary the stub admittance. In Reference [126], the stubs are composed of a set of digital MEMS varactors in a parallel combination that serves as a digital capacitor bank. In [127], the stubs are composed of shorted transmission lines loaded with MEMS switches. Both designs use a large number of MEMS switches to realize the required impedance coverage (8 switches in [126] and 12 switches in [127]).

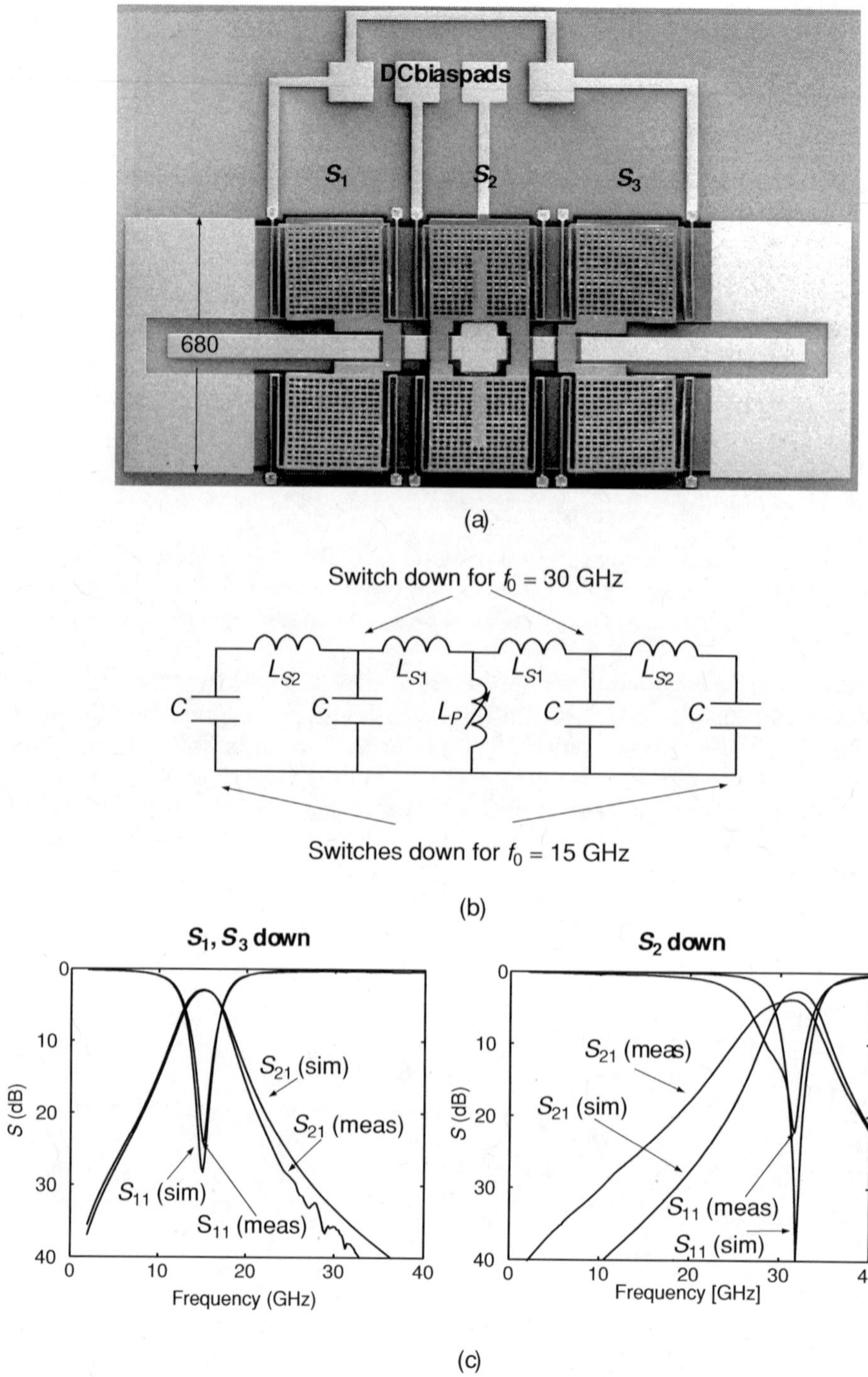

Figure 3-20 Bandpass filter: (a) layout (dimensions are in μm); (b) simplified equivalent circuit; and (c) simulated and measured response.

A 6-20 GHz impedance tuners based on triple-stub topology is reported in [128]. Triple-stub tuners can theoretically provide larger impedance coverage than double-stub tuners [131]. The stubs are shorted transmission lines loaded by MEMS capacitors similar to those used in [127]. Eleven switches are used in this particular design. Similar designs utilizing a combination of capacitively loaded transmission line with stub-based topology have also been reported for tuners in different frequency bands [129, 130].

Although these previously reported tuners have provided, in general, very good impedance coverage, they have not considered the issue of power handling. They all use a large number of MEMS switches in the designs which suffer from a) dielectric charging, b) stiction and welding; b) low-power handling; and c) expensive packaging requirements. Consequently, tuners based on MEMS switches are not an optimal solution for load-pull applications where high power handling is desired. Additionally, the interconnection lines [126] and the bias lines [128, 130] for the switches in these tuners also occupy a large chip space and they could limit the tuner's performance in terms of frequency bandwidth and tuner loss.

In this section a novel impedance tuner is presented that is simultaneously optimized for impedance range and power handling [132]. Instead of MEMS switches the design employs contact-less MEMS varactors with a very wide tuning range (3:1) that can reliably handle more than 3.5 W of RF power (hot-tuning) with no hermetic package. Furthermore the placement of the MEMS varactors is optimized so each varactor exhibits approximately the same voltage swing. This ensures that self-actuation due to high RF power [81] is minimized for the whole tuner.

3.4.1 Impedance Tuner Design

3.4.1.1 Basic Design First consider a two-capacitor impedance tuner connected to a 50 Ω load (Figure 3-21). The tuner's impedance coverage is determined by the capacitor tuning range and transmission line t-line 2. In the interest of using the minimum number of MEMS varactors, each capacitor is implemented by a single MEMS varactor. This varactor was presented in Chapter 2. It will be treated here as a variable capacitor with a continuous (analog) tuning range of 0.1–0.3 pF.

The length of t-line 2 is traditionally chosen as $\lambda/4$ to achieve maximum impedance coverage. It serves as an admittance transformer that transforms a relatively high-admittance value of G_A to a lower admittance G_B (see Figure 3-21). Varactor 2 thus has more effective tuning and results in a larger coverage on the Smith Chart. The $\lambda/4$ line also enables a more uniform impedance coverage since the changes of C_1 and C_2 move Z_L in orthogonal directions on the Smith Chart (Figure 3-21(b)) [127].

3.4.1.2 Power Handling The tuner's power handling capability is determined by 1) the maximum power each varactor can individually handle and 2) the varactor's self-actuation voltage when placed in the tuner. As shown in chapter 2 the employed capacitors can reliably handle at least 3.5 W when measured individually. Maximum power handling of the tuner as a system can be achieved by ensuring that the maximum voltage swings across the varactors are equalized. These voltage swings can be calculated using the equivalent circuit of Figure 3-21.

Figure 3-22 shows the result of the calculation for the voltage ratio $|V_2/V_{\text{load}}|$ as a function of the electrical length θ_2 and the capacitance C_1. It can be seen that at $C_1 = 0.3$ pF and $\theta_2 = 90°$ (the optimal value of θ_2 if only impedance coverage is considered) the rms voltage of varactor 2 is about twice of the rms voltage at varactor 1 and the 50 Ω load. This is highly undesirable since varactor 2 is much more likely to self-actuate. As a result, the tuner's power handling capacity is primarily limited by varactor 2. Reducing the transmission line length from 90° can reduce the voltage ratio. Nevertheless, it will also reduce the impedance coverage. Fortunately, the curve in Figure 3-22 also shows that the slope of $|V_2/V_{\text{load}}|$ versus θ_2 is quite high around $\theta_2 = 90°$. This

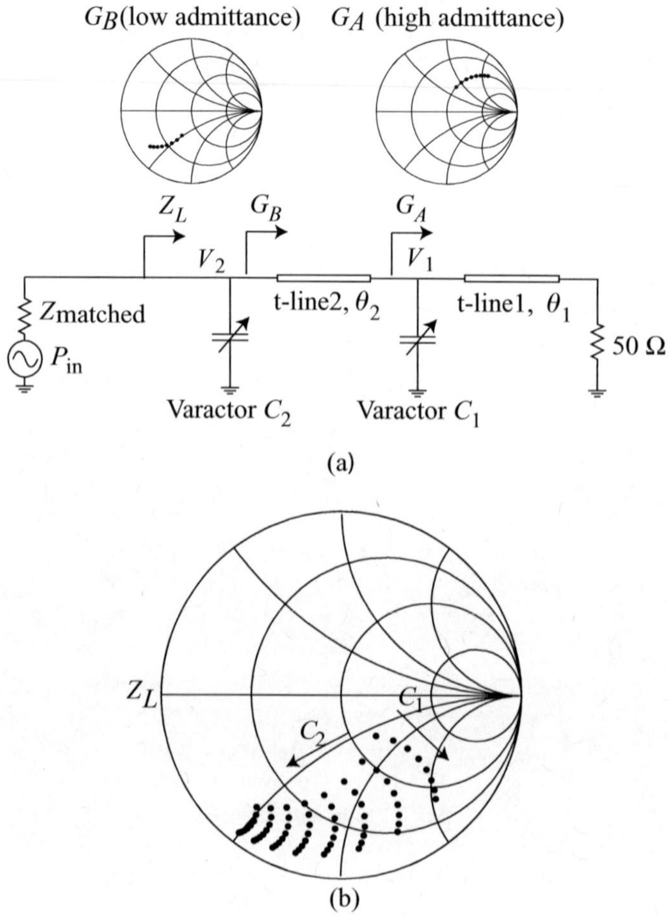

Figure 3-21 A two-capacitor tuner: (a) topology; and (b) its simulated impedance coverage at $f = 25$ GHz. The arrows indicate the directions of increase for C_1 and C_2. Both are varied from 0.1 pF to 0.3 pF.

suggests that a relatively small change of the line length can significantly decrease $|V_2/V_{load}|$ while at the same time, maintain a reasonable impedance coverage. An optimized design can then be obtained based on the trade-off between impedance coverage and voltage ratio.

The analysis on the rms voltage so far has been based on a matched power source assumption (Figure 3-21). If the power source has a mismatched impedance, the power delivered to the load will be less due to reflection. However, the ratio of the rms voltages of the varactors with that of the load depends only on the tuner's configuration and does not change with the matching condition. The previously calculated voltage ratio relation is still valid and the rms voltage on the varactors reduces accordingly with that on the load. Therefore, the self-actuation of the varactor will occur at a higher P_{in} level as compared to the matched case. This is important since in practice, we can only use a 50 Ω power source for the experimental test of the tuner's power handling capability. It should be noted that in this condition, the varactors can survive higher input power (from the power source) than in the matched case before self-actuation occurs.

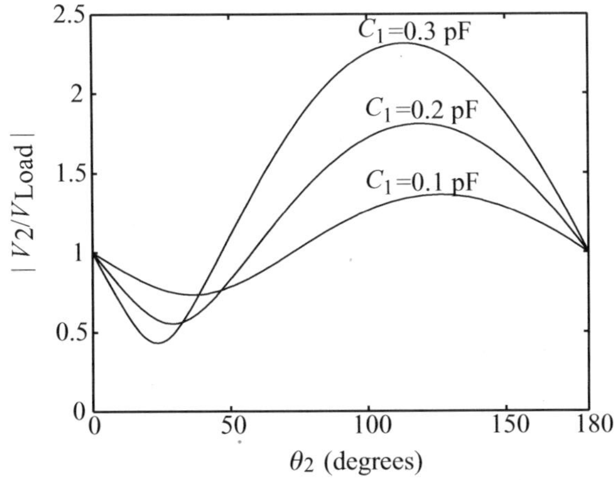

Figure 3-22 Magnitude of the voltage ratio V_2/V_{load} versus the electrical length of t-line 2, θ_2, for various C_1 values.

3.4.1.3 Improving Power Handling and Impedance Coverage Following a similar procedure we can also show that if a load with admittance of $G_0 + jb$ is used instead of just G_0, the varactor's voltage swing can be further reduced. An example is shown in Figure 3-23 where a small $jb = j0.5$ (normalized) is added to the matched load. The voltage ratio for $\theta_2 = 90^o$ and $C_1 = 0.3$ pF has been reduced from 2.13 to 1.82. Based on these calculations, the optimized jb value can be found for different values of θ_2. The inclusion of such a susceptance can be easily accomplished by adding a shunt MIM capacitor at the end of the tuner as shown in Figure 3-24(a).

It is worth noting that adding the extra MIM capacitor also increases the tuner's impedance coverage. Figure 3-24(b) shows the simulated impedance coverage of the tuner with a 0.18 pF

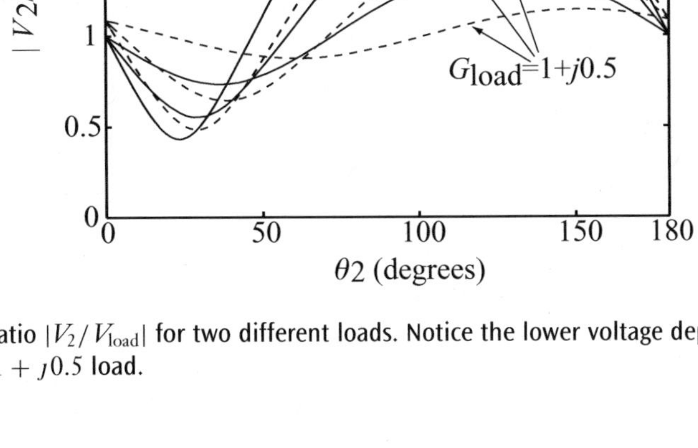

Figure 3-23 Voltage ratio $|V_2/V_{\text{load}}|$ for two different loads. Notice the lower voltage dependance on θ_2 for the $G_{\text{load}} = 1 + j0.5$ load.

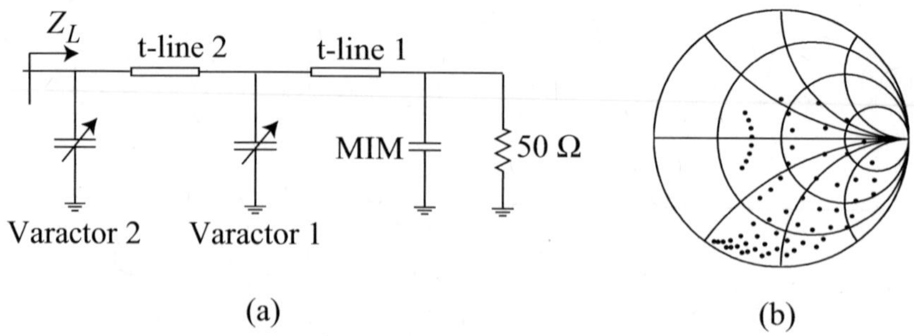

Figure 3-24 Two-capacitor tuner: (a) circuit with an additional MIM capacitor; and (b) tuner's impedance coverage at $f = 25$ GHz with a shunt 0.18 pF MIM capacitor (compare with Figure 3-21(b)).

MIM capacitor. Both t-lines are $\lambda/4$ lines as in Figure 3-21. An improvement of the Smith Chart coverage by about 80% is achieved as compared to Figure 3-21. The reason for a better coverage is that the MIM capacitor transforms the 50Ω load to a lower impedance value.

3.4.1.4 Multi-Capacitor Tuner As seen in Figure 3-24, the two-capacitor tuner covers ideally about two quadrants of the Smith Chart. To increase the impedance coverage, two such tuners can be cascaded together as shown in Figure 3-25. The tuner is optimized to achieve a large impedance coverage and at the same time have a uniform maximum RMS voltage on each of the four varactors. The design optimization steps are summarized below.

1. An MIM capacitor is placed at the end of the tuner to optimize the coverage and the varactor rms voltages.
2. The transmission lines used in the tuner are initially $\lambda/4$ lines. However, the final lengths of the lines are chosen such that the maximum rms voltages of the four varactors are approximately equalized. This of course, requires a trade-off with the impedance coverage.
3. Each capacitance is implemented with only one MEMS analog varactor which can be electrically tuned to provide about 0.1–0.3 pF of shunt capacitance. Therefore a total of 4 MEMS varactors are used in this design.

Table 3-3 summarizes the parameters used in the tuner shown in Figure 3-25. Its impedance coverage is shown in Figure 3-26(a). The varactor capacitance values are set as [0.1 0.14 0.18 0.22 0.26 0.3] pF in the simulation to produce this figure.

Figure 3-27 presents the ADS simulated time-domain varactor voltages for all possible bias voltages. These waveforms are normalized to the voltage of the load so they are independent of

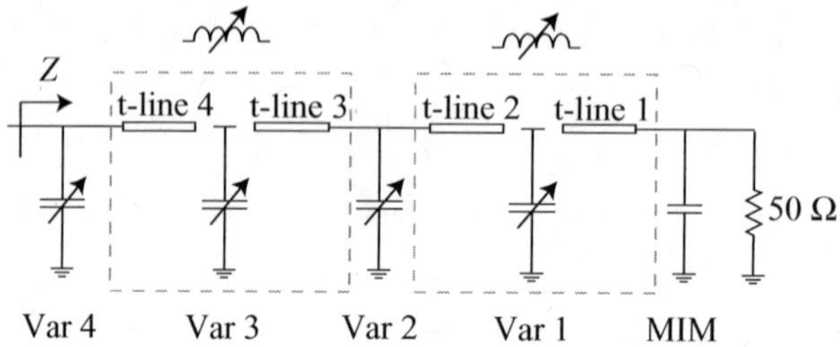

Figure 3-25 Two two-capacitor tuners cascaded together to increase the tuner impedance coverage.

Table 3-3 The parameters used in the tuner design in Figure 3-25 on Si substrate

Description	Value
MIM cap	0.25 pF
MEMS varactor	0.1–0.3 pF
t-line 1	0.19 λ
t-line 2	0.21 λ
t-line 3	0.18 λ
t-line 4	0.13 λ

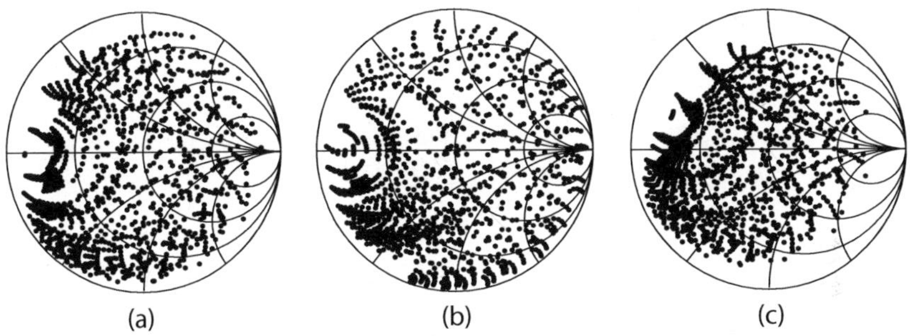

Figure 3-26 Simulated impedance coverage of various four-capacitor tuners at 25 GHz: (a) Design 1—Optimized for impedance coverage and power handling as shown in Figure 3-25 (lossless); (b) Design 2—Conventional approach optimized only for impedance coverage (lossless); and (c) Design 1—Impedance coverage with loss included in the simulation.

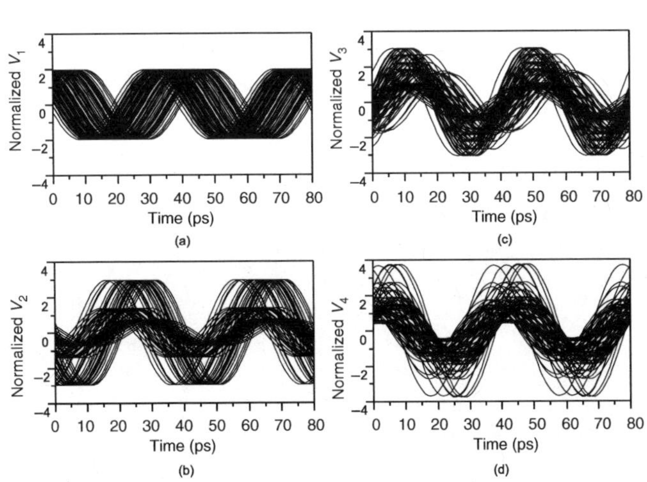

Figure 3-27 Simulated time-domain voltage waveforms for the four varactors of the tuner for all possible bias voltages. Each waveform is normalized to the amplitude of the voltage at the load. The maximum amplitude for each varactor is listed in Table 3-4 (Design 1).

Table 3-4 Maximum rms voltage of the four MEMS varactors. Design 1 is the optimized design whose parameters are given in Table 3-3 and design 2 is a comparison design which does not optimize the varactor rms voltages

Maximum RMS voltage (normalized to that of the 50 Ω load)	Design 1	Design 2
Varactor 1	2.0	1.0
Varactor 2	3.0	2.6
Varactor 3	3.1	5.1
Varactor 4	3.7	7.8

the input power level. The maximum voltage swing for each varactor of this design is summarized and compared to a conventional design in Table 3-4. As a comparison, Table 3-4 also presents the rms voltage of a conventional design (design 2 in Table 3-4) which uses exact 90° transmission lines (except the line between varactors 3 and 4 which is 75°) and no MIM capacitor. Although design 2 has a wide impedance coverage (Figure 3-26b), the maximum rms voltage on the varactors are not optimized. For example, the maximum voltage swing on varactor 4 is about 8 times of that on varactor 1, and 3 times of that on varactor 2. This means that varactor 4 may experience much higher voltage swing at some certain configurations and cause self-actuation. For instance, with varactors of self-actuation voltage of 24 V, the tuner can only deliver 0.19 W power to the load before self-actuation occurs for varactor 4. The optimized design, on the other hand, has close maximum rms voltage on each of the four varactors so no particular varactor would have high risk of failure due to self-actuation. With the same varactors, the maximum power that the tuner can deliver to the load increases to 0.84 W. This is about 4.5-times improvement in terms of the power that the tuner can deliver.

3.4.1.5 Tuner Loss The loss due to the MEMS varactors is pretty low [107] and can be neglected compared to the transmission line loss. A 100/120/100-μm CPW line with 2 μm gold on a 10, 000-Ωcm Si substrate was chosen to be used for our impedance tuner. The resultant transmission line has characteristic impedance of 55 Ω and a measured attenuation of about 0.7 dB/cm at 25 GHz. With the transmission-line loss included in the simulation, the resultant tuner impedance coverage is reduced as shown in Figure 3-26(c).

3.4.2 Experimental Results

The impedance tuner shown in Figure 3-25 was fabricated on a 10,000-Ωcm Si substrate. The fabrication process flow is the same as reported in [107]. Figure 3-28 shows a picture of the fabricated tuner. Due to a condition change of the sputter tool [132], the deposited lower beam

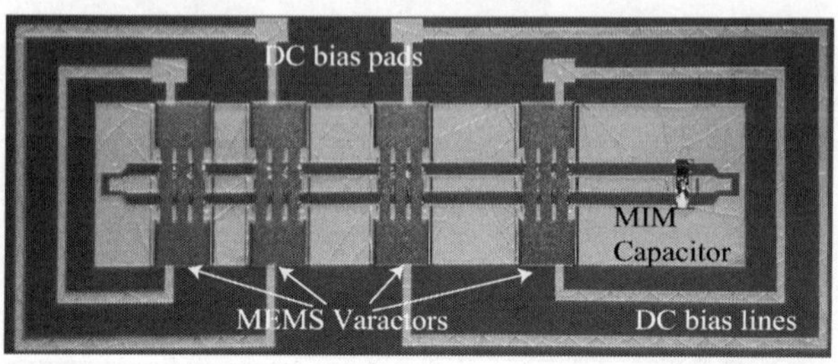

Figure 3-28 Microphotograph of the fabricated impedance tuner. The tuner occupies an area of 0.49×0.12 mm^2.

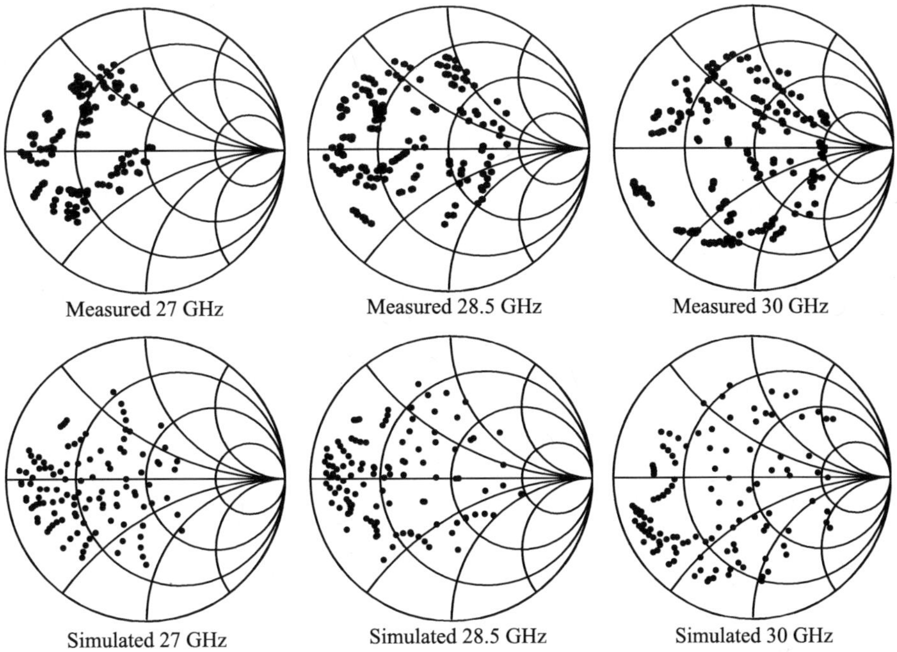

Figure 3-29 Measured and simulated tuner impedance coverage at 27 GHz, 28.5 GHz, and 30 GHz.

had a somewhat higher residual stress level than expected. This resulted in a reduced capacitor range of the MEMS varactors because the varactor's minimum capacitance was increased to about 0.15 pF instead of the nominal 0.1 pF. Due to these fabrication issues, the tuner presented its optimum performance at 30 instead of the originally designed 25 GHz.

Figure 3-29 presents the simulated and experimental results for various frequencies. At frequencies lower than the design frequency the tuner's functionality is primarily limited by the reduced impedance coverage due to the reduction of the transmission lines' electrical lengths as well as the reduction of the capacitors' admittances. Limited coverage is not a major concern at high frequencies unless the frequency is significantly higher than the design frequency. Figure 3-30 proves this for frequencies up to 35 GHz. At high frequencies, however, the maximum rms voltages increase rapidly. Table 3-3 summarizes this data.

Table 3-5 Simulated tuner loss and maximum rms voltages of the varactors at various frequencies. The maximum power is calculated with varactor self-actuation voltage of 24 V.

	maximum rms voltage of varactors (normalized to that of the 50 Ω load)				
Freq. (GHz)	**Varac 1**	**Varac 2**	**Varac 3**	**Varac 4**	**max. power (dBm)**
22.5	1.6	1.5	2.6	2.7	32
25	2.0	3.0	3.1	3.7	29.2
27.5	2.4	5.0	8.6	6.9	23.8
30	2.7	7.3	17.4	24.7	12.8
32.5	3.0	9.7	29.2	58.1	5.3

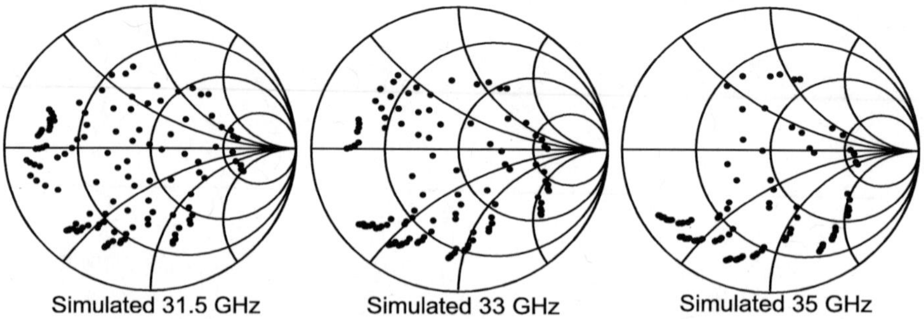

Figure 3-30 Simulated tuner impedance coverage at 31.5 GHz, 33 GHz, and 35 GHz.

Please note that this is an analog tuner and it is not feasible to measure all possible configurations. The simulated data in Figures 3-29 and 3-30 are presented for six distinct capacitance values per varactor. About 190 experimental points were collected from the continuous range. As a comparison between simulated and measured data, Figure 3-31 shows the S-parameters of the tuner with all varactors in the up and down states. Good agreement is observed between the designed and the acquired values.

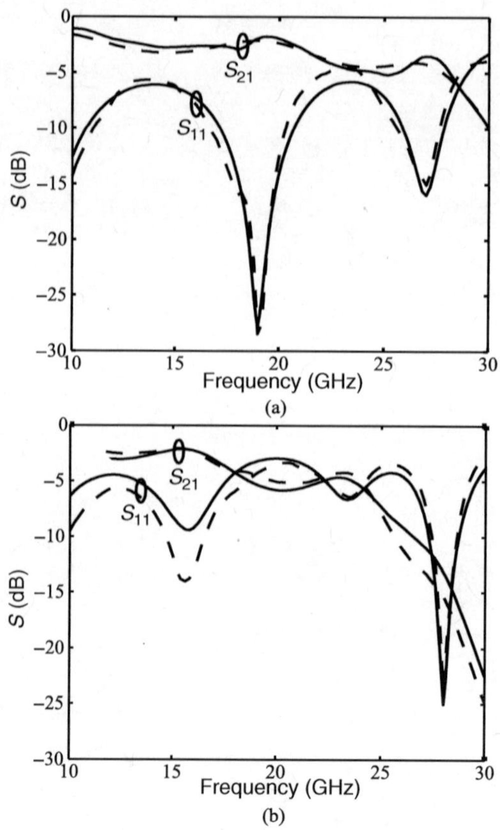

Figure 3-31 Measured and simulated S-parameters of the tuner when all varactors are in the (a) up and (b) down position.

3.5 TUNABLE POWER AMPLIFIERS

Design of efficient RF power amplifiers is a critical concern, because the efficiency of the RF front-end power amplifier dominates the battery life and heat dissipation of the RF front-end transmitter. In particular, the Class E power amplifier, invented in 1975 by Sokals [133], has become a promising candidate for high-frequency and high-efficiency applications especially with the dramatic developments of the solid-state device technology [134]. Class E mode operation requires a precise load impedance to be presented at the drain end of the device. Nonetheless, in practice a variety of factors including a) the variations of the antenna impedance connected to the power amplifier particularly when employed in wireless handheld devices, b) the nonlinearity of the transistor output capacitance and lack of accurate nonlinear transistor models, c) the batch-to-batch fabrication variations, d) the uncertainty of interconnection parasitics cause considerable difficulties in realizing the precisely needed output impedance for every power transistor. Therefore it would be beneficial to have a tunable matching network that can cover a certain impedance range and achieve performance optimization after the circuit is fabricated. Such a matching network is unlikely to be implemented using solid-state devices like varactor diodes since the relatively high loss of these devices significantly degrades the amplifier efficiency.

In this section, a design methodology is presented for an X-band power amplifier operated in Class E mode with a double stub output impedance tuner consisting of MEMS analog varactors. The results show that it is feasible to have a high-power and high-efficiency power amplifier that maintains optimum efficiency for load impedance variations that exceed $\pm 100\%$.

3.5.1 Mems Double-Stub Tuner with Analog varactors

The analog varactor employed in this design is presented in detail in Chapter 2. Figure 3-32 shows the exact measured C-V curve of the particular varactor used for this design. The varactor provides a capacitance of 97 fF when no bias is applied and a maximum capacitance of 272 fF at a bias voltage of 32 V, resulting in an analog tuning ratio of 2.8:1.

A double-stub impedance tuner with the developed MEMS analog varactor is shown in Figure 3-33. For a given stub spacing, d, the tuner transforms a 50 Ω load (Z_0) to an impedance Z_L.

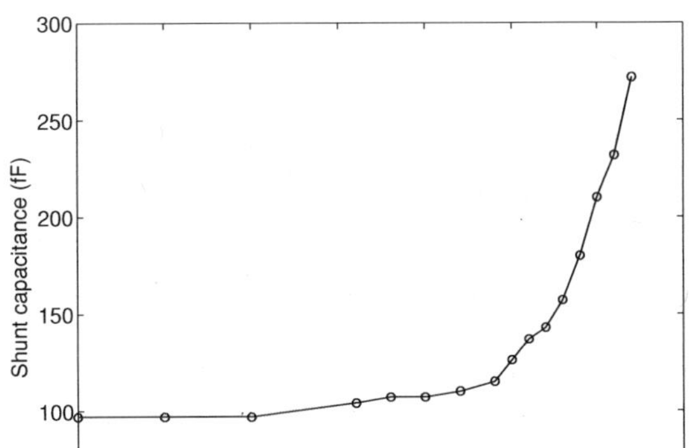

Figure 3-32 Measured C-V curve of the varactor in Figure 3-34.

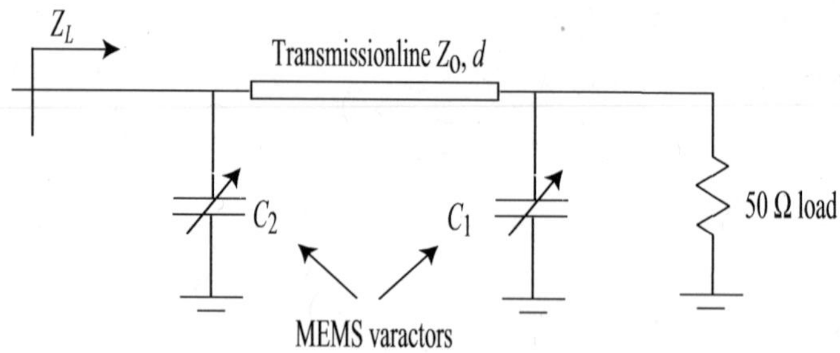

Figure 3-33 Schematic of the double-stub output tuner.

The real part, G_L, and imaginary part, B_L, of its admittance are given as:

$$G_L = G_0 \frac{G_0^2 + G_0^2 t^2}{(G_0 - B_1 t)^2 + G_0^2 t^2} \tag{3.36}$$

$$B_L = B_2 + B_1 \frac{G_0 - B_1 t - G_0 t^2}{(G_0 - B_1 t)^2 + G_0^2 t^2} \tag{3.37}$$

where $t = \tan \beta d$, β is the propagation constant. B_1 and B_2 are the admittances of the two capacitors ($B = 2\pi f C$) and G_0 is the characteristic admittance ($G_0 = 1/Z_0$). The tuner's coverage is therefore determined by the capacitance tuning range of the varactor, the spacing d and the operation frequency. While this is a robust and simple design, unless it is implemented with MEMS devices with large tuning range (>2:1), an excessively large transmission line section will be required with obvious undesirable consequences on the tuner's loss and bandwidth. If needed, more sophisticated tunable-matching networks can be designed as shown in Section 3.4.

3.5.2 Class E Power Amplifier Design

To achieve high efficiency, a Class E amplifier drives the transistor into switch-mode operation where the overlap of nonzero voltage and current waveforms of the transistor is minimized (and is ideally zero). Consequently the power dissipated by the transistor is minimized and the efficiency is substantially improved. To achieve the required current and voltage waveform, however, the load impedance for this mode of operation has to be a specific complex number, Z_L, as shown in Equation 3.38 [135,136]. At harmonics, very high-load impedances are required to prevent power from getting delivered to the load:

$$Z_L = \frac{0.28}{2\pi f C_{\text{out}}} e^{j49^\circ}. \tag{3.38}$$

A number of practical issues hinder the implementations of real-world Class E power amplifiers due to the need for accurate knowledge of the amplifiers' load impedance. However, the exact value of the output capacitance C_{out} is typically not available from the device large signal model in most cases. In addition, C_{out} is nonlinear and varies with the drain voltage swing as well [137]. Furthermore, the impedance of the antenna typically connected to the power amplifier varies widely depending on the environment around it, particularly in wireless handheld units. In this perspective, an output matching tuner provides the possibility of postprocessing tuning to

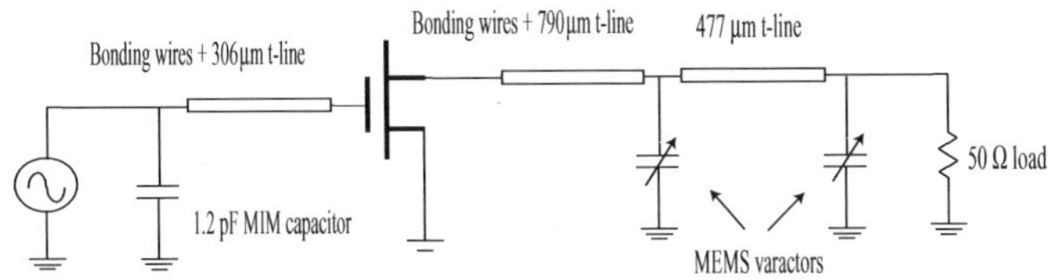

Figure 3-34 Circuit diagram of the Class E power amplifier design.

compensate for those nonideal factors and may considerably reduce the time and cost of the actual design.

Raytheon's 12 finger PHEMT transistor [138] is chosen as a vehicle to demonstrate this design. It is an 1.2 mm device and provides maximum current of 700 mA. The C_{out} is estimated to be around 0.3 pF. At $V_{DS} = 7$ V, the device operates in a suboptimal Class E mode at 10 GHz [139]. From Equation (3.38), Z_L is calculated as $14.63 + j16.84\ \Omega$ at 10 GHz. Load-pull and source-pull simulations are carried out in the Agilent ADS circuit simulator [115] with the large-signal device model developed by Raytheon [140]. In this model, a two-terminal-voltage-dependent gate charge formulation is used to improve the large signal accuracy in high-power and high-frequency applications [140]. The simulations provides the optimal nominal load and source impedances as $11.85 + j17.1\ \Omega$ and $2.35 + j8.9\ \Omega$ respectively. The output- and input-matching networks are designed based on these simulation results. The bonding wire's parasitics have a significant effect at the design frequency and have to be included as part of the matching networks. A set of test structures for wire bonding were fabricated and bonding wires with different lengths are made on these test structures. The *S* parameters of the test structures were measured from 2 GHz to 40 GHz and deembedded to get the *S* parameter of the bonding wires. These deembedded *S* parameters were then used in the circuit simulation of the amplifier. Figure 3-34 shows the final circuit schematic of the amplifier. The bonding wires have been included into the matching networks. The completed circuit is shown in Figure 3-35.

3.5.3 Measurements and Discussion

3.5.3.1 Double-Stub Tuner Impedance Coverage Figure 3-36(a) shows the measured return loss of the tuner at 35 combinations of the varactor settings at 10 GHz. Figure 3-36(b) shows the simulation results from Agilent ADS [115]. The measurements agree very favorably with the simulated results. The real part and imaginary part of the impedances extracted from the S parameter are shown in Figure 3-37. The tuner effectively transforms a 50 Ω load to any impedance with resistance of about 10-30 Ω and reactance of about -3–9 Ω. Consequently this matching network can absorb impedance variations in the order of ± 100%.

3.5.3.2 Power Performance of the Class E Power Amplifier The power amplifier's performance is measured by connecting the input port to a HP 83624A Synthesized Sweeper and the output port to a HP 8564E Spectrum Analyzer. The loss of the test setup is calibrated out by replacing the circuit by a through line and measuring the loss of the connecting cables and probes (about 5.5–6.5 dB total for 9.5–10.5 GHz). The amplifier was biased at $V_{GS} = -1\ V$ and $V_{DS} = 7\ V$ and the output matching network was tuned to get optimal circuit performance. Figure 3-38 shows the output power and Power Added Efficiency (PAE) of the amplifier from 9.5 GHz to 11.5 GHz with input power of 19 dBm. The peak PAE is measured at 10.5 GHz of 52%. Figure 3-39 shows the output power and PAE versus input power measured at 10.5 GHz. Table 3-6 summarizes the power measurement results. As compared to other X-band Class E power

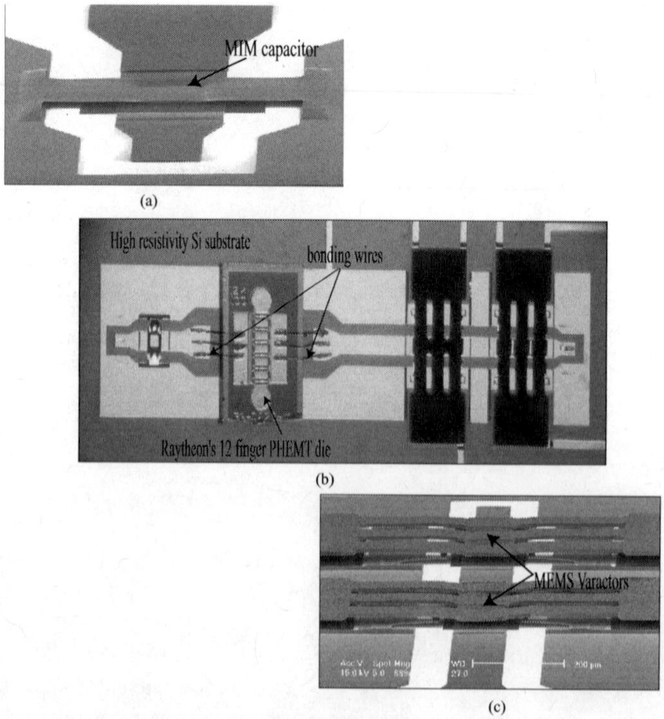

Figure 3-35 Pictures of the fabricated Class E amplifier: (a) detail of input matching network; (b) the complete circuit; and (c) detail of the output-matching network.

amplifiers , this amplifier has achieved high output power with comparable high efficiency. However, to the best of the authors' knowledge this is the first time that such performance has been presented from a power amplifier integrated with a MEMS tuner. The integration of the MEMS devices in the power amplifier provides circuit tunability for postprocessing optimization without degrading circuit's power efficiency and output-power level. It is also worth noticing that all measurements were performed in a normal laboratory environment with no packages for the

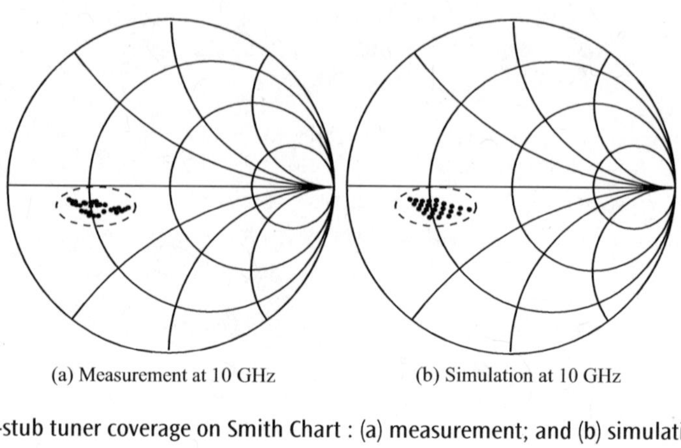

Figure 3-36 Double-stub tuner coverage on Smith Chart : (a) measurement; and (b) simulation.

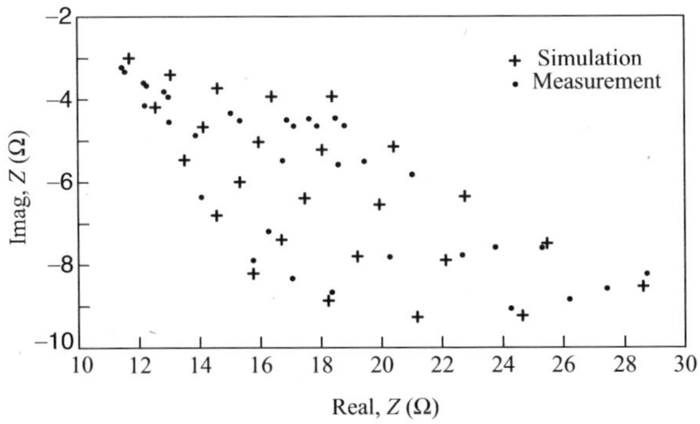

Figure 3-37 Measured and simulated impedance coverage of the tuner.

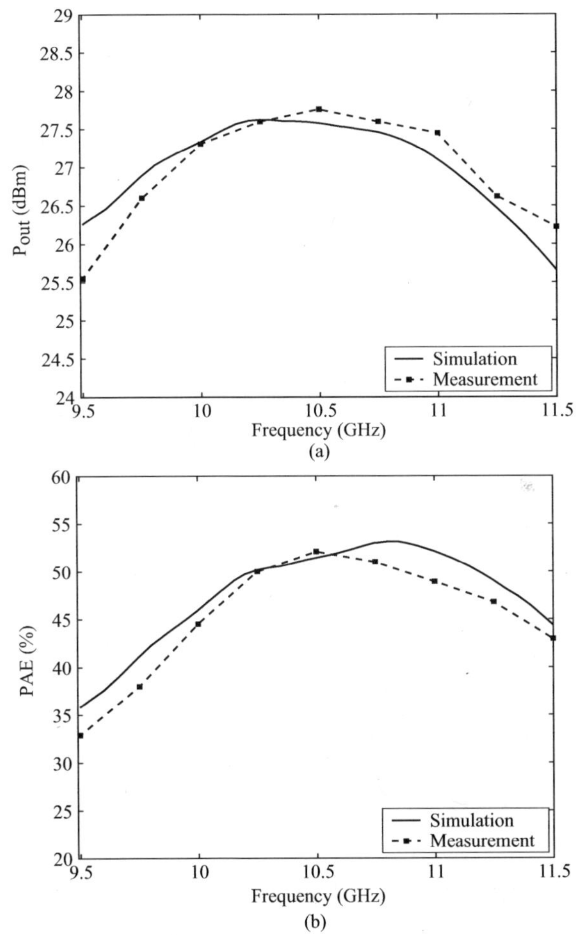

Figure 3-38 Performance of the Class E power amplifier versus frequency: (a) output power; and (b) power added efficiency (PAE).

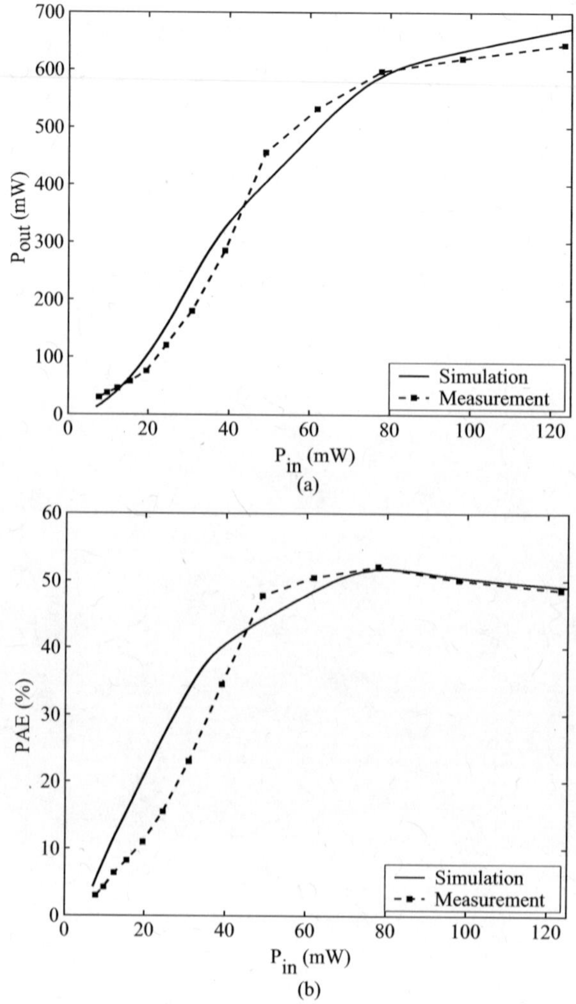

Figure 3-39 Performance of the Class E power amplifier versus input power: (a) output power; and (b) power added efficiency (PAE).

Table 3-6 The performance of the Class E power amplifier.

Frequency (GHz)	Implementation	P_{out} (dBm)	PAE (%)	Gain (dB)	Reference
10.5	Hybrid	27.8	52	8.8	This work
10	Hybrid	20.1	56	7	[141]
10	Monolithic	21	65	11	[142]
10.6	Monolithic	24	63	10	[143]

MEMS devices or special nitrogen environment. This is a significant advantage of using analog contact-less varactors.

3.6 RECONFIGURABLE SLOT ANTENNAS

In the previous sections a variety of tunable circuits using MEMS switches and varactors were considered. MEMS however can be also particularly beneficial for tunable antennas. In modern wireless systems spread spectrum signals are used to suppress the effects of the interference from other users who share the same channel and self-interference due to multipath propagation. Also spread spectrum signals are used for securing the message in the presence of unintended listeners and alleviating the effects of communication jammers. One common feature of spread spectrum signals is their relatively high bandwidth. This is specifically true for frequency-hopped spread spectrum communications system. In a frequency-hopped spread spectrum system a relatively large number of contiguous frequency slots spread over a relatively wide bandwidth are used to transmit intervals of the information signal. The selection of the frequency slots for each signal interval is according to a pseudo-random pattern known to the receiver.

Signal propagation over large distances and in urban and forested environment can take place at UHF and lower frequencies. At these frequencies, the size of broadband and efficient antennas is considerable. Techniques used to make the antenna size small, usually renders narrow-band antennas. In realizing miniature antennas compatible with a frequency-hopped spread spectrum system, a reconfigurable narrow-band antenna can be considered that dynamically follows the pseudorandom pattern of the frequency-hopped modulation. In this section the design aspects of compact, planar, and reconfigurable antennas are considered and the feasibility of such designs is demonstrated by a planar reconfigurable slot antenna operating at UHF [144].

Compared to broadband antennas, reconfigurable antennas offer the following advantages: 1) compact size, 2) similar radiation pattern and gain for all designed frequency bands, and 3) frequency selectivity useful for reducing the adverse effects of co-site interference and jamming.

In recent years, reconfigurable antennas have received significant attention for their applications in communications, electronic surveillance and countermeasures by adapting their properties to achieve selectivity in frequency, bandwidth, polarization, and gain. In particular, preliminary studies have been carried out to demonstrate electronic tunability for different antenna structures. It has been shown that the operating frequency or bandwidth of resonant antennas can be varied when a tuning mechanism is introduced. Several interesting approaches are presented by Sengupta [145, 146] and Guney [147]. In the literature, tuning is accomplished using varactor diodes [148, 149], or by the application of electrically [150] and magnetically tunable substrates [151, 152] with the use of barium strontium titanate (BST) and ferrite materials respectively.

Tuning of printed dipole or slot antennas has also been considered since they share the same advantages of portability, low profile and compatibility in integration with other monolithic microwave integrated circuits (MMICs). Kawasaki and Itoh [153] presented a 1-λ slot tunable antenna loaded with reactive FET components. Although the radiation pattern properties were preserved in all resonant frequencies, the tuning range of the resulting antenna was very limited. Second-resonance cross-slot antennas were also presented by Forman *et al.* [154] in a mixer/phase detector system. A varactor diode was used in the microstrip feed line and the resonance could be electronically tuned over a 10% bandwidth. The bandwidth was increased to 45% when mechanical tuning was used by varying the feed-line length. Dipole-tunable antennas were proposed by Roscoe et al. [155] where printed dipoles in series with PIN diodes were studied. The dipole length was varied from $\lambda/2$ to 1λ depending on whether the diodes were off or on. The operating frequencies were selected from 5.2 to 5.8 GHz, while matching of only 4–5 dB was achieved.

The slot antenna discussed in this section utilizes shunt switches that effectively change the electrical length of the antenna over a very wide bandwidth. To demonstrate the technique a reconfigurable slot antenna capable of operating at four different resonant frequencies over a bandwidth of 1.7:1 is presented [144]. Measurements of the return loss indicate that excellent impedance match can be obtained for all selected resonant frequencies. No especial matching network is used and the matching properties are solely determined by the placement of the switches. The loading effect of the PIN diodes in the antenna is also characterized by a full wave analysis and transmission line theory and comparisons between the real and ideal switches are also studied. In relation to the design goals, it is demonstrated that the reconfigurable slot antenna has the same radiation pattern at all frequencies. Also, the measured radiation patterns agree with the theoretical ones. The polarization characteristics and the efficiency behavior of the antenna as a function of frequency are investigated using both theoretical and experimental data. Finally, several design guidelines are provided and possible design improvements are discussed.

The strict requirements of a constant input impedance, gain, radiation pattern and polarization can only be met, if both the passive structure and the tuning mechanism are carefully designed and effectively integrated into the final design. Therefore, these issues are discussed separately in the following.

3.6.1 Passive Antenna Design

The antenna size at UHF and lower frequencies becomes critical and therefore special consideration is required. A compact planar geometry is best suited since three-dimensional large and bulky structures are in general undesirable. Furthermore, some miniaturization techniques have been applied to reduce the size. This section focuses on the passive slot antenna design issues emanating from the above principles.

First, the miniaturization capabilities provided by a high dielectric constant substrate were investigated in [144]. Inasmuch as an accurate characterization of its effect is needed, a commercially available moment method code [92] was employed. First, simple slot antennas were simulated at 600 MHz and their resonant lengths were determined as a function of the substrate thickness and dielectric constant, see Figure 3-40. This analysis suggests that even at low frequencies where the substrate is very thin compared to the wavelength, a miniaturization factor of about 2:1 is possible, if a high dielectric constant substrate is employed. However, the standard commercially available substrates are electrically thin at UHF and below and therefore the 2:1 factor seems to be a limit difficult to exceed even for substrate permittivities as high as 10.

In an effort to further decrease the total area occupied by the antenna, the slot configuration was altered from its standard straight form to an S shape. From the simulated equivalent magnetic current distribution on the straight and S-shape slots (Figure 3-41, antennas (a) and (b)), it is obvious that both closely follow a sinusoidal pattern with the maximum current concentrated in the middle of the slot. As a result, the two antennas share similar properties and only differ in their polarization orientation. Antenna (a) is horizontally polarized, while antenna (b) is slant-linear polarized. Other more complicated geometrical shapes can also be used, but the S-shape slot does not contain any segments supporting opposing currents, which would considerably deteriorate the radiation efficiency. It should also be mentioned that, although the total area of the antenna is greatly reduced by this geometrical change, the resonant length remains almost unchanged. For example, a resonant length of 136 mm for a straight slot is slightly increased to 139 mm for an S-slot antenna at 600 MHz for a substrate with $\epsilon_r = 10.2$ and thickness of 2.54 mm.

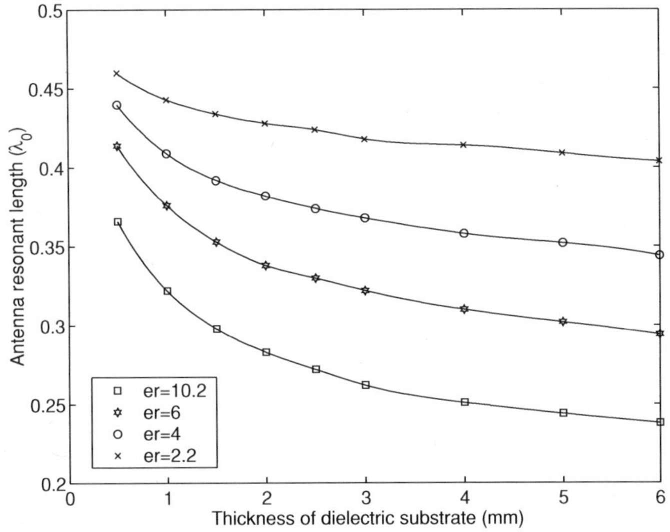

Figure 3-40 Resonant length at 600 MHz for straight slot antenna (in terms of free-space wavelength) as a function of substrate thickness and dielectric constant.

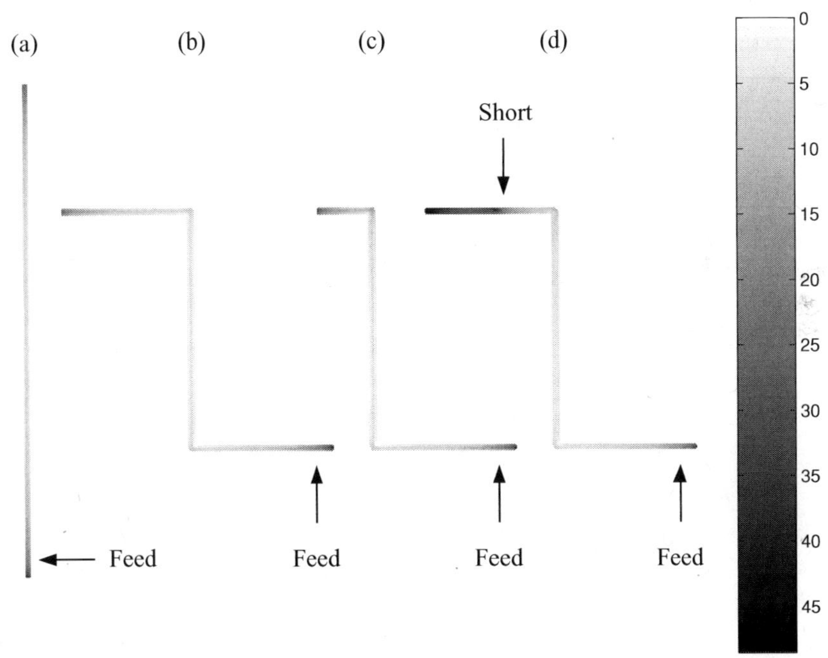

Figure 3-41 Computed magnetic current distributions: (a) 600 MHz straight slot antenna; (b) 600 MHz S-shape slot antenna; (c) 700 MHz S-slot; and (d) 600 MHz S-slot with a short circuit 21 mm above its bottom edge.

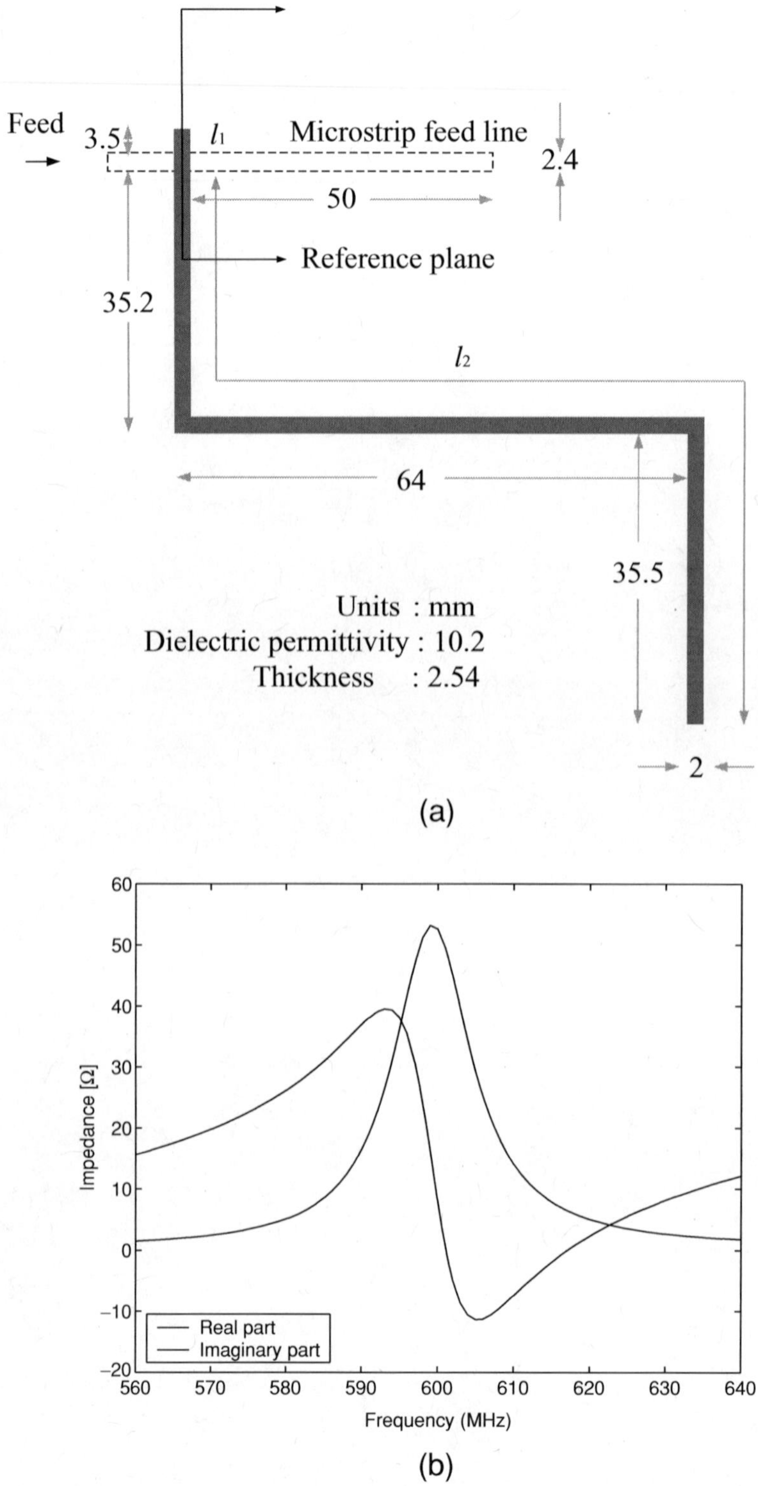

Figure 3-42 S-shape slot antenna: (a) layout with microstrip feed line; and (b) the real and imaginary parts of the input impedance as a function of frequency.

The standard microstrip feed for the simple slot can also be used for the S-shaped slot antenna. Figure 3-42(a) shows the slot antenna with its feed line, while Figure 3-42(b) presents the input impedance at the feeding point as a function of frequency. To achieve a good match to a 50 Ω line, the microstrip feed line must be moved close to one end of the slot antenna. This implies that the antenna input impedance is not very sensitive to small changes in the length of the longer segment (l_2, see Figure 3-42(a)). This property greatly simplifies the design of the tunable slot and its feeding network and results in minimum complexity and maximum reliability for the final antenna. This property of the slot antenna makes it an attractive choice as a reconfigurable structure, since most other antennas (such as dipoles) would require a specially designed matching network.

The resonant frequency of the above structure can be tuned by changing the electrical length of the slot. This may be readily accomplished by introducing a short circuit at a specific location. Then the slot will appear to be shorter and therefore the antenna will resonate at a higher frequency. The three S-shape slots in Figure 3-41 demonstrate these concepts. Slot antenna (b) resonates at 600 MHz with a resonant length of 139 mm. Antenna (c) is 21 mm shorter and is designed to resonate at 700 MHz. Finally, antenna (d) is obtained by modifying (b). In particular, antenna (b) is short circuited at 21 mm above its lower end. The simulated return losses for these three slot antennas are shown in Figure 3-43. It is also important to note that the microstrip feed line remains unchanged in all three cases. That is, the distance between the top end of the slot and the feed-line cross point remains constant and is equal to 3.2 mm. This means that, although the resonant frequency is shifted by 100 MHz, very good matching is achieved for both (c) and (d) slot antennas without needing to modify the feeding network. In addition, slot antennas (c) and (d) have almost identical resonant frequencies. The small difference in the resonant frequency comes from the fact that antenna (d) appears somewhat electrically longer than (c) due to the parasitic effects of the short circuit. Therefore, tunability is possible by introducing these short circuits with no special matching network. Although Figure 3-41 illustrates the basic concept of reconfigurability on a dual band antenna, it is obvious that it can be extended to antennas with several bands of operation. The number of these bands depends on the number of switches on the antenna. For

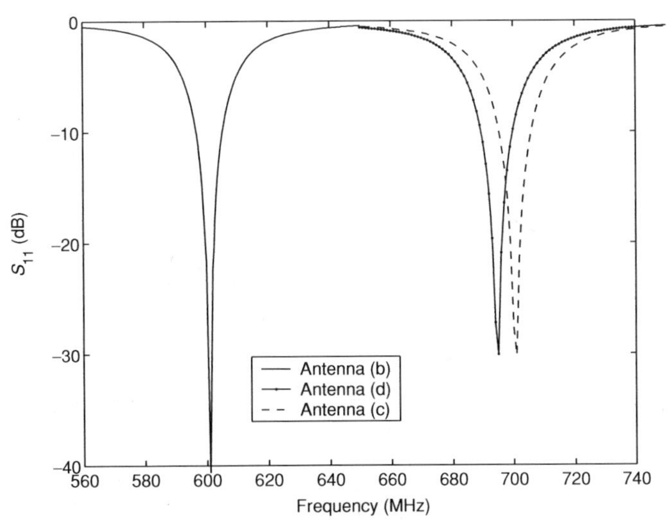

Figure 3-43 Simulated results for the return loss of the S-shape slot antennas presented in Figure 3-41.

example, a four-band antenna is presented in Section 8.3.3 and it is demonstrated that the resonant frequency can be digitally controlled by an array of four switches.

3.6.2 Modeling and Design of Active antennas

In the previous section the basic principle of controlling the antenna resonant frequency was presented. It was also shown that even when a perfect short circuit is used, the parasitic effects of the short can slightly affect the antenna performance and particularly the resonant frequency. The parasitic effects become worse when a switch with finite isolation is used. This section addresses the issues related to the design of a suitable solid-state switch and on the characterization of its effects on the antenna performance. Finally, the complete reconfigurable antenna design is presented at the end of the section together with its theoretical performance.

3.6.2.1 Switch Design To implement the electronic reconfigurability, the ideal shunt switches must be replaced with PIN diodes. This antenna was intended for use in an open uncontrolled environment. Therefore, PIN diode's reliability and high switching speed made it most appropriate for the application at hand. Packaged RF MEMS switches, however, could also be used without changing any of the design principles presented in this section. The RF equivalent circuit of the diode is shown in Figure 3-44(b) for both the on and off states. The reactive components C_p and L_p model the effect of the package, while the others come from the electric properties of the diode junction in the on and off positions [156]. Typical values are also given for the HSMP-3860 diode [157] used in this chapter. The computed isolation (defined as $1/|S_{21}|^2$) for the circuit shown

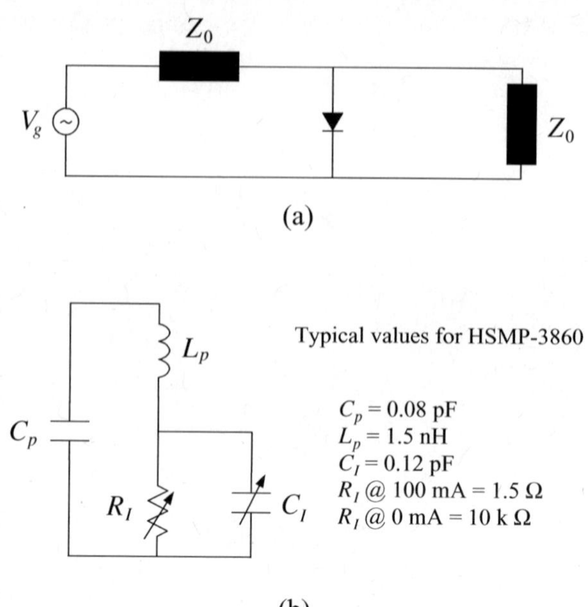

Figure 3-44 PIN diode switch: (a) PIN diode connected as a shunt switch in a transmission line; and (b) RF equivalent circuit for PIN diode including packaging effects.

in Figure 3-44(a) is given by [157]:

$$\alpha = 10\log\left[\frac{\left(\frac{R_d Z_0}{R_d^2 + X_d^2} + 2\right)^2 + \left(\frac{X_d Z_0}{R_d^2 + X_d^2}\right)^2}{4}\right] \tag{3.39}$$

where $Z_d = R_d + jX_d$ is the equivalent impedance of the diode and Z_0 is the characteristic impedance of the line. In the example considered here, the characteristic impedance of the line is approximately equal to 60 Ω, which is calculated by a moment method electromagnetic code [92] for a slot line with a width of 2 mm, a finite ground plane of 60 mm (on both sides of the slot) and a substrate permittivity $\epsilon_r = 10.2$ (RT/Duroid) [158].

The switch bias network is presented in Figure 3-45. An inductor of 470 nH and three 10 pF capacitors are used to improve the RF-DC signal isolation. These values were chosen based on the bias network RF equivalent circuit shown in Figure 3-45(b). The simulated performance for the on and off states is presented in Figure 3-45(c). The RF-DC isolation is better than 30 dB for both states and the return loss is less than −20 dB for the off state. Finally, the RF-RF isolation is greater than 16 dB for frequencies lower than 500 MHz and degrades to 11 dB at 1 GHz due to the diode parasitic elements.

3.6.2.2 Switch Loading on the Antenna Although the switch isolation is important since it determines the frequency selectivity of the antenna, the switch loading on the antenna is equally important inasmuch as it affects its resonant frequency and input impedance. The loading effects must be taken into account for an accurate prediction of the antenna resonant frequency and input impedance, especially when more than one switch is used for multifrequency operation.

A transmission line equivalent circuit that models the loading effect of one diode on the antenna is shown in Figure 3-46. The transverse resonant technique [159] states that:

$$Z_R(z') + Z_L(z') = 0 \tag{3.40}$$

where $Z_R(z')$ and $Z_L(z')$ are the input impedances on the right and left of the reference point respectively. For the unloaded transmission line in Figure 3-46(a), Equation (3.40) simplifies to:

$$\tan(\beta l_L) + \tan(\beta l_R) = 0 \tag{3.41}$$

or

$$\beta(l_L + l_R) = n\frac{\pi}{2}, \quad n = 1, 2, 3\ldots \tag{3.42}$$

which is the well-known formula for these resonant antennas. Now it is important to see what happens in the simplest case of having one switch on the antenna. Figure 3-46(b) shows the equivalent circuit of a transmission line loaded with one switch in the off position. Equation (3.40) then becomes:

$$[Z_0\omega_R C - \cot(\beta l_{R2})]\,[\tan(\beta l_L) + \tan(\beta l_{R1})] = 1 + \tan(\beta l_L)\tan(\beta l_{R1}). \tag{3.43}$$

Equation (3.43) can of course be solved numerically and an iterative method can be employed for finding the unknown lengths until the desired resonant frequency (f_R) has been achieved. A similar procedure can be followed if more than one switch is used on the slot, but the process becomes a

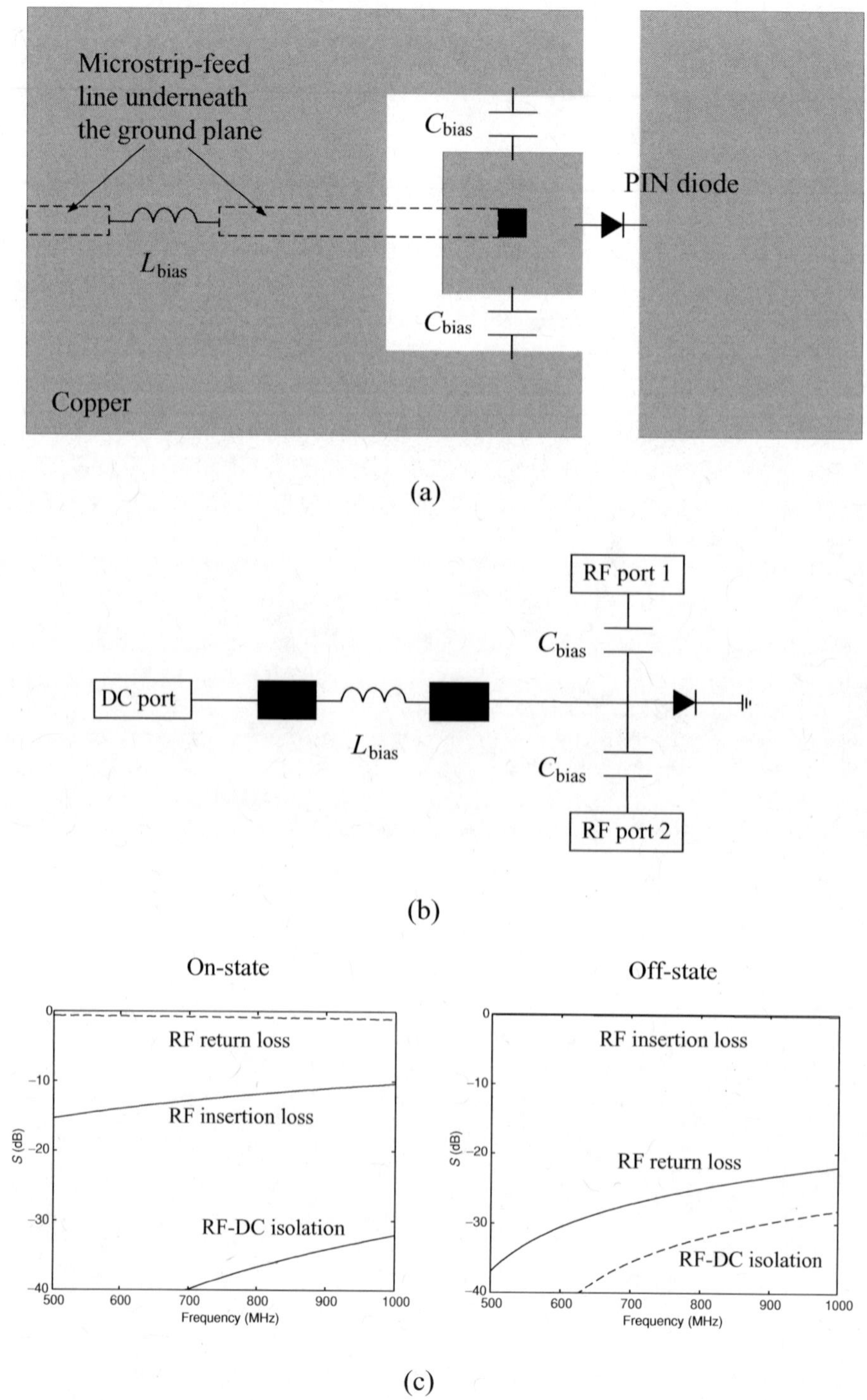

Figure 3-45 Switched antenna: (a) layout of switch-biasing network; (b) RF equivalent circuit; and (c) on and off-state simulated RF performance.

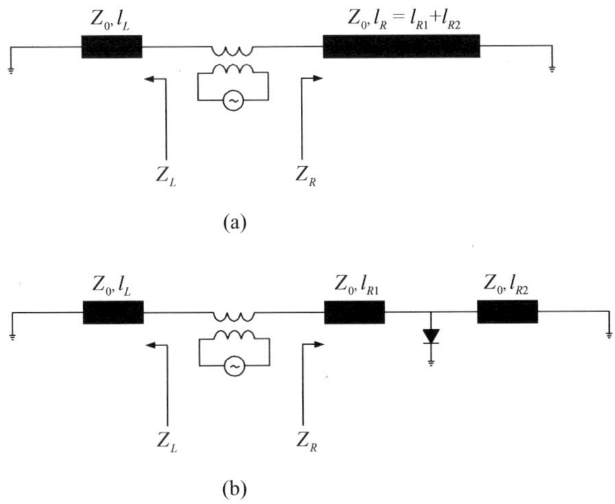

Figure 3-46 RF equivalent circuits for determining the resonant frequency of (a) unloaded and (b) loaded with a single switch slot antenna.

little more complicated if all resonant frequencies are to be specified. It should also be noted that Equation (3.43) does not include any packaging effects, but these can be readily incorporated in the model, resulting in a more accurate computation.

Only the loading effects when the switch is in its off state have been discussed up to now. Nevertheless, the small on-state resistance also affects the antenna performance and particularly its input impedance. Full-wave analysis was used to model these effects. For a first-order approximation, the resistance was modeled as a thin-film resistor on top of the slot and the packaging parasitic elements were neglected in this analysis. The parasitic element effects in the on state can be important especially at the highest frequencies, see Figure 3-45(c). Figure 3-47(a) shows the simulated geometry of an S-shape slot antenna loaded with a resistive film, which is fed by a microstrip line and Figure 3-47(b) shows the simulated return loss versus the switch on-state resistance for four different cases between 0 to 5.6 Ω. In all four cases the position of the 50 Ω feed line was kept unchanged. It is obvious that the matching level deteriorates rapidly as the resistance value increases, and for resistance values above 1.5 Ω the matching level becomes unacceptable.

However, this degradation can be avoided to some extent by elongating the upper end of the slot as the resistance is increased. Figure 3-47(c) shows the improvement on the antenna matching when the slot length is adjusted. It is found that in all three cases, only a very small segment of line needs to be added in order to improve the input impedance of the antenna. Even for a resistance value of 5.6 Ω the required line segment length is less than 3% of the total slot length, resulting in only a small change in the resonant frequency. This method of maintaining a good impedance match will be utilized later for the design of the reconfigurable antenna by placing additional switches (matching switches) on the slot above the feed-line and synchronizing them together with the switches at the other end of the slot (frequency switches). However, it should be noted that the matching switches will not represent perfect shorts and they will introduce an extra loading effect. Nonetheless, this effect is negligible and matching levels of better than −20 dB can be achieved, as will be seen next. Therefore, the matching properties of the reconfigurable antenna will solely depend on the position of an array of switches on the slot and no matching network will be necessary as frequency changes.

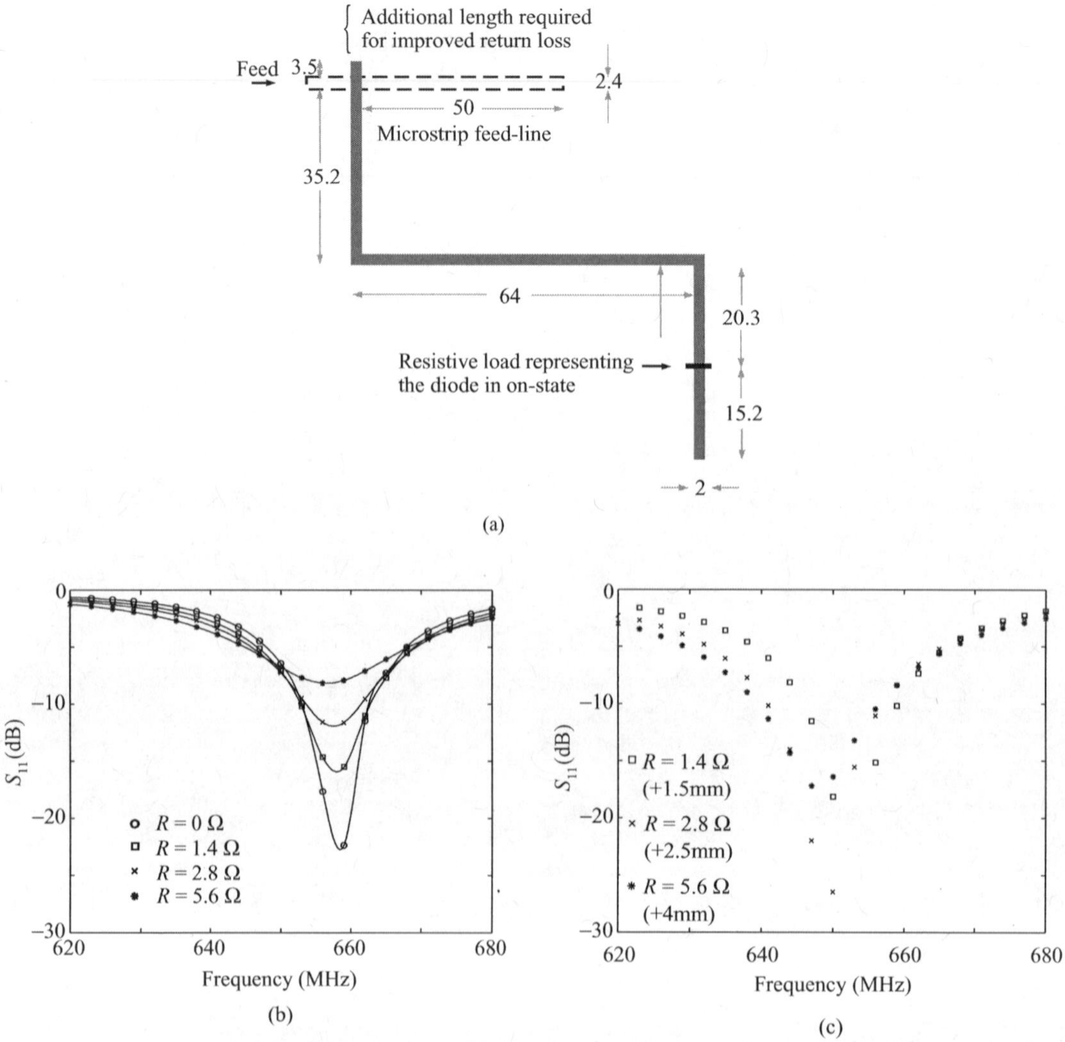

Figure 3-47 Slot antenna: (a) layout with resistive load representing actuated switch (units are in mm); (b) return loss for different values of switch resistance; and (c) improved return loss with minor adjustments (< 4 mm) in the slot length above the feeding point.

Having discussed the loading effects of the switches on the matching properties of the antenna, their effects on the radiation characteristics of the antenna need to be found as well. Ideally, the radiation efficiency should be that of the half-wavelength dipole, since the antenna behaves effectively as a $\lambda/2$ resonant slot at each of its operating frequencies. However, the on-state resistance of the switches will obviously result in power dissipation and finally degradation in the antenna efficiency. The dissipated power obviously depends on the diode's on-resistance and on the number of switches on the antenna. Table 3-7 shows the computed efficiency for the antennas previously discussed in Figure 3-47. Dielectric loss has been included in all cases. This explains the nonideal efficiency when $R = 0\ \Omega$.

Table 3-7 Computed efficiency for slot antennas with a single switch versus on-state resistance value.

R (Ω)	0	1.4	2.8	5.6
Efficiency (%)	71.8	55.6	45.6	33.9

The above antenna efficiency analysis shows that even for a small series resistance of $R = 1.4\ \Omega$ the antenna gain will be approximately 2.5 dB lower than that of an ideal half-wavelength dipole. This is an inherent drawback of using PIN diode switches. However, microelectromechanical (MEMS) switches are becoming increasingly important and are now a viable alternative as they offer very low power consumption and they come in even smaller packages. It has also be shown that capacitive-type MEMS switches exhibit very low ohmic losses and therefore can be used to maximize antenna efficiency. However, the required on-capacitance values renders them impractical for UHF frequencies. Hence metal-to-metal contact switches, which have no cutoff frequency should be considered in such a design.

3.6.3 Final Reconfigurable Antenna Design and Properties

Based on the previously discussed design principles, a reconfigurable slot antenna design, shown in Figure 3-48(a), is presented here. Four switches are used in order to tune the antenna over a range of 540 to 900 MHz.

Both full-wave analysis and the transmission line model were used in the design process. In this design three frequency switches and a single matching switch are used. Table 3-8 summarizes the calculated resonant frequencies and the conditions of all four switches for each resonant frequency. The transmission line model (or circuit model) has the advantage of allowing fast and accurate (as will be shown later) computation of the resonant frequencies and can easily incorporate diode parasitics. However full wave analysis is essential when an accurate prediction of the antenna input impedance is needed. In the moment method code, the diodes were simulated as metal-insulator-metal (MIM) capacitors and as thin-film resistors in the off and on states respectively and as a result the packaging parasitics were ignored. This explains the 5% differences observed in the computed resonances between the two models. Figure 3-48(b) shows the calculated return loss where a matching level of better than −20 dB has been achieved for all operating frequencies.

Since at every operating frequency the antenna radiates as a $\lambda/2$ slot, the radiation pattern remains unchanged when the frequency is shifted. The same holds for the antenna directivity. The E and H planes of a typical calculated antenna pattern are shown in Figure 3-49(a). Since

Table 3-8 Theoretically calculated resonant frequencies using full wave analysis and transmission line model.

F_R (MHz) (TLM)[a]	F_R (MHz) (MM)[b]	Switch configuration
542	561	4 = ON 1,2,3 = OFF
596	627	1,4 = ON 2,3 = OFF
688	711	2,4 = ON 1,3 = OFF
1002	950	3 = ON 1,2,4 = OFF

[a] Transmission line model
[b] Moment method (full-wave) model

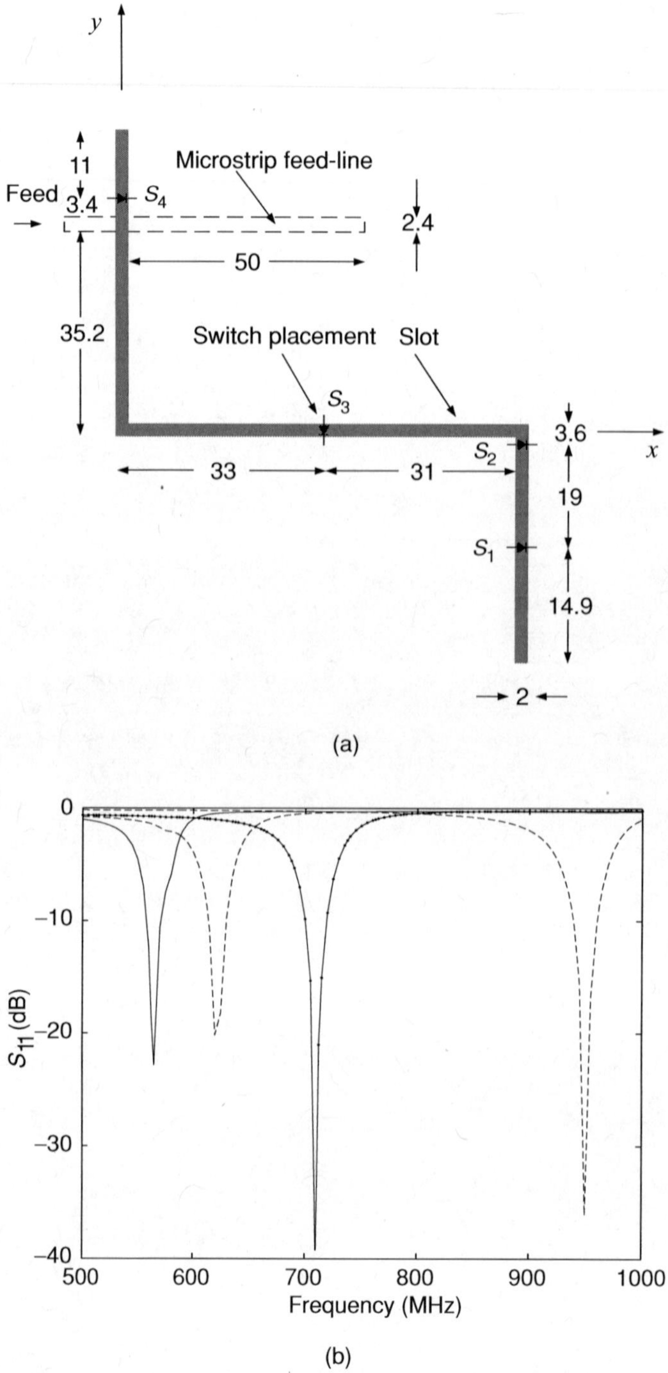

Figure 3-48 Reconfigurable slot antenna: (a) layout, (dimensions are in mm); (b) simulated return loss for the four resonant frequencies; (c) typical radiation pattern; and (d) simulated gain the four resonant frequencies.

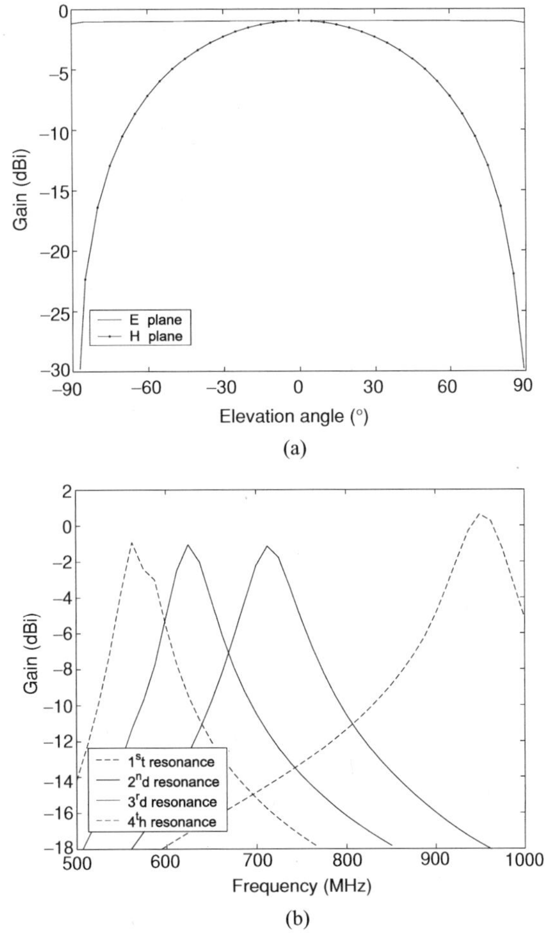

Figure 3-49 Reconfigurable slot antenna shown in Figure 3-48: (a) typical radiation pattern; and (d) simulated gain the four resonant frequencies.

the antenna has been designed for an electrically thin substrate (at UHF) the radiation pattern is symmetric on the two sides of the slot. However, the efficiency and the gain will be reduced compared to a half-wavelength dipole due to the resistive losses introduced by the diodes. Figure 3-48(b) shows the calculated gain using the moment method full-wave electromagnetic analysis [92]. The gain is approximately -1 dB for the lowest frequencies and increases to about 0.7 dB for the highest one. Similar results hold for the antenna efficiency.

The reference angle of 0° in the previous graphs represents the direction normal to the antenna ground plane. Although the S-shape pattern considerably reduces the area occupied by the antenna, it has the inherent drawback that the polarization does not remain constant as the frequency is changed. However, as Table 3-9 shows, the polarization does not change considerably (variation of about 30°). This is due to the fact that the antenna polarization (always slant linear) is dominated by the orientation of the middle segment of the slot from which most radiated field emanates. Therefore, if the orientation of the receiving antenna does not follow that of the transmitter as the

Table 3-9 Calculated polarization for the reconfigurable antenna.

f_R [MHz]	561	627	711	950
Angle (°)	60	70	60	40

frequency is changed, a maximum polarization mismatch of 25% will be incurred. The orientation of linear polarization reported in Table 3-9 is with respect to the x axis, see Figure 3-48(a).

3.6.4 Measurements and Discussion

The reconfigurable antenna designed in the previous section was fabricated on a 2.54 mm (100 mil) thick RT/Duroid substrate (with $\epsilon_r = 10.2$) [144]. The size of the ground plane was 125×125 mm square (5×5 in square).

An HP8753D vector network analyzer was used for the S-parameter measurements. The biasing voltage for the switches was provided by a DC voltage source. After calibrating the network analyzer the antenna return loss was measured for different combinations of activated switches. The measured data are presented in Figure 3-50, where a return loss of better than −13 dB is observed at all resonances. The measured resonances are shown in Table 3-10 together with the necessary biasing conditions. Satisfactory agreement between theoretical, Table 3-8, and experimental, Table 3-10, data is observed. In addition, in comparison to the method of moments simulation the transmission line model yields slightly better results—except at the highest frequency—mainly because the parasitic reactive elements have been included in this model and not in the moment method technique. However, this is not true for the characterization of the antenna configured for the highest resonant frequency where an error of 13% exists between the transmission line model and the measurement. This discrepancy can be attributed to the fact that the properties of

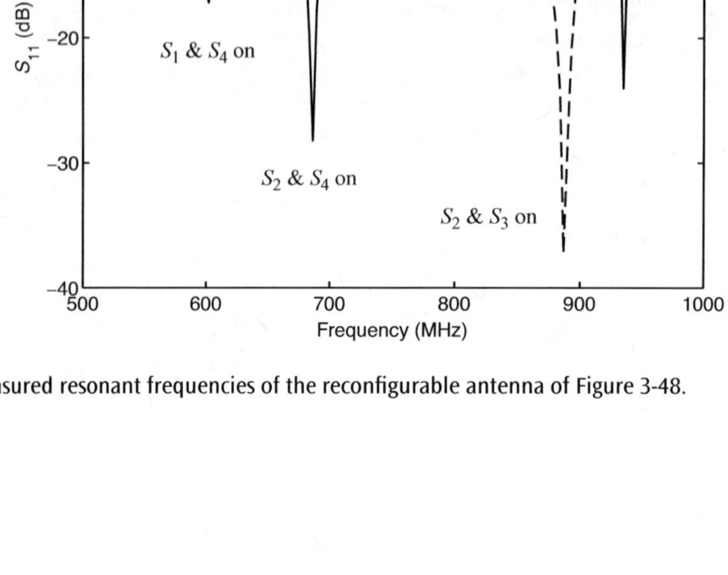

Figure 3-50 Measured resonant frequencies of the reconfigurable antenna of Figure 3-48.

Table 3-10 Measured resonant frequencies and the necessary bias voltages for the switches.

	Bias voltage (V)			
F_R (MHz)	S_1	S_2	S_3	S_4
537	−20	−20	−20	1.1
603	1.1	−20	−20	1.1
684	0	1.1	−20	1.1
887	0	1.1	1.1	0.2

the diodes, and particularly the element values of its equivalent circuit, cannot be assumed to be constant up to 1 GHz.

A reverse voltage of −20 V was applied to maintain the switches in the off position and by doing so a better matching level was achieved. This is an important issue particularly when the antenna is used as the transmitter. Since the structure is a resonant structure strong electric fields are established that can turn the diodes on and off at the RF frequency and ruin the small signal design. This effect was clearly observed at the lowest resonance with an input power of 0 dBm. In this case an improvement of about 5 dB was achieved by changing 0 V bias to −20 V.

One more interesting effect was observed for the highest resonance. Notice that better matching level would occur if not only S_3 but also S_2 were forward biased. This is due to the relatively low isolation that each diode provides at these relatively high frequencies, see Figure 3-44(c). Therefore, biasing S_2 results in higher isolation and reduces the effect of leaked magnetic current in the area after the switch. The improvement in the return loss was approximately 10 dB compared to leaving S_2 unbiased for this frequency.

The far-field antenna patterns were also measured. The E and H planes were measured as well as the corresponding cross-polarization for each operating frequency. An RF signal and a DC voltage source were used with the reconfigurable antenna and a dipole with adjustable length was employed as the receiving antenna. The dipole length was appropriately adjusted for each operating frequency of the transmitting antenna until maximum received power was recorded. In order to find the E-plane, the transmitting antenna was rotated until the electric field was vertically polarized. Then the transmitter, placed on a turn table, was azimuthally rotated for measuring the E-plane cuts. The cross-polarized pattern was measured by rotating the receiver antenna by 90°. The H-plane antenna pattern measurements were conducted in a similar manner.

For slot antennas printed on a substrate it is expected that the radiated power be higher in the half-space than that include the dielectric substrate. However no appreciable difference was observed experimentally. This is easily explained since in this case the dielectric thickness is about $\lambda/200$ at 600 MHz and the size of the ground plane is small (approximately $\lambda/3$) at the same frequency. Therefore the antenna is almost bi-directional and equivalent to a dipole in free space.

The measured data are presented in Figure 3-51 for each resonant frequency. In these plots 0° denotes the direction of maximum radiated power. These measurements show that the H-plane pattern closely follows the expected sinusoidal pattern. However, some slight asymmetries near ±90° exist for almost all frequencies. These discrepancies originate primarily from two sources. First, parasitic radiation from the cables and the feeding network and second, radiation from the edges of the dielectric. These sources of radiation also affected the E-plane pattern measurements and they caused a difference of 3–4 dB between the minimum and maximum measured value, see

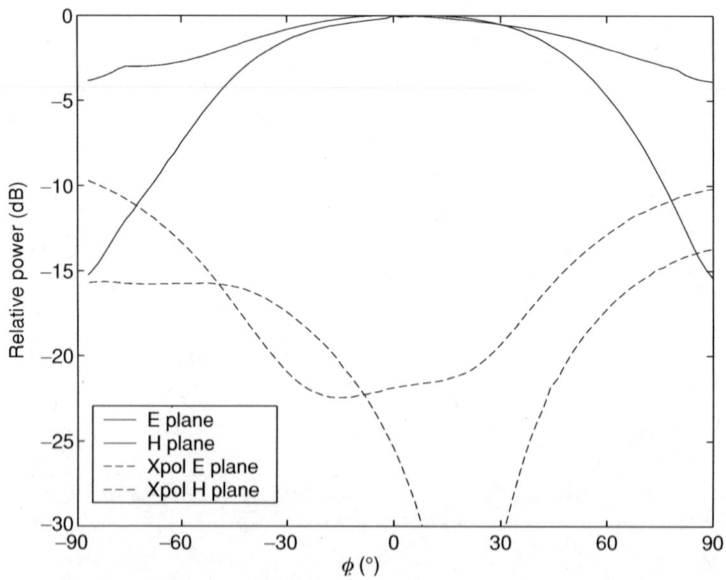

(a)

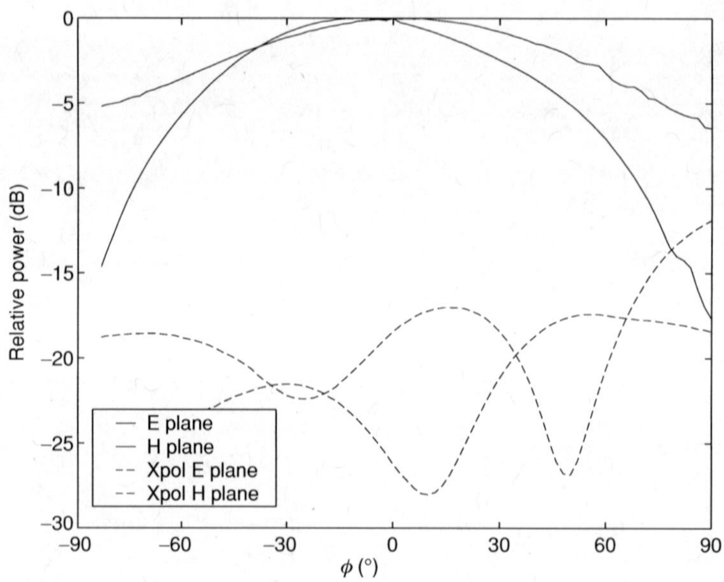

(b)

Figure 3-51 continued. Measured radiation patterns for the four resonant frequencies.

Figure 3-51. Despite these discrepancies, it is clear that the far-field pattern remains unchanged versus the frequency tuning.

Gain measurements are accomplished using the comparison method [160]. A log-periodic antenna with 6 dBi gain at 600 MHz was used as a reference antenna for these measurements. The second resonance at 593 MHz was chosen as the operating frequency of the reconfigurable

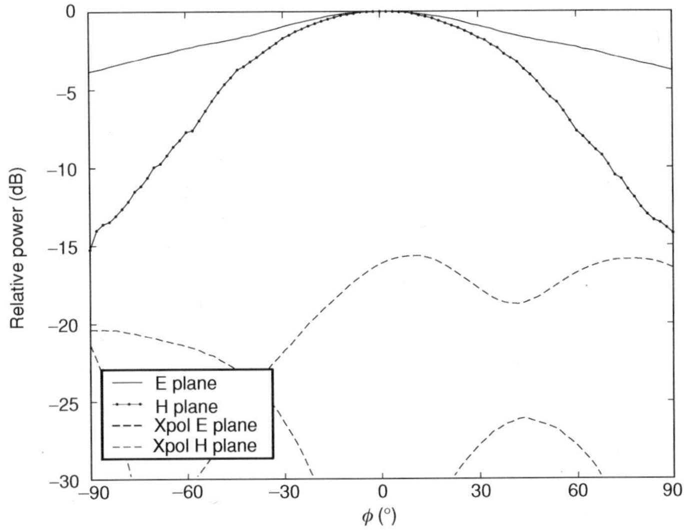

(c)

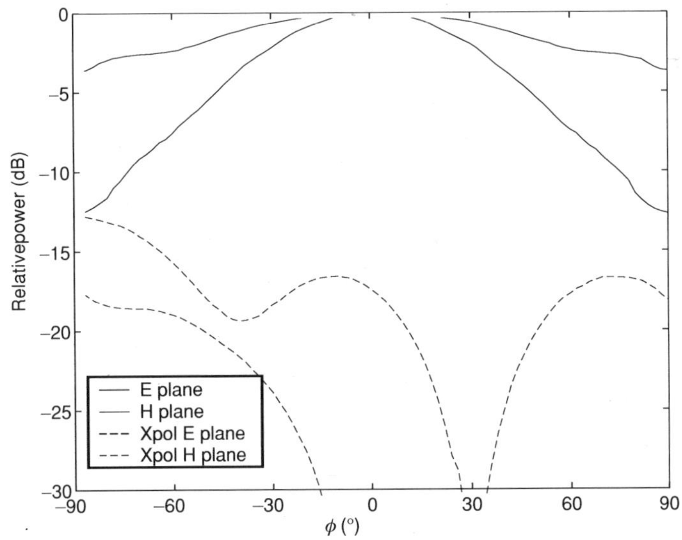

(d)

Figure 3-51 continued.

antenna, so that to make direct comparisons with the reference antenna possible. To measure the gain, the power received by the receiver dipole at 593 MHz was recorded when both the reference and the reconfigurable antennas were used in the transmitting mode inside the anechoic chamber under the same conditions. The measured gain was found to be –1.1 dBi, which corresponds to an efficiency of 47%. These results closely resemble the calculated data. It should also be pointed out that the gain of the slot antenna is reduced not only from the forward-biased diode resistance, but also from the small ground plane size. However, a comparison between the measured and

Table 3-11 Measured polarization for the reconfigurable antenna.

f_R (MHz)	537	603	684	887
Angle (°)	57	70	55	33

calculated data reveals that the dominant degrading factor in gain is the dissipated power on the diodes rather the ground-plane size.

Finally, the antenna polarization was measured and the method previously described for the pattern measurement was employed. The measured polarization orientation at each frequency is provided in Table 3-11. As discussed before, although the polarization does not remain absolutely constant as the frequency is changed, the variation range is small and comparable to the theoretical data, see Table 3-9.

3.7 CONCLUSION

Several RF MEMS applications were presented in this chapter. High-isolation switch packets using distributed π−networks showed isolation in the order of 50–60 dB limited only by substrate modes. The bandwidth of such designs can be easily controlled to reach 20–30 GHz. Similar techniques can also be employed to achieve switchable filters with tuning ratios of 2:1. Both lowpass and bandpass filters are demonstrated with excellent performance. The bandpass filter in particular preserves its bandwidth over the 2:1 tuning ratio, shows an average insertion loss of 3–4 dB, and occupies an area of less than 1 mm^2. The designs of high-performance and high-power impedance tuners and tunable power amplifiers are subsequently shown for 30 and 10 GHz respectively. Tuners have the capability to match nearly any impedance on the Smith Chart and are simultaneously optimized for maximum power handling (nearly 1W). This enables class E power amplifiers that maintain optimum output power and power added efficiency even for considerable variations of their loads. Finally, similar approaches are followed to develop reconfigurable slot antenna with tuning ratio of 1.7:1 and the ability to maintain their characteristics (input impedance, gain, polarization and radiation pattern) essentially unaffected over the entire tuning range without external tuning networks.

4

Tunable Dielectrics for RF Circuits

Robert York

4.1 INTRODUCTION

In most electronic materials the change in dielectric constant with applied electric field is an effect too small to be useful or even easily measurable. However for a special class of high-permittivity materials the effect can be quite pronounced. This chapter explores challenges and opportunities for exploiting tunable dielectrics to make reconfigurable, adaptive, frequency-agile RF devices. The chapter focuses on thin-film barium strontium titanate (BST) materials, discusses design and modeling of BST varactors, and surveys some circuit implementations and heterogeneous integration efforts.

4.1.1 BST Ceramics

The field-dependent permittivity (i.e. the "tunability") of high-permittivity dielectrics has been known for quite some time, and its potential for use in RF circuits was recognized immediately [161–167]. Barium-titanate, $BaTiO_3$ (abbreviated as BTO) and related compounds are now considered the prototypical "high-κ" or high dielectric constant materials for this purpose. In bulk ceramic form, barium-titanate is a ferroelectric at room temperature, with a ferroelectric-paraelectric transition at $T_c \sim 116°C$, the so-called Curie temperature. Figure 4-1 illustrates the behavior of a ferroelectric material around the Curie temperature. In the ferroelectric phase below T_c the material exhibits memory effects or hysteresis in the polarization-field response that can be exploited for nonvolatile embedded memories [168]. Above T_c the material is paraelectric with a very large dielectric constant and field-dependent nonlinearity. In both phases the electrical properties are strongly temperature-dependent. Although the paraelectric phase is of most interest in this work, to the material is still colloquially referred to as a ferroelectric.

The Curie temperature in bulk BTO can be easily manipulated by mixing with other materials or compounds to allow for room-temperature operation. Figure 4-2(a) shows the variation in Curie point with various additives [169]. Strontium titanate, $SrTiO_3$ (STO) is an interesting and useful choice because it also has a high permittivity, so the dielectric constant and nonlinearity (tunability) remain high as the Ba/Sr ratio is changed, but the Curie temperature decreases almost linearly with the amount of strontium as shown. The solid solution of BTO and STO is barium-strontium titanate, $Ba_xSr_{1-x}TiO_3$ (BST), with x specifying the mole fraction of barium. Three representative examples are marked in Figure 4-2(a). The Curie temperature falls below room temperature when $x < 0.7$.

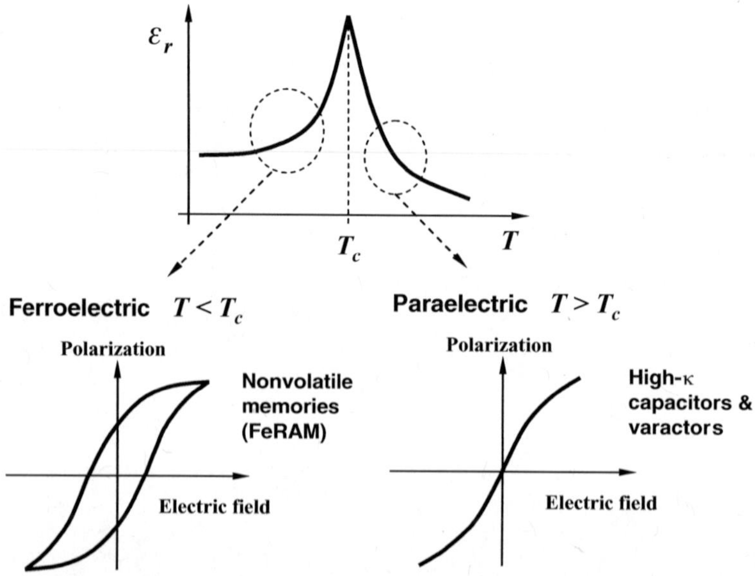

Figure 4-1 Behavior of ferroelectric materials with temperature around the Curie temperature.

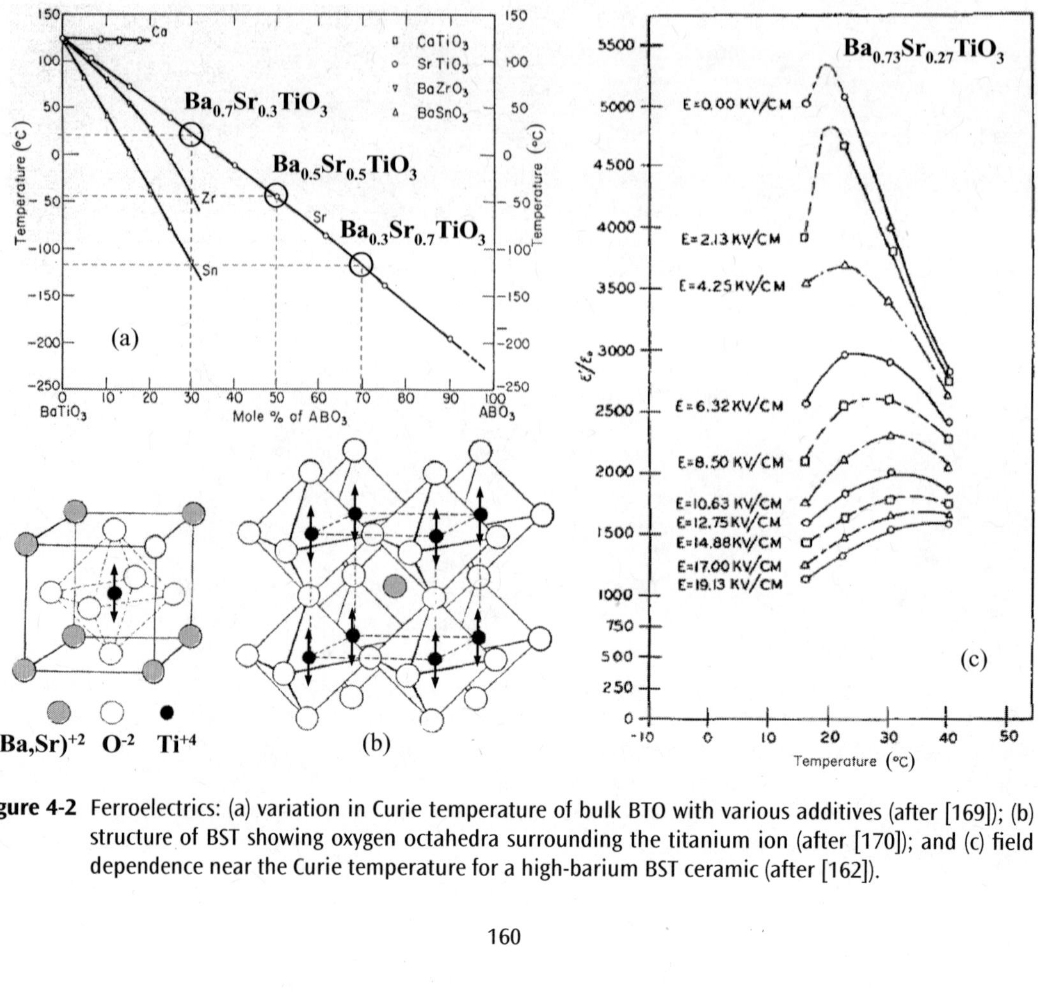

Figure 4-2 Ferroelectrics: (a) variation in Curie temperature of bulk BTO with various additives (after [169]); (b) structure of BST showing oxygen octahedra surrounding the titanium ion (after [170]); and (c) field dependence near the Curie temperature for a high-barium BST ceramic (after [162]).

BST has a cubic perovskite structure, with the large barium and strontium ions occupying the corners of the unit cell as shown in Figure 4-2(b) [170]. The oxygen ions form an octahedral "cage" surrounding the small central titanium atom. In the paraelectric phase, the high-permittivity of the material derives from the fact that the titanium ion can be easily displaced by an applied field, yielding a large induced dipole moment or polarization. This effect is further enhanced by long-range ordering effects at low temperatures. The displacement or polarization (indicated by arrows in the figure, assuming a vertical field) begins to saturate at high fields, leading to a reduction in the small-signal effective dielectric constant. The measured dielectric response for a representative bulk ceramic $Ba_{0.73}Sr_{0.27}TiO_3$ mixture is shown in Figure 4-2(c) [162].

Significant changes in the dielectric constant with applied field are apparent in Figure 4-2(c), especially near the Curie temperature, but the concurrent strong variation with temperature in bulk BST material is a source of concern for practical applications. Another limitation associated with tunability in bulk ceramics is the rather high voltage required to achieve the field strength needed for significant tunability (typically 10-20 kV/cm or higher), and associated breakdown considerations. For these and other technological reasons the field-dependence of tunable dielectrics was not widely exploited, and the effect remained a laboratory curiosity for decades. However, it is important to note that BTO and derivatives have been extensively used in high capacitance-density passive components, enjoying huge commercial success in both the ceramic capacitor and thermistor industries. Thus considerable resources have been and continue to be employed towards understanding and improving these materials for commercial use.

4.1.2 Thin-Film Versus Thick Film

In the early 1990's, two separate technological developments—the discovery of high-temperature superconductors, and evolutionary scaling issues in silicon dynamic random access memories (DRAMs)—revived interest in BST and STO for RF applications. In the case of high-temperature superconductors, low-loss tunable dielectrics were sought to integrate with the superconducting circuits to create high-performance filters [171]. For the CMOS and DRAM industry, high-κ alternatives to SiO_2 were sought to maintain the steady rate of improvements predicted by Moore's law [172, 173]. In DRAMs a high dielectric constant is desired to reduce the size of the storage cell capacitor; the capacitor dielectric is deposited in very thin films (<100 nm) and must be process-compatible with the remaining CMOS transistor circuitry. Significant resources were dedicated to developing candidate dielectrics such as BST for this demanding application.

For these new applications, methods were developed and/or refined for depositing thin-film BST by industry-standard and high-throughput techniques such as RF sputtering and metal-organic chemical vapor deposition (MOCVD), and the electrical properties of the thin films proved to be significantly different than those of bulk ceramics of similar composition. As shown in Figure 4-3, not only are the dielectric constants much lower, but more importantly the temperature dependence is drastically different, with no obvious ferroelectric-paraelectric transition and much smaller temperature coefficients. In addition, the tunability (change in dielectric constant with voltage) is quite large aided in part by the higher fields that can be sustained in thin films in comparison to bulk materials. Naturally the voltages required to achieve a given field are also much lower than for bulk materials.

These remarkable differences between thin-film and bulk BST are generally favorable as far as potential RF applications are concerned, resolving many of the limitations that discouraged early application efforts. As these properties became more widely appreciated, several high-frequency devices and circuits using thin-film STO and BST were reported [175–182] (these are just a few representative references on thin-film devices only, not thick-film or bulk ceramics). Many of these

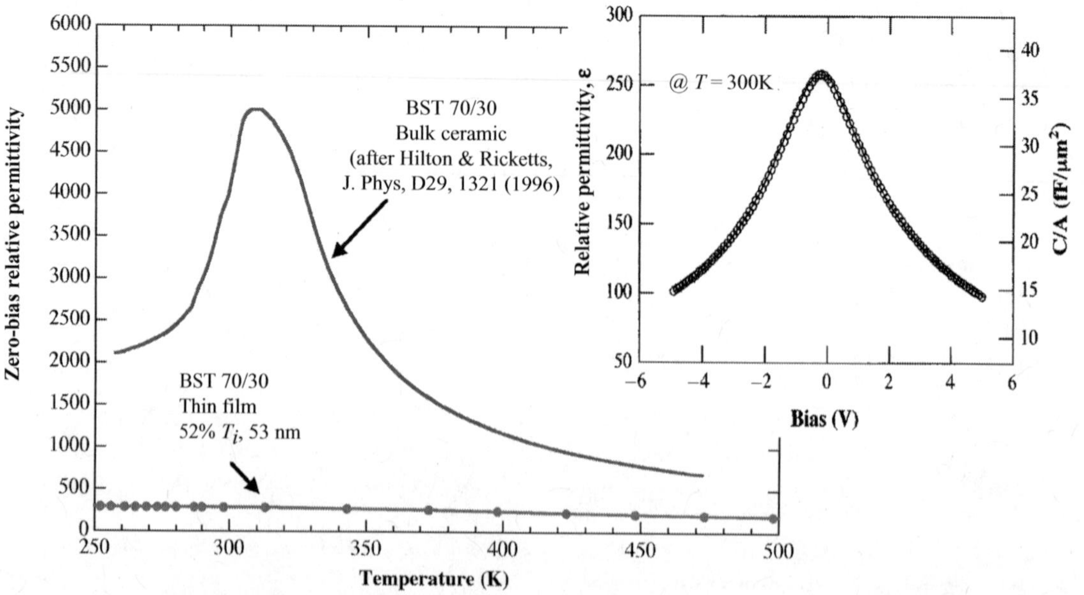

Figure 4-3 Comparison of bulk and thin-film BST of similar composition (courtesy S. Streiffer, Argonne National Laboratories), and room-temperature tuning curve for thin film on silicon substrates using Pt electrodes. [174].

early reports also used high-temperature superconductors, reflecting the evolutionary origins of the work.

In the last decade, steady progress has continued towards an understanding of BST thin films. It is now believed, for example, that the large differences observed in thin-film BST in comparison to bulk materials are probably due to a combination of factors including so-called "size-effects" [183] and residual strain in the materials arising from high-temperature growth on thermally mismatched substrates [184–186]. The relevant material deposition technology has also evolved continuously, and some progress has been made in understanding the failure mechanisms and long-term reliability of the materials (more on that later).

On the other hand, it is also now widely appreciated that many factors can influence the behavior of thin-film BST, including the method of deposition, growth temperature, composition, substrate, film thickness, processing conditions, and contact metallurgy. This chapter will focus, therefore, on behavioral or phenomenological modeling approaches that capture the general features of tunable dielectric devices, independent of the particulars of the technological implementation.

4.1.3 Potential for RF Applications

Thin-film BST has already been commercially exploited for high-capacitance density decoupling capacitors [187–189], but the focus of attention here is on voltage tunable integrated capacitors (varactors). Two questions arise: where can BST varactors be useful, and will the devices offer some compelling advantage(s) over alternative technologies

Semiconductor diode varactors have been exploited for decades [190] and a large body of work exists on the design of varactors and varactors circuits. Tunable filters, phase-shifters, linearization networks, and voltage-controlled oscillators are several important examples. Despite the different

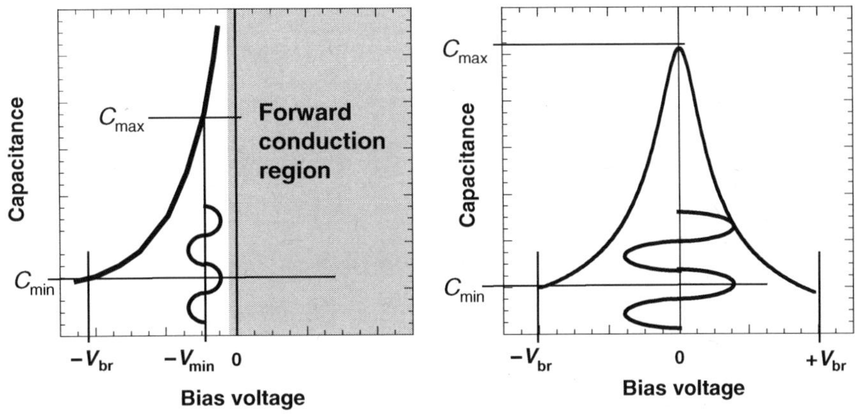

Figure 4-4 Comparison of diode and dielectric varactors characteristics. BST varactors have no forward conduction region, and hence can sustain large RF voltage swings, especially near zero volts (exaggerated here for illustrative purposes).

physical mechanisms responsible for the tuning, diode varactors and dielectric varactors have similar tuning ratios and quality factors. Dielectric varactors do have one key advantage relative to diodes: there is no forward conduction region. As illustrated in Figure 4-4 this allows for larger RF voltage swings, improved power handling, and also somewhat simpler biasing compared with diodes. Dielectric varactors also require less sophisticated processing for a given Q-factor and operating frequency, and can be easily integrated with other high-Q passive components, factors that tend to lower overall implementation cost. On the other hand, diode varactors are an established and mature technology, and at present it is unclear whether the linearity and cost advantages of dielectric varactors will be sufficient to displace diode varactors.

Another emerging technology for wireless applications is RF Micro-Electro-Mechanical systems (MEMS). MEMS devices seem best suited to a switching function, although analog varactors have been realized using MEMS techniques. In comparison to MEMS, dielectric varactors have the following advantages: they are smaller; require lower control voltages; require less processing and less complicated packaging (lower cost); have a fast intrinsic response time; and most importantly have excellent RF power-handling characteristics, allowing for hot-switching without degradation. The only real negative in relation to MEMs is the nonlinearity of the device, but this can be managed with techniques discussed later.

So relative to diode varactors and MEMS, there is at least a potential opportunity for dielectric varactors in applications requiring large RF voltage swings and low cost. The analog front end of mobile wireless communication devices is one such application, and might potentially benefit from frequency agile filters, matching networks, tunable antennas, and phase-shifters that can be realized with tunable dielectrics.

4.1.4 Doping, Composites, and Other Tunable Dielectric Materials

Most work in tunable dielectrics for RF applications has centered on BST and STO because of the unique combination of high tunability and good RF loss characteristics, but there are other candidate materials [191]. Many are structurally similar perovskite structures of the form ABO_3, where the B-site cation is typically small in comparison to the A-site cation and often one of the following three: 1) titanium, e.g., lead titanate, $PbTiO_3$ (PTO), and lead zirconium titanate, $Pb(Zr_xTi_{1-x})O_3$ (PZT); 2) niobium, e.g,. lithium niobate, $LiNbO_3$ and potassium niobate, $KNbO_3$ (KNO);

or 3) tantalum, e.g,. potassium tantalate, $KTaO_3$ (KTO). These materials are either ferroelectric (like BTO) or incipient ferroelectrics (like STO), displaying high permittivity and tunability and similar temperature dependencies. PTO and PZT are technologically important materials for piezoelectric and embedded memory applications, but the RF losses are typically high in comparison to BST [192, 193], limiting their usefulness in varactor applications at high frequencies. As with BTO, the addition of STO to lead titanate results in a solid solution $(Pb,Sr)TiO_3$ (PST) with a Curie temperature that depends on the Pb/Sr ratio [194]. It has recently been shown that strontium-enriched PZT, e.g,. $Pb_{1-x}Sr_xZr_{0.52}Ti_{0.48}O_3$ (PSZT) has improved characteristics as compared with PST [195]. Similarly $K(Ta,Nb)O_3$, a solid-solution of KTO and KNO, has been shown to have potentially useful properties for RF applications [191]. Clearly there are some common threads of scientific thought at work in the development of these materials.

Another emerging material worthy of special mention is bismuth zinc niobate [196], $Bi_{1.5}Zn_{1.0}Nb_{1.5}O_7$ (BZN). In contrast to BST and many of the other materials mentioned earlier, BZN is a nonferroelectric material with a cubic pyrochlore structure, exhibiting a relatively large dielectric constant (150-200) and a low loss tangent ($<10^{-4}$ at 1 MHz). BZN thin films show a significant field-dependence of the permittivity, with more than 2:1 change in the dielectric constant at field strengths of $\sim$2.4 MV/cm [196, 197]. Although BZN bulk ceramics exhibit a dielectric relaxation and high losses at microwave frequencies [198], we have recently shown that dielectric losses of thin-film BZN capacitors remain comparatively low at least up to 20 GHz [197, 199]. Hence BZN thin films appear to be attractive for microwave tunable applications [200–202].

In most cases of practical interest the thin-film materials will be polycrystalline, with defect and grain-boundary structures that depend on growth temperature, choice of substrate and electrodes, and impurity concentrations. It is well known from the ceramic capacitor industry that oxygen vacancies are a problem in these materials, often linked with leakage, reduced lifetime, and bias-induced performance degradation. A time-honored solution to these problems is the addition of small amounts of compensational "dopants" [203]; this has been shown to also work well with thin-film STO [204] and BST [205–208]. Since the precise amount of compensation doping is difficult to ascertain a priori, amphoteric dopants such as yttrium [203,209,210] and erbium [211] appear to be a promising solution for thin-film ferroelectrics.

From a fundamental standpoint high losses are an unavoidable companion of high permittivity and tunability, and hence there is always an inherent tradeoff between loss and tunability. Another approach to manipulating the properties of tunable dielectrics is to add relatively large amounts of nontunable compounds, usually a low-loss/low-permittivity linear dielectric (e.g,. MgO). The resulting 'composite' material has a reduced dielectric loss tangent, but of course this comes at the expense of reduced tunability which, as will be seen, is fundamentally linked to the zero-field permittivity. It can be shown by relatively simple electrostatic considerations [212] that one can never increase the figure-of-merit (often defined as a product of tunability and Q-factor) of a tunable dielectric using such composite mixtures. Composites are also more prone to structural defects and contamination, making them undesirable for a robust and reproducible high-volume thin-film manufacturing process. Fortunately there are more effective techniques to address dielectric losses and linearity concerns, these are discussed later.

4.1.5 Summary

No materials have yet demonstrated characteristics superior to BST when all of the critical variables are considered. However, as noted earlier, BST is itself a complex material with electrical properties that are dependent on many physical factors including Ba/Sr ratio, grain size, dopants, film thickness, temperature, frequency, and field strength. BST films are sensitive to the methods used in material preparation and device processing. Electrical properties of BST devices are also strongly dependent on the substrates and contact metallurgy. For these reasons and in combination with its technological importance, BST is sometimes referred to as the "BeaST" of electroceramics.

The goal of this section is the development of a simple but general framework for modeling tunable dielectric devices and circuits with respect to important independent variables (frequency, applied electric field, film thickness, and electrode size) in a way that is transparent to the technological details. The approach is largely phenomenological in order to obtain a simple closed-form modeling of these electrical properties. The focus is exclusively on thin-film, parallel-plate capacitors, since these are the most important structures from a practical standpoint. No attempt will be made to understand or model the observed properties of the materials or devices from a detailed consideration of the underlying physics. The reader is referred to [191] and [213] and references therein for a more in-depth overview of materials, deposition, and processing technology associated with tunable dielectrics.

4.2 LOW-FREQUENCY MEASUREMENTS AND MODELING

Device characterization at RF and microwave frequencies (>100 MHz) is complicated by several factors that stem from the high-permittivity of the films, as well as difficulties inherent in making accurate high-frequency measurements on high-Q reactive components. Low-frequency (<100 MHz) measurements can therefore be especially useful for basic material characterization of tunable dielectrics, and much of what has been learned about BST devices is based on low-frequency characterizations.

4.2.1 Test Structure Design and Impedance Analyzer Measurements

At low-frequencies, relatively simple large-area test structures can be used and fabricated with a minimal amount of processing, increasing the turnaround time between material growth and characterization—an important consideration for circuit development. With carefully chosen device sizes and frequencies, certain parasitics (such as electrode resistance and inductance) can be safely neglected simplifying data analysis. Most importantly, highly accurate measurements can also be made over several orders of magnitude in frequency using balanced-bridge I-V methods [214], and such broadband dielectric spectroscopy is especially helpful for developing accurate circuit models [215]. The Agilent 4294A impedance analyzer is one example of an instrument that uses this technique, can be easily configured for on-wafer probe measurements.

A useful test structure for thin-film BST characterization is shown in Figure 4-5. After depositing the BST film on a metalized substrate, the BST film is subsequently patterned to form a "mesa," and a second metal contact layer is deposited on the film and bottom electrode simultaneously. Since the BST films are usually quite thin (100–500 nm), the variation in height between the top and side contacts is negligible, thus the structure can be directly probed by GSG (ground–signal–ground) coplanar probes as shown, provided the top contact area is larger than the probe tips. In practice, devices as small as 20 μm × 20 μm have been characterized using this method. Although simple two-pronged (GS) probes could also be used, a GSG test structure has the advantage that the series resistance arising from the bottom electrode is reduced by a factor of two (more on that later). Furthermore, GSG probes have been shown to yield more accurate calibration and measurements.

A variety of substrates, contact metals, and process conditions can be used and these will affect the results in various ways. In much of the work reported here the BST films for low-frequency characterization were deposited by RF magnetron sputtering onto platinum-coated (100–200 nm) c-plane sapphire substrates. The BST films were then wet-etched using a buffered HF solution,

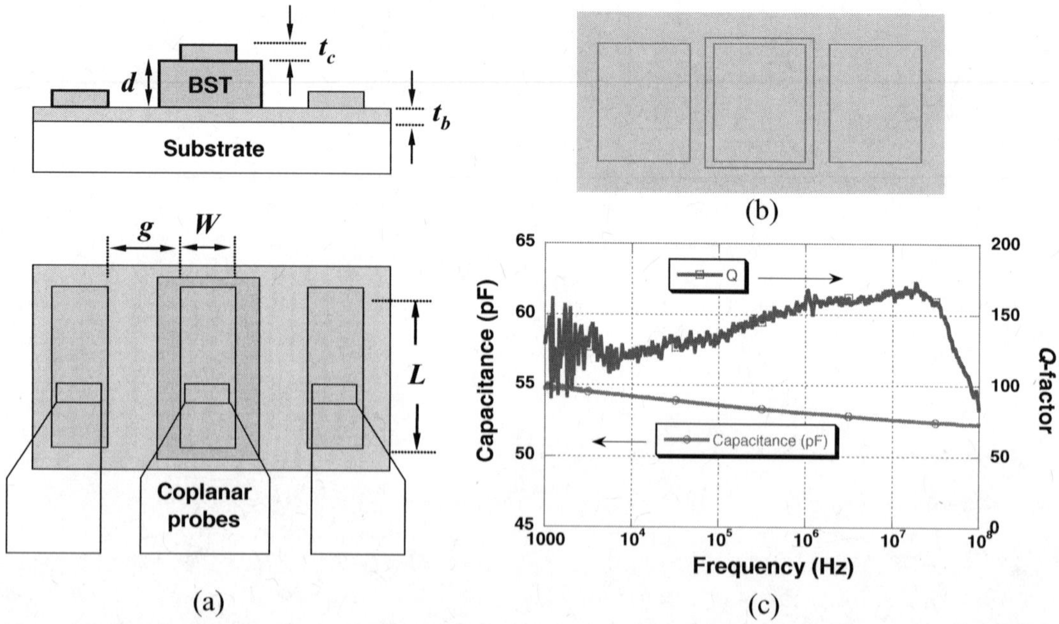

Figure 4-5 Low-frequency characterization: (a) simple test structure for electrical characterization of BST films showing contact by coplanar probes; (b) photograph of a representative structure; and (c) typical capacitance-frequency and Q-frequency data.

and Pt(100 nm)–Au(200 nm) top contacts were then formed using a lift-off procedure. More details on the processing can be found in [186] and [216].

Representative data on a 160 nm thick, nominally stoichiometric 50/50 BST film ($Ba_{0.5}Sr_{0.5}TiO_3$) is shown in Figure 4-6. This data was recorded using an Agilent 4294A analyzer and RF probe station; the capacitance data assumes a parallel G-C (admittance) model for the device, and the Q-factor (inverse loss tangent) is computed as $\mathrm{Im}\{Y\}/\mathrm{Re}\{Y\}$. The frequency was swept from 40 Hz to 110 MHz (the limits of the instrument) at different DC bias voltages from 0 to 10 V in 1 V increments. The instantaneous catastrophic breakdown voltage for this particular film was >22 V, corresponding to a breakdown field of >1.3 MV/cm.

Clearly there is a complicated dependence on frequency and bias for both the capacitance and the Q-factor (which is an indication of loss). The following sections will develop a fairly complete phenomenological model for these dependencies. It is important to note that the particular device measured in Figure 4-6 was a large-area device, nominally $2.5 \times 10^4\ \mu m^2$, such that peripheral effects (discussed later) are expected to be negligible. It is also important to mention that the impedance analyzer calculates device admittance as $Y = \Delta I/\Delta V$, where ΔI is the complex AC current (the in-phase and quadrature components) that flows upon the application of an AC voltage ΔV [214]. If the AC voltage is small enough, the data approximates the desired small-signal admittance $Y = \partial I/\partial V$. The choice of oscillation amplitude is thus dependent on how strongly the material properties change with voltage, which in turn depends on material composition, film thickness, etc. If the oscillation amplitude is too large, the calculated capacitance will differ from the small-signal value, and the loss will appear to increase as a result of nonlinear frequency conversion (harmonic generation) and self-heating effects. In the data set of Figure 4-6, a 200 mV oscillation amplitude was used.

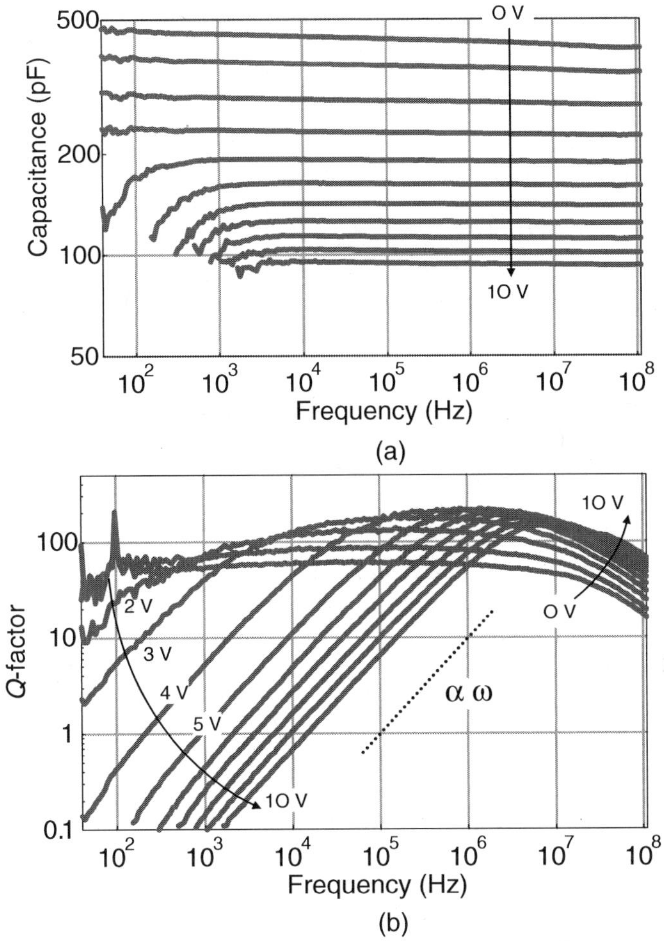

Figure 4-6 Low-frequency characterization: (a) typical capacitance-frequency; and (b) Q-frequency data from 0 to 10 V in 1 V increments. Data taken on 160 nm $Ba_{0.5}Sr_{0.5}TiO_3$ films using an Agilent 4294A impedance analyzer with a 200 mV test signal amplitude.

4.2.2 Capacitance-Voltage Relationship

In Figure 4-6(a) it is seen that at any one frequency the capacitance decreases monotonically as the bias field is increased. For a symmetrical device (having identical top and bottom contacts) the result is independent of the polarity of the bias field. At zero field the capacitance starts at some maximum value $C_{\max}$ (dependent on electrode area, film thickness, frequency, temperature, etc.). As the applied DC field increases the small-signal capacitance dQ/dV decreases monotonically. At some voltage V_τ the capacitance is reduced to $C_{\min}$; we define the tunability, τ, as the ratio of maximum-to-minimum capacitance at this voltage:

$$\tau \equiv \frac{C_{\max}}{C_{\min}} \tag{4.1}$$

The tunability thus defined is dependent on the choice of V_τ. In this analysis, V_τ can be chosen arbitrarily, but it is shown later that the "2:1" voltage V_2 is an obvious choice.

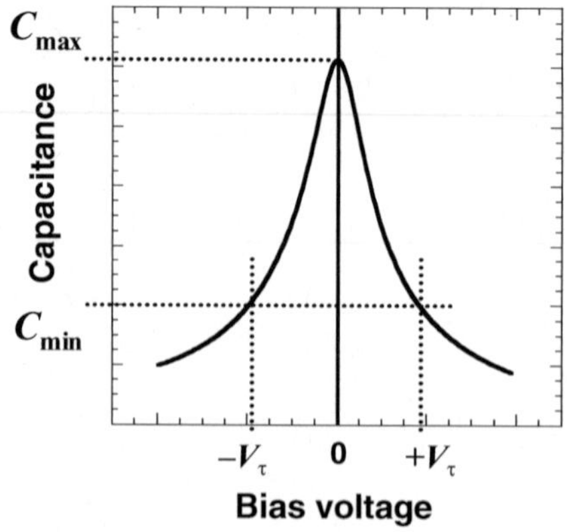

Figure 4-7 Definitions for modeling the capacitive nonlinearity.

Recently a simple closed-form expression has been derived for the $C(V)$ relation [217], beginning with a power-series expansion for the field-polarization relation of the form [161, 174, 184]:

$$E = \alpha_1(T)D + \alpha_3 D^3 + \cdots \tag{4.2}$$

where $\alpha_1(T)$ is the inverse of the zero-bias permittivity (temperature-dependent) and α_3 describes the nonlinearity of the material. In the context of ferroelectric films Equation (4.2) is called the Landau–Devonshire–Ginzburg (LDG) model. Irrespective of any physical justification for this model, there is a simple phenomenological basis: the even symmetry of the $C(V)$ relationship for dielectric varactors means that there can only be odd terms in a power-series expansion for $E(D)$, and higher-order terms can be neglected because dielectric breakdown is usually encountered before these terms become significant.

For an ideal capacitor (having no interfacial layers or space-charge) it can be assumed that the E-field and flux density D are uniform throughout the film, and relate to the external applied voltage and charge through

$$E = V/d \qquad D = Q/A \tag{4.3}$$

where d is the capacitor thickness and A is the area. This transforms Equation (4.2) into

$$V = \frac{\alpha_1 d}{A} Q + \frac{\alpha_3 d}{A^3} Q^3. \tag{4.4}$$

The small-signal capacitance is defined by

$$C(V) = \frac{\partial Q}{\partial V}. \tag{4.5}$$

Using Equations (4.4) and (4.5) and the definition of tunability in Equation (4.1), it can be shown [217] that the capacitance nonlinearity can be modeled as

$$C(V) = \frac{C_{\max}}{2\cosh\left[\frac{2}{3}\sinh^{-1}\left(\frac{2V}{V_2}\right)\right] - 1} \tag{4.6}$$

where

$$V_2 = \frac{4V_\tau}{(\tau + 2)\sqrt{\tau - 1}} \tag{4.7}$$

is the "2:1" voltage at which $C(V_2) = C_{\max}/2$, an easily measured quantity. Experimentally there are only two parameters that define the ideal $C(V)$ curve: $C_{\max}$ and V_2. Once V_2 for a given device is known, Equation (4.6) can also be used to determine the voltage required to achieve a desired tunability.

Although the explicit $C(V)$ relation in Equation (4.6) has only recently been derived, the formula for the 2:1 voltage, Equation (4.7), has been known for some time [161]; in fact this relationship defines an implicit C-V curve since, using Equation (4.1), it can shown that

$$V = \frac{1}{4}V_2\left(\frac{C_{\max}}{C(V)} + 2\right)\sqrt{\frac{C_{\max}}{C(V)} - 1}. \tag{4.8}$$

This is quite useful for predicting the voltage at which a certain capacitance value is reached, an issue relevant to the control circuit design in frequency-agile networks. For completeness it should also be noted that the original LDG expansion, Equation (4.2), has now been successfully inverted in a simple closed from

$$D(E) = \frac{3}{2}\varepsilon_b E_2 \sinh\left[\frac{1}{3}\sinh^{-1}\left(\frac{2E}{E_2}\right)\right] \tag{4.9}$$

where $E_2 = V_2/d$ is the field at which the permittivity changes is reduced by a factor of 2, and $\varepsilon_b = 1/\alpha_1$ is the zero-bias permittivity. This is a potentially useful result for use in electromagnetic simulators.

The model presented here compares quite favorably with data on large area devices (large area-to-periphery ratios) where edge effects can be neglected. Figure 4-8 shows a representative example, with excellent agreement out to voltages in excess of 8 V_2. This is somewhat remarkable given that there are only two parameters in the model.

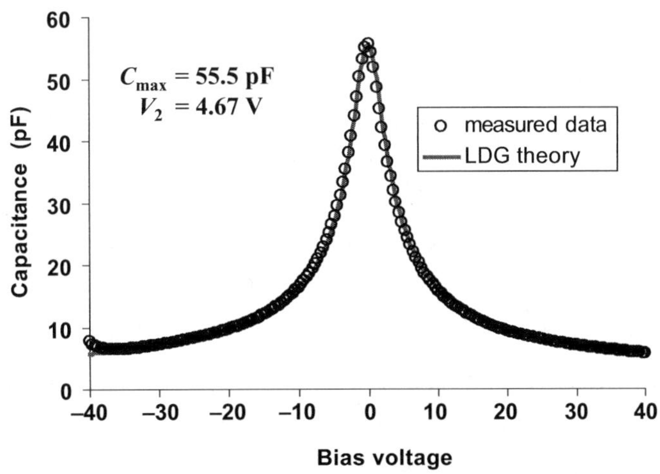

Figure 4-8 Comparison of Equation (4.6) with measured data at 1 MHz for a large-area device (2000 μm^2) fabricated on 140 nm thick, 50/50 BST with Pt electrodes [217].

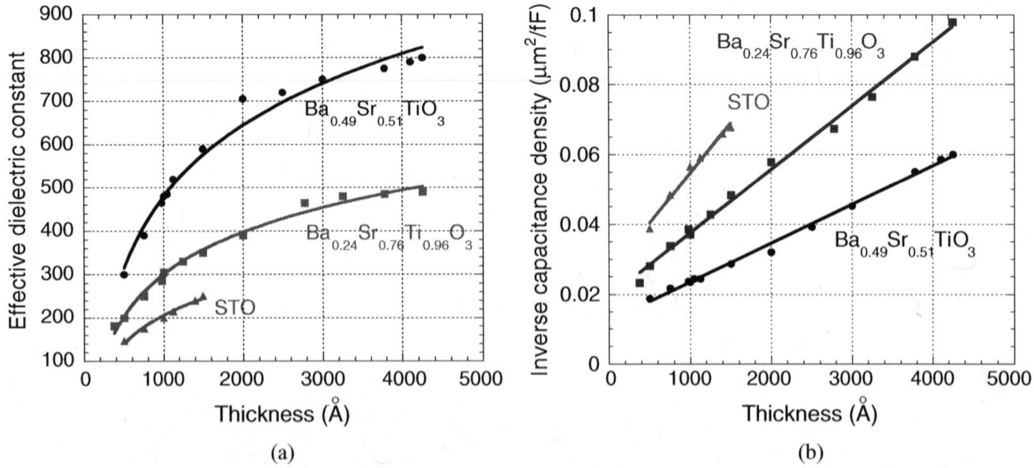

Figure 4-9 Thickness dependence of the effective dielectric constant: (a) calculated from zero-bias capacitance data at 1 MHz for various film compositions (sapphire substrates and Pt electrodes); and (b) corresponding plot of inverse capacitance density [186].

4.2.3 Effect of Interfaces and Breakdown on Tunability

The field-dependent tunability is ultimately linked to the inherent nonlinearity of the material, the third-order term in Equation (4.2), and not surprisingly the tunability is dependent on material composition and deposition conditions. However, the tunability is also affected by the electrode-BST interface, and also depends indirectly on the breakdown field, both of which lead to a thickness-dependent tunability.

Figure 4-9 shows the measured dielectric constants (calculated from the zero-bias capacitance data C_{max} at 1 MHz) as a function of film thickness for three different material compositions on sapphire substrates. For very thick films the apparent dielectric constant saturates; this can be explained by the presence of a nontunable interfacial capacitance [184, 218] associated with the electrode-BST interfaces. These are sometimes called "dead" layers since they do not respond to the field in the same manner as the remainder of the film. There is no general agreement yet on the exact origin of this capacitance, but possible candidates are: 1) stiffening of the material's optical phonon mode near the interface [219]; 2) an effective interfacial capacitance due to field penetration into the electrodes [220]; or 3) an interfacial capacitance associated with near-surface charge traps [221]. In any case, the interfaces are modeled as suggested in Figure 4-10 by a fixed

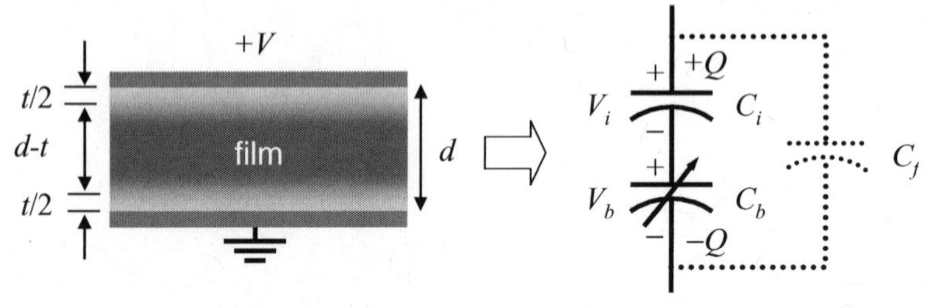

Figure 4-10 Illustration of the interfacial capacitance and equivalent circuit (after [217]).

capacitance C_i in series with the remaining bulk film capacitance C_b such that

$$\frac{1}{C(V)} = \frac{1}{C_i} + \frac{1}{C_b(V)}. \tag{4.10}$$

Using the parallel-plate capacitor formula the dependence on electrode area can be eliminated and write

$$\frac{1}{\varepsilon(V)} = \frac{t}{\varepsilon_i} + \frac{d-t}{\varepsilon_b(V)} \tag{4.11}$$

indicating that a plot of the inverse dielectric constant versus film thickness should approach a straight line for $d \gg t$, and the resulting y-intercept gives the interfacial capacitance density (the slope yields the bulk permittivity ε_b). From Figure 4-9(b) it is seen that interfacial capacitances are typically on the order of 40-60 fF/μm^2. Note that this effect may be present in any thin-film capacitor, but for low-permittivity materials there is a negligible impact because the interfacial capacitance is so large in comparison to the bulk capacitance density. On the contrary, for high-permittivity materials the bulk and interfacial capacitance densities can be comparable in value, particularly for very thin materials that are desired for low control voltages, and thus the effect of the dead layers can be significant.

Using Equation (4.10) it has been shown [217] that the composite $C(V)$ relationship has the same functional form derived earlier, but with thickness-dependent parameters $C_{\max}(d)$ and $V_2(d)$. For device optimization it is helpful to make the thickness dependence more explicit, especially with regards to the tradeoffs between control voltage and power handling or linearity. Once the interfacial capacitance density is determined, it is only necessary to measure the maximum capacitance and 2:1 voltage at some nominal material thickness d_0, then the general thickness-dependent tuning parameters become

$$C_{\max}(d) = \frac{1}{\dfrac{1}{C_i} + \dfrac{d}{d_0}\left[\dfrac{1}{C_{\max}(d_0)} - \dfrac{1}{C_i}\right]} \quad \text{and} \quad \frac{V_2(d)}{V_2(d_0)} = \sqrt{\frac{d_0}{d}\frac{C_{\max}^3(d_0)}{C_{\max}^3(d)}}. \tag{4.12}$$

The $C(V, d)$ functional is then uniquely determined for any film thickness by the specification of C_i, $C_{\max}(d_0)$, and $V_2(d_0)$.

Figure 4-11 shows the data for three devices of identical electrode area, processed from three different thicknesses of low-barium BST with Pt electrodes. From a thickness series an interfacial capacitance density of approximately 32 fF/μm^2 and a relative bulk permittivity of $\varepsilon_b = 420$ were computed. The data was extracted from broadband RF data taken on a network analyzer as described later. Using these parameters, and using the 575 nm material as the reference, the theoretical curves for each device were generated from Equations (4.6) and (4.12). Excellent agreement is observed using the dead-layer model for the thickness dependence.

Figure 4-12 is an example of the dependence of tunability on deposition conditions, in this case the Argon/Oxygen gas mixture during sputter deposition (see [222] for details). It has been shown [223–225] that varying the gas mixture and background pressure influences the stoichiometry of the film, and this in turn has a strong influence on the zero-bias permittivity ε_b. A key observation is that regardless of the initial dielectric constant, all curves tend to approach a similar asymptotic permittivity at high fields. Thus the tunability is not only determined by the zero-field permittivity, it is also dependent on the maximum field that can be sustained by the material.

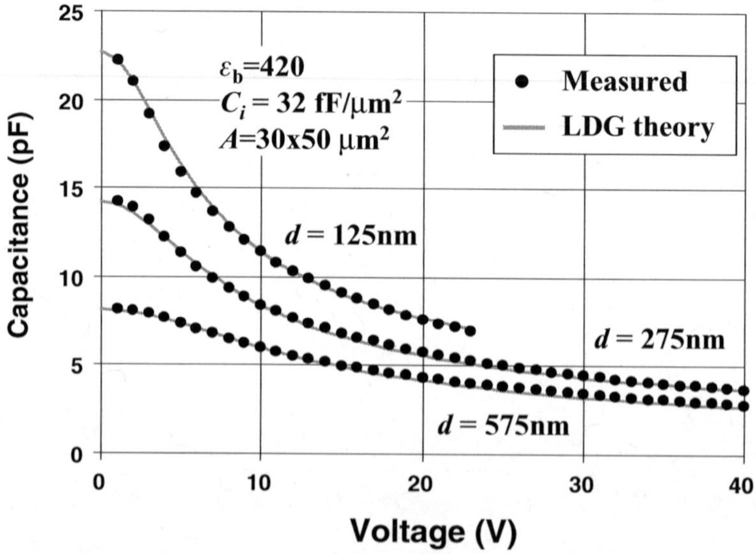

Figure 4-11 Comparison of thickness-dependent $C(V)$ model and 1-MHz data (after [217]).

The breakdown field E_{br} governs the maximum voltage $V_{br} \approx E_{br}d$ that can be applied to the material. In simple experiments, breakdown is usually taken to mean an instantaneous, catastrophic device failure; thus published C-V curves like those in Figure 4-12 are usually terminated at a field or voltage just below breakdown. Using this definition, the dielectric strength of thin-film BST is typically 1.5-2 MV/cm. Although this number can vary somewhat with thickness, deposition and processing conditions, it is generally consistent with the general downward trend in dielectric strength observed in many materials as the dielectric constant increases, see Figure 4-13.

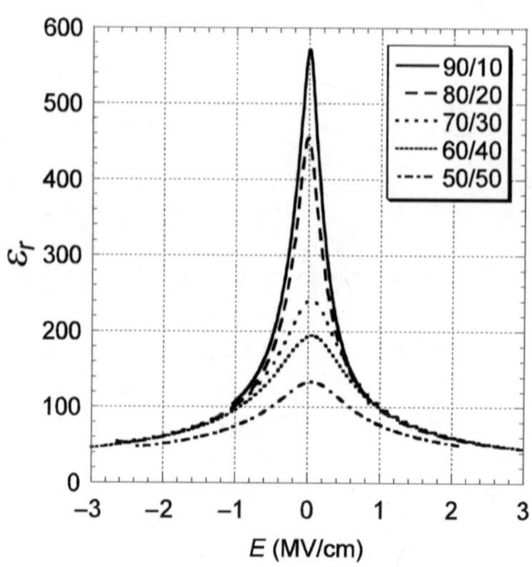

Figure 4-12 Dependence of tunability (1 MHz) on material deposition parameters, in this case the Ar/O_2 gas ratio during magnetron sputtering [222].

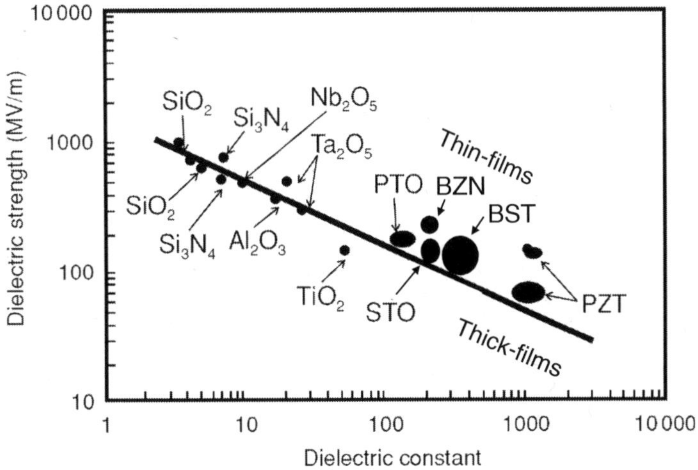

Figure 4-13 Dielectric strength versus dielectric constant (adapted and modified from [226]).

The next logical question that arises is what maximum field can be safely applied to the materials for long periods of time. It is well known that the lifetime of thin-film capacitors is degraded by prolonged exposure to high fields [227]; in high-κ ceramics this is due to many effects such as accumulated charge injection, migration of charged defects (such as oxygen vacancies), self-heating, etc. [183, 228]. The safe operating fields for tunable RF devices can therefore only be determined by careful, long-term reliability studies. A small but growing body of literature exists on this subject for BST (e.g. [229–234]), and although there is no clear consensus, it is believed that bias fields should be kept below 500–600 kV/cm for long-term operation at temperatures up to 85°C.

Thus if the field strength is contrained to some maximum value based on reliability considerations (typically well below the catastrophic failure field), then the maximum voltage and hence tunability becomes thickness dependent. This is illustrated in Figure 4-14 for some representative field strengths, using the data from Figure 4-11 and the model developed in (4.12). For example, an application requiring 2:1 tunability at a maximum field of 500 kV/cm would require a 250 nm film using this particular material. Since the control voltage also depends on thickness, this example highlights a tradeoff between the control voltage and tunability as a result of both breakdown and

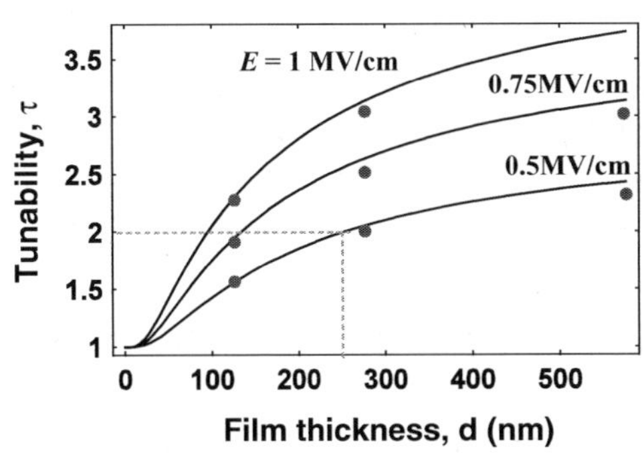

Figure 4-14 Maximum achievable tunability as a function of film thickness. Dots represent the data in Figure 4-11, solid lines are the model (4.12).

interfacial capacitance considerations. This is important in some applications where low control voltages are desirable, such as battery-operated devices.

It is interesting to note that the dielectric strength of thin-film BST is quite a bit larger than what is often quoted for bulk ceramics of similar composition. A simple empirical model for thickness-dependent breakdown in bulk materials is given in [235] as

$$E_{br} \approx 0.988\ (d\ [\mu\text{m}])^{-0.39}\ \ [\text{MV/cm}] \tag{4.13}$$

where the fit was determined from samples on the order of a few millimeters thickness. It is almost absurd to expect this formula to hold for thin-film materials that are 4 orders of magnitude thinner, but in fact the predictions are in reasonable agreement with measurements on high-quality samples such as those shown in Figure 4-12. Assuming a similar relationship applies to the maximum fields determined from long-term reliability studies, then we can conclude that thin-film materials will always have higher tunabilities than bulk materials of similar composition, because they can sustain much larger fields. This observation is consistent with most of the results reported in the literature.

4.2.4 Low-Frequency Loss and Dispersion

Turning attention back to the measured Q-factor in Figure 4-6(b), we can identify some important trends with frequency and voltage: At the lower frequencies the Q-factor decreases with applied field, approaching a linear frequency dependence at moderate to large bias fields; In the middle of the frequency range the Q-factor is initially frequency independent at zero bias, and then increases slightly with bias field; At the upper end of the frequency range the Q-factor begins to roll off but also increases with bias field.

Below 10 MHz the behavior can be understood using a parallel conductance model, where the conductance G can be related to the measured Q in Figure 4-6(b) as

$$G = \frac{\omega C}{Q} = \omega C \tan\delta. \tag{4.14}$$

This is shown in Figure 4-15. At very low frequencies the conductance is frequency-independent (constant) under bias, increasing significantly with voltage; this is associated with electronic

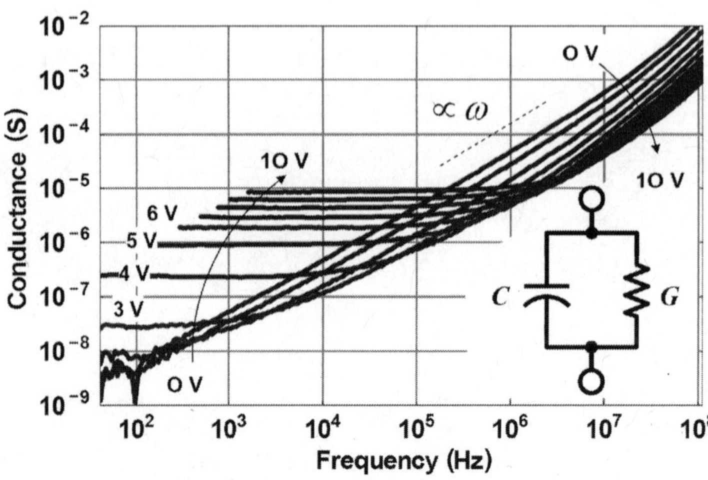

Figure 4-15 Parallel conductance versus frequency at different bias voltages, computed from Equation (4.14) and the date in Figure 4-6.

conduction (leakage currents) through the device. As the frequency is increased, a different loss mechanism starts to dominate yielding a frequency-dependent conductance that decreases with applied field. Over most of the frequency range this second loss mechanism, which is generally associated with AC losses within the BST material, varies almost linearly with frequency. Thus as a starting point for a simple model the conductance can be written as

$$G = G_{\mathrm{DC}}(V) + G_{\mathrm{AC}}(\omega, V). \tag{4.15}$$

The leakage term $G_{\mathrm{DC}}(V)$ can be easily determined by a simple DC current-voltage measurement using a picoammeter. Although DC leakage is linked with long-term reliability issues, it does not usually impact the RF loss in the device under normal operating conditions. In addition, accurate modeling of leakage requires a detailed treatment of several different conduction mechanisms and the results depend critically on the choice of electrodes, process conditions, and other factors. For these reasons leakage will not considered here and the reader is referred instead to [236–238] and references therein for an excellent treatment of leakage in thin-film BST capacitors. It is important to note, however, that any bias-dependent AC measurements can be influenced by leakage in the device at high fields.

From Equation (4.14) it can be seen that the AC loss term, modeled as a conductance with a nearly linear frequency dependence, is consistent with a constant loss tangent over several orders of magnitude in frequency. This behavior is observed in many materials and is sometimes called "universal relaxation" [239] in which the complex permittivity follows the power-law

$$\varepsilon(\omega) - \varepsilon_{\infty} \propto (j\omega)^{n-1} \tag{4.16}$$

where ε_{∞} is a high-frequency asymptote, and n is an exponent that is usually close to unity and can be fundamentally linked to measurements of time-dependent depolarization currents [240, 241] (in the time domain, universal relaxation is called Curie–von Schweidler behavior). Separating out the real and imaginary parts of the complex permittivity and converting to capacitance and Q-factor gives [242]

$$\begin{aligned} C(f) &= C_{\infty} + C_0 \left(\frac{f}{f_0}\right)^{n-1} \\ Q(f) &= \frac{C(f)}{C(f) - C_{\infty}} \tan\left(\frac{n\pi}{2}\right) \approx \tan\left(\frac{n\pi}{2}\right) = Q_f \end{aligned} \tag{4.17}$$

where f_0 is some suitably chosen reference frequency (usually 1 Hz) such that the fitting parameter C_0 has the units of capacitance. Thus the Q in this intermediate frequency range is intimately linked with the observed dispersion in the capacitance versus frequency curves, as required by causality and the Kramers–Kronig relations [243].

After curve fitting the capacitance data to determine the exponent n, the Q or loss tangent can be determined. Typical exponents for BST capacitors are in the range of $0.990 < n < 0.998$, so the change in capacitance is often slight, requiring measurements over a wide range of frequencies in order to determine accurately. Since capacitance measurements are generally more accurate than loss measurements for low-loss materials, the relationship in Equation (4.17) can be exploited to characterize the loss and identify possible extrinsic contributions, or simply to establish confidence in a direct measurement of the loss. Figure 4-16(a) shows a typical result of such a calculation, with error bars based on the uncertainty in the fitting parameters. Similarly, Figure 4-16(b) compares the calculated Q at 1 MHz versus the measured Q using this technique, for materials of different thickness (from 72 to 380 nm), at different temperatures (from 150 K to 325 K), and different

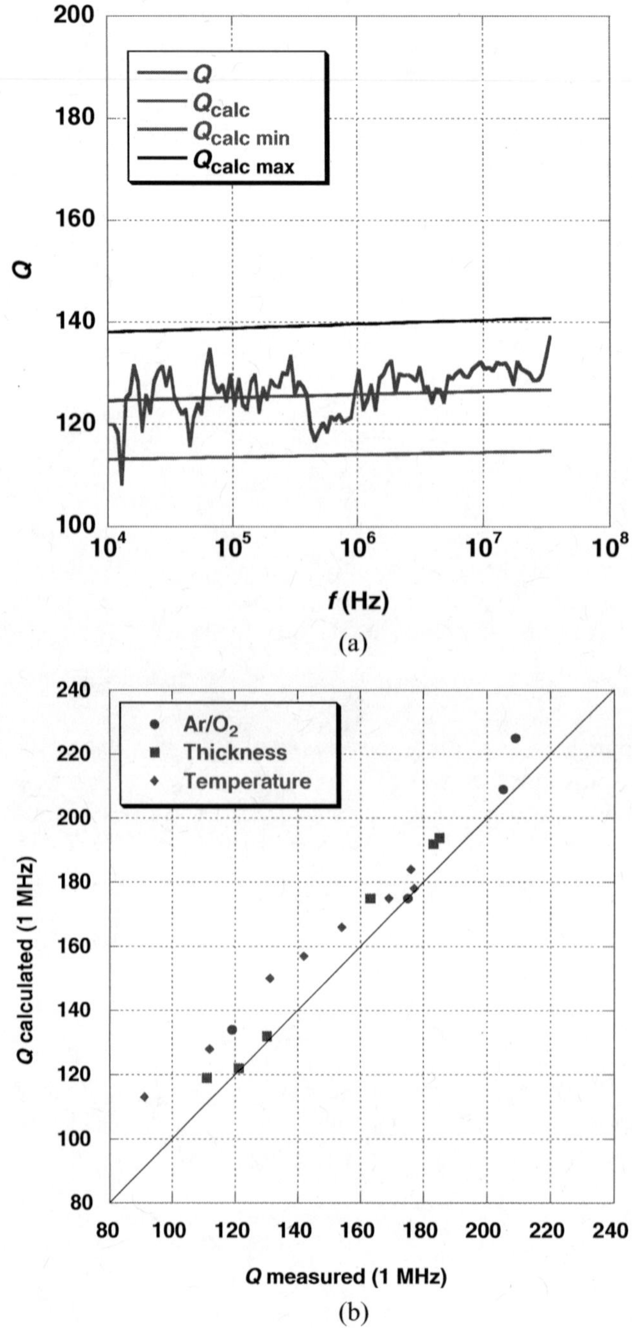

Figure 4-16 Determination of Q: (a) comparison of Q computed from measured capacitance dispersion [244]; and (b) computed Q compared to directly measured Q.

deposition parameters (varying Ar/O_2 ratio in sputtering). The calculated Q-factors are usually slightly higher than the measured Q, suggesting some small extrinsic contribution, but overall there is excellent agreement for a wide range of films and temperatures.

In the spirit of simple modeling, we can take the capacitance and film Q Q_f can be taken to be roughly constant in the megahertz range, so that

$$G_{\mathrm{AC}}(\omega, V) \approx \omega C(V)/Q_f(V) \tag{4.18}$$

and all that remains is to characterize the voltage-dependence of the film loss, $Q_f(V)$. Since device leakage makes such characterization difficult at low-frequencies, this will be examined in the section on high-frequency measurements.

4.2.5 High-Frequency *Q* Roll-Off

At the upper edge of the data presented in Figure 4-6(b) the Q begins to roll off. In most cases the roll-off tends to asymptotically approach a $1/f$ dependence. There are at least three possible explanations for this: 1) a frequency-dependent loss tangent, arising from some high-frequency relaxation processes in the material; 2) a series resistance due to electrodes or other extrinsic effects; 3) measurement errors due to limitations of the instrument (impedance analyzer) or calibration technique at the upper end of its measurement range. In most cases there is likely to be some combination of all three effects going on simultaneously.

It is important to understand the difficulty in distinguishing between the first two effects without additional experimentation. The parallel GC circuit in Figure 4-17 can be represented as a series RC equivalent circuit as shown, where

$$R' \approx \frac{G}{\omega^2 C^2} \text{ and } C' \approx C \text{ when } \tan\delta \ll 1. \tag{4.19}$$

Remembering that the material losses can be represented as $G = \omega C \tan\delta$, it can be seen that a loss tangent with a linear frequency-dependence can be modeled by a constant series resistance, which in turn leads to a Q that rolls off as $1/f$ since for the series circuit we have

$$Q = \frac{1}{\omega R' C'} \tag{4.20}$$

and thus a frequency-dependent loss tangent is indistinguishable from a series resistance arising from the metal electrodes. Note that a linear (or close to linear) frequency dependence for the loss

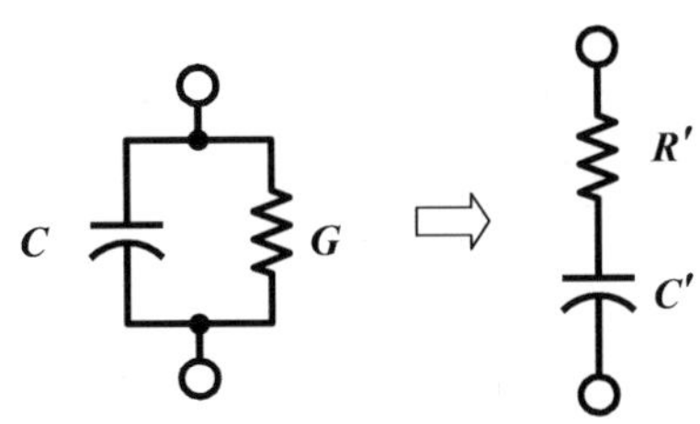

Figure 4-17 Parallel–series circuit transformation.

is predicted by many simple relaxation processes such as the Debye law [243]

$$\varepsilon(\omega) = \varepsilon_b + \frac{\Delta\varepsilon}{1 + j\omega\tau} \overset{\omega\tau \ll 1}{\Rightarrow} \tan\delta \approx \omega\tau \frac{\Delta\varepsilon}{\varepsilon_b}. \tag{4.21}$$

Frequency-dependent loss tangents also imply a corresponding change in the frequency-dependence of the capacitance. This is not usually observed in thin-film capacitors, where the capacitance generally appears to follow the weak power law predicted by universal relaxation Equation (4.17) up into the gigahertz range [241]. However, tunable materials are somewhat unique in that they have very large dielectric constants, such that the relaxation process contributing to the loss may have a negligible impact on the overall capacitance. This is equivalent to assuming that $\Delta\varepsilon/\varepsilon_b \ll 1$ in Equation (4.21). In fact, the small series inductance in a real device is often sufficient to mask the additional capacitive dispersion that would accompany a frequency-dependent loss tangent.

Regardless of the physical origin, it seems that we can always model the high-frequency roll-off reasonably well by adding a series resistance to the model, and then the question becomes how this term scales with geometry and applied field. The geometrical dependence of ohmic losses are relatively easy to measure and quantify for a given device structure; for example, using a distributed-circuit model [216] the series resistance for the simple test structure shown in Figure 4-5 can be written as

$$R_{\text{electrode}} \cong \frac{r_b}{2L}\left(g + \frac{W}{6}\right) + \frac{1}{3}\frac{L}{W}\left(r_t + \frac{r_b \parallel r_2}{2}\right) \tag{4.22}$$

where r_b is the sheet resistance of the bottom electrode, and r_t is the sheet resistance of the top electrode. In most cases the first term is dominant, since the bottom electrode is usually a thin refractory metal. It is relatively easy to characterize the sheet resistance of each metal layer using suitably designed process monitors, but the actual device resistance can also be estimated more directly by fabricated short-circuited devices (no dielectric) alongside the test structures. This will not include some of the distributed effects accounted for in Equation (4.22), but these are often second-order effects.

Frequency-dependent material losses, on the other hand, should scale with area in the same way as a contact resistance term,

$$R_{\text{material}} \approx \frac{r_c(V)}{A} \tag{4.23}$$

where r_c involves the loss tangent, has the units of specific contact resistivity ($\Omega \cdot \mu m^2$), and is voltage dependent. (A similar expression would arise from interfacial contributions to the series resistance, like that discussed in [221]). The total series resistance is then a sum of $R_{\text{electrode}} + R_{\text{material}}$. As will be seen later, the different geometrical dependencies of each term can be helpful in identifying sources of loss.

4.2.6 Other Geometrical Effects on Tunability and *Q*

It is important to appreciate that until this point, all the devices and results that have been considered were based on relatively large-area capacitors. This was intentional, because there are some important geometry effects that influence the electrical properties of the device as the capacitor area shrinks. Dielectric varactors have a high capacitance density, up to 100 times that of conventional integrated capacitors using SiO_2 or SiN dielectrics. The typical electrode areas are therefore

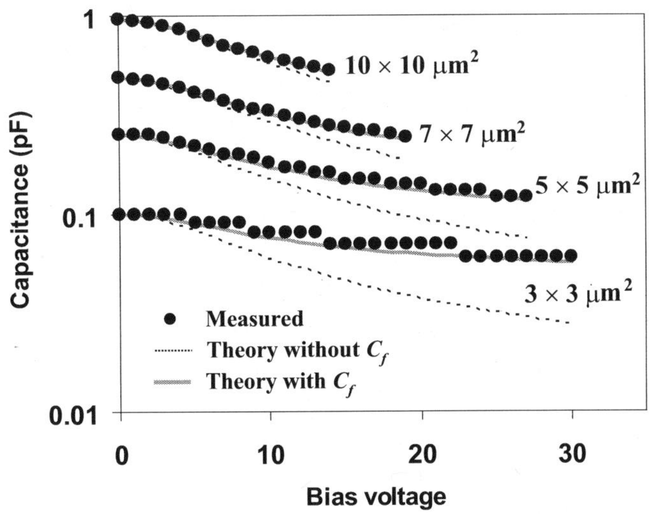

Figure 4-18 Dependence of tunability on area [217].

much smaller by comparison, and the periphery-to-area ratios are much higher for a given total capacitance. This is especially true for tunable RF applications because relatively small devices are required for circuit designs in the gigahertz range.

Experimentally we observe that smaller capacitors have a reduced tunability compared with larger devices on the same wafer. This appears to be well modeled by a nontunable peripheral or "fringing" capacitance C_f in parallel with the tunable device, as was shown in Figure 4-10. As the device size is reduced this contribution represents an increasing fraction of the overall capacitance, and the tuning curves are observed to level off prematurely. The data collected is consistent with a modified $C(V)$ relationship of the form

$$C(V) = \frac{C_{\max} - C_f}{2\cosh\left[\frac{2}{3}\sinh^{-1}\left(\frac{2V}{V_2}\right)\right] - 1} + C_f. \tag{4.24}$$

The fringing capacitance scales with periphery and seems to have a weak thickness dependence. Figure 4-18 shows the tuning curves for several small-area devices and a comparison to the theoretical model with and without the fringing correction [217]. The devices were made using sputtered 30/70 BST with Pt electrodes. The dashed curves in Figure 4-18 were generated from Equation (4.6) using $V_2 = 13$ V, a value determined experimentally from larger area devices on the same wafer. The solid curves were generated from Equation (4.24) using a 2.8 fF/μm fringing capacitance density. The data is shown on a log-scale in capacitance for clarity (note that the data was rounded to the nearest 0.01 pF, which is apparent in the data for the smallest device).

The fact that there is a fringing capacitance is not too surprising, but simple electrostatic modeling of the device suggests that an ordinary fringing contribution should have a negligible impact on the tuning curves, in contrast to what is observed experimentally. Figure 4-19 shows the results from a 2-D numerical model of a dielectric varactor biased at $V = V_2$, assuming a zero-bias permittivity of $\varepsilon_b = 250$; here the spatial variation of permittivity within the material is being shown, based on a computation of the electrostatic field that included the material nonlinearity

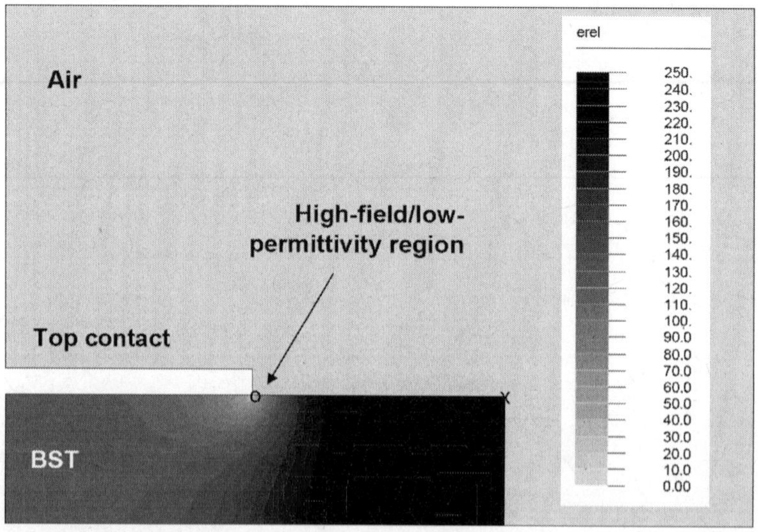

Figure 4-19 Numerical contour plot of permittivity in the region of the top edge for a BST varactors based at $V = V_2$. A 2-D PDE solver was used.

described by Equation (4.9). Near the contact edge we can see the relative permittivity changes, from its intermediate value of 125 in the active region beneath the top contact, to its unbiased value of 250 in the shelf away from the top contact edge. The high fields near the top contact edge reduce the permittivity even further. According to this analysis, the fringing capacitance should tune with bias voltage and should have a negligible impact on the tunability curve, other than to increase the total capacitance accordingly. Thus other explanations must be sought for the physical origin of this fringing term.

Another geometrical effect that has been observed experimentally is an apparent area-dependent Q in the megahertz region [244], as shown in Figure 4-20(a). For an ideal parallel-plate capacitor there can never be an area dependence for material-related losses, because the area dependence always cancels out in the expression for Q. Assuming this is not a measurement artifact, the only

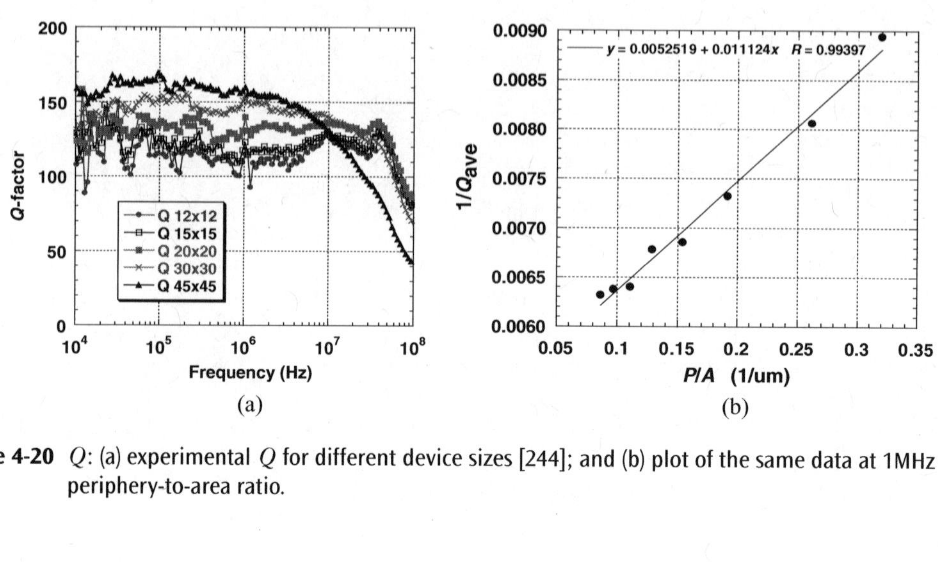

Figure 4-20 Q: (a) experimental Q for different device sizes [244]; and (b) plot of the same data at 1MHz versus periphery-to-area ratio.

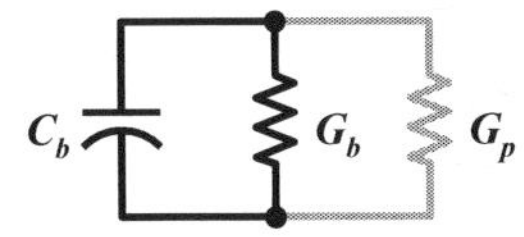

Figure 4-21 A periphery-dependent conductance can model area-dependent Q-factors.

way to explain the data is by introducing a periphery-dependent loss term, much as was done for the fringing capacitance. This is shown in Figure 4-21. Adding a conductance of the form $G_p = \sigma' P$, where σ' is an effective conductivity and P is the device periphery, it can be shown that the inverse of the total device Q for this circuit is

$$\frac{1}{Q} = \frac{G_b + G_p}{\omega C_b} = \tan\delta + \left(\frac{\sigma'}{\omega c_d}\right)\frac{P}{A} \tag{4.25}$$

where c_d is the capacitance density of the film. Hence a plot of the inverse Q versus periphery-to-area ratio should yield a straight line, with an intercept that gives the bulk loss tangent, and the slope gives information on the conductivity of the periphery. Figure 4-20(b) is a plot of the inverse Q data from Figure 4-20(a) taken at 1 MHz, showing excellent agreement with this simple phenomenological model. The physical origin of this peripheral loss term is not clear, nor is it important from the standpoint of modeling the effect, but it may be linked with surface conduction on the BST shelf around the top electrode [245].

4.3 HIGH-FREQUENCY MEASUREMENTS AND MODEL

4.3.1 RF Device Structure and Network Analyzer Measurements

Characterization of BST varactors at RF is complicated by several factors. At these frequencies, series inductance associated with the electrode geometry introduces a self-resonant frequency that limits the useful measurement and operating bandwidth. The high-capacitance density of the films also means that capacitors intended for use in this frequency range (typically on the order of 0.1–10 pF) will have rather small electrode areas, often smaller than the tips of on-wafer probes. This has at least three important ramifications: first, an additional dielectric cross-over layer is required to make an external connection to the device; secondly, the bottom electrode must be patterned to allow for the external connections; and third, the device must be integrated into a structure with large electrical contacts, introducing additional parasitics that must be de-embedded from the measurements. Thus the fabrication of RF test structures and the measurement procedures are always more difficult in the gigahertz range. Further complicating the issue is the intrinsic limitation of network analyzer measurements, a notoriously inaccurate technique for highly reactive devices (impedances on the outer rim of the Smith Chart). Measurement interpretation is also more difficult, as new loss mechanisms become significant in the gigahertz range that may not be apparent at lower frequencies (dielectric relaxation processes, interfacial losses arising from electron transfer between the electrodes and surface states, and skin-effect losses in the electrodes).

A basic RF device structure that has been used in most of the work here is shown in Figure 4-22 [216, 246, 248]. For the most part this is a standard integrated-circuit capacitor structure [187, 188] where the top contact to the capacitor dielectric (BST) is defined by a window in a second (usually low-κ) dielectric. This second "interlayer" dielectric serves two important roles: first, as a cross-over layer to separate the top contact away from the edge of the bottom electrode,

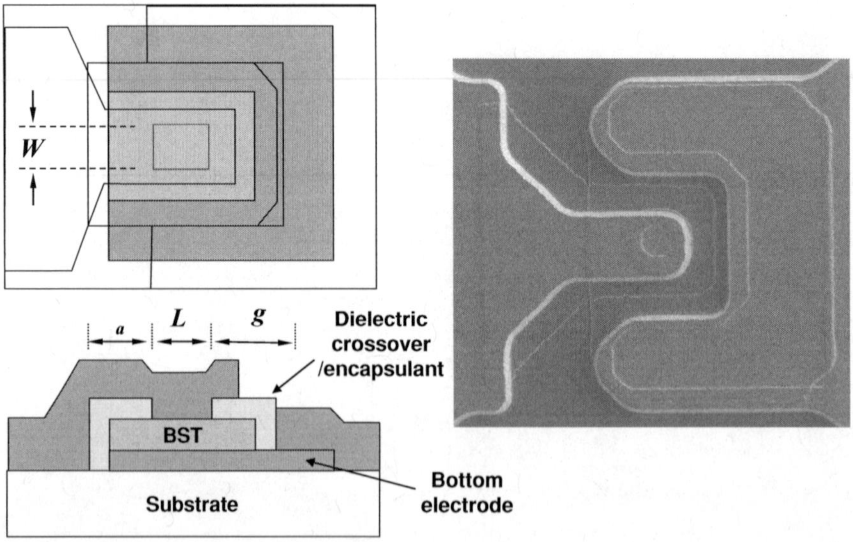

Figure 4-22 Simple RF device layout and photo of a finished structure. The small contact area requires a small access finger that must cross over the bottom electrode, necessitating the use of an air-bridge or dielectric cross-over layer. A U-shaped "collector" is used to minimize resistance associated with the bottom electrode [216, 249].

a problem region for premature breakdown; secondly, as an environmental encapsulant to protect the BST film from exposure to subsequent contamination in processing or operation. In the latter role the layer is also sometimes referred to an a "passivation" layer, suggesting that it may be helpful in controlling surface states in the film, but this effect has not been widely studied yet. Several materials have proven adequate for this encapsulant, such as SiO_2, SiN [187, 246] and Al_2O_3 [247]. Some process-related considerations are discussed in [216, 246, 248]. The bottom electrode is patterned (by lift-off, dry-etching, or ion-milling techniques), and is often a thin, refractory metal such as Platinum. The only slightly unusual aspect of this device design is the "U"-shaped connection to the thick metal interconnect layer, which is designed to minimize the series resistance associated with the bottom electrode layer [216,249]. This is similar to ideas used in high-speed Schottky diode layouts [250].

For on-wafer characterization the device must be embedded in a "probe-friendly" structure; Figure 4-23 illustrates a simple and commonly used scheme for one-port RF device characterization using GSG probes [251, 252]. In order to remove the influence of the large CPW probe structure on the device measurement, two additional structures are fabricated alongside the device-under-test (DUT) in which the DUT is replaced by an open- or a short circuit. If the probe pads are represented electrically by the L-network of Z_s and Y_p, then the short-circuit measurement yields Z_s, and the open-circuit measurement then yields Y_p. Once these are known, their contribution to the measured impedance of the DUT can be mathematically removed. It has been shown in [252] that inconsistent placement of the probe tips between the various test structures can lead to measurement errors; these can be compensated to some extent, but a simple safeguard is to include some reference marks in the test structure as shown in Figure 4-23.

Usually the smallest probe-pitch (separation between the probe tips) is used for optimal accuracy. Somewhat simpler test structures are made possible if two-conductor GS (ground–signal) probes are used instead of GSG probes [253], but accurate calibration is a more difficult challenge

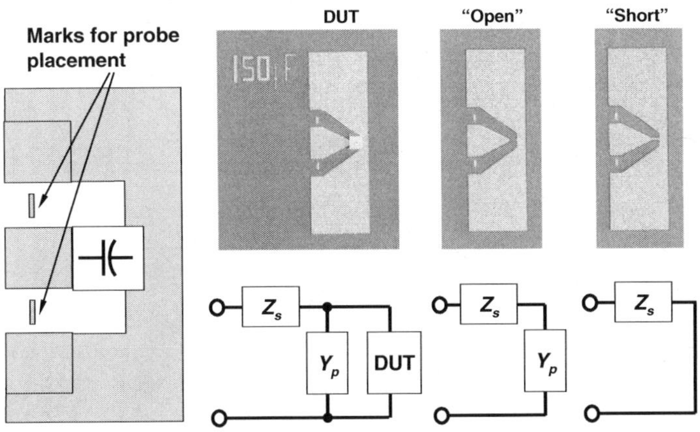

Figure 4-23 Simple CPW test structures and equivalent circuits for deembedding device characteristics from one-port network analyzer measurements.

with GS probes and will not be considered here. For one-port network analyzer measurements, a standard short-open-load (SOL) technique is typically used, with calibration "standards" provided by the probe manufacturer. Unfortunately this technique is highly sensitive to the accuracy of the impedance standards [254], especially for highly reactive and low-loss devices (reflection coefficients close to unity). Furthermore, even with perfect standards, the intrinsic measurement accuracy of the network analyzer is also suspect for such devices [255]. Two approaches that can improve measurement accuracy are the use of additional on-wafer lumped-element structures [256] as part of an over-determined calibration procedure, and the use of two-port structures [257] as shown in Figure 4-24. Two port methods help in two ways: first, they allow for more advanced calibration methods such as the through-short-delay (TSD) and line-reflect-match (LRM) techniques, which do not require perfectly characterized standards; and secondly by bringing the impedance trajectory in closer to the center of the Smith Chart(close to 50 Ω) where the measurement accuracy of the network analyzer is significantly higher.

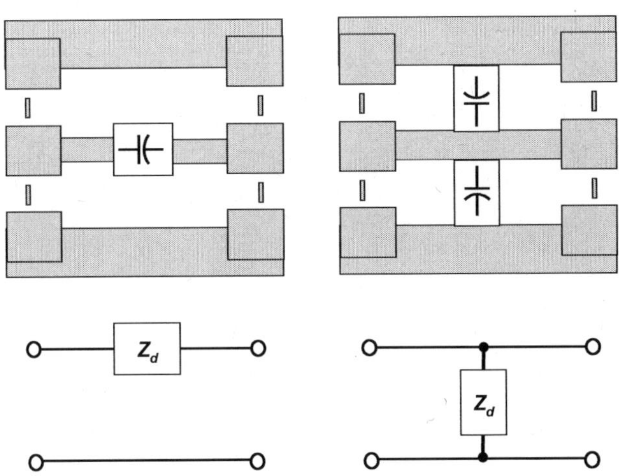

Figure 4-24 Two-port test structures for RF device characterization and equivalent circuits.

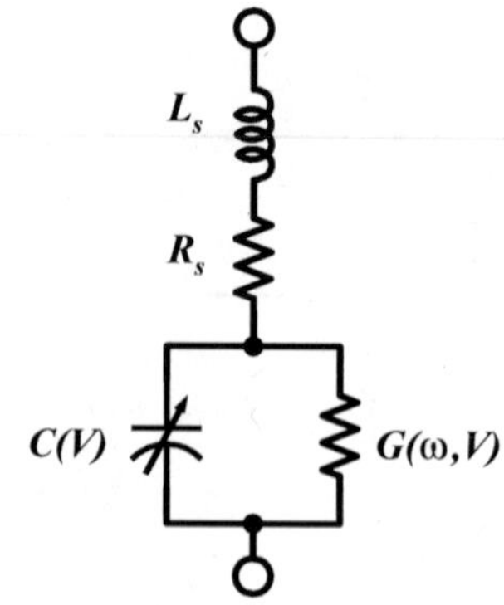

Figure 4-25 RF Model for a dielectric varactor.

In a two-port measurement the DUT can be embedded in series or in shunt. The CPW shunt arrangement has the disadvantage of requiring two devices which must be assumed identical, but has the advantage of simpler biasing than the series configuration. The series configuration tends to have better accuracy when the DUT impedance is low, so it tends to work well at higher frequencies for capacitive devices; conversely the shunt configuration has better accuracy at the low end of the band when the device impedance is high. In practice, measurements taken from a combination of series and shunt structures will yield the best accuracy over a wide bandwidth.

4.3.2 Curve Fitting to a Circuit Model

Although swept-frequency network analyzer measurements are not as accurate as some other techniques (notably high-Q resonator methods), they provide information on the behavior of the devices over a broad frequency range, which is invaluable for developing models and exploring various contributions to the device impedance.

Based on our low-frequency measurements and physical expectations, dielectric varactors should be reasonably well-modeled by the equivalent circuit shown in Figure 4-25. The only new addition to the model from the previous section is the series inductance, which can no longer be neglected in the gigahertz range. To estimate these model parameters the intrinsic device impedance is first computed by deembedding the probe parasitics at each frequency (using either the one-port or two-port techniques described earlier), and then a least-squares curve-fitting procedure is used following [258]. The equivalent circuit leads to an expression for the frequency dependence of the reflection coefficient

$$\Gamma(\omega) = \frac{1 + (F_1 - Z_0 F_0)\omega + F_2\omega^2}{1 + (F_1 + Z_0 F_0)\omega + F_2\omega^2} \tag{4.26}$$

where

$$F_0 = jC \quad F_1 = jRC \quad F_2 = -LC$$

and Z_0 is the reference impedance (e.g., 50 Ω). Note that the loss tangent of the material is accounted for by allowing the unknown capacitance to be complex, $C(1 - j\tan\delta)$. If measurements on each standard are made at N frequencies ω_i yielding a deembedded reflection coefficient $\Gamma(\omega_i)$, then Equation (4.26) leads to the following over-determined system

$$\begin{bmatrix} A_1 & B_1 & C_1 \\ A_2 & B_2 & C_2 \\ \vdots & \vdots & \vdots \\ A_N & B_N & C_N \end{bmatrix} \begin{bmatrix} F_0 \\ F_1 \\ F_2 \end{bmatrix} = \begin{bmatrix} 1 - \Gamma(\omega_1) \\ 1 - \Gamma(\omega_2) \\ \vdots \\ 1 - \Gamma(\omega_N) \end{bmatrix} \tag{4.27}$$

where

$$A_i = \omega_i Z_0 \left[1 + \Gamma(\omega_i)\right] \quad B_i = \omega_i \left[1 - \Gamma(\omega_i)\right] \quad C_i = \omega_i^2 \left[1 - \Gamma(\omega_i)\right].$$

This equation can be solved for a least-squares best fit for the three unknowns F_0, F_1, and F_2, and hence the model parameters from Equation (4.26). Typically a singular-value decomposition (SVD) algorithm [259] is used. A wide frequency range and large number of measurement points are preferred for a good model fit. Note that a frequency-dependent loss mechanism (such as skin-effect losses) can also easily be included in this model if desired.

Figure 4-26 shows the results of this procedure using broadband on-wafer RF data measured from 50 MHz to 40 GHz on an Agilent E8364A PNA-series network analyzer [216]. After a standard on-wafer SOL calibration, the device impedance was determined by first de-embedding the probe pad parasitics, and then the fitting procedure was applied, with the best-fit model parameters shown. Excellent agreement is observed between the data and the simple model of Figure 4-25. The influence of the small series inductance is clearly visible. Using the simple model in Figure 4-25, all of the high-frequency roll-off in Q is attributed to the series resistance and inductance, such that the parallel-conductance term is represented by a constant loss tangent.

4.3.3 RF Loss Modeling and Electrostrictive Resonances

At the low end of the RF frequency range the Q asymptotically approach loss tangents that are consistent with observations and models developed at lower frequencies, taken as a frequency-independent constant $Q_f = 1/\tan\delta$ in the model. As shown earlier in connection with Equation (4.19), a frequency-dependence in the loss tangent can effectively be accounted for in the series resistance term, R_s. This combination of loss terms in the model seems to adequately account for the observed losses in the low gigahertz range.

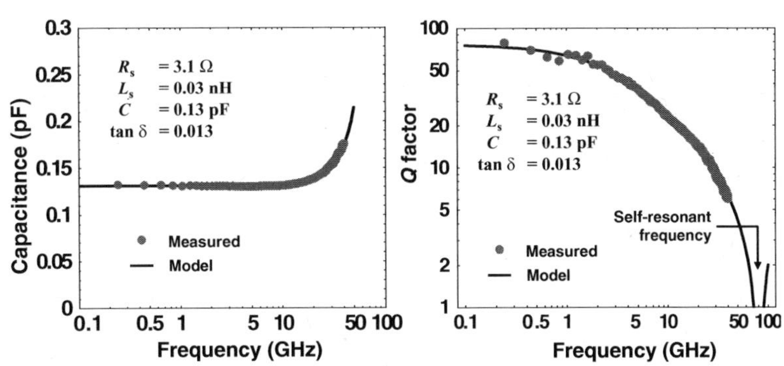

Figure 4-26 Example curve fit to network analyzer data from 200 MHz-40 GHz using the equivalent circuit of Figure 4-25 [216].

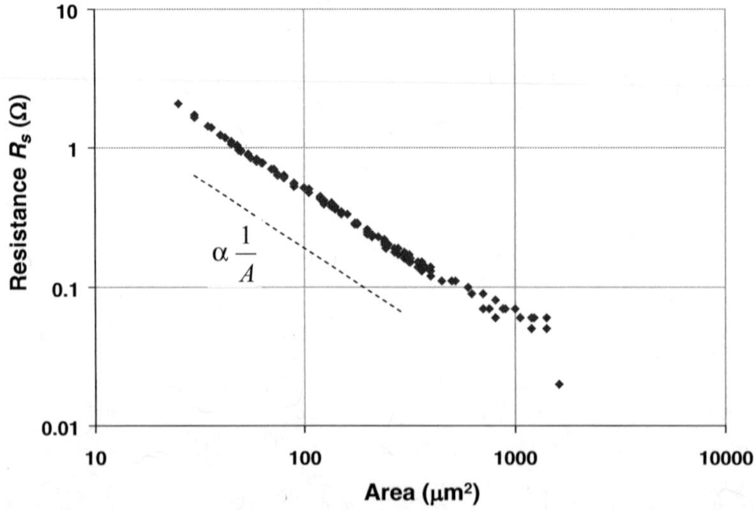

Figure 4-27 Zero-bias data (series resistance, R_s) for 144 different devices of different areas and aspect ratios.

The model parameter that immediately jumps out of the example in Figure 4-26 is the excessive series resistance R_s. This number is significantly larger than what would be expected from electrode contributions alone, suggesting another dominant loss term. One simple approach to identifying possible loss mechanisms is to explore the dependence on various geometrical factors. For example, it can be expected that certain losses would scale with aspect ratio (e.g., top electrode resistance), or periphery (e.g., bottom electrode and/or fringing effects), or area (e.g., material or interface effects). So a series resistance of the following form can be assumed:

$$R_s \approx r_s \frac{\ell}{w} + \frac{r_p}{P} + \frac{r_c}{A} \tag{4.28}$$

where the coefficients for each term are determined by experiment. Figure 4-27 shows the net series resistance determined by model-fitting of broadband RF data from 200 MHz to 5 GHz, using a large matrix of 144 devices with rectangular contacts varying in both length and width from 5 μm to 40 μm, and aspect ratios (L/W) varying from 40/5 to 5/40. For this particular material and process technology there seems to be an inverse area dependence for these devices, which would suggest either a frequency-dependent loss tangent or interfacial contribution that is large in comparison to other terms. This is consistent with observations of other researchers [260–264]. Without speculating further on the origin of the loss, simply note that the zero-bias RF loss seems to be well modeled by a term of the form $R_s \propto r_c/A$, where the effective contact resistance appears to depend on several technological variables such as grain size and/or defect concentration. With careful optimization of the material and device structure this term can be reduced. Relatively high RF Q have been demonstrated from well-designed devices [265].

Under bias, both the low-frequency asymptotic loss tangent and the effective series resistance vary with applied field as shown in Figure 4-28. The variation in series resistance (which further argues for a material or interface contribution) is relatively weak and can be ignored without too much error, or fitted to a linear dependence if desired. The background loss tangent seems to obey

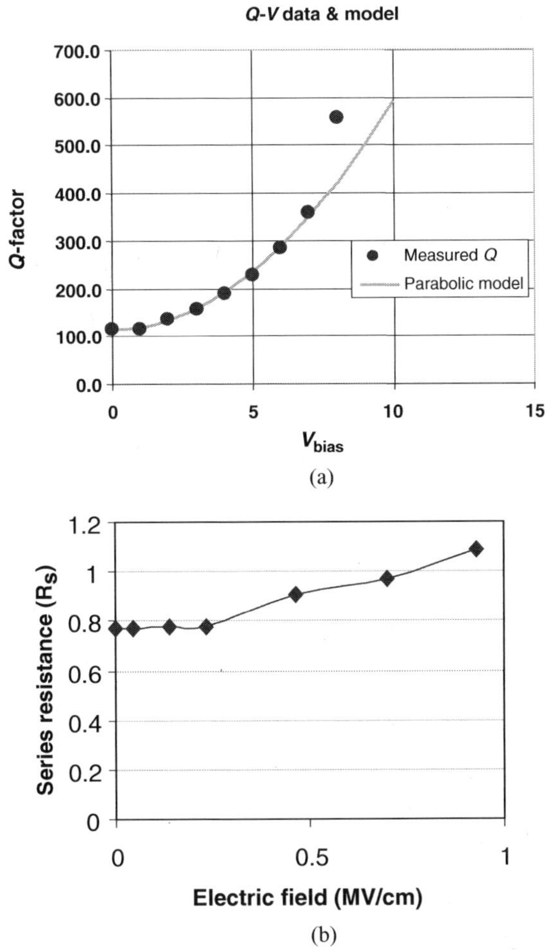

Figure 4-28 Variation of (a) the asymptotic quality-factor Q_f; and (b) the effective series resistance R_s with the applied field or bias.

a parabolic model reasonably well such that

$$Q_f(V) = Q_0\left[1 + (V/V_q)^2\right] \tag{4.29}$$

where Q_0 is the zero-bias asymptotic Q, and V_q is a fitting parameter (note that the voltage could be replaced by field if desired). Q_0 is essentially the same as that determined by low-frequency impedance analyzer measurements, and may display similar dependencies on film thickness and device geometry.

At high fields the data and Equation (4.29) diverge; it is difficult to ascribe any significance to this observation because of the inherent inaccuracy in measuring such high Q on a network analyzer, and also because of the possible influence of leakage on the data at extremely high fields (see previous section). Fortunately this is a nonissue because as Q_f becomes large, its impact on the model predictions becomes small and the series resistance term is dominant. So Equation (4.29) works reasonably well in a practical sense.

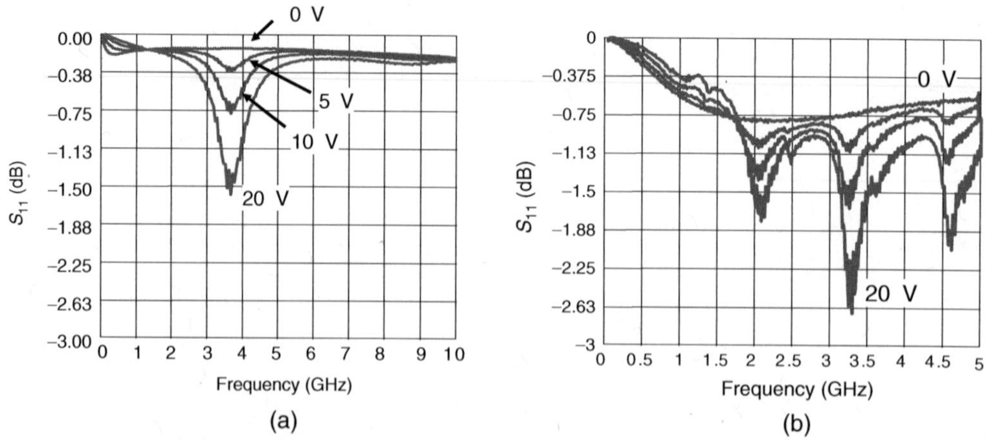

Figure 4-29 Voltage-induced resonances in two different BST capacitors: (a) a $30 \times 30\,\mu$m device with 100nm Pt tope electrode; (b) $10 \times 30\,\mu$m device with 100 nm/1 μm Pt/Au top electrode.

Most importantly, a new effect appears under bias that was not apparent at low-frequencies: resonant dips appear in the reflection coefficient data, with a depth that varies with bias, and at frequencies that depend on the device layer structure. Figure 4-29 shows examples from two devices. This is now understood to be a consequence of the electrostrictive property of the material, which has been known for some time [161, 266] but seemed to have escaped notice in thin-film research community until recently [267, 268]. The extent to which this effect is observable (note the scale in Figure 4-29) depends on the film thickness, electrode area, and applied field.

This electrostrictive property is akin to a piezoelectric effect, but depends on the applied field so it can be called "field-induced piezoelectricity". The piezoelectric coupling coefficients for the material can be quite large and seem to vary approximately linearly with applied field. If the piezoelectric coupling constants are known, the BST varactor can be modeled as a thickness-mode bulk-acoustic-wave (BAW) device using standard models such as the KLM model [269] or the Mason model [270]. The KLM model is attractive because each of the various layers in the vertical cross section of the device is modeled as an equivalent transmission line. Figure 4-30

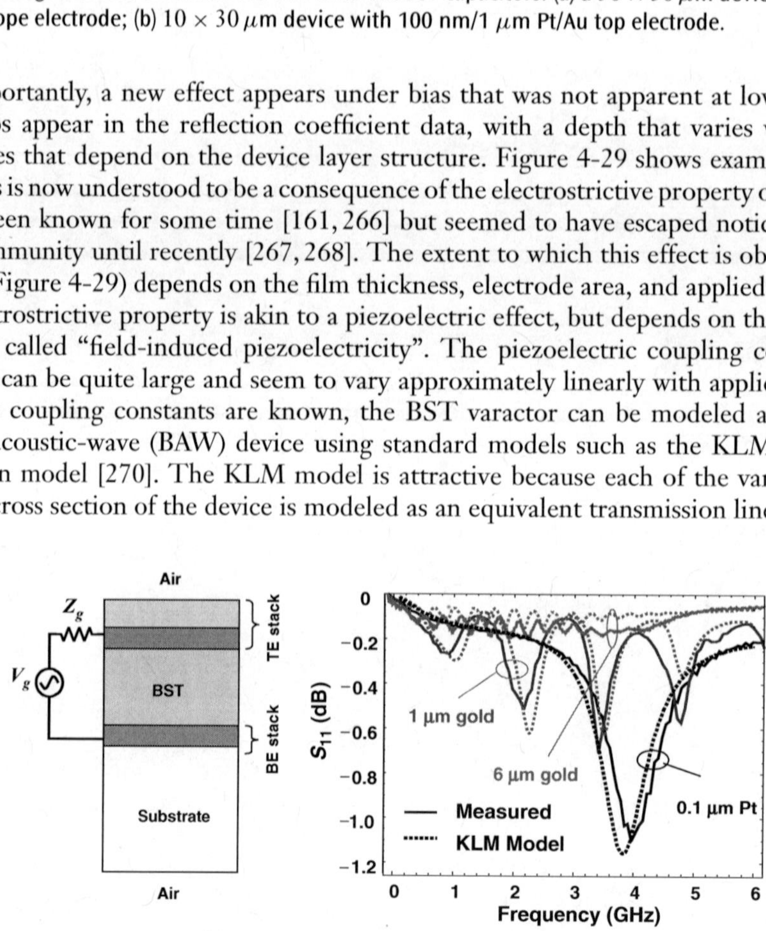

Figure 4-30 Measured reflection coefficient of a $31 \times 32\,\mu$m device, 160 nm BST, at 20 V bias, for three different top electrode stacks, illustrating the damping effect of a thick gold layer on the electrostrictive resonances (courtesy of Agile Materials & Technologies).

shows a comparison of the modeled reflection coefficient for a typical varactor structure using three different top electrodes.

Although this property of STO and BST is quite interesting and potentially quite useful for high-Q resonators, it will not be considered it further in this chapter since the focus is on tunable capacitors. As Figure 4-30 indicates, the resonances can be effectively suppressed to a large extent by piling on thick layers of interconnect metal, which acts like a mechanical damper on the structure. There is still an adverse residual effect on the loss properties of the device, but this can effectively be accounted for by existing terms in the simple loss model.

4.3.4 Design Optimization

With the various dependencies of loss and tunability on geometrical variables articulated, we can begin to focus on designs that optimize certain performance variables.

Our analysis already began with a device structure that minimizes electrode resistance (Figure 4-22) with a thick metal collector surrounding the top contact. As a result, the net Q for the device was dominated by mostly area-independent effects, so further dramatic improvements in Q-factor can not be expected with device design, except perhaps by decreasing the sheet resistance of the electrode layers [265]. Focusing then on secondary effects, two small contributors to the high-frequency roll-off are the top electrode resistance, which includes the "access" resistance associated with the length ℓ_a in Figure 4-22, and the series inductance L_s, each of which scale with the aspect ratio L/W; this tends to argue against long, narrow top contacts. The inductance contributes to high-frequency roll-off through the selfresonant frequency, given by

$$\omega_r \approx \frac{1}{\sqrt{L_s C}}. \tag{4.30}$$

Data indicates that the series inductance is roughly area independent, so maximizing the self-resonant frequency requires the smallest possible capacitance. Small-area capacitors are therefore expected to perform best with respect to high-frequency roll-off in Q, and this is indeed borne out by experiment. On the other hand, it has also been observed that the constant background loss tangent (associated with parameter Q_f) may have an area-dependence such that it increases with area-to-perimeter ratio. In addition, the tunability also increases with area-to-perimeter ratio because of the nontunable fringing capacitance; this effect is shown in Figure 4-31.

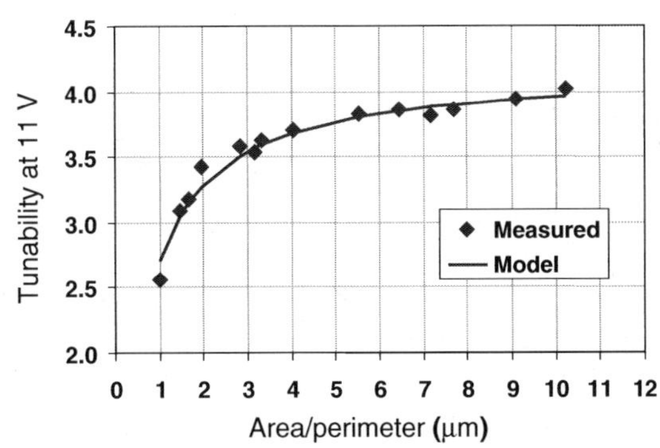

Figure 4-31 Comparison of tunability at a fixed field (voltage) versus device size, and comparison to our fringing capacitance model, Equation (4.24).

A partial solution to these competing demands is to use a number of small-area devices in parallel to achieve a given capacitance, and to use a top-electrode geometry for each unit cell that maximizes the area-to-perimeter ratio for a given area; this argues for circular top electrodes. Circular electrodes have also shown to be better than rectangular electrodes from the standpoint for breakdown and long-term reliability in thin-film capacitors.

Another reason for using multiple parallel devices is to minimize the current density in the electrode layers [249]. This can be important for applications requiring large RF voltage swings. It is well known from studies in integrated circuit technology that the current density must be kept below some critical value J_c to prevent failure:

$$J < J_c. \tag{4.31}$$

The critical current density depends on the type of metal and its thermal environment, but is typically on the order of 10^6 A/cm^2. The amount of AC current that will flow in the capacitor is a function of the RF voltage and impedance. If the peak AC voltage swing is denoted by $V_{\max}$, the peak AC current through the capacitor is

$$I_{\max} = j\omega C V_{\max}. \tag{4.32}$$

For a single device with a round top electrode of radius r, top interconnect thickness t_c and bottom electrode thickness t_b, then Equation (4.31) implies the following approximate inequalities for the top and bottom metal layers:

$$\frac{I_{\max}}{t_c 2r} < J_c \qquad \frac{I_{\max}}{t_b \pi r} < J_c \tag{4.33}$$

Since we usually have $t_b \ll t_c$, the second inequality is more restrictive. Using Equation (4.32) and writing $C = c_d \pi r^2$ we find

$$r < \frac{J_c t_b}{\omega c_d V_{\max}}. \tag{4.34}$$

As an example, consider an application at 1 GHz with an 10 Vp-p signal swing (corresponding to 1 W in a 50 Ω system). If we assume a bottom electrode thickness of $t_b = 300$ nm, a critical current of 10^6 A/cm^2, a capacitance density of $c_d = 20$ fF/μm^2 gives $r < 2.4\ \mu$m, corresponding to a capacitance of about 0.4 pF. If the application requires a capacitance of 2 pF, this would have to be realized using parallel combinations of small-area devices.

4.3.5 Linearity and Control Voltage Trade-offs

The current-handling analysis leading to Equation (4.34) indicates that a low capacitance density is desirable. Similarly it has been found (Figure 4-14) that films of thickness in excess of 100–200 nm are also desired to maximize tunability for a given maximum field strength. Low-capacitance density also helps from a processing standpoint, especially for high frequencies where the required capacitance values are small and the electrode areas become small. The downside of thicker films, however, is the need for a larger DC control voltage. Some applications, especially in wireless handheld devices, would prefer to have low control voltages commensurate with the battery technology, which may be as low as 2.7 V.

The tradeoff with control voltage is especially important with respect to linearity concerns in the RF front-end. The capacitance change with voltage occurs almost instantly in tunable dielectrics, so a large RF voltage swing across the device will modulate the capacitance and generate harmonic distortion, including third-order products (inter-modulation distortion, or IMD) if the waveform is modulated. IMD is a particular concern in modern wireless systems with close channel separation such as GSM and CDMA. Devices with low control voltage will generally produce more harmonic distortion because, by definition, they generate a larger capacitance change per volt. So the desire for low control voltage is always at odds with the desire for high power handling and high linearity.

One solution to this problem is to use a simple DC–DC boost-converter or charge-pump to step up the available control voltage to a much higher value, large enough to drive a dielectric varactor with sufficient power-handling for the application. This is in fact a reasonably attractive option for some applications, because dielectric varactors draw no appreciable current and hence consume almost no power from the control circuit (only during switching transients is there any significant current draw). DC–DC converters can be quite small and inexpensive when they do not have to source much current, and can be designed for quite large step-up ratios.

However, the linearity specifications for many applications simply require film thicknesses and control voltages that are too large to be practical, and hence a different solution is needed. The best solution to this problem appears to be the use of a series combination of capacitors, as shown in Figure 4-33 [271].

In this figure, the resistors are all large enough that they are effectively not part of the RF circuit, so the total RF voltage is divided more-or-less equally among each of the capacitors. However, the biasing circuit is designed such that the full applied DC control voltage appears across each capacitor. There are a number of possibilities for accomplishing this, depending on whether the varactor is mounted in series or shunt in the RF circuit, and depending on whether the bias voltage can be superimposed on the RF signal. Other straightforward extensions of this concept (not shown) might include a large-value DC blocking capacitor to isolate the bias circuit from the RF circuit, or perhaps using inductive chokes instead of resistors. Vertical stacking has been demonstrated with both interdigital and parallel-plate capacitors [272, 273].

The vertical stacking concept can be scaled to arbitrarily large numbers of devices in series, so there is great flexibility in manipulating linearity and power-handling with respect to control voltage. Note that if there are a total of N capacitors C_s in series, then the net capacitance of the structure is reduced to C_s/N, so as the stacking increases, the unit cell capacitance must also increase. In order to maximize the Q and current-handling, it is advantageous to combine the parallel device concept described in Figure 4-32 with the series stacking described by Figure 4-33. An example of such a device is shown in Figure 4-34.

There are at least two practical challenges with the use of stacked capacitors; first, the need for a number of large value resistors, which can occupy significant substrate area unless a high

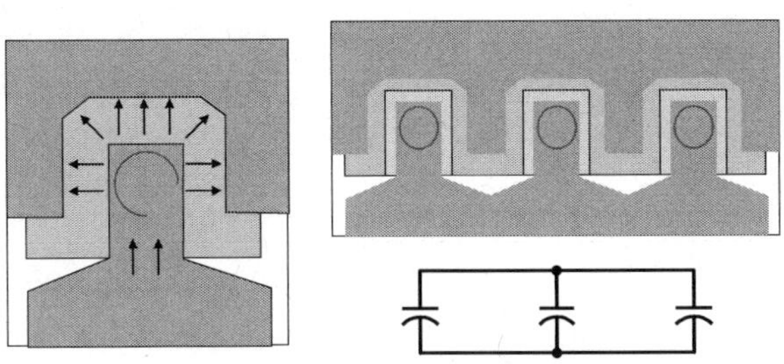

Figure 4-32 Current flow through top and bottom electrodes on basic device, and illustration of parallel-finger designs to optimize Q, tunability, and current handling [249].

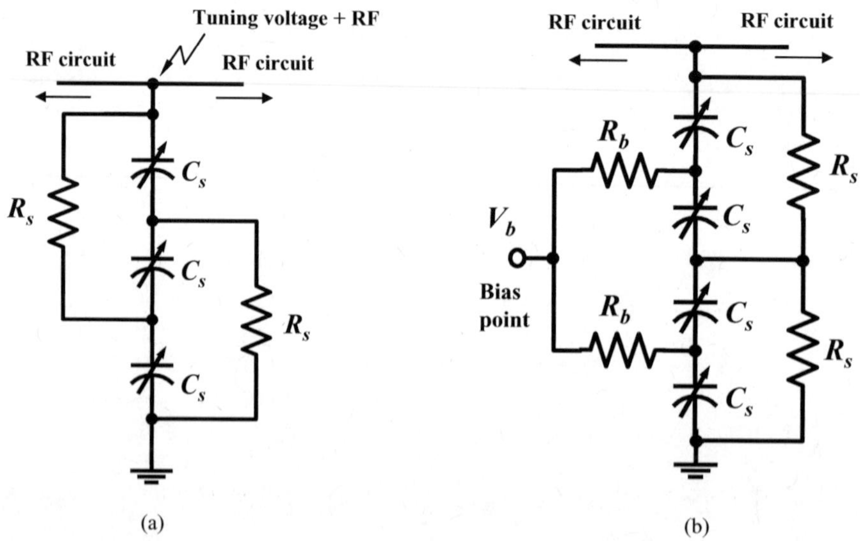

Figure 4-33 Examples of vertical device stacking to enhance power handling and linearity with low control voltage. (a) Two-terminal structure with bias and RF superimposed at one terminal; and (b) three-terminal arrangement with separate bias and RF circuits.

sheet-resistance process is available. Secondly, as the number of devices grows, the settling time with respect to a step-function control voltage increases, because the inner devices in the stack must be charged through an increasing number of resistors. This tends to favor smaller resistors, but small resistors decrease the overall Q and increase the leakage currents, so there is an inherent tradeoff between Q and settling time. This can be partially addressed using alternative biasing schemes that will not be discussed here.

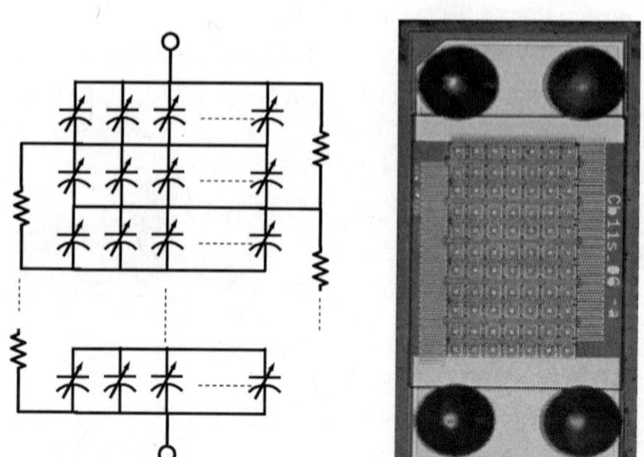

Figure 4-34 A high-linearity varactor design using stacking and parallelism for high power handling and low control voltage (photo courtesy of Agile Materials and Technologies Inc.).

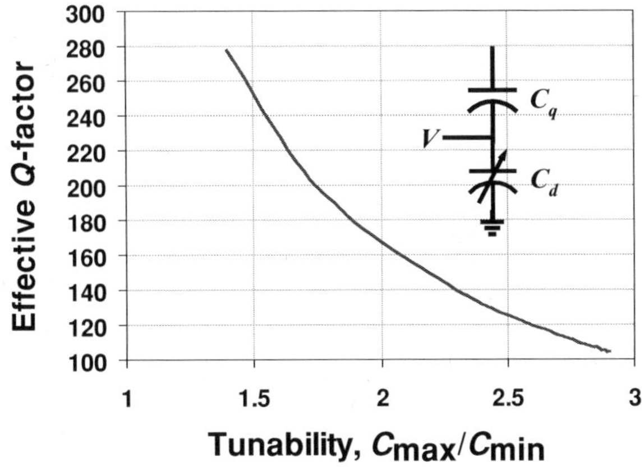

Figure 4-35 Effective Q-factor versus tunability for the series combination of a high-Q nontunable capacitor and a low-Q tunable capacitor ($Q_d = 100,\ Q_q = 500,\ \tau_d = 3,\ C = 1\text{pF}$).

4.3.6 *Q* Enhancement

Tunability and material Q are fundamentally linked, such that the choice of material always involves a compromise between the two factors. But occasionally it is more effective to use circuit techniques to manipulate this tradeoff. An example is a series combination of high-Q nontunable capacitor with a low-Q tunable capacitor. Many thin-film processes include some kind of low-κ dielectric layer that can be used for a high-Q fixed capacitor, making this a very practical method for enhancing the overall Q at the expense of some tunability.

In the following the high-Q nontunable capacitor is represented as C_q with a Q of Q_q, in series with a BST capacitor C_d of Q Q_d. If the BST capacitor has a tunability of τ_d over some control voltage range, then we can show that the overall Q and tunability are of the series combination is given by

$$Q = \frac{Q_d Q_q (C_q + C_d)}{C_q Q_q + C_d Q_d} \qquad \tau = \frac{C_d + \tau_d C_q}{C_d + C_q}. \tag{4.35}$$

The effective Q versus tunability for the series combination is shown in Figure 4-35, assuming a BST tunability of 3:1, a BST Q of 100, a nontunable device Q of 500, and an overall net capacitance of 1 pF. We can see that this technique can be effective for Q enhancement in situations where Q is at a higher premium than tunability. It should be noted that similar ideas have been used to enhance Q using layered materials rather than distinct capacitor structures [224]; recently, Yan et al. [275] have also demonstrated this concept with a layered structure of BZN and BST dielectrics.

4.4 PHASE-SHIFTERS AND DELAY LINES

Early interest in thin-film BST focused on the potential use of this technology in phase-shifters for low-cost phased-array antennas. Compared with conventional MMIC processing, BST varactor circuits have a potential for fewer processing steps, cheaper substrates, and larger lithographic dimensions, all factors that reduce costs. In addition, BST-based phase-shifters provide continuous analog phase control with a single control voltage, negligible control power dissipation, and fast response times. Such analog functionality is especially useful in beam-scanning applications for improving beam-pointing accuracy and compensating for temperature or aging drift. BST-based phase-shifters circuits do not require hermetic packaging and are compatible with flip-chip manufacturing techniques. Collectively these features make BST phase-shifters an attractive alternative to competing MMIC or MEMs technologies.

Key requirements for a viable BST phase-shifter technology are low insertion loss, wide bandwidth, and small die size. This section compares some possible design approaches with respect to these criteria.

4.4.1 Periodically Loaded Transmission-Lines

Traveling-wave phase-shifters were among the first types of structures to be examined using BST materials [276–283]. Here a transmission line is loaded with BST material in such a way that the phase velocity is dependent on the permittivity of the material, and hence the propagation delay along the structure is controlled by the DC applied field to the material. Early circuits typically used a CPW or microstrip arrangement, where the BST material continuously loads the transmission line. This creates a difficult impedance matching problem because of the high permittivity of the material.

A more attractive solution is shown in Figure 4-36, where discrete BST varactors periodically load a transmission line (a CPW line is shown). This was first demonstrated using Schottky diode varactors [282]. A design optimization procedure is described in [282] and summarized here. It is helpful to define a "loading factor" x, which is the ratio of the loading (varactor) capacitance to the unit-cell capacitance of the unloaded line:

$$x \equiv \frac{C_d(V)/\ell_{\text{cell}}}{C'}. \tag{4.36}$$

The impedance and phase velocity of this structure are then given by

$$Z_L = \frac{Z_i}{\sqrt{1+x}} \qquad v_p = v_i\sqrt{1+x} \tag{4.37}$$

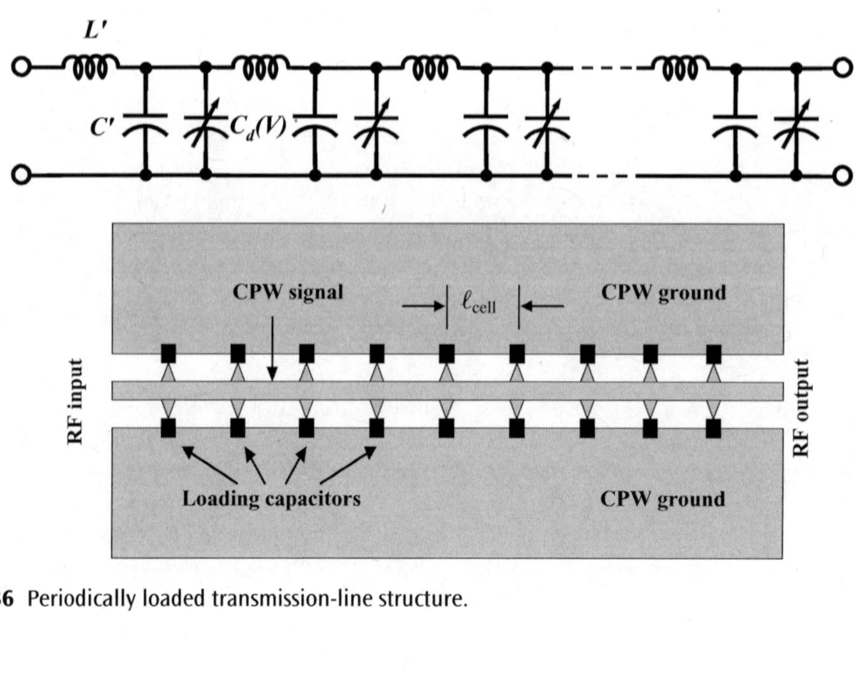

Figure 4-36 Periodically loaded transmission-line structure.

where Z_i and v_i are the impedance and velocity of the unloaded transmission line, which can be related to L' and C', the inductance and capacitance per unit length as

$$v_i = \sqrt{L'C'} \qquad Z_i = \sqrt{\frac{L'}{C'}}. \tag{4.38}$$

The periodic nature of the structure introduces a cutoff frequency called the Bragg frequency:

$$f_{\text{Bragg}} = \frac{v_i}{\pi \ell_{\text{cell}} \sqrt{1+x}}. \tag{4.39}$$

In writing Equation (4.37) we basically assumed that the loading capacitance is spread uniformly over the unit cell, so these expressions only work well below the Bragg frequency. In this case, the maximum differential phase-shift per unit cell is given by

$$\Delta\phi = \omega \frac{\ell_{\text{cell}}}{v_i} \left(\sqrt{1+x} - \sqrt{1+x/\tau} \right) \tag{4.40}$$

and the total loss per unit cell can be written as

$$\alpha \ell_{\text{cell}} = \pi Z_L C_d^{\max} \left(\frac{f}{Q_f} + \frac{f^2}{f_c} \right) + \alpha_i \ell_{\text{cell}} \frac{Z_i}{Z_L} \tag{4.41}$$

where α_i is the attenuation factor of the unloaded line. Thus a device cutoff frequency f_c due to ohmic loss in the BST varactor can be defined as

$$f_c = \frac{1}{2\pi R_s C_d^{\max}}. \tag{4.42}$$

Combining Equation (4.40) and Equation (4.42) to give a figure-of-merit (FOM) as

$$\text{FOM} = \frac{2\pi f \left(\sqrt{1+x} - \sqrt{1+\frac{x}{\tau}} \right) /8.686}{\frac{\pi x}{\sqrt{1+x}} \left(\frac{f}{Q_{\text{film}}} + \frac{f^2}{f_c} \right) + \alpha_i v_i \sqrt{1+x}} \quad [^\circ/\text{dB}] \tag{4.43}$$

and the overall length of the structure that is required to achieve a 360 degree phase delay at some frequency f_0 is given by

$$L_{\text{total}} = \frac{v_i / f_0}{\sqrt{1+x} - \sqrt{1+\frac{x}{y}}}. \tag{4.44}$$

For a given set of design variables the FOM increases with loading factor, and the total length decreases with loading factor. The design frequency is chosen as the frequency which maximizes the FOM. In order to maintain a good impedance match, the loaded impedance must remain close to 50 Ω over all bias states. For large loading factors, this requires that the unloaded line impedance be quite large. In addition, the return loss begins to increase dramatically near the Bragg frequency, so the operating frequency must be chosen well below the Bragg frequency, usually around 25–50% lower.

Early designs using coplanar waveguide [283–289] were limited to small loading factors because of the limited range of CPW impedances that can be realized with low loss. Some impressive FOM

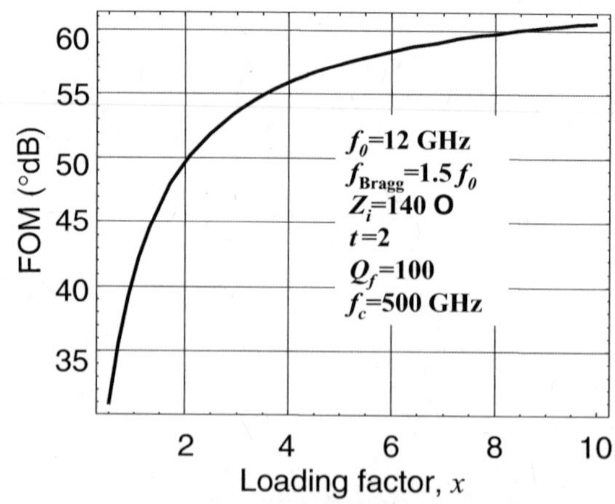

(a)

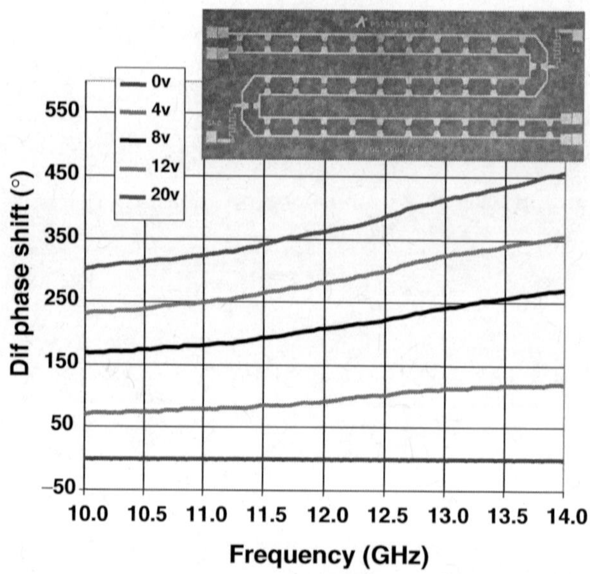

(b)

Figure 4-37 Phase-shifter figure of merit (FOM): (a) FOM calculation versus loading factor; and (b) measured differential phase result for a loaded CPS delay line (courtesy of Agile Materials and Technologies Inc.).

results (approaching 90°/dB) were achieved [288, 289], but these structures were quite large. A solution to this problem is to use coplanar-stripline (CPS), the dual of CPW. CPS can be realized with quite large unloaded impedances and with relatively low attenuation, allowing for high loading factors and hence more compact designs. Figure 4-37 shows the FOM calculation and a measured result for a 12 GHz CPS-based design. This device measured 1 × 3 mm overall, considerably smaller than CPW designs. Similar structures were used recently in a 24 GHz wafer-scale phased-array system [290].

For high tunability and/or high loading factors, the loaded impedance of the phase-shifter can vary significantly. A simple partial solution to this problem involves using smaller varactors in the first and last sections, effectively providing a small amount of impedance transformation. Another approach using a different unit cell design is described in [291].

4.4.2 LC Ladder Structures

The wide bandwidth and parameter-insensitive design make distributed delay-lines attractive, but there are two serious drawbacks: first, the designs are physically large, increasing the unit cost and complicating array designs; secondly, both the CPW and CPS structures require off-chip baluns for proper functioning in the microstrip environment common for a host circuit.

One way to reduce the size of the structure is to implement the transmission line as a synthetic LC ladder network, using lumped-element inductors for the series element. Again, this technique has been implemented using Schottky varactors [292, 293] as well as BST varactors [294, 295]. Figure 4-38 shows two small BST-based phase-shifter structures using this LC ladder approach. These designs yielded approximately 90° phase shift with ~2 dB of loss at the design frequency of 10 GHz, corresponding to an FOM of approximately 45°/dB, with excellent return loss over the bias states as shown. Similar structures were used recently in an adaptive amplifier linearization network, reported in [296, 297].

The principal drawback of using spiral inductors is their increased loss compared to transmission lines, but optimization with electromagnetic field solvers can be carried out to find the inductor geometry with the lowest series resistance. Note that the optimization procedure for the overall FOM is somewhat different here than for the truly distributed structure, because the assumption of constant line loss regardless of Bragg frequency is no longer true. A lower f_{Bragg} design using a smaller number of higher valued inductors may incur more loss than a design using a larger number of smaller inductors.

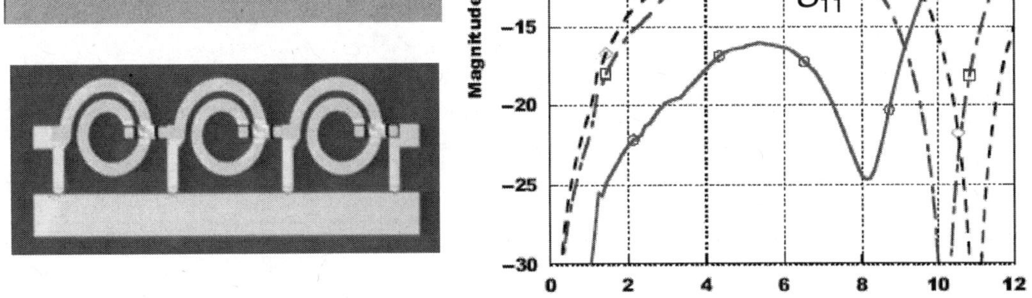

Figure 4-38 Two small variable delay-line structures using an LC ladder network, and representative S-parameters at various bias states (after [295]).

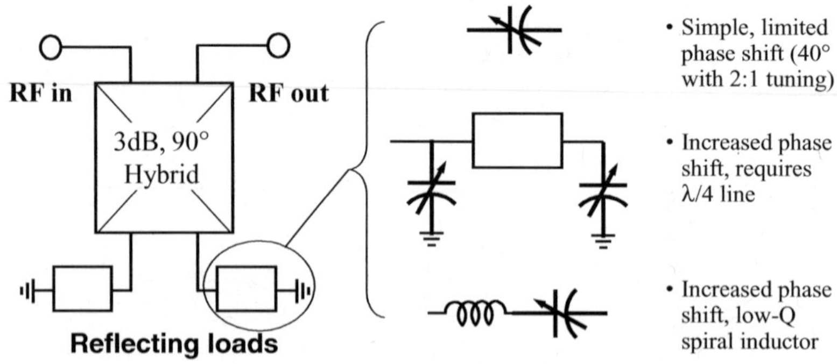

Figure 4-39 Reflection phase-shifter concept [298] and some possible reflecting loads.

4.4.3 Reflection Phase-Shifters

There are many other types of phase-shifters that have been developed to exploit varactors [298]. One example is a so-called reflection phase-shifter, shown in Figure 4-39 [299, 300]. In this circuit, a quadrature hybrid circuit is used with two of the ports terminated in reflecting loads. The hybrid splits the input signal, sending it to the reflecting loads, and the reflected signals then add coherently at the output port. The phase is varied by implementing the reflecting loads as variable reactance networks, some examples of which are shown.

Series LC resonators with spiral inductors are attractive for compact circuits. In this case, phase-shifters can be made with far fewer varactors than is needed for the delay-line structures described earlier. For a given varactor value C_v and tunability τ, it can be shown that the phase-shift and the required inductor value is given by

$$\Delta\phi = 4\tan^{-1}\left[\frac{1}{2\omega_0 Z_0 C_{v0}}\left(\sqrt{\tau}-\frac{1}{\sqrt{\tau}}\right)\right]$$
$$L = \frac{1}{2\omega_0^2 C_{v0}}\left(\sqrt{\tau}+\frac{1}{\sqrt{\tau}}\right) \tag{4.45}$$

where $C_{v0} = C_{\max}/\sqrt{\tau}$ is the geometric average of the maximum and minimum varactor capacitance, and Z_0 is the characteristic impedance looking into the ports of the branch line coupler at the location of the reflecting loads. Large phase-shifts can theoretically be realized given a large enough tunability, but in practice the inductor values can become too large, so phase-shifts from 90–200° are realistic for simple resonators.

The quadrature hybrid can be implemented in a number of ways. Using a Lange coupler [301] is one common method. Figure 4-40 shows a reflection phase-shifter using a Lange coupler and BST varactors [294, 302]. This particular network achieved a maximum of 120° of phase with a little over 3 dB of insertion loss at around 11 GHz, corresponding to a FOM of approximately 40°/dB. The return loss was excellent over the entire 8-12 GHz range. The circuit occupied a die area of 3.3 mm × 0.65 mm, dominated of course by the Lange coupler which must be on the order of a quarter-wavelength long. Similar designs have also been reported using inter-digital BST varactors [303].

A branch line coupler is another possibility for implementing a quadrature hybrid, but this is even larger than the Lange coupler so it is not attractive for compact monolithic circuits. An interesting alternative is to implement the hybrid using lumped-elements [304, 305]. A diode

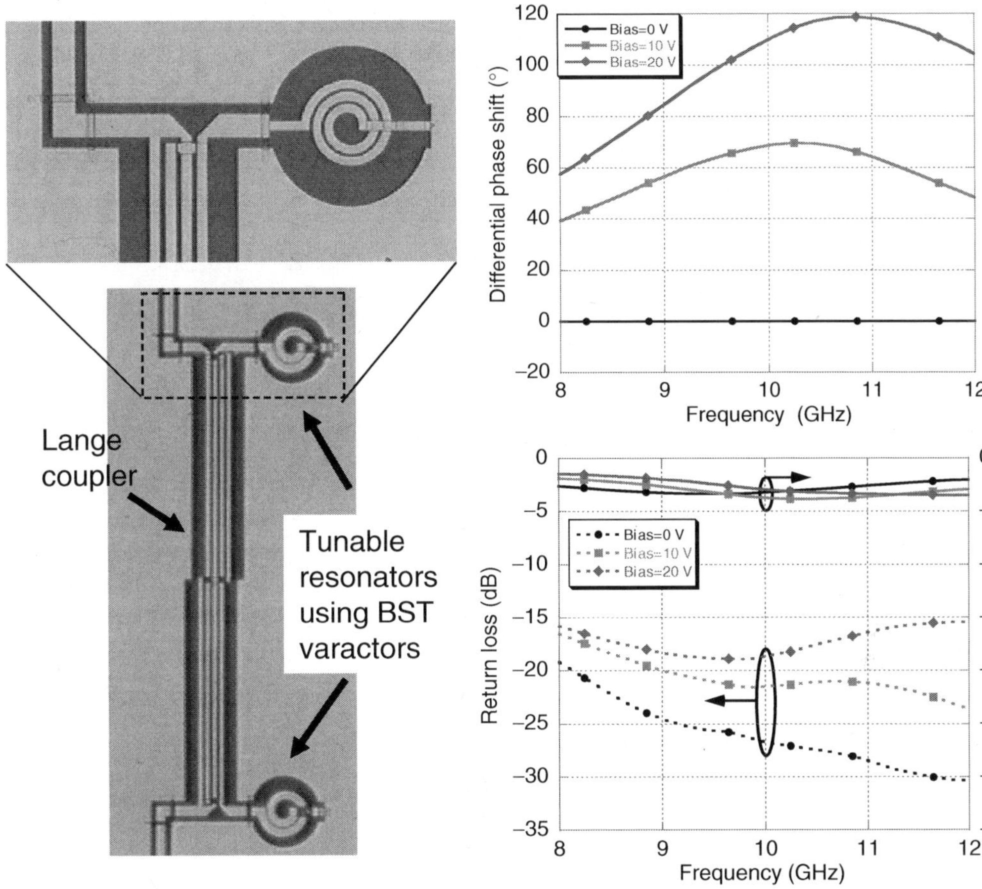

Figure 4-40 A reflection phase-shifter using a Lange coupler and series resonators [307].

varactor implementation using a lumped-element hybrid was described in [306], demonstrating that extremely compact designs could be realized with this approach. As a further refinement, note from the design equations, Equation (4.45), that the phase-shift scales inversely with the characteristic impedance of the ports; if an impedance-transforming hybrid design is used [305], the load ports can be designed for low impedance while maintaining a good impedance match at the RF input and output. This has the further advantage that smaller inductor values can be used, which in turn makes the circuit smaller and reduces some of the spiral parasitic capacitances that conspire to reducing the phase shift. This technique was adopted in [302, 307], and resulted in an extremely compact phase-shifter, shown in Figure 4-40. This device measured 0.6 mm×0.6 mm, and produced over 200° of phase shift.

A drawback of the lumped-element hybrid implementation in comparison to the Lange coupler example of Figure 4-40 is a reduced bandwidth and higher loss (lower figure-of-merit). However, the small size may make this a more attractive approach in some applications. In both cases, the reduced number of varactors needed in comparison to the distributed delay-line architectures may be attractive from a yield and reliability standpoint.

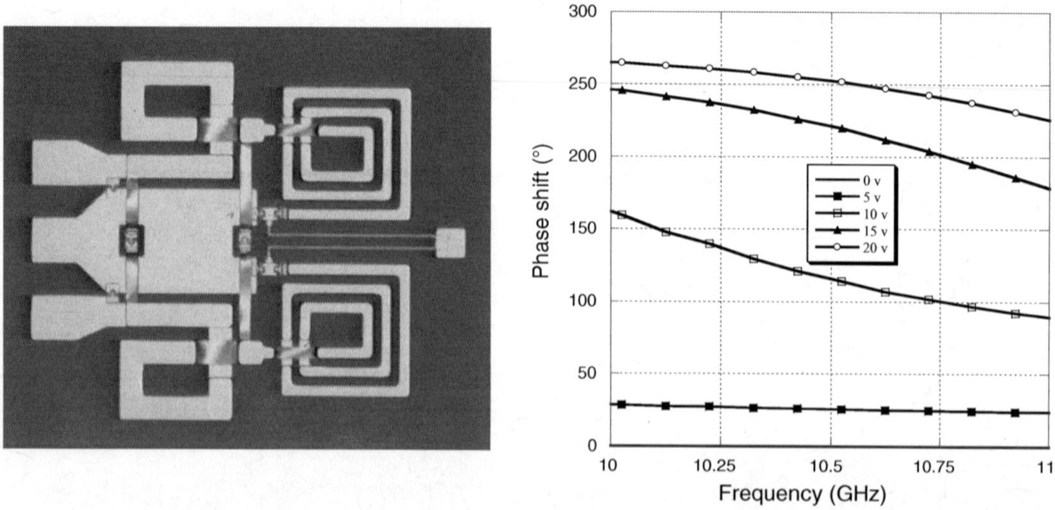

Figure 4-41 A very compact reflection phase-shifter using BST varactors and an impedance-transforming lumped-element hybrid design [302, 307]. This circuit measured 0.6 mm × 0.6 mm.

4.4.4 All-Pass Networks

The last kind of lumped-element phase-shifter considered here is based on 2 nd-order all-pass networks as shown in Figure 4-43. If we note that the LC ladder network described earlier is basically a low-pass filter, the Bragg-frequency (or cutoff frequency) is the source of some trouble, leading to a strong increase in insertion loss and limiting the amount of phase-shift that can be achieved from each unit cell. The all-pass network solves this problem by adding a bridging element to counteract this increase in loss. If designed properly, this network can give a flat amplitude response over all frequencies, with a greater phase variation than either a simple low-pass or high-pass network. Lumped-element values that achieve this are shown in the figure for the ideal case, where R is the desired impedance level (usually 50 Ω) and ω_0 is the center-frequency at which the maximum phase variation occurs.

The all-pass network has historically found use in phase equalization networks, and has also been used in varactor-based phase-shifters using diodes [308], GaAs FETs [309, 310], and more recently with tunable dielectric materials [311, 312]. An example of a simple bridged high-pass structure is shown in Figure 4-43. Nonidealities in the circuit, principally losses in the varactor

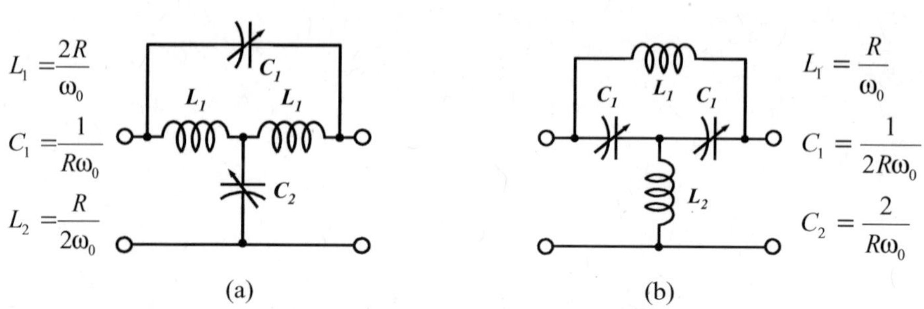

Figure 4-42 Bridged–tee circuits and design equations for realizing Second-order all–pass transfer functions: (a) bridged lowpass tee; and (b) bridged highpass tee.

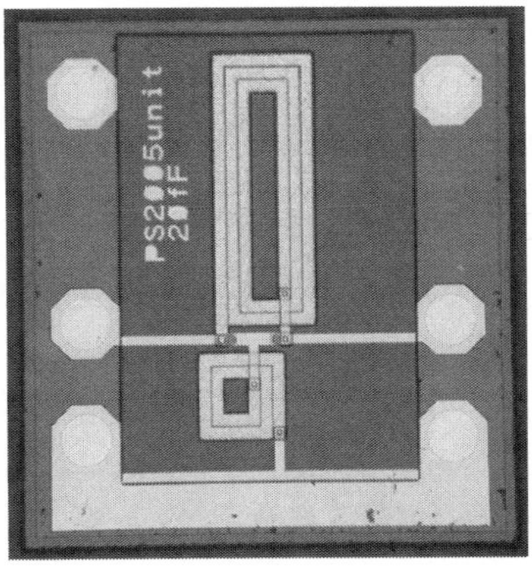

(a)

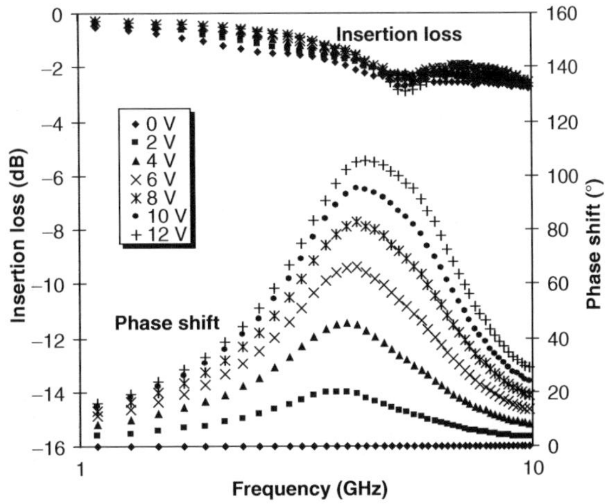

(b)

Figure 4-43 Simple all-pass phase-shifter structure (0.25 mm^2) and measured characteristics (courtesy of Agile Materials and Technologies Inc.).

and spiral inductors, lead to a slight dip in the insertion loss near the center frequency (around 5 GHz in this example), but the small variation in insertion loss over the bias states is an improvement compared with phase-shifters discussed earlier.

Like the lumped-element reflection phase-shifter this is a narrowband design, but the all-pass structure is very attractive in terms of giving a large phase-shift for a given size and varactor tunability, is extremely easy to design, and uses a minimum number of varactors.

4.5 TUNABLE FILTERS AND MATCHING NETWORKS

4.5.1 Tunable Bandpass Filters

The possibility of using BST varactors in tunable filters has been recognized for quite some time, but this application space has not been as well developed or investigated as phase-shifters, although it could be argued that many of the phase-shifter structures discussed in the previous section are essentially tunable filters themselves. The reason is two-fold; first, high-quality filters—that is, high-order reactive circuits with sharp passband skirts and large out-of-band rejection—are very difficult to design and implement, being quite sensitive to parasitic effects in the components. Secondly, most filter applications place a very high premium on insertion-loss in the pass-band, and the technology for BST varactors was simply not mature enough to satisfy the difficult requirements on Q and tunability.

Nevertheless, some good progress has been made in this area. Some of the first efforts to demonstrate tunable bandpass filters with BST technology were reported by researchers active in the high-temperature superconductor field, and typically used structures involving transmission-line resonators, usually with high-T_c superconducting electrodes but with some reports of good performance at room temperature with normal conductors [181, 313–316]. More recent circuit demonstrations [317–320] have focused on filters using integrated thin-film varactors and have continued to improve on the early results.

High-quality filters require high-Q resonator structures, and this involves some direct and indirect design challenges. First, even if the BST varactor Q is high, it must always be combined with an inductance of some kind (lumped or distributed), lowering the overall Q. For on-chip inductors such as spirals, the Q are usually limited to 30–40 in the gigahertz range; this then becomes the dominant loss in the resonator and filter structure. Distributed (transmission-line) resonators have higher Q, but not usually in excess of 100 due to a combination of ohmic loss, parasitic substrate modes, and radiation (some completely shielded resonators have Q approaching 200 but are large). Transmission-line structures also become prohibitively large below a few gigahertz. A practical solution in this frequency range is to use discrete off-chip air-core inductors with Q in the range of 100–200.

A second issue, sometimes overlooked, is that high-Q resonators tend to amplify the RF voltage across the varactor (or current depending on whether it is a series or parallel resonator), thus raising significant power-handling and linearity concerns. These can be addressed to some extent by the ideas developed in earlier sections, but generally involve tradeoffs with control voltage and Q, among other considerations. Lastly, since only the capacitors are tuned and not the inductors, it becomes difficult to maintain a desirable passband shape and out-of-band rejection as the device is tuned, and careful optimization of the circuit is usually required.

With careful attention to detail in the design of the structures, some impressive results can been obtained [320]. Figure 4-44 shows an example of a lumped-element IF filter using air-core inductors and BST varactors. Within each separate band the frequency tuning is accomplished with an analog voltage control to the varactors. In order for the filter to operate over multiple IF bands, three different varactors sizes are used, which are connected to the RF circuit using a simple switching matrix. An advantage of off-chip inductors is that they can be mechanically adjusted to optimize the response shape after assembly. Using tapped inductor matching at the input and output of the circuit also allows for the resonator impedance to be optimized somewhat independently of the external matching considerations.

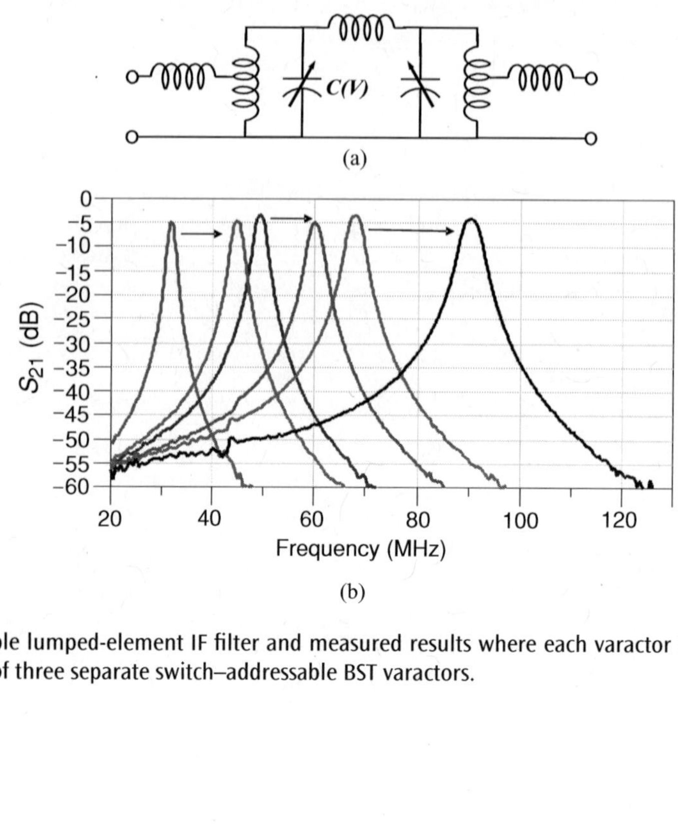

Figure 4-44 A simple lumped-element IF filter and measured results where each varactor is implemented as a bank of three separate switch–addressable BST varactors.

4.5.2 Impedance Matching Networks

The use of BST varactors in impedance-matching networks is a more recent suggestion that may prove useful in RF front-ends for optimizing amplifier and antenna efficiency. In modern wireless systems the efficiency of the RF power-amplifier has a strong influence on battery-life. Ordinarily the amplifier is designed to operate with maximum efficiency at its maximum power level, but in a typical wireless system the amplifier is more commonly operating at a much lower power and efficiency. The solution is to use a dynamic load-line that can be adjusted to maintain high power-added efficiency at different power levels, thus keeping the current draw from the battery to a minimum.

Figure 4-45 is a simple impedance transforming network that illustrates the concept [321]. The circuit uses two BST varactors and an off-chip inductor. The measured data is taken using a 50 Ω load, and shows that the input impedance is can be varied from 13–29 Ω using a 2:1 change in the capacitance, and remains on or near the real axis over the measurement frequency range.

Figure 4-46 shows a similar impedance matching network, designed for matching to a small helical antenna in the 420–490 MHz range [321]. The problem being addressed in this example is two-fold: first, it is ordinarily difficult to create an efficient match to an electrically-small antenna over a wide frequency range, due to the large reactance and small radiation resistance of the structure and fundamental Bode–Fano limitations [121]. So, using a tunable matching network allows for a less complicated and more efficient narrow-band matching network to be used, which can then be tuned to operate at various channels within the band of interest. The measured response for various combinations of tuning voltages is shown in Figure 4-46, and indicates that a good (< -10 dB) match can be achieved over the range of 420–490 MHz as desired.

A second problem that is addressed with a tunable antenna matching network is the potential variation in the antenna impedance as its operating environment is changed. Figure 4-47 shows how the impedance of a simple helical antenna can vary as it is moved from the head to a table (see Smith Chart in Figure 4-47). The accompanying data shows how the measured return loss can vary if the impedance match is fixed, and how it can be subsequently tuned to improve the match.

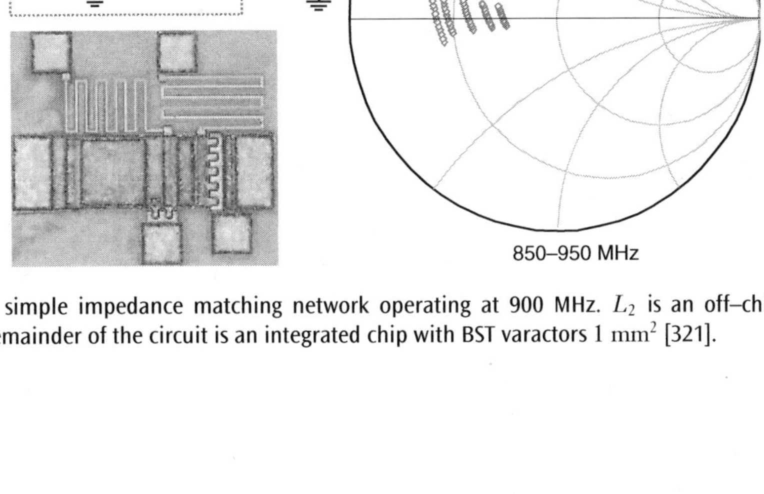

Figure 4-45 A simple impedance matching network operating at 900 MHz. L_2 is an off–chip inductor, the remainder of the circuit is an integrated chip with BST varactors 1 mm^2 [321].

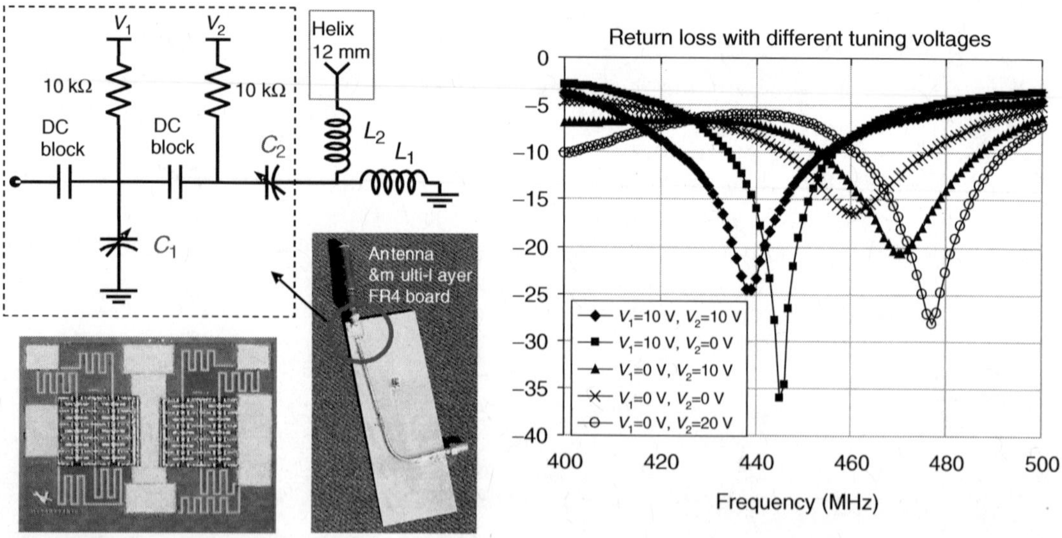

Figure 4-46 An impedance-matching network for antenna tuning (chip size $0.5\ \text{mm}^2$) [321].

The ability to compensate for antenna mismatch could significantly enhance the performance of any RF front-end.

4.6 HETEROGENEOUS INTEGRATION

4.6.1 Integration with Semiconductor Electronics

Some interesting and potentially useful circuits have been demonstrated with thin-film BST technology, but the potential application space could be even greater if the films are successfully integrated with other technologies. Some significant progress has already been made in this direction. We have noted earlier that some of the initial motivation for developing thin-film BST was

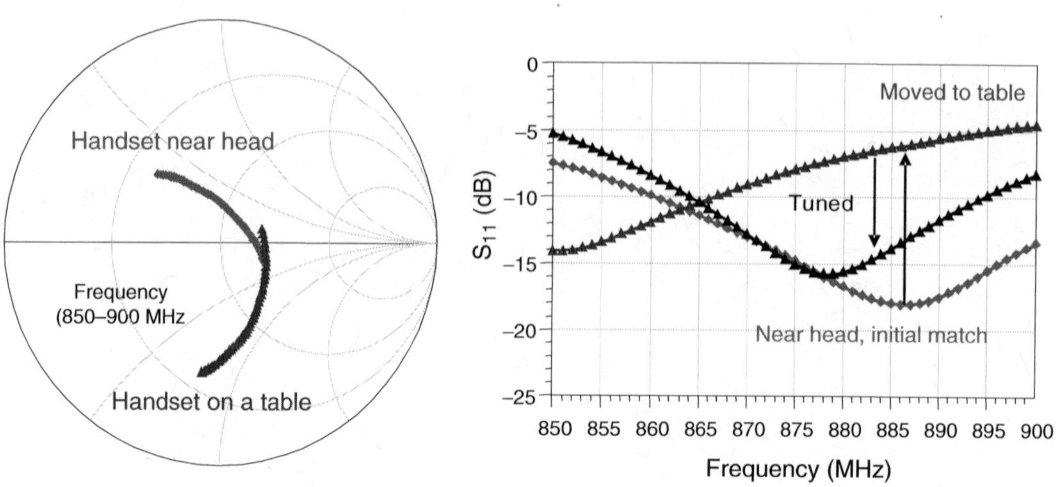

Figure 4-47 Use of the adaptive matching to compensate for changes in antenna impedance, using the circuit shown in Figure 4-46 [321].

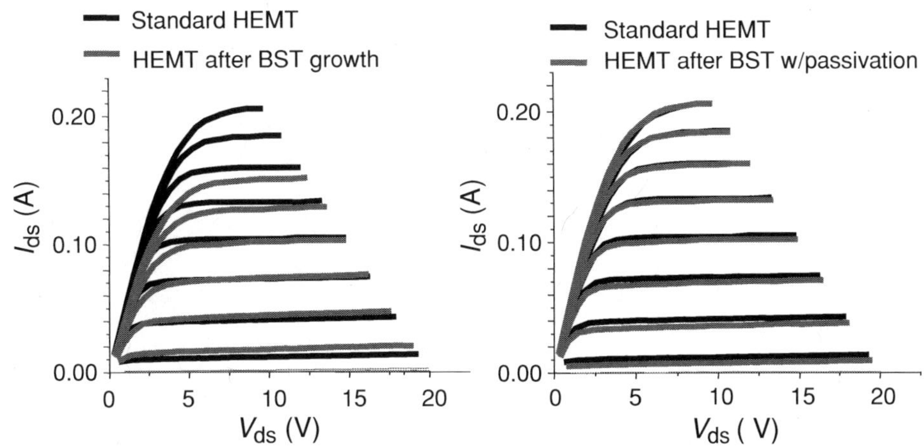

Figure 4-48 GaN HEMT I–V curves showing the importance of the surface encapsulation/passivation step prior to BST growth by magnetron sputtering.

in connection with the silicon CMOS and DRAM industry [172, 173], and processes were successfully developed there to integrate BST and STO films into Si integrated circuits as gate-oxides and DRAM capacitors [121, 322–324].

One of the key integration challenges is the high processing temperature of BST and other related ceramic materials. Ordinarily the films are deposited and/or annealed at temperatures in the range of 600–700°C or higher, so the process must be sequenced in such a way that any temperature-sensitive steps are done after the ferroelectric deposition. Another challenge is that the tunable dielectrics are often grown in a highly-oxidizing environment, which is necessary to minimize oxygen vacancies in the material. This oxidizing environment can adversely affect exposed semiconductor surfaces, so some kind of encapsulation layer is required to protect the epitaxial semiconductor layers. A process for integrating BST films with GaAs electronics has been reported [189] which addresses both of these concerns by first encapsulating the GaAs surface with silicon nitride, depositing and annealing the BST film, and finally removing the SiN encapsulant to fabricate the GaAs transistors and BST capacitors. Several circuits have been reported using this or a similar process [187, 188], and it is believed that a substantial number of chip-sets have been sold that exploit this technology.

A similar process was recently developed for the integration of BST films with gallium nitride (GaN) electronics [325]. In this case a sacrificial SiO_2 layer was used to protect the GaN epitaxial layers during BST growth. Figure 4-49 shows the measured I-V curves of device fabricated with and without this encapsulation layer, showing the importance of protecting the active layers during BST growth. Some progress has also been made in using BST as a gate oxide in GaN electronics to reduce gate leakage [326]. Figure 4-49 shows an example of an RF circuit that combines a GaN transistor and BST capacitor. In this particular circuit, the BST capacitor is only used as an AC bypass for bias decoupling, but results in a substantial size reduction (a SiN capacitor of similar capacity would occupy an area as large as the entire circuit of Figure 4-49).

4.6.2 Integration with MEMS

A final simple example of heterogeneous integration is combining high-κ materials like BST with RF microelectroMechanical systems (MEMS). This is potentially interesting for two reasons.

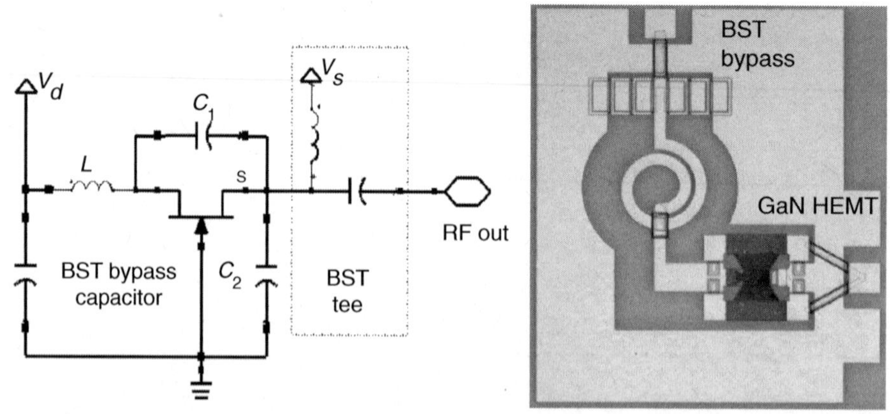

Figure 4-49 Example of an oscillator circuit using heterogeneous GaN/BST integration [327].

First, since BST is inherently an analog (continuously-variable) varactor technology, there are certain circuit functions that can not be implemented easily using BST varactors, and hence there are a limited number of circuits that can be created using BST varactors alone. A low-loss and high-isolation switch is one example, something for which MEMS is particularly well-suited. Thus the combination of a MEMS switch and a BST varactor opens up a large number of new potential applications. Secondly, the MEMS devices themselves might benefit from the availability of a high-κ material like BST. For example, in the simple electrostatic switch shown in Figure 4-50, the dielectric layer plays an important role in determining the on-off capacitance ratio according to

$$\frac{C_{\text{on}}}{C_{\text{off}}} = 1 + \varepsilon_r \frac{g}{d} \tag{4.46}$$

where d is the dielectric thickness. It can be seen that a high permittivity film can dramatically increase the on-off ratio and hence the performance of the device

This performance enhancement has been demonstrated in [328] and also recently in [329]. Figure 4-51 shows the improvement in switch isolation using a BST film in contrast to a more conventional SiN-based structure. There are some other technological issues such as dielectric breakdown strength that make the use of BST in this application somewhat challenging, but the example suggests some of the potential merits of marrying BST and MEMS technologies.

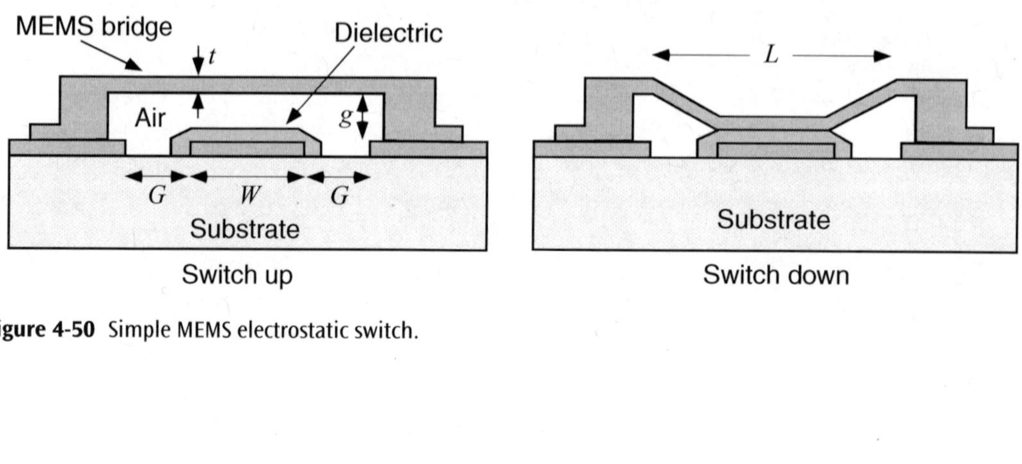

Figure 4-50 Simple MEMS electrostatic switch.

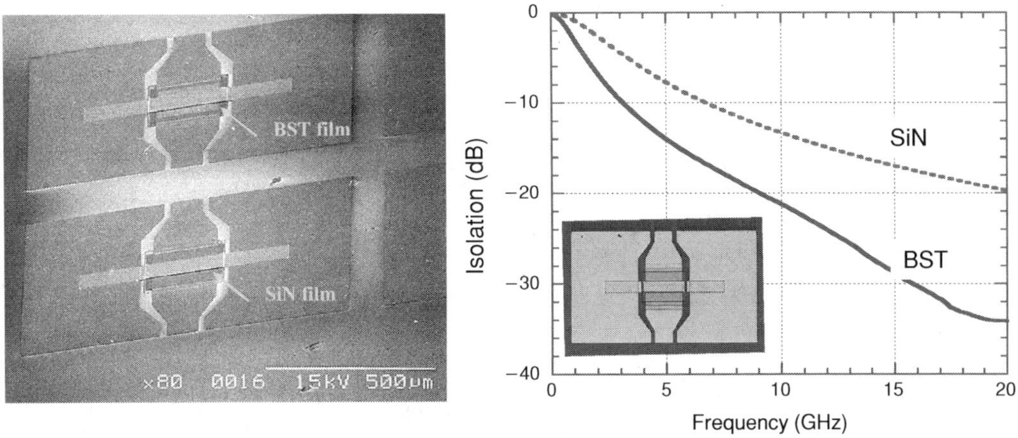

Figure 4-51 Simple MEMS electrostatic switch and improvement in down–state isolation using BST films [328].

4.7 CONCLUSION

Tunable dielectric materials appear to have a strong potential for future wireless systems. The technology is cost-effective, compact, and with suitable device design can have excellent power-handling and linearity at low control voltages.

Presently the materials are reasonably well understood from a modeling and technology standpoint, and a variety of circuit demonstrations have been made. Phase-shifters, tunable matching networks, filters, phased-arrays, VCOs [330], switches [331], and a number of other proof-of-principle circuits have been reported, some of which have been reviewed here.

Much of the work presented in this chapter was based in whole or in part on work carried out at the University of California at Santa Barbara (UCSB) over the course of several years, and with the help of many students and collaborators. The author wishes to acknowledge several former students that were responsible for the development of this technology: Dr. Amit Nagra and Dr. Baki Acikel for their work in process development and distributed phase-shifter circuits; Dr. Troy Taylor and Dr. Peter Hansen for material growth and characterization; Dr. Yu Liu for MEMs integration; Justin Serraicco for work in compact phase-shifters; Dr. Hongtao Xu for RF device optimization and integration with GaN; and Dr. Nadia Pervez for work in the understanding and modeling of losses in BST capacitors. The author also acknowledges the contributions of Dr. James Speck and Dr. Susanne Stemmer in the Materials Dept. at UCSB, and many fruitful discussions with Dr. Stephen Streiffer at Argonne National Laboratories. Lastly, the authors would like to thank Dr. Chris Elsass, Dr. Vicki Chen, and Roger Forse at Agile Materials and Technologies Inc. in Goleta, CA for providing some of the materials and data used here.

In addition to the MARRS MURI support, the work reported in this chapter was also supported by DARPA through the FAME (Frequency Agile Materials for Electronics) program, contract DABT63-98-1-0006; DMEA through the Center for Nanoscience Innovation for Defense (CNID), contract DMEA90-02-2-0215; and the Office of Naval Research through the Center for Advanced Nitride Electronics (CANE MURI), contract N00014-01-1-0764.

5

Adaptive Circuit Approaches for Microwave Transmitters

Peter M. Asbeck, Yu Zhao, Dongjiang Qiao, Mingyuan Li, Jeremy Rode and Tomás O'Sullivan

5.1 MOTIVATION FOR ADAPTIVE TRANSMITTER CIRCUITS

As wireless communication systems evolve toward greater sophistication, their RF sections are faced with increasingly challenging requirements. These typically include handling multiple carrier frequencies, widely varying RF power levels, differing modulation formats and linearity specifications, and varying impedance provided by the antenna. These challenges are particularly severe in mobile units where small form factor and high efficiency are needed. This chapter discusses approaches to meet these challenges for the RF transmitter section, chiefly the power amplifier. The ability to integrate very large numbers of transistors together is enabling progressively smarter transmitters and amplifiers. The system components needed for adaptation include, however, not only digital transistors, but also microwave components that can provide switching or impedance variation without introducing nonlinearity or dissipation; and often devices for measuring the electromagnetic characteristics of the "environment". This chapter summarizes the motivations for adaptive power amplifiers, the potential benefits they can provide, and their general architecture; and it provides various examples of transmitter systems and subsystems currently under development whose characteristics can be adapted. In a number of the examples provided here, the focus has been on application to commercial wireless systems, such as CDMA, to provide a concrete example with real system requirements. However, the techniques described may be applied in a wide range of contexts, and over a wide range of frequencies.

Adaptive amplifiers provide the capability of reconfiguring their characteristics to meet different operational specifications or changing environments. Representative examples of varying specifications include the following:

Variable microwave output frequency. Communication systems have proliferated to occupy numerous spectral bands, characterized by different usable transmission bandwidths and propagation characteristics. Table 5-1 lists some of the more popular personal communication bands for commercial use.

For military applications there is an even greater diversity, dictated by considerations of line-of-site versus non-line-of-sight communications, antenna size, differences in data rates supported, and potential interference sources. In the future it is envisioned that single radios may be capable of operating with a wide variety of frequencies and waveforms [332], for example, within the Joint

Table 5-1 Characteristics of various commercial wireless communication signals.

System	Peak-to-average ratio (dB)	Center frequency (GHz)	Power control dynamic range (dB)	Bandwidth (MHz)	Access type
GSM	0	0.9, 1.8	30	0.2	TDMA
GPRS	0	0.9, 1.8	30	0.2	TDMA
EDGE	3.2	0.9, 1.8	30	0.2	TDMA
CDMA ONE	5.5–12	0.85, 1.8	73	1.25	CDMA
UMTS	3.5–7	1.8	80	5	CDMA
CDMA 2000	4–9	1.8	80	1.25	CDMA
802.11 a/g	8–10	2.4, 5.2, 5.8	25	20	TDMA

Tactical Radio System (JTRS). An even more attractive idea is that of allowing the radio itself to scan the frequency spectrum and select open regions of the spectrum for a particular transmission (cognitive radio) [333]. It is highly desirable to have power amplifiers that can operate on any chosen frequency band. However, as dictated by Fano's Theorem [334] the performance metrics of an amplifier (gain, efficiency or linearity) are invariably better over a narrow band than over a wide bandwidth. For multimode systems with signals transmitted sequentially at different frequencies, it is thus desirable to build amplifiers that can achieve the high performance associated with a narrowband design, and apply this performance at the different frequencies by retuning, rather than using a single wideband amplifier (as shown schematically in Figure 5-1).

Variable output power level. The output power required of a communication transmitter can vary over many orders of magnitude, depending on the proximity of the receiver and the shadow or multipath-fading characteristics of the wireless "channel". It is highly desirable to change the output power of the transmitter to correspond to the varying needs of the link—both to conserve power, and to avoid excessive interference to other users of the spectrum. Figure 5-2 shows a representative probability distribution of the output power required of a CDMA cell phone, detailing how variation over seven orders of magnitude is required. By controlling the emitted power to what is necessary to complete the link, and no more, the overall time-average transmitted power can be reduced by more than an order of magnitude. However, in order to maximize the benefit from this strategy, the power amplifier should must retain high efficiency at the different power levels (which is not the case for conventional Class A or Class AB amplifiers).

Variable modulation formats. As the output signal is changed between CDMA, OFDM, GMSK (used in GSM) and 8PSK (used in EDGE), for example, the linearity required of the amplifier is

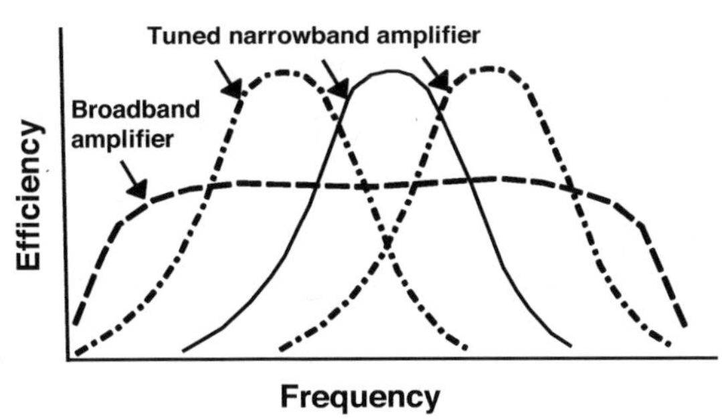

Figure 5-1 Schematic variation of efficiency with frequency for narrowband and broadband matching.

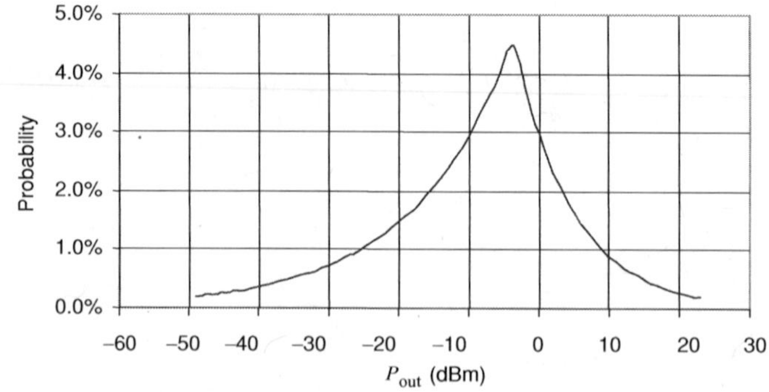

Figure 5-2 Representative probability distribution of output power for CDMA radios in urban environments.

changed. The amplifier should change in order to make the most favorable tradeoff of linearity and efficiency (which can make differences of a factor of two or three in battery life).

Variable antenna impedance. In handheld units the impedance provided by the antenna is seldom the nominal 50 Ω design value. Proximity of metal or even the user's hand or body changes the impedance significantly, as shown by data in Figure 5-3. It is necessary to design for a 4:1 voltage standing wave ratio for power amplifiers in cell phones. Typically, the output power, efficiency and linearity of the amplifier are all degraded in the presence of such large mismatches. A related example occurs in antenna array systems, for which the antenna impedance depends on the beam power and angle. Isolators can be used to minimize the effect on amplifier linearity, but there is an associated penalty in terms of output power, efficiency, size, and cost.

Variable interference sources and constraints. As the electromagnetic spectrum becomes more crowded, the effect of intentional or unintentional interference on wireless links becomes more problematic. It is desirable for receivers to be able to filter out "jammers" that are detected at various frequencies. Similarly, transmitters must filter out their own out-of-band components in order to avoid co-site interference (to other friendly users) or even to avoid self-jamming (which is a real threat in frequency division duplex systems, which transmit at the same time as they receive).

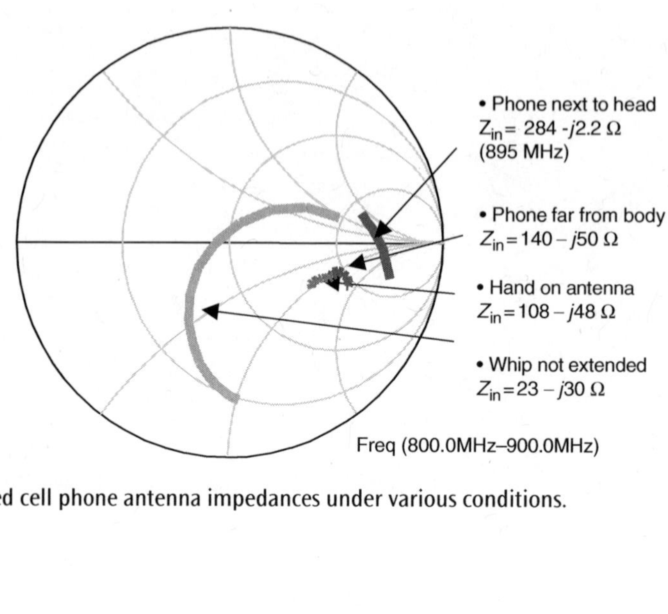

Figure 5-3 Measured cell phone antenna impedances under various conditions.

Leakage of the transmitted signal through the diplexer into the receive channel can cause in-band distortion via the cross-modulation process if there is an interfering signal near the receive band.

The feasibility of amplifiers which can vary their characteristics to meet requirements of this type has been under investigation for over a decade, and already some of the desired adaptations can be carried out. This experience, together with research initiatives such as the MARRS MURI and DARPA's Intelligent RF Front-End program, suggests that there can be further benefits of adaptive amplifiers, including the following:

1) Amplifiers could reconfigure themselves to address different mission requirements (such as switching from communication to radar applications); this can be a major benefit in spacecraft or UAVs where resources are at a premium.
2) Amplifiers could tune themselves after manufacturing, thereby eliminating the cost of manual tuning and relaxing component tolerances as well as making the amplifiers far less susceptible to temperature variations or to aging.
3) Design of the amplifier no longer would have to correspond to operation in the *worst case* scenario; instead, the tuning and bias conditions selected can correspond to *nominal* operating conditions, and the amplifier itself can adjust if the demands at a given moment are greater than usual. This strategy can significantly improve the overall efficiency and linearity of the amplifier.

5.2 ADAPTATION MECHANISMS FOR RF POWER AMPLIFIERS

Power amplifiers typically consist of input impedance matching networks, various amplification stages with associated interstage matching, an output matching network, and possibly, a power combiner. The amplifier is typically connected to the antenna through fixed switches or filters. Circuit components for tuning these networks are challenging to implement since they must have very low loss and handle large voltages and currents. The demands are particularly stringent on the output side of the power amplifier. A variety of device technologies have emerged which approach the required performance; a number of them are shown in Figure 5-4.

MEMS varactors and switches have demonstrated exciting characteristics for this application (with low insertion loss and wide frequency response), and are expected to have a major future impact, as detailed elsewhere in this book. MEMS reliability and packaging have made significant strides, although these have yet to be verified for power levels above 5 W [335–337]. Switches based on FETs or PIN diodes are also used, although to alter the tuning on a fine scale, a large number must be employed. Continuously tunable variable capacitors (varactors) have been made with semiconductor p-n junctions, and more recently with barium strontium titanate thin films [321] which provide capacitance versus voltage relationships as shown in Figure 5-4. For these last two approaches it is necessary to carefully design the circuit environment so that the microwave signal itself does not vary the capacitance also (which would result in a nonlinear response). In addition, circuit techniques may be employed, such as turning on or off various transistors (or portions of the transistors) by control over their bias; varying the impedance levels via purely electronic techniques such as active load pulling as in the Doherty amplifier; and through variation of the power supply voltage. The application of these methods is illustrated in the following sections, which detail specific approaches to adapt transmitters to meet different requirements.

5.3 ADAPTATION TO POWER LEVEL VARIATIONS

The power level needed for transmission in wireless communication systems varies over time, as a result of the signal modulation in many cases (for example, QPSK, OQPSK, Edge, or OFDM

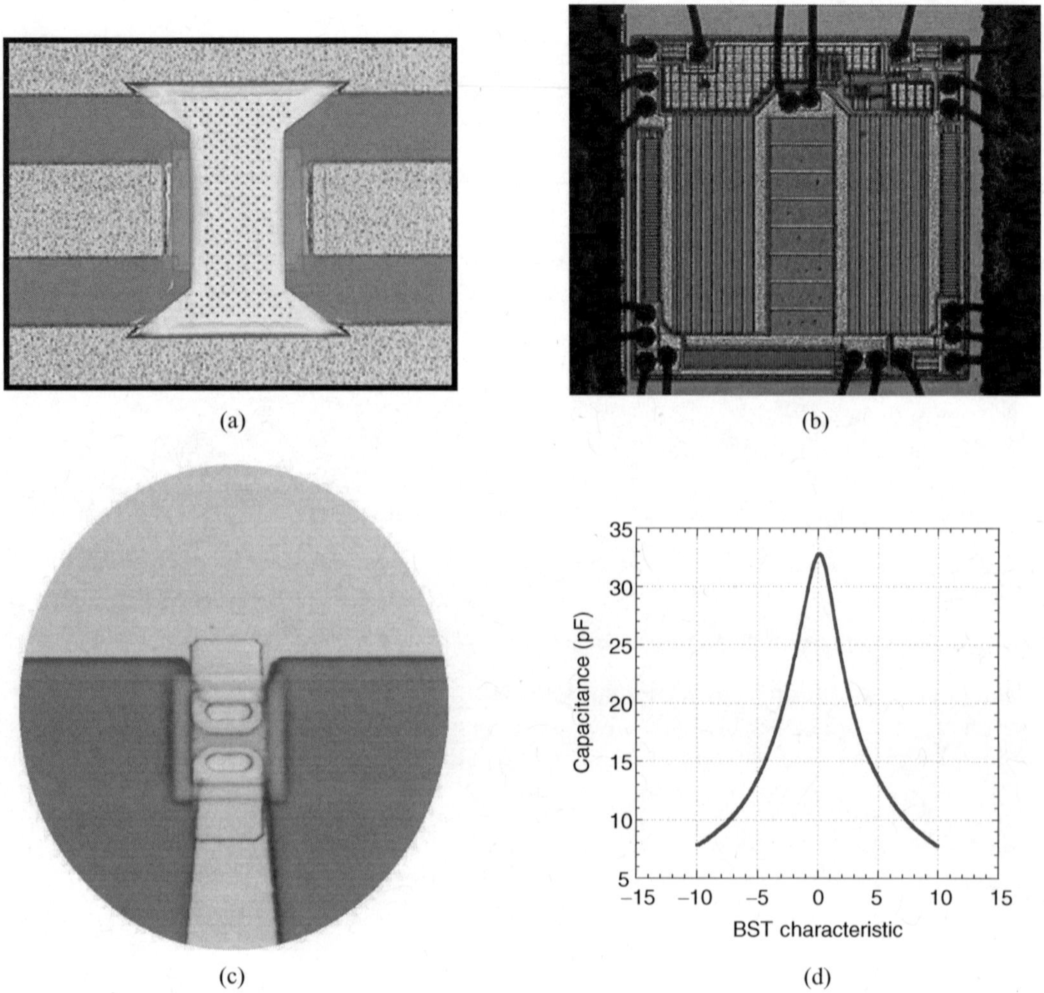

Figure 5-4 Representative components for varying microwave impedance: (a) MEMS switch; (b) FET switch; (c) BST varactor; and (d) BST characteristic.

modulation); it also varies as result of the time-varying distance between transmitter and receiver, and changing fading characteristics of the channel. In multimode transmitters it can also vary as a result of the changing air interface standards that are used. For an overall efficient system, the power amplifier efficiency must remain high at the various output power levels. This requirement has proved to be challenging for RF power amplifiers. The efficiency of classical amplifiers such as Class A and Class AB amplifiers, often used in wireless systems, drops off as the output power is backed off from its maximum value, as shown in Figure 5-5. As a result, the average efficiency for realistic wireless signals can be quite low (for ideal Class A amplifiers, efficiency is limited to 1.5% for the power probability distribution shown in Figure 5-2). It is desirable to use amplifier approaches that have higher efficiency at low output power relative to the peak power of the amplifier. Architectures that have been under development for this purpose include the envelope elimination and restoration and the closely-related envelope tracking architectures; the Doherty

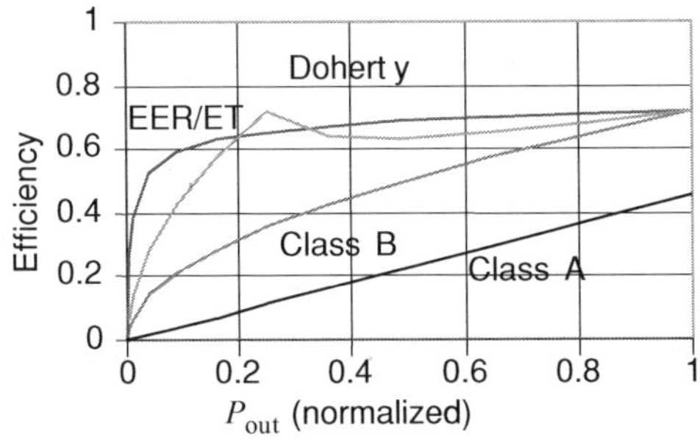

Figure 5-5 Efficiency of various amplifier architectures versus output power (normalized to the peak output power).

architecture, and the outphasing architecture (also known as the Chireix or LINC architecture) [338]. Recent publications detail significant progress in each of these areas. In this chapter the characteristics of the Doherty system are described in detail. Results that are presented show that very high efficiency and good linearity can be obtained over a wide power range for handset power amplifiers (with peak output power of about 1W), in systems in which DSP is used to optimize the amplifier operation. Similar benefits can be obtained with a Doherty amplifier in which a BST-based phase-shifter is used for linearization.

5.3.1 Doherty Amplifier

The Doherty amplifier architecture was introduced in the 1930's [339] and has been utilized recently for a variety of communications applications [340, 341]. The Doherty structure employs two amplifiers, main and auxiliary, as shown in Figure 5-6. The Doherty concept provides a means by which the load impedance seen by the main amplifier automatically adapts to a value that maximizes efficiency as the power level is changed. The auxiliary amplifier is biased in Class C, and operates only under high output power conditions. Under low-power conditions, the main amplifier is operated with a relatively high load impedance which is conducive to high efficiency. With increasing power levels, the auxiliary amplifier turns on, and the load impedance seen by the main amplifier decreases, which allows the amplifier to stay out of deep saturation. The output of

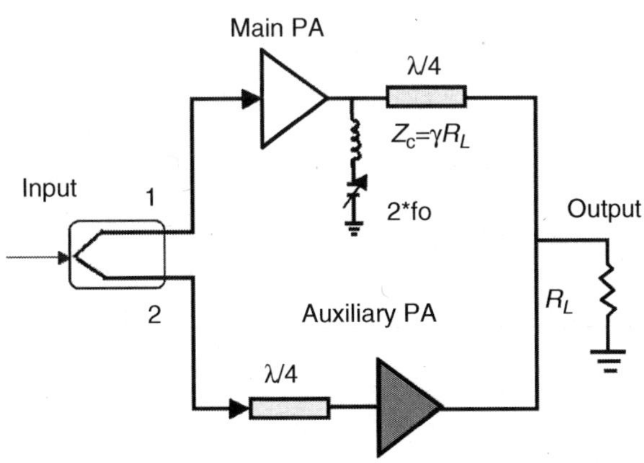

Figure 5-6 Doherty amplifier topology.

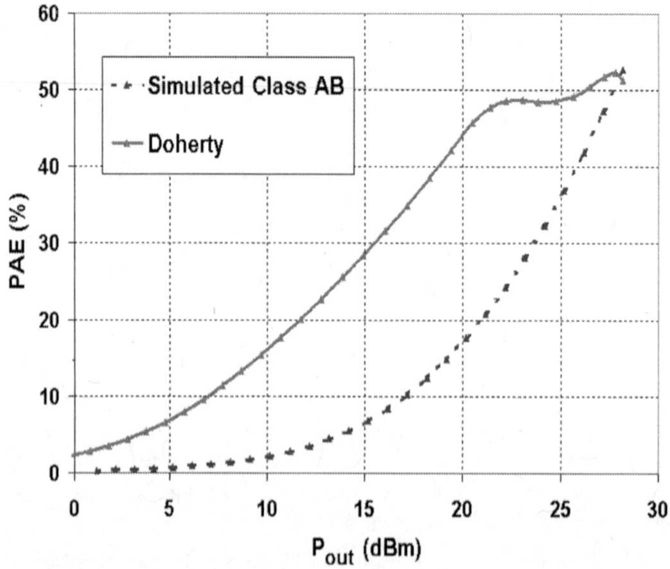

Figure 5-7 Measured single-tone PAE versus output power of a MESFET-based Doherty amplifier and a simulated Class AB amplifier.

the main and auxiliary amplifiers continue to increase, providing higher power at high efficiency, until the auxiliary amplifier goes into compression. The Doherty architecture is advantageous in achieving high average efficiency over a wide range of output power levels, which is highly desirable for RF communication systems. Various publications have described Doherty amplifiers used for basestation applications, including the design tradeoff encountered between efficiency and linearity [321,335–338]. In this chapter a Doherty amplifier oriented toward CDMA handset applications is described. Here DSP control is used to improve its linearity. This amplifier exhibits outstanding average efficiency (taking into account the probability distribution of output power level) as well as good linearity.

The active devices used were GaAs MESFETs, with an MwT-971 for the main amplifier and an Infineon CLY5 for the auxiliary amplifier. Figure 5-7 shows the measured power added efficiency (PAE) of the Doherty amplifier, obtained with CW tones, prior to linearization. The figure also includes, for comparison, the simulated PAE of a Class AB amplifier constructed with the MwT-971 FET and having similar output power and quiescent bias. High PAE (>40%) is obtained over an approximately 9 dB wide region for the Doherty amplifier, and a significant efficiency improvement is obtained over a wide power range. The large power range of high efficiency is influenced by the ratio of the transmission line characteristic impedance Z_c to R_L (γ), which has the value $\gamma = 4$ in our design ($Z_c = 16\ \Omega$, and $R_L = 4\ \Omega$). The circuit was constructed on a printed circuit board, employing surface mount components and microstrip transmission lines.

When the circuit was optimized solely for efficiency, the gain and phase were found to vary significantly with output power in the high power region. Correspondingly the adjacent channel power ratio (ACPR) behavior of the amplifier tested with CDMA signals was unsatisfactory. The gain and phase data measured with CW signal were used as input to a behavioral model [339] in MATLAB. The calculated ACPR of the Doherty PA is compared with the measured results in Figure 5-8. Also shown in Figure 5-8 are ACPR values calculated with only gain or only phase distortion contributions (that is, calculated using ideal flat phase or flat gain in the behavioral

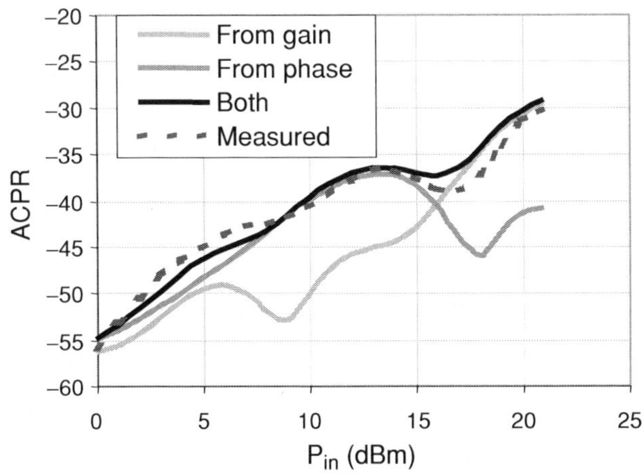

Figure 5-8 Contributions to ACPR from gain and phase distortion of Doherty amplifier.

model). It can be seen that phase variation with power (AM-PM distortion) is a major source of ACPR in the amplifier, and cannot be neglected in the linearization approach chosen.

To improve the amplifier linearity, DSP was introduced. The linearization strategy is twofold: (a) dynamically adjust the gate bias of the auxiliary amplifier at the rate of the signal envelope to obtain gain flatness (i.e., AM-AM distortion correction); and (b) use DSP to adjust the phase of the input signal at baseband (i.e., AM-PM distortion correction). The gate bias of the auxiliary amplifier (V_{gg2}) has a significant influence on both the gain of the Doherty amplifier and on its efficiency. DSP was used to dynamically change the bias voltage of the auxiliary amplifier according to the instantaneous envelope of the input signal in order to make gain flat, while at the same time maximizing the efficiency. Digital predistortion of the phase was also applied to the input baseband signal (based on phase measurements after taking into account dynamic gate bias).

The same Doherty amplifier described previously was tested using an IS-95 signal with the DSP linearization applied. Figure 5-9 shows the reduction of ACPR between the original and

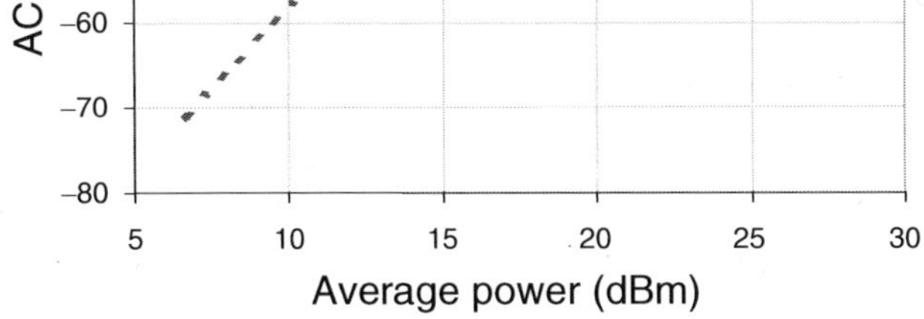

Figure 5-9 ACPR at 885 kHz offset at the right shoulder versus CDMA average output power of uncorrected and DSP-corrected PA.

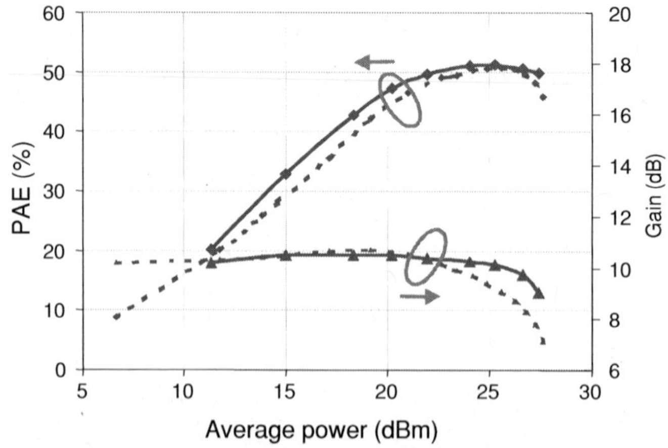

Figure 5-10 Power-added efficiency (PAE) versus CDMA average output power of the uncorrected and DSP-corrected power amplifier.

the DSP-controlled amplifier. Examination of the frequency spectrum of the output with CDMA input signal at 23 dBm average power with and without DSP-control shows that there is 10 dB improvement of adjacent channel power by utilizing the DSP techniques. The ACPR specification for IS-95 (at −42 dBc) is met over the operating range, up to the output power of 26.5 dBm, at which point the auxiliary and main PAs saturate. To reach higher output power, a redesign with slightly larger transistors may be employed. The high PAE of the Doherty amplifier is maintained even with IS-95 signal input, as seen in Figure 5-10. The average efficiency of the Doherty PA (and of the Class AB PA for comparison) can be calculated based on an assumed probability distribution function (PDF) for the output power with IS-95 signals. The PDFs can differ according to the environment of the cellular network. We chose a probability distribution that is intermediate between urban and suburban environments. The average PAE for the Doherty is 20%, which is more than four times of the average PAE for the Class AB PA, which is 5%.

5.3.2 Doherty Linearization with a BST Phase-Shifter

An alternative way to linearize the Doherty amplifier was demonstrated using a $BaSrTiO_3$ (BST)-based phase-shifter [342]. This phase-shifter is controlled by DSP to provide phase predistortion in the RF signal path before the PA. Predistorting in this fashion provides advantages by avoiding extra complexity in the transmitter.

The block diagram of the Doherty PA with analog phase control is shown in Figure 5-11. Here a Doherty amplifier with relatively flat gain and very nonflat phase is used. The phase variation was found to be as much as 50°. Consequently the nonlinearity of the Doherty PA is considered to be mainly from the nonflat phase. The baseband signal is generated in the form of I and Q and then up-converted to RF. After being amplified by a preamplifier the signal is split by a 1:2 ratio Wilkinson power divider into the main and auxiliary amplifier paths. DSP generates another voltage signal which is applied to the BST phase-shifter to control the phase behavior of the RF signal. Normally gain control through V_{gg2} at the gate of the auxiliary PA is required. However here the gain characteristic is relatively flat (there is little AM-AM distortion) and gain control is not required. DSP was only used to correct the phase of the amplifier. Predistorting the phase in the analog fashion at RF can be advantageous as it reduces complexity in the transmitter. By using an analog phase-shifter, one can avoid the need for a high resolution, high bandwidth DAC for the I and Q signals, as well as avoid difficulties with the channel filter (since the predistorted

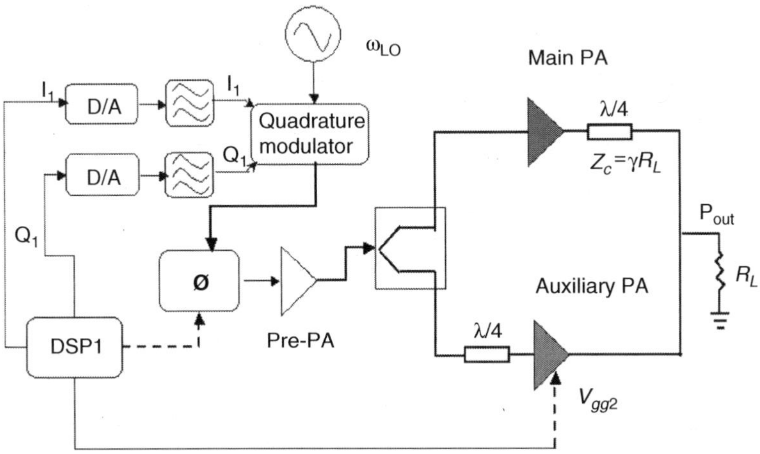

Figure 5-11 Doherty architecture with DSP control using a BST phase-shifter.

signals have expanded bandwidth relative to the signal), and reduce the complexity of the whole up-conversion chain.

The analog phase-shifter was implemented using voltage-controllable BST-based capacitors in a lumped LC transmission line. A schematic layout of the transmission line phase-shifter with BST capacitors, and a packaged BST-based phase-shifter are shown in Figure 5-12. Phase data from the amplifier was analyzed to develop the phase predistortion strategy. From these curves and the calibration curve of the phase-shifter, a signal-dependent control voltage was derived, and generated (via PC and suitable DAC) during the amplification of an IS-95 CDMA signal.

To verify the efficacy of the described phase predistortion, the performance of the Doherty power amplifier was measured in terms of PAE, gain, and ACPR for an IS-95 signal. Shown in Figure 5-13 is the measured PAE and gain versus output power of the Doherty amplifier with a CDMA IS-95 test signal prior to phase correction. High PAE (>40%) is obtained over an approximately 6 dB wide region. It is estimated that, when averaged over the possibility distribution for output power usage in a CDMA system, the overall efficiency of this amplifier is higher by more than a factor of 2.5 than that of a Class AB amplifier. The largest improvement of ACPR achieved by this phase predistortion is 5.2 dB at 22 dBm output power, which is particularly significant because it allows the amplifier to meet the linearity specifications of −42 dBc ACPR for the IS-95 standard. It is expected that further improvement will be possible by combining gain predistortion (obtained by dynamically biasing the gate of the auxiliary or peaking amplifier).

5.4 ADAPTATION TO LOAD IMPEDANCE VARIATIONS

Mobile handsets are used in a variety of configurations and positions, by users who manipulate the handset particularly the antenna, in ways that are difficult to predict. While a nominal antenna provides an input impedance of 50 Ω, in actual usage the impedance at the antenna terminal can vary over a wide range, characterized by a voltage standing wave ratio (VSWR) of up to 10. It is a major challenge to maintain operation of the handset with such a wide range of impedances. For the power amplifier such severe impedance mismatch greatly impacts efficiency, power gain, maximum output power and linearity. In the worst case, high VSWR, or the possibly oscillation caused by the mismatch in the circuit, may damage the power amplifier. In past practice, an isolator or a VSWR protection circuitry was often inserted between the amplifier and antenna to mitigate the problem. However the added attenuation of the isolator reduces overall efficiency. Here a

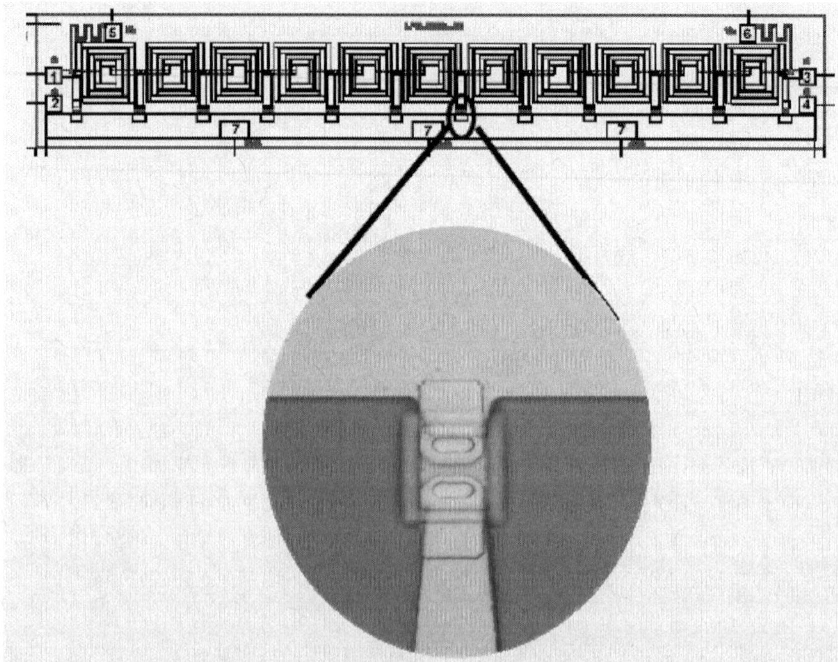

(a)

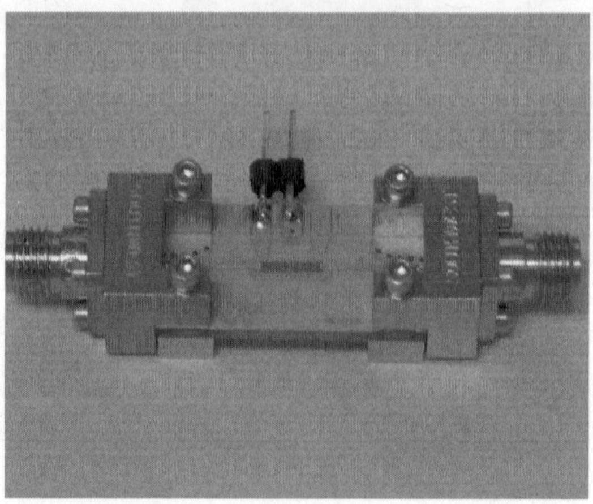

(b)

Figure 5-12 Phase-shifter-based on BST material: (a) schematic layout; (b) photograph.

simple example of an adaptive amplifier is described. In this system the antenna impedance is measured, and then this impedance is matched approximately to the desired load of a microwave power amplifier stage in accordance with the measurements [343]. The system is oriented toward the specifications of CDMA handsets, and thus the operating frequency was 850 MHz, the output power levels was up to 650 mW, and the linearity specifications were exacting (−42 dBc for ACPR).

The structure of the adaptive amplifier is shown in Figure 5-14. In this structure, a controller (analog or digital) tunes the matching network based on the measurement of the load impedance,

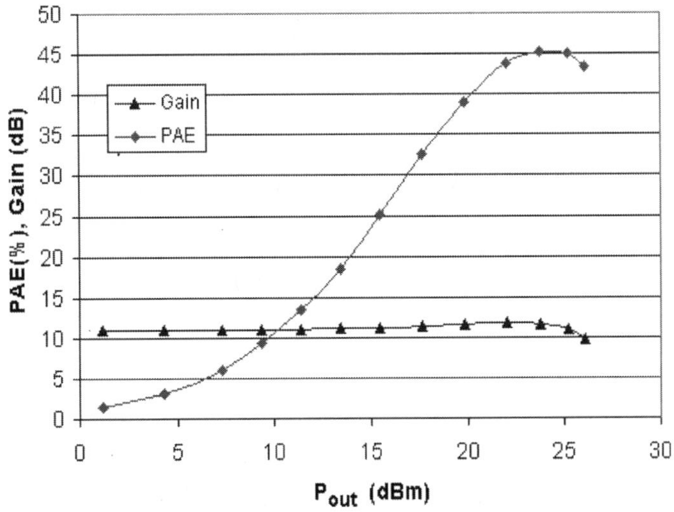

Figure 5-13 Measured gain and PAE of the BST-linearized Doherty PA tested with an IS-95 signal.

thereby keeping the handset in an optimum working condition. In practice, control of the tunable matching network in handset requires real time measurement of the load impedance. Due to their relatively large size, the traditionally used direction couplers are not suitable for handset applications. A method based on sectioned transmission lines has been proposed to measure the load impedance. This technique estimates the load impedance by measuring the voltages at multiple points of the transmission line.

5.4.1 Design of Tunable Matching Network

Silicon-on-sapphire (SOS) switches were used to implement the tunable matching network. Using a solid-phase epitaxial regrowth process [344], these SOS MOSFETs are built in a 100 nm film of silicon with transconductance and mobility comparable to that of bulk CMOS. At this thickness the transistors are symmetric three-terminal devices that can be stacked in series to withstand high voltage despite the low breakdown voltage of a single transistor. The insulating substrate also provides for low parasitic capacitance and thus low insertion loss. The matched network implemented in this paper used six SOS switches. Depending on the load impedance, the switches are turned on or off to reduce the impedance mismatch. Measurement results show that this matching network has the advantage of large tuning ratio, low-insertion loss and high linearity.

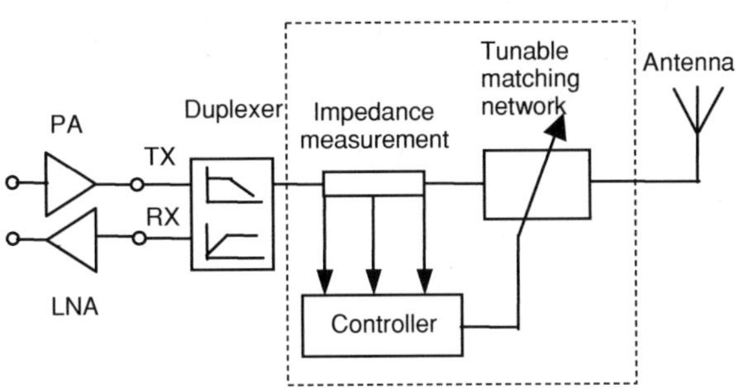

Figure 5-14 Schematic diagram of adaptive impedance tuner.

An L-topology is a commonly used design impedance matching networks. However, a particular L matching network can only match impedance in part of the Smith Chart to 50 Ω, i.e., there is a forbidden region for a particular L-topology. In this work, a cascaded L matching network consisting of two fixed inductors and two tunable capacitors was designed, see Figure 5-15(a). Simulations show if the tuning ratio of the capacitors is large enough, most of the impedance range in the Smith Chart can be matched to 50 Ω. To match all the impedance within the VSWR = 7circle of the Smith Chart, the capacitance tuning ratio required is approximately 3:1. Larger tuning ratios are required for higher VSWR correction. In order to increase the tuning range, switches with shunt capacitors are used in this paper. In this case, the tuning ratio is defined by the fixed capacitors and the parasitic capacitance of the switches. Increasing the capacitance of the fixed capacitors can increase the tuning ratio.

If the tunable parameters in the matching network only consist of switches, tuning is a digital process. In this case, the tuning precision should also be considered when designing the matching network. More switches are preferred for more accurate control but increase the cost and complexity of control. Here six switches are used which provides 64 available states. The fixed capacitors are selected to provide a relatively uniform distribution of the impedances that the matching network can match to 50 Ω in the Smith chart. For a particular mismatch circumstance the best state is selected. Another design for consideration the matching network is that there should be a state which does not change the impedance of 50 Ω, since for most of the time, the handset is operated with a 50 Ω antenna.

Figure 5-15(b) shows the matching network used in this work. The SOS switch building block is a stack of six transistors which each has a drain-source breakdown voltage (BVDSS) of approximately 2.0 V yielding a stack that can withstand a peak voltage of 12 V. Integrated 30 kΩ blocking resistors are placed between the FET gates and the control pad to provide for proper RF voltage division across the FET stack when biased in the off condition. Inside the switch core

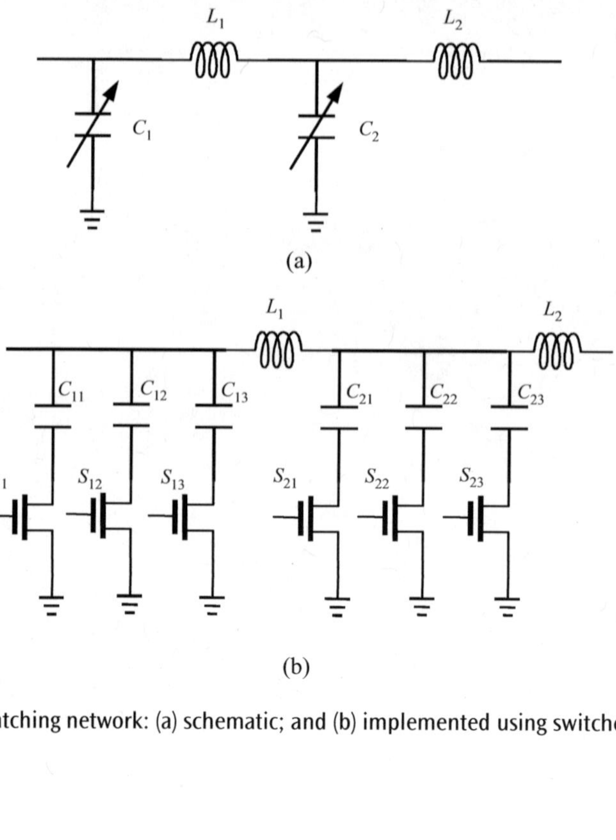

Figure 5-15 Tunable matching network: (a) schematic; and (b) implemented using switches.

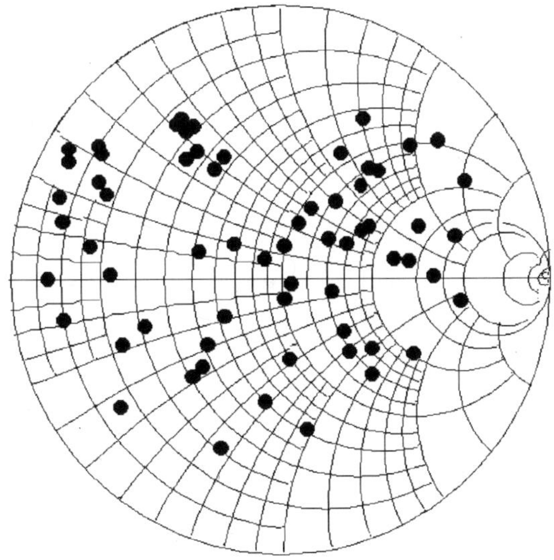

Figure 5-16 Measured range of matched load impedance using the tunable matching network.

there are six 1000 μm × 0.5 μm FETs in parallel, which are combined by wire bonding. The on resistance for the total switch is ~3 Ω and the off capacitance is 0.3 pF. ESD protection is provided from the gates to ground node of the stack. The tunable matching network was implemented on a PCB board with FR4 substrate and gold coated for wire bonding. Toko LL2012 (whose Q is about 40–90) chip inductors and ATC 100A Porcelain Superchip multilayer capacitors were selected for the fixed elements. Unpackaged SOS switches were wire-bonded to the PCB board. The size of the circuit prototype was approximately 4.3 cm × 2.7 cm. Figure 5-16 shows the matched load impedances using the tunable matching network. It is seen that the matched impedance is well distributed over the Smith chart, and there are some states that provide impedance very close to 50 Ω.

5.4.2 Amplifier with Tunable Matching Network

In this section the tunable matching network is used with an RF amplifier. Many digital communication systems, such as CDMA, require high linearity. Thus, in this section, a highly linear Class A power amplifier is used to characterize the linearity of the tunable matching network. Figure 5-17 shows the spectra of the power amplifier with and without the tunable matching network between the amplifier and a 50 Ω load with a CDMA IS-95 input. The switches were set to a state close to 50 Ω. The output power of the amplifier was adjusted to approximately 28.5 dBm to measure the ACPR. Measurements showed that even with a small mismatch, the ACPR1 of the output signal only changed from –59.5 dBc to −59.2 dBc, suggesting the tunable matching network did not obviously degrade the linearity of the system. The insertion loss associated with the tunable matching network was also measured using a CDMA IS-95 signal. It was found the insertion loss was about 0.4 dB.

5.4.3 Measurement Circuit for Antenna Load Impedance

Figure 5-18 shows a transmission line terminated in an arbitrary load Z_L with a characteristics impedance of Z_0. For two points on the transmission line, the ratio of the voltages at the two points is

$$\frac{V_{z1}}{V_{z2}} = \frac{e^{-j\beta z_1} + \Gamma e^{j\beta z_1}}{e^{-j\beta z_2} + \Gamma e^{j\beta z_2}} \tag{5.1}$$

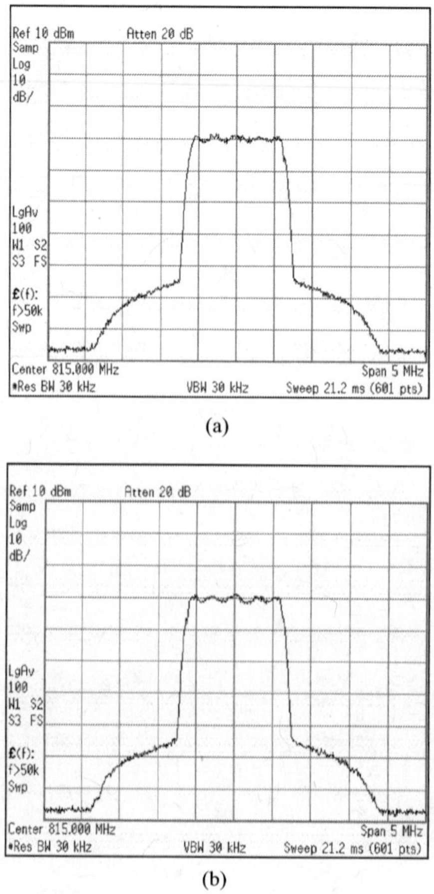

Figure 5-17 Output power spectra of a power amplifier: (a) without; and (b) with the tunable matching network.

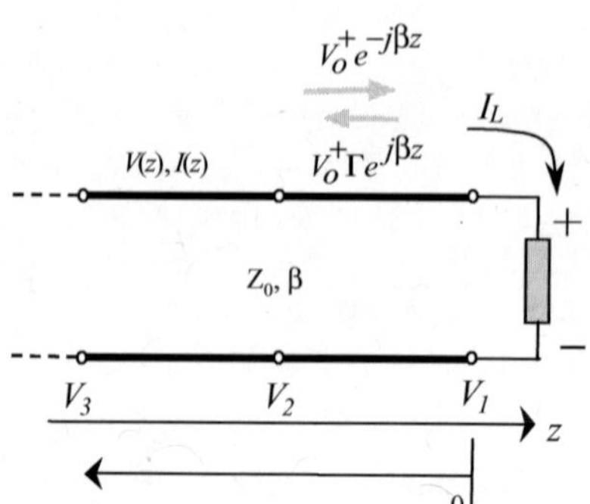

Figure 5-18 Transmission line terminated in a load showing measurement points.

where β is the propagation constant of the transmission line, and Γ is the reflection coefficient. In Equation (5.1) the voltage ratio of two points on the transmission line does not depend on the magnitude and phase of the incident wave and the source impedance, but depends on the position of the points and the load impedance. The real part and the imaginary part of Γ (and thus the antenna impedance) can be obtained by solving Equation (5.1) for two independent voltage ratios, i.e., the voltages at three different positions on the transmission line.

For easy validation, a quarter-wavelength-long transmission line is used in this section. The voltage is measured at positions of 0°, −45°, and −90° from the load toward the generator on the transmission line. Selection of measurements at these positions simplifies Equation (5.1), as well as providing relatively large differences between the measured voltages to alleviate the accuracy requirement of measurement. Theoretically, shorter transmission lines can be used.

The circuit used in this work for envelope monitoring was based on resistive taps. To measure the voltage at each position on the transmission line, a 1.9 kΩ resistor was used to couple the signal into a high impedance transmission line, which is connected to a power meter (represented by 50 Ω resistors in the schematic). The circuit was fabricated on a Rogers RO4003C PCB board. A sinusoidal signal with a frequency of 815 MHz and three different powers of 15 dBm, 18 dBm, and 20 dBm, and a CDMA IS-95 signal with a frequency of 815 MHz and at three different powers of 12 dBm, 14 dBm, and 16 dBm were used for measurement. An impedance tuner was used to produce different load impedance. Equation (5.1) was then solved numerically to obtain Γ and load impedance.

Measurements of load impedance by the transmission line approach and the actual load impedance—measured by a vector network analyzer (VNA)—show good correspondence For loads with smaller VSWR, the measured results are more accurate. Part of the large error for high-impedance loads is caused by the perturbation to the circuit by the resistors and high impedance transmission lines used for measuring the voltage. It is also found that the loads measured by the transmission line approach do not depend on the input power, in agreement with Equation (5.1). Power measurement showed the insertion loss of this circuit to be about 0.4 dB.

The load impedance measurement was used in conjunction with the impedance tuner under computer control. The result of using this now adaptive impedance tuner was a significant improvement in the amplifier performance under mismatch conditions. Figure 5-19 shows measured results for output power and efficiency under various mismatch conditions, both with and without

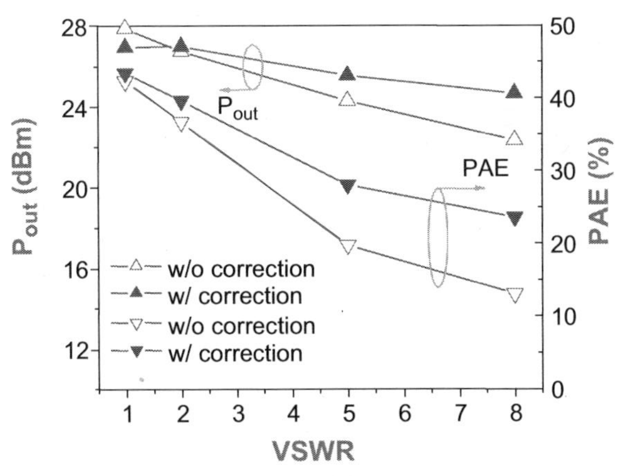

Figure 5-19 Measured improvement of output power and power-added efficiency from use of adaptive tuner.

the adaptive tuner. Improvements of several decibels of output power and a factor of two in efficiency were obtained for the highest mismatch values. The reported values correspond to averages of power and efficiency measurements obtained with different phase angles at each VSWR value.

From the above discussion, it can be seen that it is appealing to apply the tunable impedance matching network in conjunction with the circuitry for measuring antenna load impedance and so obtaining adaptive control of the CDMA transceiver. Simulation showed that the size of the circuitry for measuring load impedance can be reduced by using a shorter transmission line or even lumped elements (the size of the tunable matching network could be readily reduced since it does not need a delay for measuring waveforms). For the tunable matching network, more careful selection of the capacitors could improve the uniformity of distribution of matched impedances over the Smith Chart. Control algorithms should also be developed to dynamically tune the matching network.

In summary, a tunable matching network using SOS switches has been developed to correct for antenna impedance mismatch. The network provides high tunability of impedance, low insertion loss, and high linearity. A simple method based on a transmission line was proposed for measuring antenna impedance for wireless handset. The complex load impedance was measured by measuring scalar voltages at three points along a transmission line.

5.5 ADAPTATION TO NONLINEARITY VARIATIONS

Power amplifier nonlinearity can lead to loss of signal accuracy (as characterized by error-vector magnitude) and to generation of interference in adjacent frequency channels as characterized by ACPR. Linearization procedures are typically used to reduce these unwanted effects [345]. However, the nonlinear mechanisms of a typical amplifier change from time to time, according to power level, temperature, load impedance, etc. Accordingly, it is desirable to measure the nonlinearity and adapt the predistortion accordingly. Traditionally single tone or two-tone testing is used to extract the amplifier nonlinearity from AM-AM and AM-PM distortion or intermodulation distortion measurements [346–348]. Employing these measurements, the spectral regrowth of complex digital modulation signals can be predicted by various behavioral models [349, 350]. However, these approaches require special input signal generation, and are generally not applicable to the characterization of amplifiers during real-time operation. An alternate method is to capture the input and output time-domain waveforms to extract the nonlinearity and use FFTs to calculate the output spectrum under operating conditions [351]. This approach is intensive in computation complexity and hardware requirements. Here simple correlation techniques are demonstrated that can estimate power amplifier nonlinearity and predict the out-of-band spectral regrowth. The correlation method can be carried out without interrupting the amplifier's normal operation and can potentially be carried out with simple analog circuitry. The proposed method is applicable to multichannel direct sequence (DS)-CDMA systems and IS-95 forward-link CDMA signals are employed as an example. In the following the correlation measurement technique is first reviewed. The calculation of spectral regrowth based on amplifier nonlinearity coefficients is then described. Finally, the correlation technique for spectral regrowth estimation is demonstrated by simulation and measurement.

The basic correlation methodology is illustrated in Figure 5-20 [352]. In a multichannel DS-CDMA system, the PA's baseband input signal is the sum of several uncorrelated PN (pseudorandom noise) digital binary sequences S_i (i is the channel number). A new binary test signal S_{test} is created by forming the product of S_i for three inputs. Both S_{test} and the output signal V_{out} are nonlinear functions of the input signal. We have shown [321] that the nonlinearity of the power

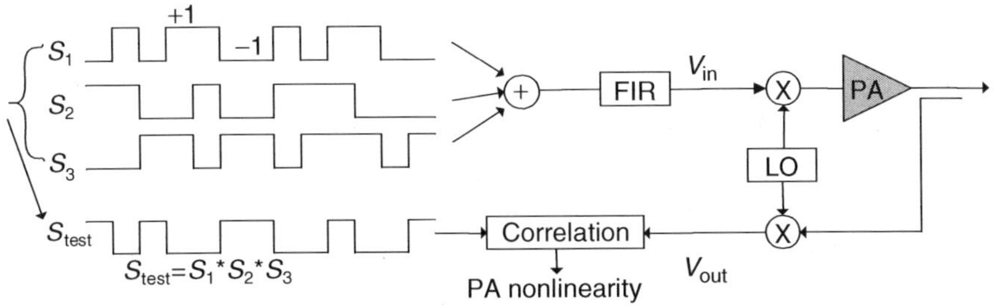

Figure 5-20 Block diagram of a transmitter showing correlation methodology.

amplifier can be extracted by the maximal correlation values between V_{out} and S_{test}:

$$\mathrm{MAX}\left(\Phi_{V_{out},S_{test}}(\tau)\right) = \mathrm{MAX}\left[\lim_{T\to\infty}\frac{1}{T}\int_0^T V_{out}(t)S_{test}(t+\tau)\right]. \tag{5.2}$$

The test signal can be inherently generated from the actual baseband CDMA signals such as IS-95 forward-link signals.

5.5.1 Nonlinearity Behavioral Model

The quasi-memoryless nonlinearity of a PA can be characterized in various ways, including: a complex polynomial model [335]; or the Saleh model [338]. Both approaches are considered here. CW single-tone measurements of a commercial wireless LAN amplifier, Intersil's ISL3990 GaAs PA, are used to obtain the model fitting parameters. Coefficients for a five-order complex polynomial were obtained by curve fitting as shown in Figure 5-21, using

$$\tilde{V}_{out} = \tilde{V}_{in}(\tilde{a}_1 + \tilde{a}_3\,|V_{in}|^2 + \tilde{a}_5\,|V_{in}|^4)\,. \tag{5.3}$$

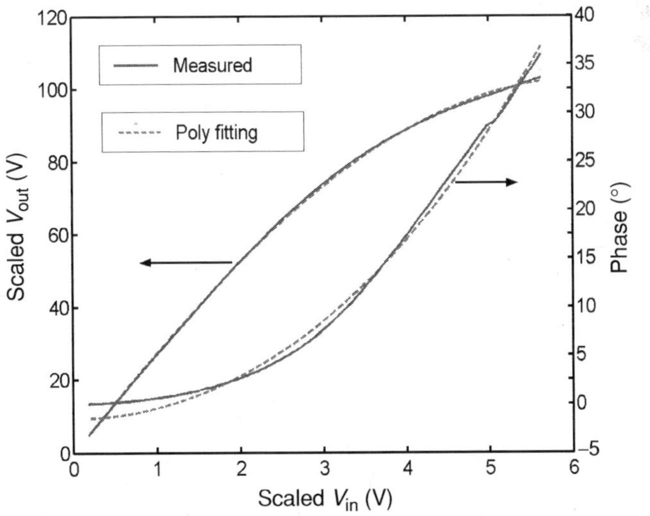

Figure 5-21 AM-AM and AM-PM curves of measured (solid blue) and fifth order polynomial fitting (dashed red).

Alternatively the Saleh model can be used in its polar form [353].

$$\begin{aligned} A(r) &= \alpha_a r/(1+\beta_a r^2) \\ \Phi(r) &= \alpha_\phi r^2/(1+\beta_\phi r^2)\,. \end{aligned} \tag{5.4}$$

This model has four nonlinearity-control parameters, as shown in Equation (5.4). Note that if the input amplitude r is very small or very large, the amplitude function $A(r)$is proportional to r or $1/r$ and phase function $\Phi(r)$approaches zero or a constant.

5.5.2 Analytical spectral regrowth formulation

Spectral regrowth can be estimated from amplifier nonlinearity together with input signal characteristics. Here the input and output signals $x(t)$ and $y(t)$ are considered in their rectangular format with polynomial nonlinear relationships. The output autocorrelation $\tilde{R}_{\tilde{y}\tilde{y}}$ can be written as a function of the input autocorrelation $\tilde{R}_{\tilde{x}\tilde{x}}$ with different orders as:

$$\tilde{R}_{\tilde{y}\tilde{y}}(\tau) = P_1\tilde{R}_{\tilde{x}\tilde{x}}(\tau) + P_3\tilde{R}^{*}_{\tilde{x}\tilde{x}}(\tau)\tilde{R}^{2}_{\tilde{x}\tilde{x}}(\tau) + P_5\tilde{R}^{*\,2}_{\tilde{x}\tilde{x}}(\tau)\tilde{R}^{3}_{\tilde{x}\tilde{x}}(\tau) \tag{5.5}$$

where $\tilde{R}^{*}_{\tilde{x}\tilde{x}}$ is the complex conjugate of $\tilde{R}_{\tilde{x}\tilde{x}}$ and P_1, P_3 and P_5 represent the coefficients of different order of input autocorrelation. P_3 and P_5 are functions of the complex polynomial coefficients in Equation (5.2) given as:

$$P_3 = K_1(|\tilde{a}_3|^2 + K_2\,|\tilde{a}_5|^2 + 2K_2\Re\{\tilde{a}_3\tilde{a}_5^*\}),\ \ P_5 = K_3\,|\tilde{a}_5|^2 \tag{5.6}$$

where $\Re$ denotes the real part and K_1,K_2 andK_3 are constant numbers.

Equation (5.6) is derived following the methods of references [349] and [354]. The output power spectrum is the Fourier transform of output autocorrelation $\tilde{R}_{\tilde{y}\tilde{y}}$. Numerical results based on this approach show that the adjacent and alternate channel leakage powers are due to the third and fifth nonlinearity coefficients, and thus are largely proportional to P_3 and P_5 respectively. As a result it is possible to estimate the ACPR values from the correlation measurements.

5.5.2.1 Simulation and Measurement Results MATLAB simulation was used to verify the similarity of spectral regrowth and power correlation values. The setup was similar to Figure 5-20 with five filtered PN sequences passing through a behavioral PA model, which can be a complex polynomial or Saleh model. The nonlinearity characteristics of PA can be changed by adjusting the parameters of its nonlinear models. Figure 5-22 shows the calculated three-dimension contours of the ACPR and the power correlation values versus coefficients $\tilde{a}_3$in Equation (5.2). The above results show that the correlation extraction method can predict the spectral regrowth phenomena quite well. The experimental test environment is shown in Figure 5-23. The baseband input signals are generated with MATLAB and downloaded to an Agilent ESG waveform generator as the RF input of the PA under test. The baseband output signals are collected by an Agilent PSA spectrum analyzer and VSA vector signal analyzer. The PA's equivalent nonlinearity characteristics can be changed by using a baseband polynomial predistorter (PD).The correlation measurements allow adjusting $\tilde{b}_3$ coefficients to minimize ACPR. The three-dimension contours of ACPR and power correlation values versus the predistorter's coefficients $\tilde{b}_3$ are shown in Figure 5-24. The similarity of these characteristics confirms the preceding simulations. Moreover, they demonstrate that the measured ACPR can be minimized by experimental minimization of the correlation values. Simple correlation techniques have been used to estimate power amplifier nonlinearity and predict its spectral regrowth. The technique is potentially applicable to the measurement of amplifier nonlinearity in real time during amplifier operation.

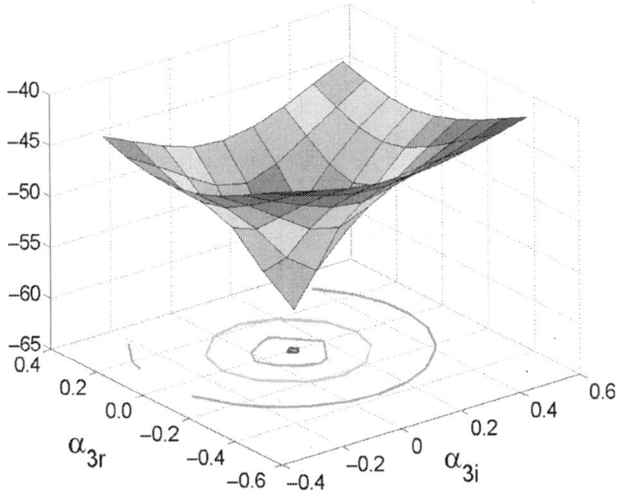

(a)

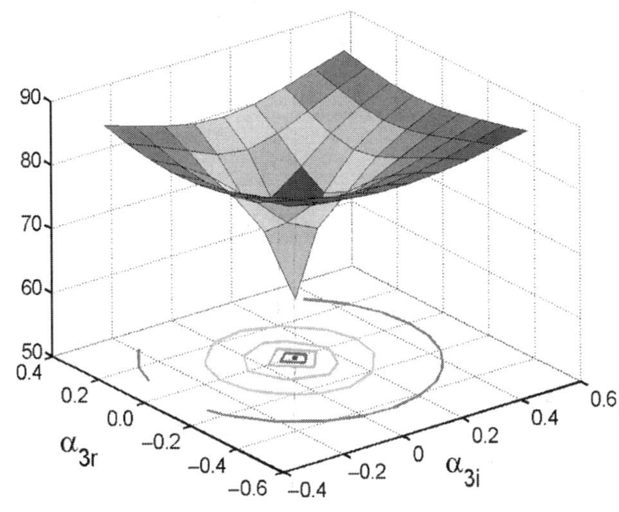

(b)

Figure 5-22 Nonlinear amplifier characteristics: (a) ACPR at 885 kHz (dBc); and (b) the third power correlation values in decibels versus third-order real and imaginary nonlinearity coefficients.

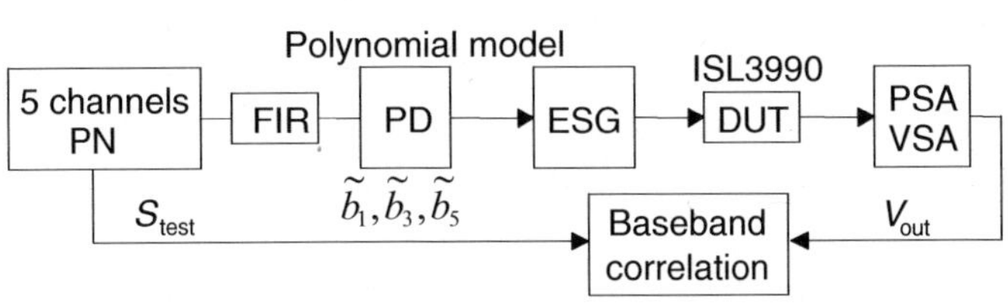

Figure 5-23 Measurement setup.

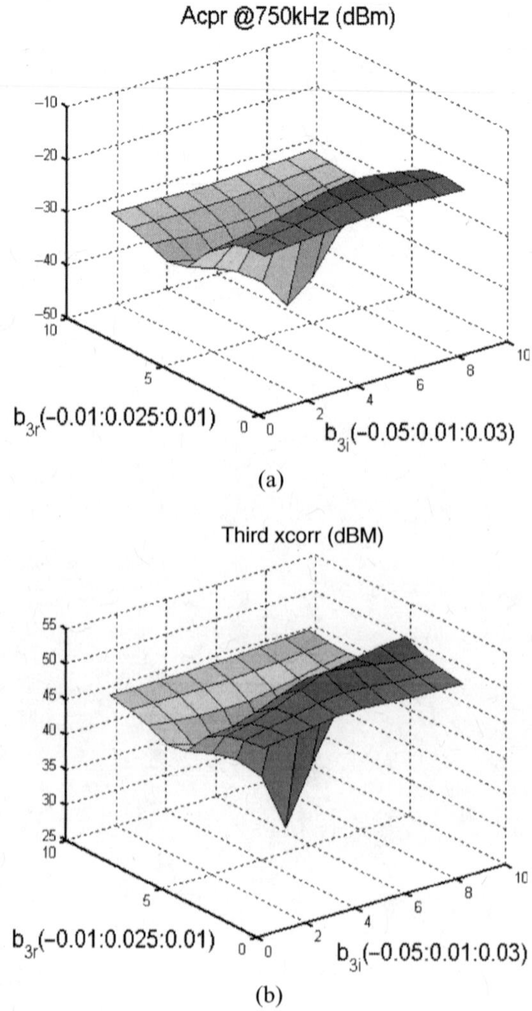

Figure 5-24 Experimentally measured values of (a) ACPR at 750 kHz (dBc); and (b) third power correlation values (dB) versus third order real and imaginary nonlinearity coefficients.

5.6 DIGITAL TRANSMITTERS

A major trend in communication systems development is the replacement of analog circuit by digital signal processing. Digital processing is inherently adaptable; their functionality can be changed on the fly in many cases, e.g., in programmable or "software radio" systems. Digital circuits can also be redesigned as the technology evolves with very little effort. In an RF context, digital transmitters are not subject to constraints from frequency planning or tuning considerations, and can be changed to employ different carrier frequencies, bandwidths and linearity requirements with relative ease. Digital circuit implementation provides many other advantages including ease of testability and decreased sensitivity to temperature, processing, and aging effects. A major added advantage of digital circuits is that they can be readily incorporated into CMOS-based systems-on-chips, and can take better advantage of process geometry shrinks than their analog counterparts.

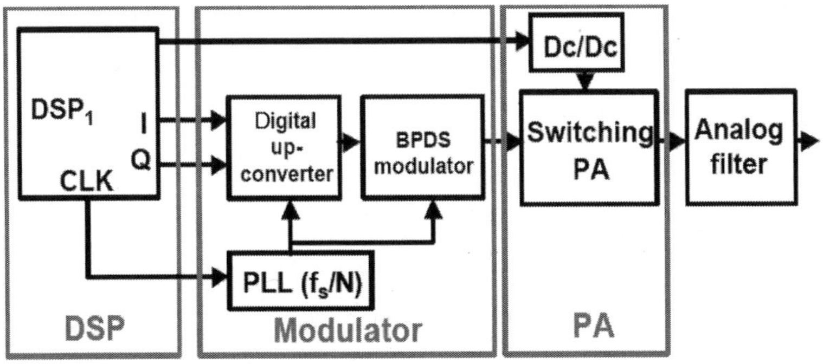

Figure 5-25 Schematic diagram of an RF transmitter using digital signal generation.

The goal of the work described here is to implement an entire transmitter section with digital circuitry. Already, digital baseband and digital IF implementations are in widespread use. The critical pieces missing in order to realize an all-digital transmitter are the upconverter, the final mixer, and the power amplification stage. The digital implementations of these blocks are becoming increasingly realistic as digital processes improve.

Figure 5-25 illustrates a block diagram of the proposed all-digital transmitter [355]. The representative application is CDMA. First, a baseband I/Q signal from a standard DSP is fed to the digital modulator section. This baseband I/Q signal is similar to that for other transmitters, except that the baseband frequency is higher than typically used. Here a baseband sampling frequency of 40–60 MHz is used in order to keep signal images resulting from up-sampling operations out of band. The representative spectrum of the baseband signal is shown in Figure 5-26(a). The baseband signal is then upsampled to four times the intended transmit frequency. A representative final sampling frequency of 3.2 GHz is used in this example (corresponding to an 800 MHz carrier frequency). The signal is then modulated in quadrature, using the following function:

$$y(n) = x_i(n) f_1(n) + x_q(n) f_2(n) \tag{5.7}$$

where $f_1(n) \equiv \{1,1,1,1,\ldots\}$ and $f_2(n) \equiv \{-1,1,1,1,\ldots\}$.

The simple form of the "LO" signals results from the fact that the sampling frequency is exactly a factor of four times higher than the desired carrier frequency. The sequences $f_1(n)$ and $f_2(n)$ are easily implemented. The representative spectrum of the upconverted signal is shown in Figure 5-26(b). The unconverted signal is then converted to a one-bit value through the use of a bandpass delta–sigma modulator [356, 357]. The block diagram of the band-pass delta–sigma modulator is diagramed in Figure 5-27. The ability to use a fast binary sequence to represent a full-blown, high dynamic range CDMA signal is surprising, but a one-bit second-order bandpass delta–sigma modulated signal has been experimentally shown to exceed CDMA IS-95 requirements.

The spectrum of the output signal from the delta–sigma modulator is indicated in Figures 5-26(c) and 5-26(d). While on a large frequency scale the quantization noise is apparent, Figure 5-26(c), on a frequency scale comparable to the CDMA transmit band, Figure 5-26(d), the noise floor is low because of the spectral noise shaping achieved in the delta–sigma modulator. The output of the modulator section is then fed into a switching amplifier. Depending on the requirements of the transmitter, the power amplifier could be as simple as a large digital inverter. The details of the switching amplifier are beyond the scope of this discussion.

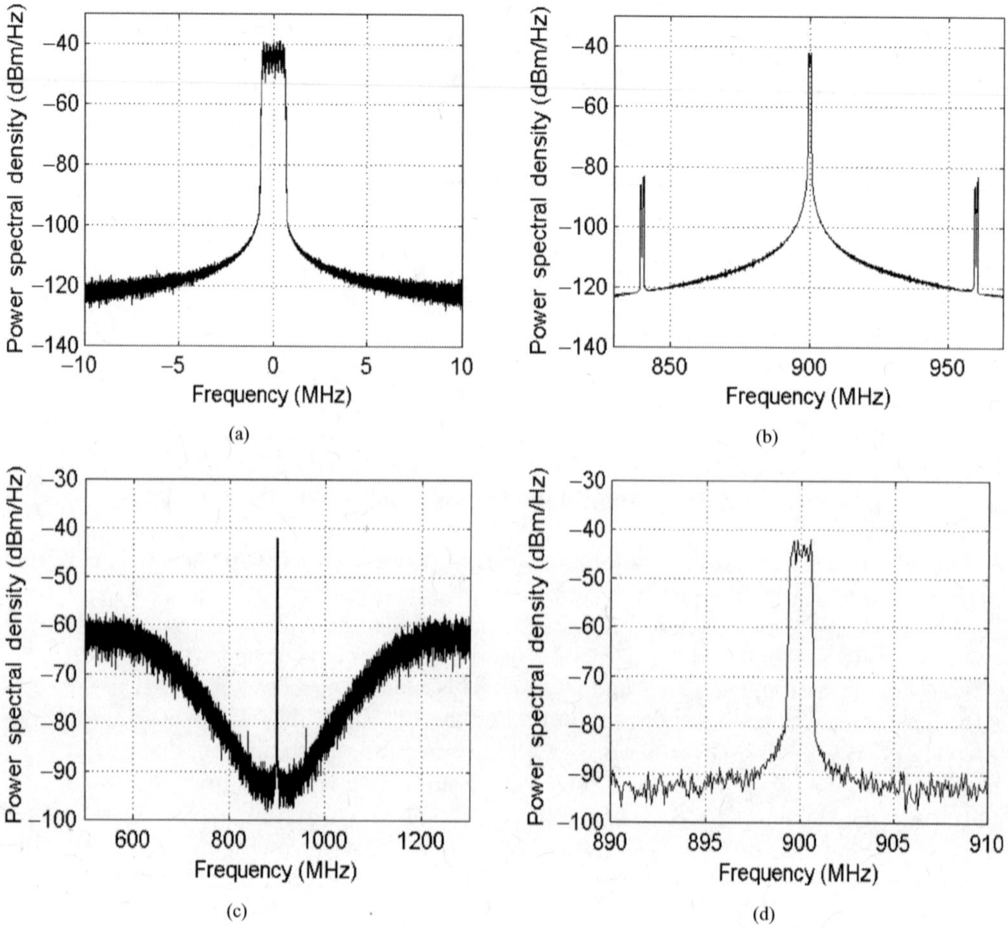

Figure 5-26 Power spectral densities of CDMA signal at various positions of the digital transmitter: (a) baseband spectrum with center frequency of 50 MHz; (b) after upconversion; and (d) after the delta–sigma modulator.

Coarse gain control of the power amplification stage needs to be done by controlling the supply voltage. This system uses a DSP-controlled DC-to-DC converter for coarse transmit power control. Power control can also be achieved by scaling down the digital input to the transmitter, but this will not lower the quantization noise floor. Lowering the power by scaling the signal degrades the signal-to-noise ratio by the same scaling factor, which is unacceptable over a large dynamic range.

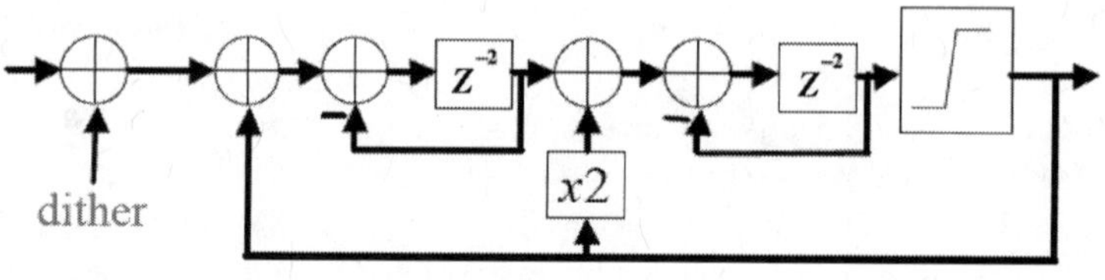

Figure 5-27 Schematic diagram of bandpass delta–sigma modulator.

Lastly, the output of the amplifier needs to be bandpass filtered to remove the out-of-band quantization noise. Experimentally it has been shown that the output signal filtered by a Toko four pole dielectric filter meets the IS-95 specification, for signals near the maximum power level of operation (additional considerations are needed in order to meet the large dynamic power range required in IS-95 systems).

5.6.1 Digital Transmitter Simulation Procedure

Simulation of the modulator operation is computationally intensive. In order to obtain accurate spectra, the input test sequence length must be adequately long. Here a sequence of 2,048 chips is used. These chips are oversampled by a factor of 32 such that the baseband sequence length is 65,536. The sequence is then bandpass filtered and serves as the CDMA baseband test sequence. The transmitter simulation is implemented in C, which offers advantages because of its speed, control over numerical representation, and memory allocation control. With a baseband sequence of 65,536 values, the output sequence length is 5,242,880. Simulating the sequence in C takes over 10^6 operations and allocates 300 MB of memory. Simulating the transmitter in an interpreted language with automatic memory management, such as MATLAB, can take hours to days. A properly optimized C implementation (provided with enough memory) can be made to run in under five seconds.

5.6.2 Delta–Sigma Modulator Implementation

In this section the performance of various delta–sigma architectures are evaluated primarily using the following criteria: spectral distribution of the test CDMA sequence and output noise; implementation complexity; and power dissipation. These follow from the key goals in designing handset transmitters: meeting specifications, lowering cost, and lengthening battery life.

5.6.2.1 A. Modulator Order Simulations demonstrated that delta–sigma modulators of order greater than two improved the noise shaping characteristics but were not necessary to meet most specifications. These implementations were rejected because of their increased power dissipation, die area, and higher potential for instability.

5.6.2.2 B. Modulator Arrangement Clocking the delta–sigma modulator at four times the transmit frequency enables all of the coefficients in the modulator to be powers of two, thus allowing their implementation to be realized by a radix shift requiring no logic. Further, delay elements in the modulator are arranged to occur immediately after major logical operations, so that the delay can be incorporated into the logical operation in the form of pipelining (in which the logical operation is split into suboperations taking several clock cycles). Using these design guidelines, the modulator implementation of Figure 5-27 was chosen.

5.6.2.3 C. Modulator Numerical Representation A signed integer-based representation (rather than floating point) was chosen based on the simplicity and speed of operations carried out with integers. In order to represent signed values, two's complement form is used based on easy adder implementation and widespread use.

The word length (or bit-depth) of the digital integrators is critical to circuit delay time, power dissipation, and die area. Longer word length not only requires more logic to compute, it also causes the carry signal to propagate through more logic. More advanced adder logic can reduce the critical path through the adder, but usually come at the expense of die area and power dissipation.

The results of detailed simulations, are shown in Figure 5-28. These show that the output spectrum is a function of the word length of the digital integrators. This bit-depth simulation used a "no-overflow" simulation, where the integrator values are stored in larger than necessary

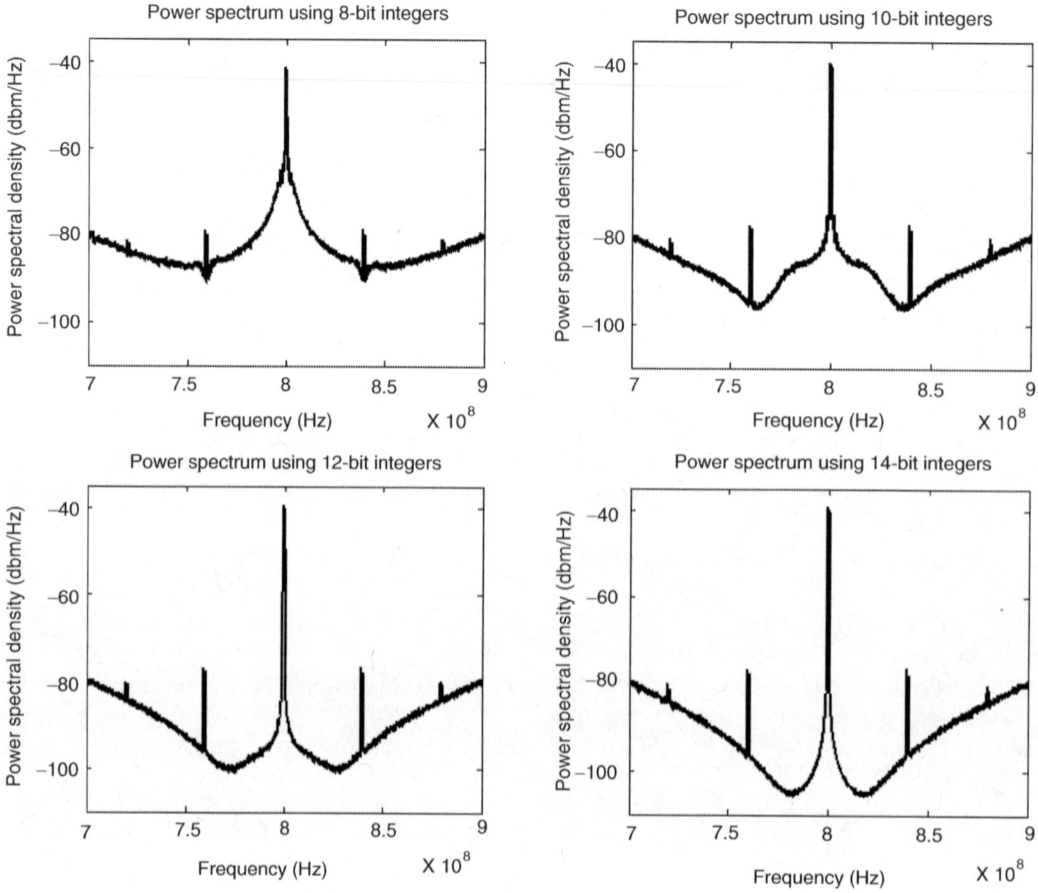

Figure 5-28 Simulated output spectra of CDMA signal using bandpass delta–sigma modulators with different integer bit lengths.

variables, and the entire value history of the variable is recorded in bin values. From the bin values, the log to base two is taken of the largest absolute value. Rounding up, this value results in the effective bit-depth of an integrator that could represent the entire range of values seen by the integrator without an overflow.

From Figure 5-28 it is found that the signal is corrupted excessively by using anything less than 10-bit representation, but that 12-bit representation provides satisfactory performance. Thus the design discussed here is based on a modulator implemented using 12-bit two's complement numerical representation.

5.6.2.4 D. Dithering Dithering is the process of purposely injecting white noise into the input signal of a delta–sigma modulator. This is necessary because delta–sigma modulators are essentially finite state machines. Thus when given an undithered periodic input, the quantization noise of the modulator will be correlated to the input sequence and spurious tones will be present in the output of the modulator [357].

Explicit dithering is essential for the operation of a delta–sigma modulator performing digital-to-analog conversion. In the analog-to-digital case, some noise is always present on the input signal, and thus explicit dithering is not always necessary. Fortunately, dithering here is typically not critical since CDMA signals already are pseudorandom in nature to begin with. The dithering operation is kept anyway in order to suppress tones during periods of malformed input, or startup transients.

Here, dithering is implemented by a linear feedback shift register (LSFR). The LSFR consists of fifteen registers in series with two XOR gates. An LSRF is a finite state machine producing a one-bit pseudorandom sequence that repeats every $2^N - 1$ cycles (where N is the number of registers in the loop). The output of the LFSR is used as the least significant bit of the input to the delta–sigma modulator.

5.6.3 Integrator Implementation

The limiting digital block for the digital transmitter is the integrator or adder within the delta–sigma modulator. This adder must operate at 3.2 GHz so that operations must take no longer than 312 picoseconds. The adder must also detect overflow conditions and override the default behavior. This is problematic since standard overflow detection algorithms XOR the last two carries to determine if an overflow has occurred. The carry out signal in adders is usually the last signal to become valid and thus any logical operations performed with it adds to the critical path delay. In order to solve this problem a scheme called "overflow look ahead" is employed. Rather than determining the overflow condition from carries, overflow look-ahead considers at the two most significant bits in each operand. The functions employed are:

$$\begin{aligned}
\text{Positive Overflow}(A, B) &= A_{\text{MSB}} \text{ AND } B_{\text{MSB}} \text{ AND } A'_{\text{MSB}-1} \text{ AND } B'_{\text{MSB}-1} \\
\text{Negative Overflow}(A, B) &= A'_{\text{MSB}} \text{ AND } B'_{\text{MSB}} \text{ AND } A_{\text{MSB}-1} \text{ AND } B_{\text{MSB}-1}\,.
\end{aligned} \tag{5.8}$$

These functions are slightly more complicated to implement than the standard overflow detection methods, however they depend only on immediately known values and thus do not impact to the critical path of the adder.

Figure 5-29 shows a block diagram of an integrator based on two-stage pipelining, along with the carry-select optimization with overflow look-ahead. The operation is broken into two clock cycle steps. During the first clock cycle each three-bit segment is added together in two separate adders with both carry conditions (except the first three-bit segment, where the carry condition is already known). Also, the overflow conditions are calculated. The second clock cycle uses multiplexers to select and propagate the correct carry combinations. Another set of multiplexers overrides the adder's result if an overflow condition is detected. The key to this architecture is parallelism; all the operations are carried out independently of each other, and come together at the end to create the final result. No operation is held back waiting for the outcome of another.

The actual CMOS implementation of the adders is in dual-rail domino logic. Domino logic was selected because of its high-speed and low-power dissipation. The dual-rail variant was selected because of its completeness; the availability of XORs is a significant benefit for adder implementation. The adder is projected to achieve the required clock rate using CMOS with gate length of 0.18 μm or shorter. The transistor count for the dual-rail domino integrator implementation is estimated to be 1,300.

5.6.4 Alternative Digital Modulator Architectures

In addition to the architecture depicted in Figure 5-27, there are various alternative architectures are under consideration for all-digital transmitters. It has been shown that frequency translation

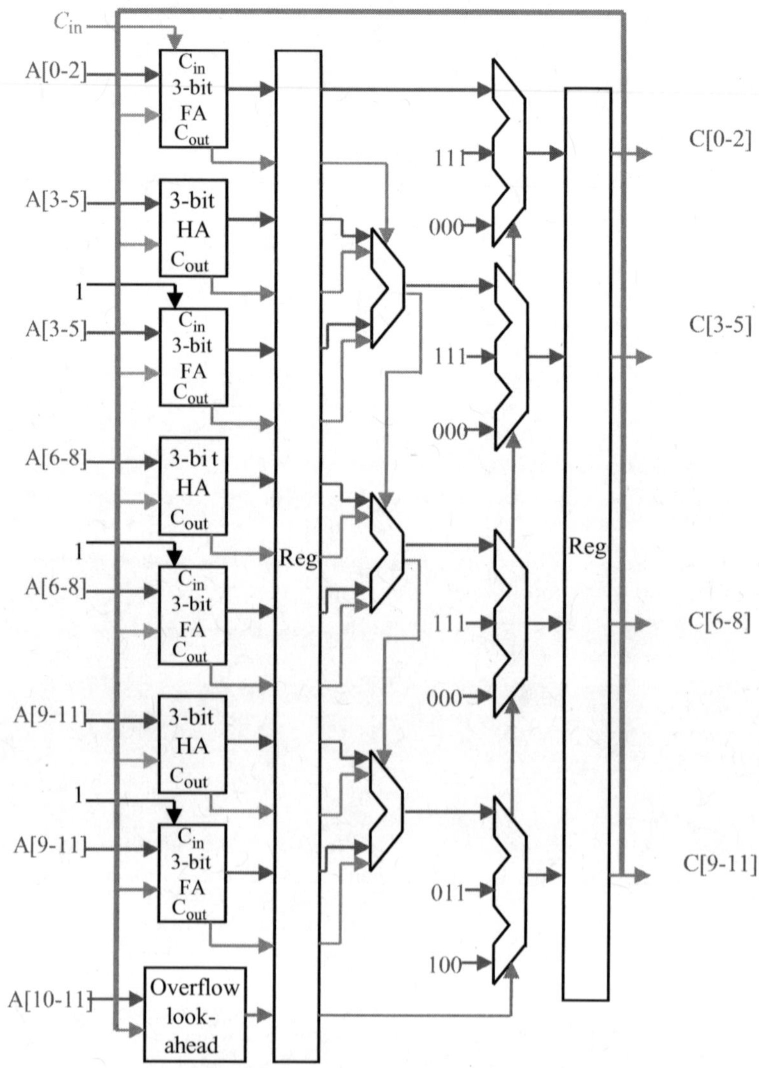

Figure 5-29 Architecture of an integrator or accumulator organized to allow rapid addition and overflow protection.

(IQ upconversion) can be simply carried out while preserving the binary nature (two-level representation) of the digital output [358]. This adds flexibility to the choice of clock frequency used for the delta-sigma modulation, such that it can be selected based on the oversampling requirements needed to achieve a desired SNR as SNR is not dependent on the carrier frequency. Another architecture, shown in Figure 5-30, is based on the equivalent of "polar modulation" in which separate signal streams are generated corresponding to the phase modulation of the output, and to the envelope of the modulation. The "phase signal" is a digital (binary) data stream at the carrier frequency, which has approximately 50% duty cycle. It can be generated by phase-locked loops in traditional analog form; by the recently introduced "digitally-controlled oscillators" [359]; or by purely digital synthesis techniques. The envelope signal is at the modulation frequency. At the

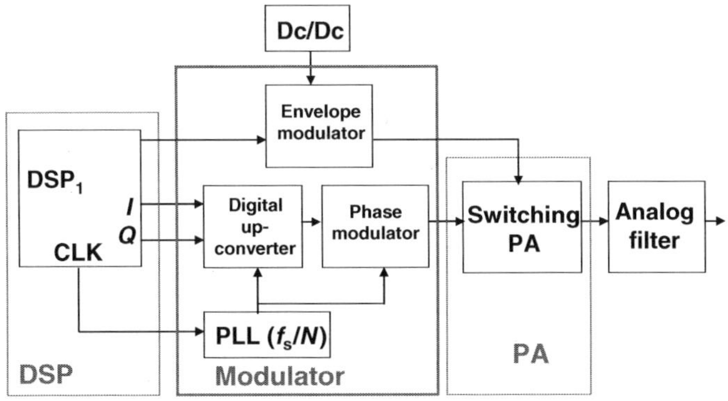

Figure 5-30 Architecture for digital transmitter employing polar modulation.

power amplifier the envelope and phase signals are recombined so that the overall output has the envelope and phase represented by the two signals, respectively.

5.7 ADAPTIVE OUTPUT FILTERING

In wireless transmitters there is often residual nonlinearity which leads to the production of spurious emissions outside of the intended frequency band. Such spurious outputs can correspond to specific frequencies as a result of carrier, IF and clock mixing (or intermodulation) products. Spurious emissions can also be broad spectrum associated with intermodulation distortion of the modulated output signal or, in the case of the digital transmitters discussed above, residual quantization noise. The spurious outputs, if radiated from the antenna, contaminate the frequency spectrum for nearby users and can cause interference in their assigned receive bands. For wireless systems that operate with frequency division duplexing, the transmitter and receiver are simultaneously active, and there can be major problems if the transmitter creates noise in the receive band. In such systems there is a duplexer filter that attenuates signals from the transmitter in the receive band before they reach the antenna or the receiver. However, the attenuation provided by the duplexer may not be sufficient to remove the unwanted components. In this section an adaptive duplexer is described that can significantly enhance the stopband attenuation of a duplexer at selected frequencies, in order to eliminate spurious emissions of the transmitter [360]. In the work described here an adaptive null in the transmission characteristics of the duplexer is achieved using feedforward techniques. Feedforward has been traditionally thought of as a means for canceling distortion in amplifiers [345], but it is also useful for noise cancelation. In the feedforward system, a portion of the input is sampled, and then adjusted in amplitude and phase to cancel the unwanted signal at a different part of the circuit. The null can be adjusted by altering the amplitude and phase of the signal as it passes through the feedforward loop, allowing the user to quickly alter the characteristics of the duplexer to "clean up" a specific communication band. To achieve perfect cancelation, however, the amplitude and phase variations through the two paths must be equal. In the presence of an amplitude mismatch ΔA and phase mismatch $\Delta\phi$, the residual signal voltage remaining after the cancelation operation for a normalized input signal is given by:

$$|V_{\mathrm{OUT}}|^2 = [1 - \cos(\Delta\phi) - \Delta A\cos(\Delta\phi)]^2 + [\sin(\Delta\phi) + \Delta A\sin(\Delta\phi)]^2 \ . \tag{5.9}$$

The effect of a group delay mismatch between the paths also needs to be considered; it can significantly reduce the bandwidth over which the feed-forward system is effective. The output

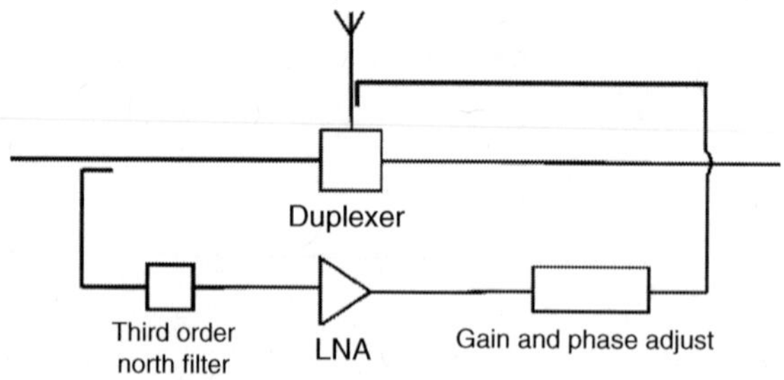

Figure 5-31 Block diagram of adaptive duplexer feedforward system.

amplitude for a normalized input signal in the presence of a delay mismatch $\Delta\tau$ between the two paths is given by:

$$V_{\text{out}} = 2\cos(\Delta\omega\Delta\tau + \pi/2)\,. \tag{5.10}$$

The adaptive duplexer design is shown in Figure 5-31. Feedforward is applied around a conventional duplexer operating between the TX (transit) and ANT (antenna) ports to enhance its performance. In the system, a fraction of the input signal is coupled off before the TX port of the duplexer. The required amplitude and phase adjustments are then applied to this signal, and the resultant is recombined with the main signal at the ANT port of the duplexer. A notch filter is added in the feed-forward loop to prevent signals in the transmit band from passing through the loop, to prevent cancellation of the desired transmit signal. The group delay difference between the added filter and the duplexer create a time delay mismatch between the two paths, which restricts the bandwidth of the system. The inclusion of the duplexer in the system allows a significant reduction in the power requirements of the system, since the duplexer provides an attenuation of the order of 40 dB. Thus the error signal amplifier shown in Figure 5-31 can be operated linearly with very low power dissipation.

The system was implemented experimentally, using an EPCOS duplexer, a hybrid LNA based on an Agilent AT-41511 transistor operated in Class A, and lumped element third-order notch filter, at a frequency band centered around 880 MHz. The phase-shifter was based on barium strontium titanate (BST) varactors, obtained from Agile Materials. The measured performance of the adaptive duplexer is shown in Figure 5-32, for three different settings of the attenuator

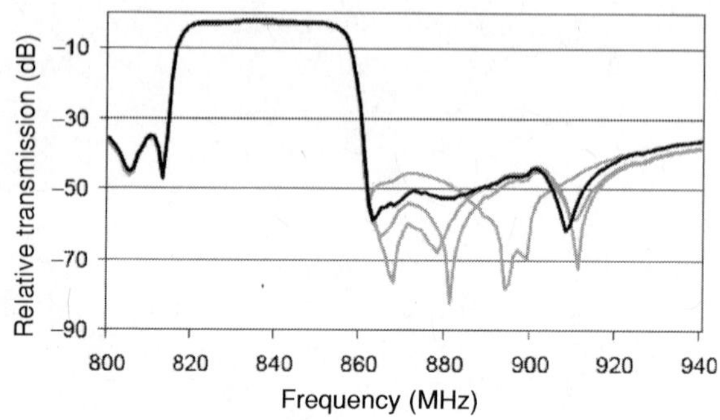

Figure 5-32 Measured adaptive duplexer performance, for three representative settings of gain and phase. The original duplexer response is also shown.

and phase-shifter. It can be seen that a null in the transmission spectrum can be produced at a frequency that can be adjusted to cover the entire receive band, with a depth greater than 20 dB. The bandwidth of the null at any particular setting is greater than 2 MHz, thus is sufficient for most communication applications. The associated insertion loss of the adaptive duplexer is low (measured to be 0.27 dB), and the associated power dissipation is also low (9.75 mW for this example). Adjustable filters of this type, operating with adaptive control loops to minimize the interference, can be expected to dramatically reduce unwanted interference from transmitters.

5.8 PROJECTIONS AND CHALLENGES FOR FUTURE

It can be expected that the adaptive techniques described here will become pervasive in the future. As the advance of digital electronics follows Moore's law it can be expected that elaborate control functions can be applied in the future to microwave power amplifiers, with little added power, space or cost. The mixed signal circuits that interface the microwaves to the digital world will also continue to make strides, and will enable increased response speed for the control functions. Development of appropriate sensors with very wide dynamic range will likely remain a challenge, requiring detailed microwave/analog design. MEMS actuators that provide low loss, low nonlinearity, wide-frequency response, and high reliability at low cost are also likely to emerge in the near future, enabling a wide variety of adaptive responses. An associated challenge, however, with be the packaging of the diverse components in a hermetic environment that allows transmission of the microwave signals with little attenuation. The result of this effort will be new generations of wireless transmitter circuits with greatly enhanced operational flexibility, efficiency, and robustness. These can be enablers of the "software radio" and "cognitive radio" systems envisioned for the future.

6

Broadband Network Design

Wael E. Fathelbab and Michael B. Steer

6.1 INTRODUCTION

Multifunctional RF systems generally require subcircuits with broad prescribed frequency response—sometimes over multiple octaves of bandwidth. The design of such subsystems is the best approached using network synthesis. The synthesis approach provides design insight, and only requires minimum optimization in a microwave circuit simulator. The reason for this is that the specified transfer function is decomposed into its constituents, i.e., the elements of the synthesized prototype, thus enabling control of all available design parameters. Aided by the use of appropriate circuit transformations, novel topologies can be created that achieve performances as close as possible to the theoretical optimum in terms of size and bandwidth. The final realization most often uses parallel coupled line (PCL) transmission line sections as lumped components, particularly inductors, are too lossy. PCL filters [361] do not necessarily require ground connections and thus are easily fabricated in planar form. They have been used to realize passband bandwidth ratios ranging from as small as 1.02:1 to as large as 3:1 [362]. Another advantage of PCL filters is that their resonators can be folded into hairpin form making them suitable for applications with space restrictions [363,364]. However, when a PCL filter is realized using inhomogeneous transmission lines such as microstrip or coplanar waveguide, it suffers from poor upper stopband performance typically having spurious passbands centered at odd harmonics of the fundamental passband center frequency, f_0.

The broadband network design concepts in this chapter exploit the intrinsic frequency-shaping properties of a PCL and development is based on novel prototypes that provide capabilities required in broadband multifunctional systems such as prescribed frequency characteristics over multiple octaves. These concepts are applied to a number of broadband networks: miniature Marchand baluns, in Section 6.3; distributed biasing networks for differential RF circuits, in Section 6.4, differential networks with broadband common-mode rejection, in Section 6.5; and networks with combined balun and matching functionality, in Section 6.6; The final development is presented in Section 6.7 with the design of bandpass filters with enhanced stopband performance.

6.2 PARALLEL COUPLED LINE (PCL) FILTERS

In this section the properties of coupled lines, and the almost universal approach to developing filters based on them, are described.

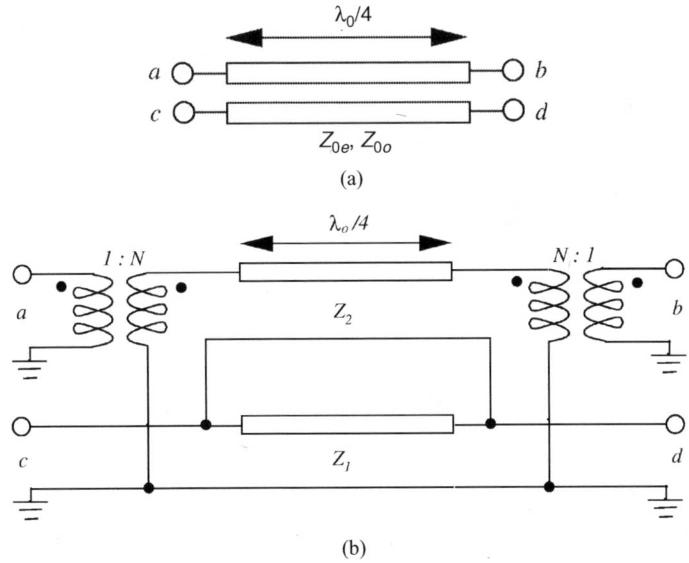

Figure 6-1 A pair of symmetrical coupled-lines: (a) physical layout; and (b) its equivalent circuit model with $N = (Z_{0e} + Z_{0o})/(Z_{0e} - Z_{0o})$.

A physical pair of coupled lines is shown in Figure 6-1(a) and propagation on these lines is partly described by the even-mode characteristic impedance Z_{0e} and the odd-mode impedance Z_{0o} with the coupling coefficient defined as

$$K = \frac{Z_{0e} - Z_{0o}}{Z_{0e} + Z_{0o}}. \tag{6.1}$$

The corresponding equivalent network model of the coupled line in Figure 6-1(a) is shown in Figure 6-1(b) with the network parameters:

$$N = \frac{1}{K} \tag{6.2}$$

$$Z_1 = \frac{Z}{\sqrt{1 - K^2}} \tag{6.3}$$

$$Z_2 = Z\frac{\sqrt{1 - K^2}}{K^2} \tag{6.4}$$

where

$$Z = \sqrt{Z_{0e} Z_{0o}}. \tag{6.5}$$

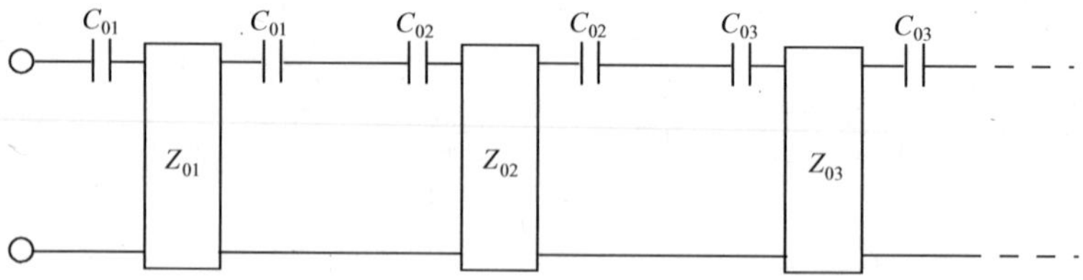

Figure 6-2 S-plane highpass prototype.

These lead to the ABCD matrix of a pair of uniform, open-circuited coupled lines in an inhomogeneous media [365]:

$$A = \frac{Z_{0e}\cot(\theta_e) + Z_{0o}\cot(\theta_o)}{Z_{0e}\csc(\theta_e) - Z_{0o}\csc(\theta_o)} = D$$

$$B = \frac{j}{2}\frac{Z_{0e}^2 + Z_{0o}^2 - 2Z_{0e}Z_{0o}\left(\cot(\theta_e)\cot(\theta_o) + \csc(\theta_e)\csc(\theta_o)\right)}{Z_{0e}\csc(\theta_e) - Z_{0o}\csc(\theta_o)}$$

$$C = \frac{2j}{Z_{0e}\csc(\theta_e) - Z_{0o}\csc(\theta_o)} \tag{6.6}$$

where Z_{0e} and Z_{0o} are the modal impedances, and θ_e and θ_o are the even- and odd-mode phase lengths. Filter design begins, generally, with lumped element lowpass filter circuits called prototypes [366]. These are transformed to obtain responses centered at radio frequencies. For PCL filters these are equated to transmission line sections by matching ABCD parameters. This transformation is eloquently captured by S, the Richards variable, defined as

$$S = j\Omega_{HP} = j\tan(\theta_{HP}) = j\tan\left(\frac{\pi}{2}\frac{f}{f_0}\right). \tag{6.7}$$

Consequently the traditional design of distributed filters is based on the S-plane highpass prototype depicted in Figure 6-2(a).

In Equation (6.7) f and S are the real and complex frequency variables respectively, and f_0 is the center frequency of the fundamental passband in the f-plane (the real world). This prototype can be synthesized in the S-plane by exact methods [367, 368] and then transformed to the f-plane using Equation (6.7). The rectangular blocks in the S-plane prototype of Figure 6-2 are known as Unit Elements (UE) [368] and become transmission lines in the f-plane whereas the capacitors represent open-circuited stubs in the f-plane. The input impedance of a UE with a load $x\ \Omega$ is $1/x\ \Omega$. The distributed elements in the f-plane are all transmission lines quarter-wave length long at f_0. Now each building block in the S-plane prototype comprises a pair of series capacitors separated by a UE and this can be replaced by a PCL section using the equivalence shown in Table 6-1. The critical property of the transition from the S-plane to the corresponding f-plane is that the f-plane response of the prototype becomes a periodic bandpass function with spurious passbands centered at odd multiples of f_0. This is an intrinsic property of the prototype of Figure 6-2. An additional complexity arises when this filter is realized on an inhomogeneous transmission media due to the inequality of the even- and odd-mode phase velocities. This effect can be investigated by utilizing the full ABCD matrix, see Equation (6.6), of each PCL section in a filter for a known even- to odd-mode velocity ratio. The inhomogeneous medium results in additional spurious passbands centered approximately at even multiples of f_0 [369–383]. In Figure 6-3

Table 6-1 Subsection equivalence. The ideal representation shown on the left-hand side has the approximate physical realization on the right. The representation is exact at the frequency with wavelength λ_0.

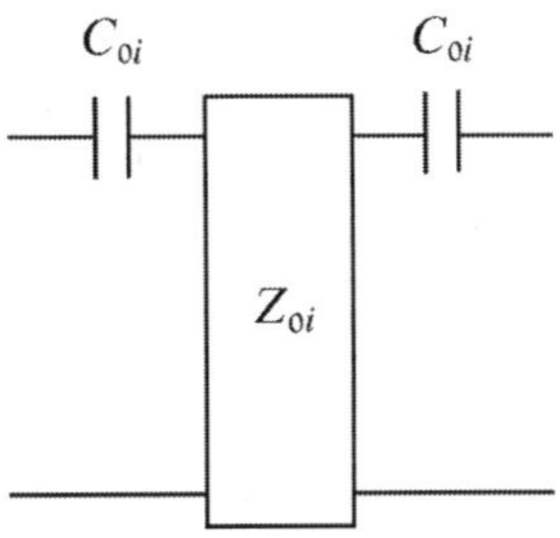

S-plane subsection comprising a pair of series capacitors separated by a UE

$$Z_{0ei} = (1/C_{oi} + 2Z_{0i}),$$
$$Z_{0oi} = Z_{0i}$$

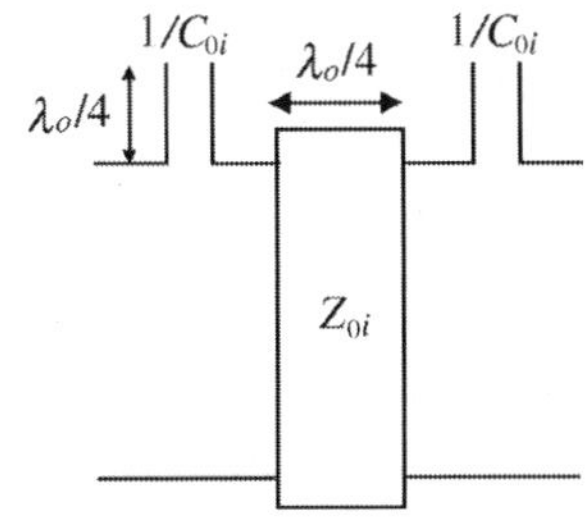

f-plane subsection comprising a pair of series open-circuited stubs separated by a transmission line

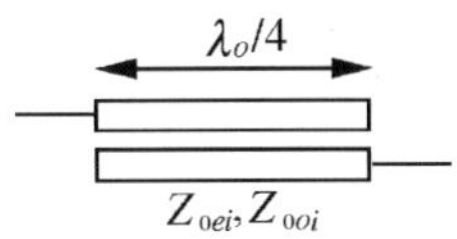

Physical equivalence

the response of a conventionally designed 5-th order Chebychev filter is shown [381]. By conventionally designed it is meant that the length of PCL sections is approximately a quarter-wavelength long.

A key concept behind all of the broadband networks presented in this chapter is the use of transmission line sections that are much shorter than a quarter-wavelength long at the center frequency of the network (usually the passband frequency).

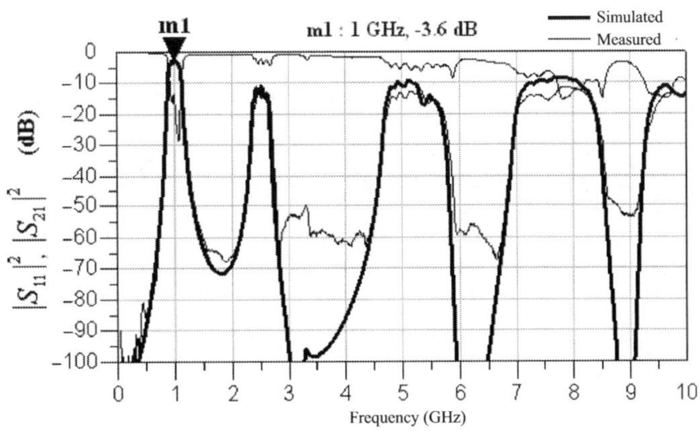

Figure 6-3 Measured and simulated wideband frequency responses of a conventional fifth-order parallel-coupled line filter utilizing optimized (Type 1) stepped-impedance resonators [383].

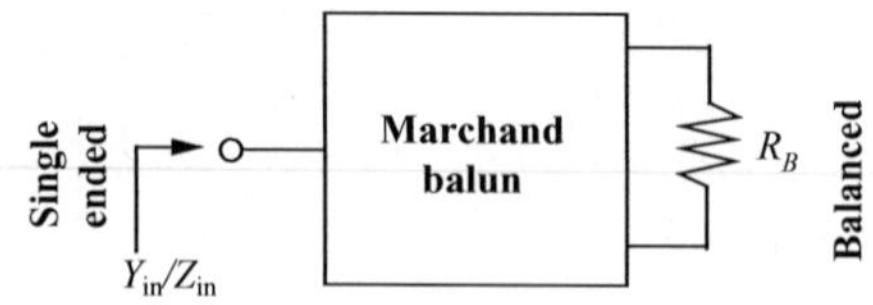

Figure 6-4 Marchand balun.

6.3 MINIATURE PLANAR MARCHAND BALUNS

Baluns [384] are key components of radio frequency (RF) and microwave communication system. They are used in balanced circuits, such as double balanced mixers, push-pull amplifiers, and frequency doublers [385]. Another application of a balun is in a system using a radio frequency integrated circuit (RFIC) where a balun transforms the differential outputs of an RFIC to unbalanced microwave circuitry. There are many types of baluns [386, 387] with the Marchand balun [384, 385, 388–390] being the most popular at microwave frequencies as it can be conveniently realized in planar or coaxial forms [384].

In this section three new classes of miniaturized Marchand balun are defined following the development presented in [391]. They are suitable for mixed lumped-distributed planar realizations with small size resulting from transmission line resonators being quarter-wavelength long at frequencies higher than the passband center frequency. Each class corresponds to an S-plane bandpass prototype derived from the specification of transmission zero locations. A tunable 50:100 Ω balun is realized at 1 GHz.

6.3.1 Marchand Balun

The most common form of microwave balun is the Marchand balun shown in symbolic form in Figure 6-4 and in planar form in Figure 6-5(a). The balun has two single-ended ports with one being identified on the left of Figure 6-5(a) and the other on the right-hand side with terminal x. The balun also has a balanced port with characteristic impedance R_B. To a limited extent the network

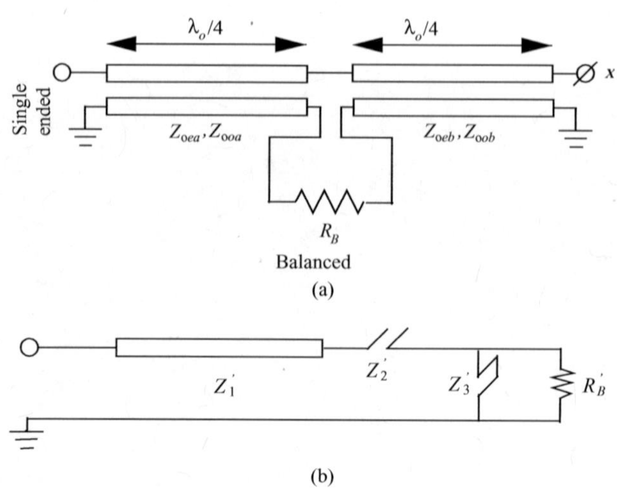

Figure 6-5 Marchand balun: (a) physical layout with balanced load impedance R_B; and (b) S-plane highpass prototype.

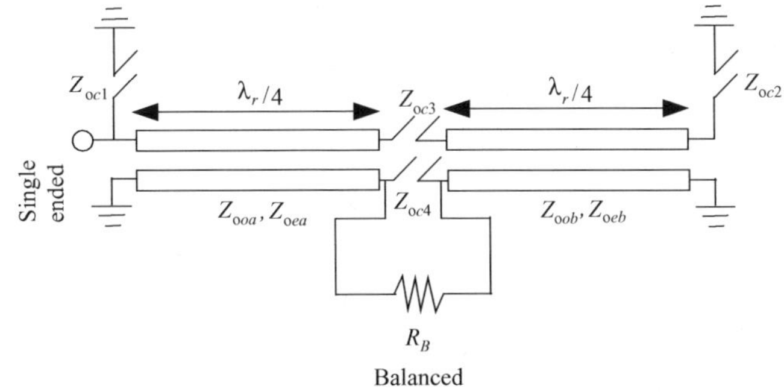

Figure 6-6 An S-plane bandpass Marchand balun with balanced load impedance R_B.

synthesis approach has been applied to the design of the Marchand balun [390]. This includes an implementation of the coaxial transmission line Marchand balun [392] and its corresponding planar coupled-line form [389]. In both approaches the balun design was based on an S-plane highpass prototype using the Richards' transformation, see Equation (6.7). Upon synthesis of such a prototype the coupled-line parameters of the Marchand balun can then be computed [392]. The balun is shown in its single-ended form as this is the form that is synthesized. In particular, the synthesis of a balun is the synthesis of a one-port network with the appropriate topology that results in a balun.

The drawback of such a design is that the center frequency of the balun is actually the resonant frequency of the transmission line resonators forming the balun. Thus the overall size of a conventional Marchand balun is determined by its center frequency. This limits the use of the S-plane highpass prototype in situations where small electrical size is critical such as in systems operating at low frequencies.

Miniaturization of the Marchand balun requires that the balun's center frequency, f_0, and the quarter-wave frequency, f_r, referred to as the commensurate frequency of the transmission line resonators, be different. To permit independent specification of f_0 and f_r, an S-plane bandpass prototype must be synthesized so that in the f-plane a periodic bandpass behavior is achieved. This basic idea was used to design a planar lumped-distributed transmission line balun [387] consisting of a power divider and a pair of bandpass phase shifters. Now Richards' transformation previously defined in Equation (6.7) becomes

$$S = j\tan\left(\frac{\pi}{2}\left(\frac{f}{f_r}\right)\right) \tag{6.8}$$

with f_r chosen to be above the passband centered at f_0. Hence a balun is reduced in size.

6.3.2 Derivation of A Bandpass Marchand Balun

The focus of this subsection is demonstrating that the procedure of element extraction from a prototype results in the most general bandpass balun topology. By appropriately loading the highpass balun of Figure 6-5(a) with open-circuited stubs, the bandpass balun of Figure 6-6 results. (The equivalent prototype of this structure can be derived in an identical fashion to that in [389] by utilizing the network model of a pair of coupled lines in homogeneous media [368].)

By substituting the network model for each of the coupled-line sections comprising the balun, the balun network of Figure 6-7(a) is obtained. Further simplification of this prototype would require that the pair of transformers in parallel with the series stub, Z_{0c3}, be of identical turns

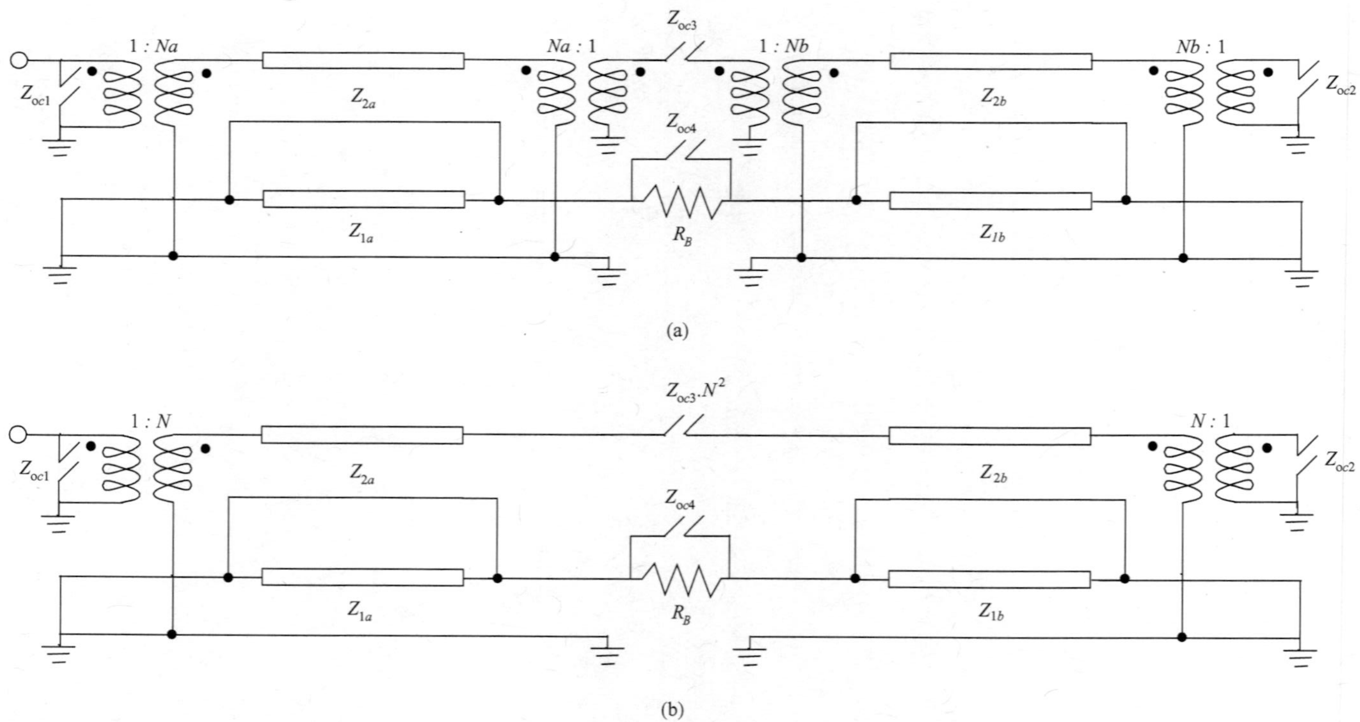

Figure 6-7 Derivation of the S-plane bandpass prototype of the balun of Figure 6-6: (a) after substitution of the network model of Figure 6-1(b) for each pair of coupled lines; and (b) after scaling up the series open-circuited stub, Z_{0c3} by N^2.

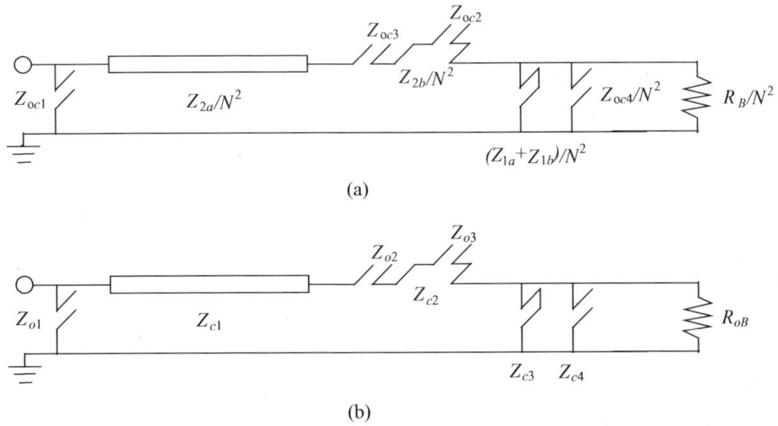

Figure 6-8 Derivation of the S-plane bandpass prototype of the balun of Figure 6-6: (a) after shifting the input transformer throughout the network and absorbing it into the load resistance; and (b) final prototype.

ratio. This leads to the loss of one degree of freedom since the pair of coupled lines comprising the balun must now have the same coupling coefficient. Proceeding with this assumption results in the circuit, Figure 6-7(b), with N_a and N_b being of equal value N; and the impedance of the series stub, Z_{0c3}, scaled up to $Z_{0c3}\,N^2$. Upon further simplification the prototype of Figure 6-8(a) is obtained after shifting the input transformer to the far right and absorbing it into the balanced load impedance. This is a redundant S-plane bandpass prototype with two zeros at $S = 0$, i.e., at DC, two zeros at infinity, and a redundant quarter-wave long transmission line as a UE. So given the impedances of the final prototype as shown in Figure 6-8(b), the PCL parameters of the balun can be extracted after solving the following equations:

$$Z_{c1} = \frac{Z_{2a}}{N^2} = Z_a\sqrt{1-K^2} \tag{6.9}$$

$$Z_{c2} = \frac{Z_{2b}}{N^2} = Z_b\sqrt{1-K^2} \tag{6.10}$$

$$Z_{c3} = \frac{Z_{1a}+Z_{1b}}{N^2} = (Z_a+Z_b)\frac{K^2}{\sqrt{1-K^2}}. \tag{6.11}$$

This gives the coupled-line parameters in terms of the prototype's circuit elements as follows:

$$K = \sqrt{\frac{Z_{c3}}{Z_{c1}+Z_{c2}+Z_{c3}}} \tag{6.12}$$

$$Z_a = \frac{Z_{c1}}{\sqrt{1-K^2}} \tag{6.13}$$

and

$$Z_b = \frac{Z_{c2}}{\sqrt{1-K^2}}. \quad (6.14)$$

Once K, Z_a, and Z_b are determined, the even- and odd-mode impedances of each individual coupled-line pair comprising the balun can be found after solving Equations (6.5) and (6.1) to give

$$Z_{0ea} = Z_a\sqrt{\frac{1+K_a}{1-K_a}} \text{ and } Z_{0oa} = \frac{Z_a^2}{Z_{0ea}} \quad (6.15)$$

$$Z_{0eb} = Z_b\sqrt{\frac{1+K_b}{1-K_b}} \text{ and } Z_{0ob} = \frac{Z_b^2}{Z_{0eb}}. \quad (6.16)$$

Also, for further clarity, the coupling coefficients used in Equations (6.15) and (6.16) are, as stated earlier, of identical values and defined using Equation (6.1) as

$$K_a = \frac{Z_{0ea} - Z_{0oa}}{Z_{0ea} + Z_{0oa}} = K_b = \frac{Z_{0eb} - Z_{0ob}}{Z_{0eb} + Z_{0ob}} = K. \quad (6.17)$$

Now the impedances of the open-circuited stub, Z_{0c4}, and the balanced load, R_B, can be found from

$$Z_{0c4} = \frac{Z_{c4}}{K^2} \quad (6.18)$$

and

$$R_B = \frac{R_o B}{K^2}. \quad (6.19)$$

Finally, without any further scaling, the impedances of the remaining stubs are found to be

$$Z_{0c1} = Z_{01}, \quad Z_{0c2} = Z_{03}, \text{ and } Z_{0c3} = Z_{02}. \quad (6.20)$$

Although the synthesis equations developed above are based on a specific prototype corresponding to a bandpass balun topology, the procedure is general and is adaptable to other classes of baluns.

6.3.3 S-Plane Bandpass Prototypes

The development of generations of prototypes culminating in the final implementation form uses the transformations shown in Figure 6-9. A logical step in deriving miniaturized baluns is to apply the half-angle transform [393] to the synthesized S-plane highpass prototype [367, 389, 392] of Figure 6-5. This transform is shown in Figure 6-9(a) for an open-circuited stub. To elaborate on the mathematics involved in this transform, consider the impedance of an open-circuited stub [393]:

$$Z_{0c} = \frac{Z_0}{S}S = -jZ_0\cot(\theta) = j\left(\frac{Z_o}{2}\right)\tan\left(\frac{\theta}{2}\right) - j\left(\frac{Z_0}{2}\right)\tan\left(\frac{\theta}{2}\right) \quad (6.21)$$

where $\theta = (\pi/2)(f/f_0)$. Equation (6.21) describes the series connection of short- and open-circuited stubs having characteristic impedances of $Z_0/2$ with half the original electrical length.

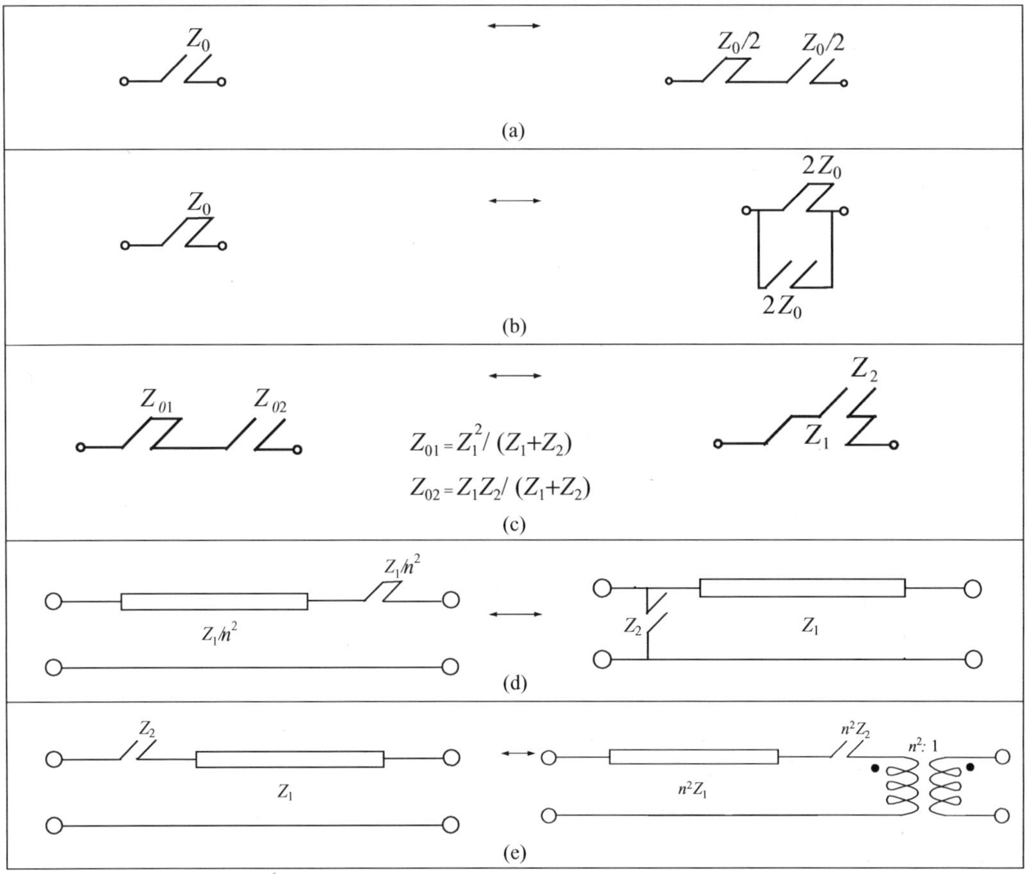

Figure 6-9 Relevant circuit transformations: (a) open-; (b) short-circuited stubs after application of the half-angle transform from [393]; (c) equivalence between series connection of an open- and short-circuited stubs and stepped impedance transmission line from [368]; (d) first Kuroda transformation of a series short-circuited stub; and (e) second Kuroda transformation of a series open-circuited stub from [368].

This implies that the resulting transmission line resonators are quarter-wavelength long at $2f_0$, i.e., they are an eighth-wavelength long at f_0. Through a similar analytical treatment the short-circuited stub has the equivalence shown in Figure 6-9(b). Application of this transform to the highpass prototype of Figure 6-5 leads to the transformed prototype of Figure 6-10(a) with the input UE now split into two equal sections. While the half-angle transformation does not affect the transmission characteristics of the network, it actually changes the distribution of transmission zeroes of the prototype. Now the prototype possesses two zeros at DC, two zeros at infinity, and two non-redundant UEs, i.e., it is now a bandpass prototype with a commensurate frequency $f_r = 2f_0$. Thus the balun equivalent of this prototype, shown in Figure 6-10(b), is reduced in size by almost a factor of 2 compared to a conventional balun design. However the series open-circuited stub is difficult to realize in planar technology such as microstrip, and there is also scope for further size reduction if the open-circuited stubs are approximated by lumped capacitors. This is shown in Figure 6-10(c). This prototype is therefore identified as Class A. In an attempt to

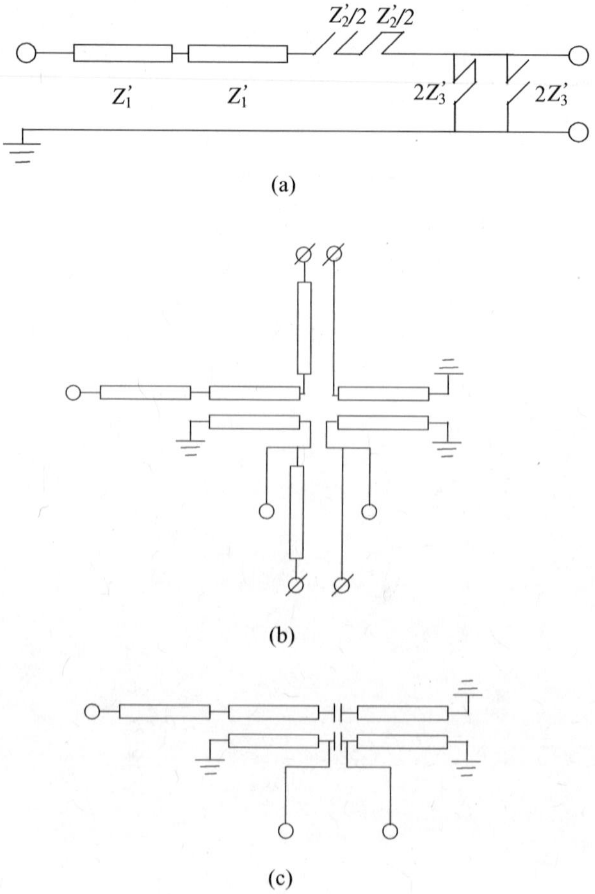

Figure 6-10 Class A balun: (a) prototype derived by transforming the highpass prototype of Figure 6-5(b) using the half-angle transformation; (b) physical topology loaded with open-circuited stubs, and (c) miniaturized mixed lumped-distributed physical topology. (The distributed prototype has two zeros at DC, two zeros at infinity, and two nonredundant UEs).

derive other prototypes suitable for realization as Marchand balun structures, different sets of transmission zeros were tried and various prototype networks were synthesized. Application of the classical synthesis technique in the transformed variable [394, 395] for an arbitrarily selected f_0, and f_q with specific sets of transmission zeros, results in the prototype Classes B and C shown in Figure 6-11 and Figure 6-12 respectively. In fact Class A prototypes can also be directly synthesized using the same technique [394, 395].

In subsequent design steps the synthesized network prototypes undergo several network transformations in the S-plane. This allows scaling of different elements of the prototype in order to obtain realizable even- and odd-mode impedances after using Equations (6.12)–(6.14). These circuit transformations also lead to various balun topologies and allow flexible designs to be achieved. Essential circuit transformations are shown in Figure 6-9. As an example, transformed Class B prototypes are presented in Figure 6-13. Similar transformed Class A prototypes may also be obtained in the same way.

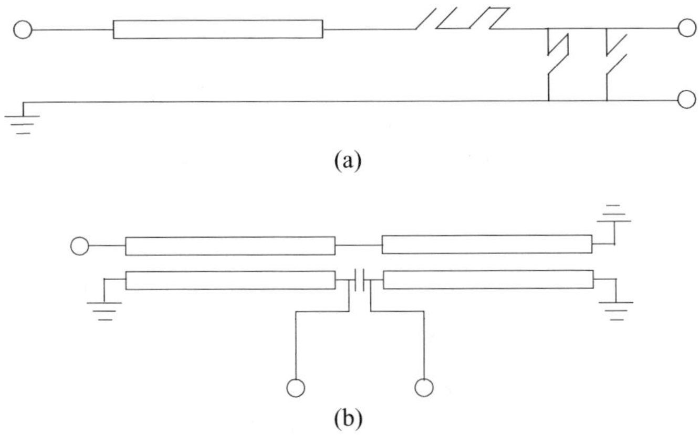

Figure 6-11 Class B balun: (a) prototype with two zeros at DC, two zeros at infinity, and a redundant UE matching the single-ended source impedance; and (b) miniaturized mixed lumped-distributed physical topology.

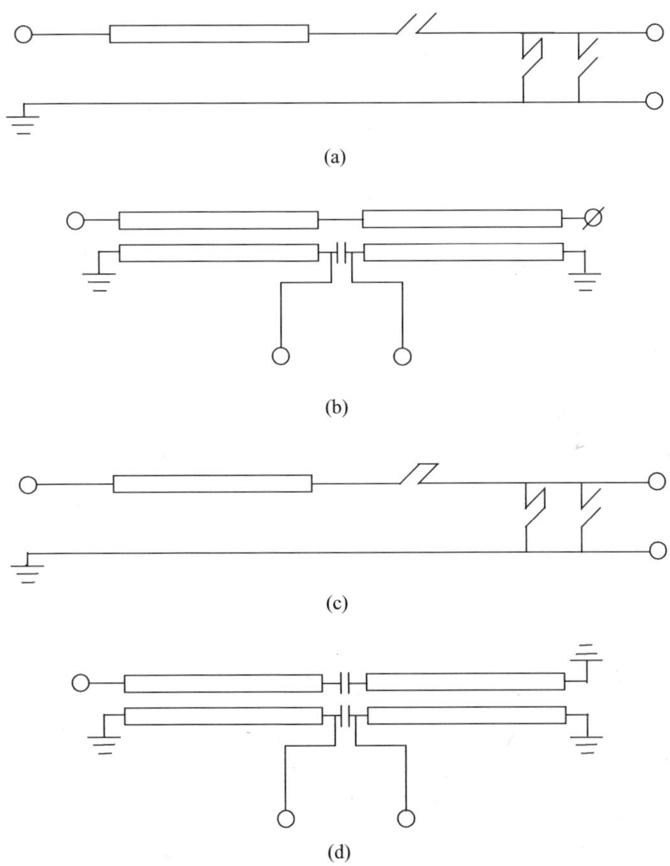

Figure 6-12 Class C baluns: (a) C1 prototype with two zeros at DC, a zero at infinity, and a nonredundant UE; (b) its corresponding miniaturized mixed lumped-distributed physical topology; (c) C2 prototype with a zero at DC, two zeros at infinity, and a nonredundant UE; and (d) corresponding miniaturized mixed lumped-distributed physical topology.

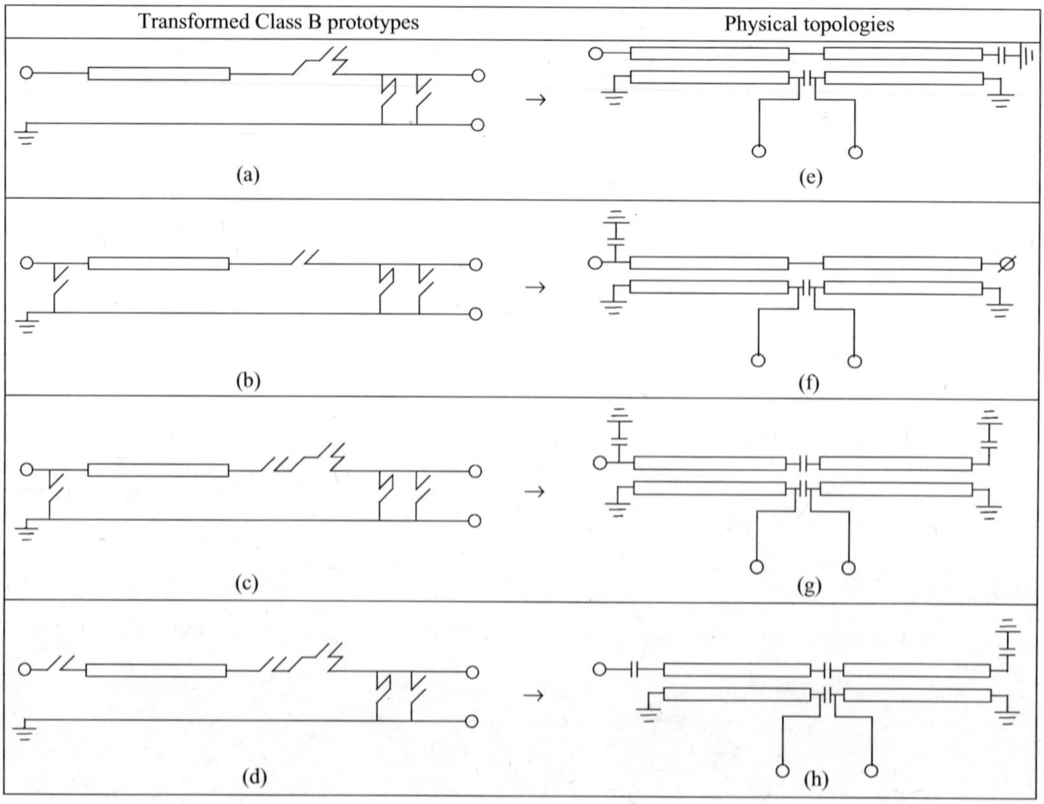

Figure 6-13 Transformed Class B prototypes of Figure 6-11(a): (a) after application of the stepped-impedance transmission line transformation to the series branch; (b) after application of the first Kuroda transformation; (c) after splitting the series connection of the open- and short-circuited stubs, application of the first Kuroda transformation to part of the series short-circuited stub and subsequently transforming the rest of the series branch to a stepped impedance transmission line; (d) after splitting the series connection of open- and short-circuited stubs, application of the second Kuroda transformation to part of the series open-circuited stub and subsequently transforming the rest of the series branch to a stepped-impedance transmission line (and then absorbing the input transformer into the load impedance); and (e)–(h) corresponding miniaturized mixed lumped-distributed physical topologies.

So the design strategy is to first classify the S-plane bandpass prototypes from a set of transmission zeros, apply circuit transformations, and then convert the specific prototypes to their physical balun realizations. The last step is performed by inspection in a backward manner as was done in Figure 6-7. Subsequently the open-circuited stubs loading the balun are all approximated by lumped capacitors at the center of the passband in the f-plane to obtain mixed lumped-distributed baluns. The next section demonstrates this process using a step-by-step numerical example.

6.3.4 Example: Class B Balun

Descriptions of network synthesis can be involved and so here a design example is presented to illustrate the procedure.

A 50:100 Ω Class B balun with a center frequency, f_0, of 1 GHz, and a commensurate frequency, f_r, of 2 GHz is synthesized here. The objective is to design a balun with a return loss greater than

15 dB at the single-ended port, and a bandwidth ratio of 1.54:1 giving lower and upper passband edge frequencies of 0.82 GHz and 1.26 GHz respectively. The purpose is to implement the balun on an FR4 Printed Circuit Board (PCB) with the following specification:

$$\begin{aligned}\text{Substrate thickness}: &\ 62\text{ mil }(1.57\text{ mm})\\ \text{Relative dielectric constant}: &\ 4.7\\ \text{Metal thickness}: &\ 2.2\text{ mil }(0.05\text{ mm})\\ \text{Minimum spacing}: &\ 7\text{ mil }(0.17\text{ mm})\\ \text{Loss tangent}: &\ 0.016.\end{aligned} \tag{6.22}$$

Now the characteristic polynomial, $K(S)$, of the prototype may be constructed for a passband ripple of 0.05 dB using the procedure described in [394] (or [395]) with two zeros at DC, and two zeros at infinity leading to[1]

$$K(S) = \frac{(0.7232261254S^4 + 1.868909798S^2 + 0.9448913438)}{(2S^2)} \tag{6.23}$$

from which the square of the magnitude of the reflection transfer function is evaluated using

$$|S_{11}(S)|^2 = \frac{|K(S)|^2}{1+|K(S)|^2}. \tag{6.24}$$

$S_{11}(S)$ may then be found with the knowledge of

$$|S_{11}(S)|^2 = S_{11}(S) \,.\, S_{11}(-S) \tag{6.25}$$

leading to

$$S_{11}(S) = \frac{(0.7232261254S^4 + 1.868909798S^2 + 0.9448913438)}{\left(\begin{array}{l}0.7232261254S^4 + 1.611778003S^3 + 3.664909798S^2\\ +1.842294719S + 0.9448913438\end{array}\right)}. \tag{6.26}$$

The input impedance is then evaluated in a 1 Ω system from

$$Z_{in}(S) = \frac{1+S_{11}(S)}{1-S_{11}(S)} \tag{6.27}$$

which is then synthesized using standard element extraction. Upon synthesis the resulting prototype is scaled up to suit a 50 Ω system and a 50 Ω UE was then added after the source impedance resulting in the prototype shown in Figure 6-14(a) where the circuit parameters come from Equations (6.9)–(6.11):

$$Z_{c1} = 50\ \Omega, \quad Z_{c2} = 44.863\ \Omega \text{ and } Z_{c3} = 60.434\ \Omega. \tag{6.28}$$

---

1 High precision of the numerical values must be retained throughout the synthesis process.

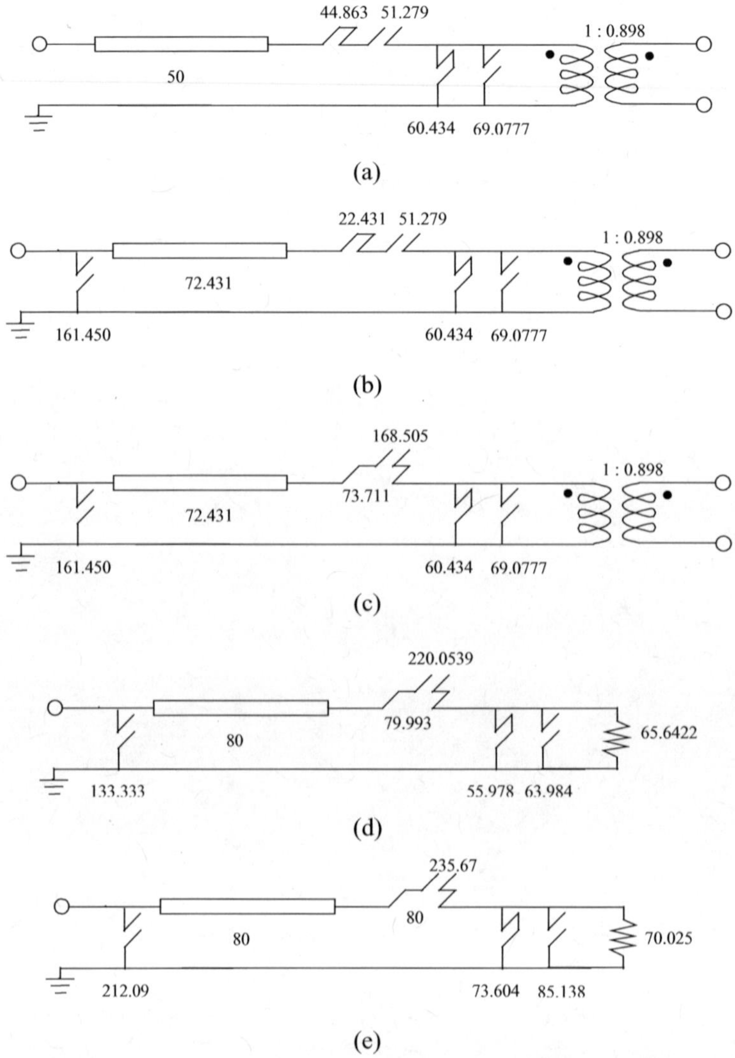

Figure 6-14 Relevant transformations for a 50:100 Ω, 1 GHz balun: (a) Class B prototype with passband ripple of 0.05 dB; (b) after equally splitting the series short-circuited stub and applying the first Kuroda transformation; (c) after transforming the series branch into a stepped impedance transmission line; (d) resynthesized and retransformed Class B prototype with passband ripple of 0.0806 dB; and (e) after manual optimization applied to the prototype of (d) to obtain a return loss of approximately 20 dB.

Substituting these into Equations (6.12)–(6.14) gives the coupled-line parameters

$$K_a = K_b = K = 0.624, \quad Z_a = 63.974, \text{ and } Z_b = 57.401 \ \Omega. \tag{6.29}$$

Now the parameters in Equation (6.29) are substituted in Equations (6.15)–(6.16) yielding

$$Z_{0ea} = 132.915 \ \Omega, \quad Z_{0oa} = 30.792 \ \Omega \tag{6.30}$$

and

$$Z_{0eb} = 119.259\ \Omega, \quad Z_{0ob} = 27.628\ \Omega. \tag{6.31}$$

Conversion of the above set of even- and odd-mode impedances into physical dimensions using ADS [115] yields width, w_i, and spacing, S_i, of the strips:

$$w_1 = 29.88\ \text{mil}\,(0.75\ \text{mm}), \quad S_1 = 3.17\ \text{mil}\,(0.08\ \text{mm}) \tag{6.32}$$

and

$$w_2 = 38.15\ \text{mil}\,(0.96\ \text{mm}), \quad S_2 = 2.49\ \text{mil}\,(0.06\ \text{mm}). \tag{6.33}$$

Unfortunately, see Equation (6.22), the spacing between the coupled lines violates the minimum fabrication spacing. This is when the circuit transformations become useful. By transforming the prototype of Figure 6-14(a) to that of Figure 6-14(b) and making use of the stepped impedance transmission line equivalence, the transformed prototype of Figure 6-14(c) results. Performing similar manipulation to that above yields new coupled-line parameters:

$$Z_{0ea} = 157.76\ \Omega, \quad Z_{0oa} = 47\ \Omega \tag{6.34}$$

and

$$Z_{0eb} = 160.548\ \Omega, \quad Z_{0ob} = 47.836\ \Omega. \tag{6.35}$$

This results in new physical dimensions of

$$w_1 = 17.14\ \text{mil}\,(0.43\ \text{mm}), \quad S_1 = 8.79\ \text{mil}\,(0.22\ \text{mm}) \tag{6.36}$$

and

$$w_2 = 16.16\ \text{mil}\,(0.41\ \text{mm}), \quad S_2 = 9.01\ \text{mil}\,(0.22\ \text{mm}). \tag{6.37}$$

The above dimensions are realizable but it is always desirable to obtain identical coupled-line parameters to construct a physically symmetrical balun structure. This would allow the division of the output load resistance to be halved enabling feasible measurements. In terms of circuit synthesis it is possible to obtain a prototype with elements values such that this condition is satisfied. This is done by iterating on the synthesis cycle for different values of passband ripple until the evaluated values of the resulting even- and odd-mode impedances of the pair of coupled lines become identical. Figure 6-14(d) shows another synthesized and transformed prototype achieving a passband ripple of 0.0806 dB, i.e., a return loss of approximately 17.37 dB with the same bandwidth ratio of 1.53:1. It is now seen from the figure that the impedances of the UE and the input series stepped transmission line are virtually equal ($80 \approx 79.993\ \Omega$)—this is just the necessary required condition.

Now for Figure 6-14(d) (referencing Equations 6.9–6.11 and Figure 6-8(b))

$$Z_{c1} \cong Z_{c2} = 80\ \Omega \text{ and } Z_{c3} = 55.978\ \Omega \tag{6.38}$$

and substituting these into Equations (6.12)–(6.14) yields the coupled-line parameters:

$$K_a = K_b = K = 0.5091, \quad Z_a = Z_b = 92.946\ \Omega. \tag{6.39}$$

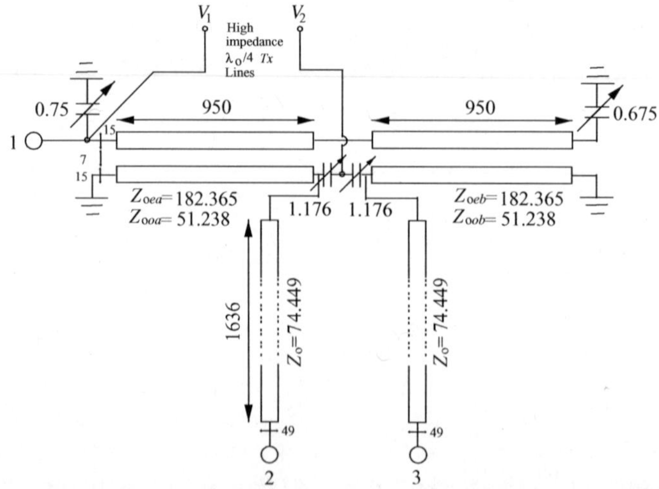

Figure 6-15 The physical balun layout based on the prototype in Figure 6-14 with lumped capacitors (in pF) and additional matching quarter-wavelength lines for impedance matching scaling the balanced load to 100 Ω (The dimensions are in mils.).

Evaluating Equations (6.15)–(6.16) using Equation (6.39) gives identical coupled-line parameters of values

$$Z_{0ea} = Z_{0eb} = 162.966 \ \Omega, \quad Z_{0oa} = Z_{0ob} = 53.011 \ \Omega \,. \tag{6.40}$$

At this stage a manual optimization step improves the return loss level resulting in the circuit of Figure 6-14(e) with a return loss of approximately 20 dB. However, the values of the impedances of the UE and input series stepped transmission line must remain fixed (80 Ω) to guarantee the condition of physical symmetry. The prototype of Figure 6-14(e) results in the physical balun layout of Figure 6-15 with even- and odd-mode impedances as shown in the same figure. In a similar step as before, ADS [115] was utilized to obtain physical dimensions:

$$w_1 = w_2 = 10.2 \ \text{mil}\,(0.25 \ \text{mm}), \quad S_1 = S_2 = 9.08 \ \text{mil}\,(0.23 \ \text{mm}). \tag{6.41}$$

The shunt open-circuited stub at the balanced output, and the balanced load, R_B, must now be scaled accordingly using Equations (6.16) and (6.19) with K evaluated as 0.561 to give 270.51 Ω and 221.707 Ω respectively. The balanced load impedance is then scaled to 100 Ω by a pair of transmission lines of impedance 74.449 Ω each. The remaining stubs are left un-scaled according to Equation (6.20).

Since a bandpass prototype is being used it is possible to approximate any or all of its open-circuited stubs by lumped capacitors with very little deterioration of the circuits passband performance. Now each open-circuited stub is approximated by equating its input admittance to that of a lumped capacitor using the relationship

$$j\left(\frac{1}{Z_i}\right)\tan\left(\frac{\pi}{2}\left(\frac{f_0}{f_r}\right)\right) = j2\pi f_0 C_i \tag{6.42}$$

where Z_i is the impedance of the open-circuited stub and C_i its corresponding capacitor value. This approximation is valid over the vicinity of the passband of the filter but will affect the upper stopband characteristics. However, in many cases such a step will lead to better stopband performance suppressing upper undesired passbands. This point will be elaborated upon in the next section. Performing the approximation using Equation (6.42) gives values of capacitors as shown in Figure 6-15. A final minor optimization step is required to adjust the return loss level giving lines that are 15 mil (0.38 mm) wide, 7 mil (0.17 mm) apart, and 950 mil (24.13 mm) long. Also each output matching line is 49 mil (1.24 mm) wide, and 1636 mil (41.55 mm) long.

It is interesting to note that this example shows yet another transformed Class B balun topology not included in Figure 6-13.

6.3.5 Implementation of the Class B Balun

The layout of Figure 6-15 (without the high impedance transmission lines) was implemented and its measured and simulated performance, using high quality lumped capacitors with Q of 600 are shown in Figures 6-16 and 6-17. In Figures 6-16(a) and 6-16(b), the measured mid-band insertion loss is about 3.6 dB. The excess 0.6 dB of loss is mainly attributed to substrate and conductor losses. The maximum measured amplitude error, from Figure 6-17(a), is about 0.2 dB over the passband

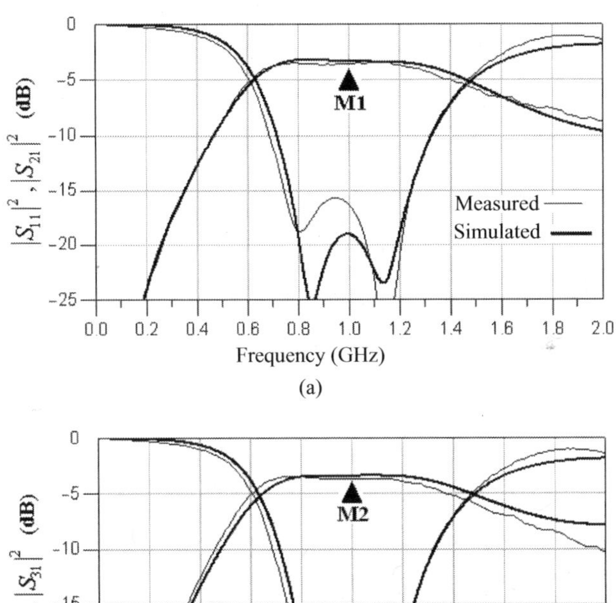

(a)

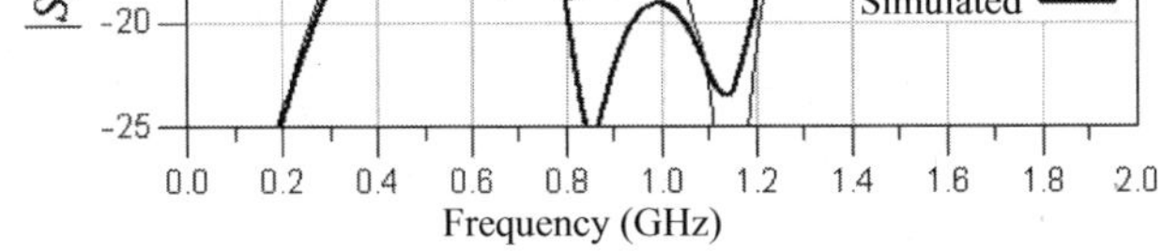

(b)

Figure 6-16 Measured and simulated balun performance of the network in Figure 6-14(f) without bias lines and with high-Q lumped capacitors: (a) $|S_{21}|^2$ and $|S_{11}|^2$ (**M1**: 1 GHz, -3.6 dB); and (b) $|S_{31}|^2$ and $|S_{11}|^2$ (**M2**: 1 GHz, -3.5 dB).

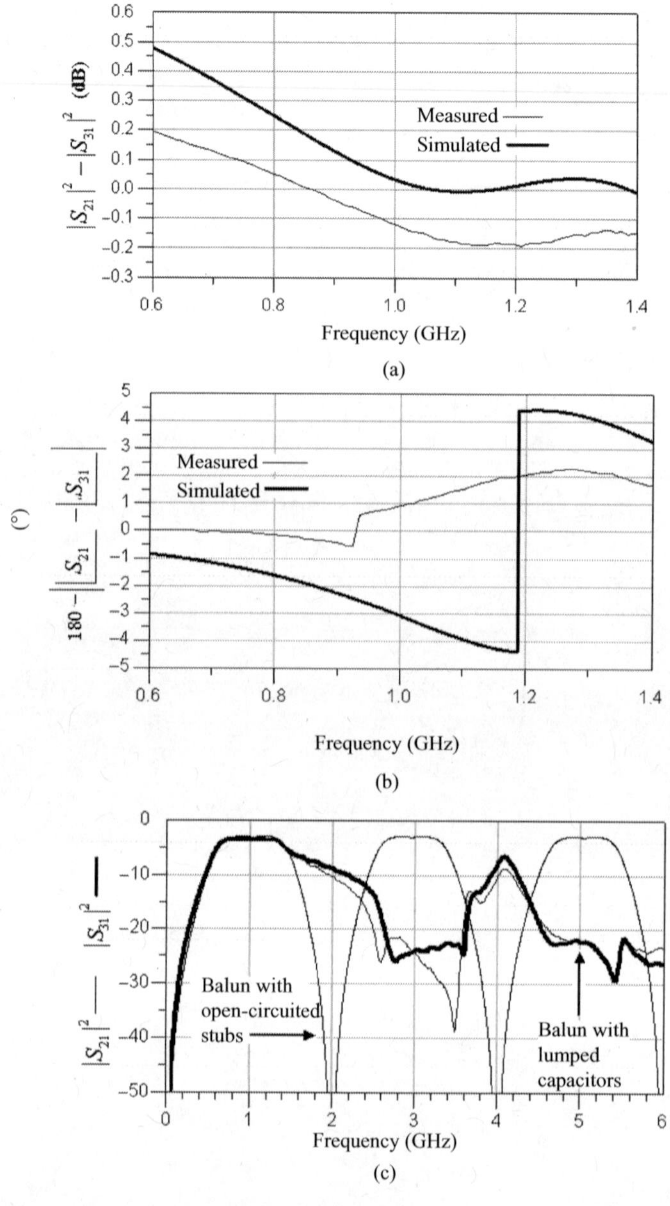

Figure 6-17 Measured and simulated balun performance using the configuration of Figure 6-14(f): (a) transmission coefficient amplitude error; (b) transmission coefficient phase error; and (c) broadband transmission coefficients.

which is just slightly better than the simulation by about 0.05 dB. Figure 6-17(b) shows a plot of transmission phase error over the passband indicating a maximum measured phase error of approximately 2° at 1.2 GHz in contrast to approximately 4° according to simulation. It is fair to note at this point that the performance of the balun is excellent. The practical capacitor values used were 0.8, 0.7, and 0.6 pF as opposed to the simulated values of (from Figure 6-15) 0.75, 0.675, and

0.588 pF respectively. The differences being due to the microstrip bias that were assumed ideal in the simulation, and also as a result of adjusting the amplitude and phase responses to counteract the intrinsic difference between the even- and odd-mode phase lengths of PCLs.

The new baluns introduced in this section lend themselves to flexible designs since the design approach is based on bandpass prototypes with selection of the commensurate frequency, f_r. Figure 6-17(c) shows the simulated f-plane response of the circuits of Figures 6-14(a) to 6-14(e) i.e., before the approximation of the open-circuited stubs by lumped capacitors. The first higher stopband is centered at f_r (2 GHz) while the following stopbands occur periodically at nf_q, where $n = 2, 3, 4, \ldots$. The second passband occurs at $2f_r - f_0$ also recurring periodically at $(2f_r - f_0) \times m$, where $m = 2, 4, 6, \ldots$. Also shown in Figure 6-17(c) is the measured performance of the balun with all the open-circuited stubs approximated by lumped capacitors. Now it is clear that the approximation step benefits the stopband rejection of the balun since the second passband now occurs at about 4 GHz with wider stopband performance around 3 GHz. Thus size reduction and control of the stopband performance are simultaneously made possible through the design approach based on bandpass prototypes.

6.3.6 Implementation of a Class B Tunable Balun

A second version of the above design was implemented with the objective of constructing an electronically tunable balun. This was another reason behind the choice of the specific sequence of circuit transformation used in the above subsection. It is generally known that passive circuits with wide bandwidths require strong coupling between their circuit elements. This is a problem particularly if the desired implementation requires coupled lines since the physical dimensions become unrealizable. A planar S-plane highpass balun is no exception [389]. A very attractive feature of the S-plane bandpass balun presented here is the feasibility of tuning its passband using variable capacitors. This implies that a moderate bandwidth balun may be designed and then tuned up or down in frequency to cover a wider bandwidth.

These goals were achieved using the layout of Figure 6-15 with the pair of quarter-wavelength bias lines (at 1 GHz) connected to low impedance voltage sources. Varactors with tunable capacitance ratios of 10:1 [396] replaced the high-Q lumped capacitors. The Q of the tunable capacitors was measured as roughly 12 at 1 GHz. This low Q has little effect on the in-band Insertion Loss due to the original design of the balun with a bandwidth ratio of 1.54:1. The passband of the balun was tuned down from 1 GHz (State 1) to 0.58 GHz (State 2). This gives a tunable frequency range of 1.7 (i.e., $1/0.58$) with approximately 6:1 capacitance ratios, see Figure 6-18.

At 0.58 GHz the Q of the varactor-diodes naturally improves to about 30 but there will be additional reactances at the biasing points due the limited bandwidth of the bias lines originally designed at 1 GHz. This contributed to some loss resulting in a measured mid-band insertion loss at 0.58 GHz of 4.0 dB. The measured maximum amplitude and phase errors in State 2 were 0.2 dB and 1° respectively. At this stage it is worth mentioning that a conventional design based on an S-plane highpass prototype would not have made tunability of the center frequency possible.

6.3.7 Summary

In this section new classes of baluns were introduced that are based on an established exact synthesis technique that is widely used in filter design. The balun prototypes were synthesized from a specification of a set of transmission zeros with bandpass characteristics and possess various features such as small overall size, and controlled broad stopband rejection. In many RF communications applications the functions of filtering and balun transformations can now be realized by a single network thus reducing the overall system losses compared to two separate realizations. The measured performance of the implemented Class B balun agrees very well with the presented theory.

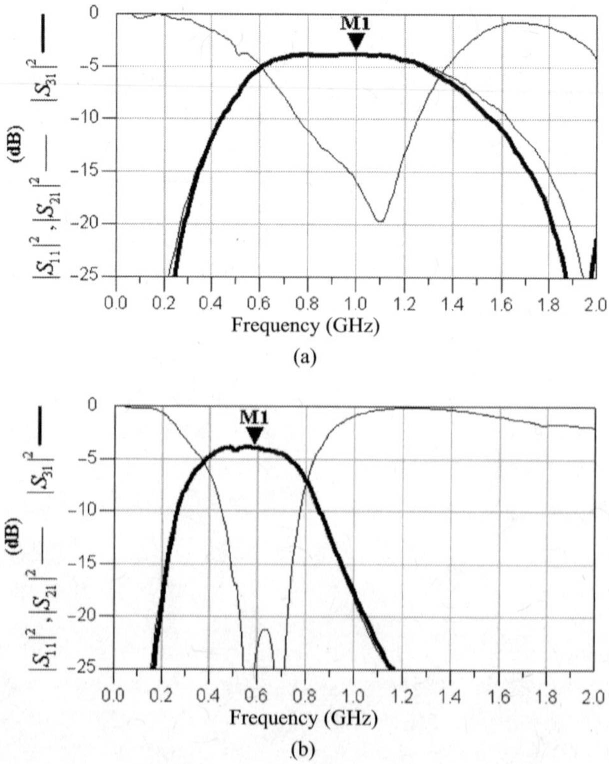

Figure 6-18 Measured performance of tunable balun of Figure 6-14(e) with bias lines and low Q varactor diodes in: (a) State 1 (**M1**: 1 GHz, −3.82 dB); and (b) State 2 (**M1**: 0.58 GHz, −4.04 dB).

Appropriate circuit transformations Classes A and B can be made tunable. This was demonstrated with a Class B tunable Marchand balun. This is a tunable balun suitable for use in wideband applications or in multifunctional RF and microwave systems covering various frequency bands.

6.4 DISTRIBUTED BIASING OF DIFFERENTIAL RF CIRCUITS

In this section a distributed balun-like section is presented as an alternative to inductor-based biasing of active differential circuits. The new scheme has broad bandwidth with low loss and is suited to broadband multifunctional RF and microwave circuits. Additionally the distributed biasing circuit discriminates between differential and common-mode signals leading to high common mode rejection. radio frequency (RF) circuits commonly utilize differential signal paths with the benefit of noise immunity and reduced sensitivity to component variations. On-chip RF power amplifiers are often pseudo-differential where the common current source typically used in fully differential circuits is sacrificed to enable a larger voltage swing, see Figure 6-19. Here the inductors present high RF impedance to the transistors (represented as transconductances) while providing a low impedance path for bias currents. With a sufficiently high-Q inductor, RF energy is delivered to the load rather than being dissipated in the bias circuitry. With the relatively low Q of on-chip planar inductors, inductance peaking near selfresonance is utilized and this leads to narrow band operation. Thus in critical situations off-chip inductors are sometimes used in conjunction with on-chip active devices. In addition, inductive biasing of pseudo-differential circuits, presents the

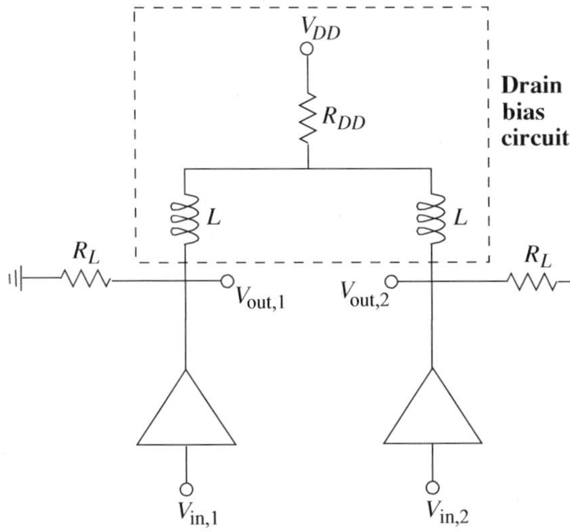

Figure 6-19 A pseudo-differential circuit without a constant-current source, bias inductors L at the drains, parasitic supply resistance R_{DD}, and single-ended load impedance R_L.

same environment to common- and differential-mode signals so that the common mode rejection ratio (CMRR) is 1.

This section presents a biasing scheme for broadband pseudo-differential circuits leading to high CMRR. The biasing circuitry consists of a coupled-line structure analogous to that of a Marchand balun [384], see Figure 6-20. The balun structure is comprised of transmission lines connected to a pair of coupled lines that normally transforms a single-ended impedance at Port 1 to a balanced output at the ports 3 and 4. The capacitors in Figure 6-20 result in miniaturization since the distributed transmission lines are quarter of a wavelength long at a frequency well above the passband of the balun. In Section 6.4.1 the equations describing CMRR are first developed, as

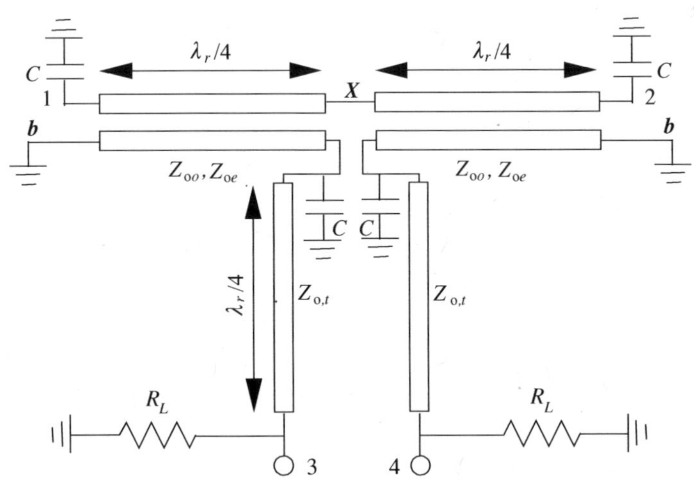

Figure 6-20 Marchand balun-like biasing circuit with single-ended load resistance R_L (DC bias may be applied at ports B using decoupling capacitors).

maximizing CMRR in pseudo-differential circuits is a key design objective. The design equations in the subsequent sections enable the design of the biasing circuit network with specified characteristics. In Sections 6.4.2–6.4.5 the use of the Marchand balun structure as a biasing circuit is investigated and analysis of the even- and odd-mode impedances of the balanced ports is presented. Practical implementations are presented in Section 6.4.6.

6.4.1 Conventional Biasing

Multifunctional systems require broadband circuits and, for differential circuits, this requires large differential gain A_d. At the same time it is desirable to minimize the common-mode gain A_c as the resulting high CMRR provides immunity to substrate induced noise. With transistor transconductance g_m, and total even- and odd-mode impedances Z_{even} and Z_{odd} presented to the drains of the transistors, the gains are approximately

$$A_d = g_m Z_{\text{odd}} \text{ and } A_c = g_m Z_{\text{even}} \tag{6.43}$$

and so

$$\text{CMRR} = \frac{A_d}{A_c} = \frac{Z_{\text{odd}}}{Z_{\text{even}}}. \tag{6.44}$$

The desired amplifier characteristics are thus obtained by synthesizing the even- and odd-mode impedances.

Modal analysis of the inductor biasing circuit results in Figure 6-21 from which the total even-mode impedance is

$$Z_{\text{even}}(s) = \left(sL + \frac{R_{DD}}{2}\right) // R_L \tag{6.45}$$

where s is the Laplace operator, // indicates a parallel connection, and the total odd-mode impedance is

$$Z_{\text{odd}}(s) = sL // R_L. \tag{6.46}$$

Thus, unless the inductance is very large, there will be strong frequency variation of Z_{odd} and Z_{Even}. Also, since R_{DD} is often negligible, the CMRR is 1. In differential-mode the gain is maximized over a broad frequency band for a specific single-ended load impedance R_L. However, if the common-mode gain has prescribed frequency characteristics then the CMRR will be inversely proportional

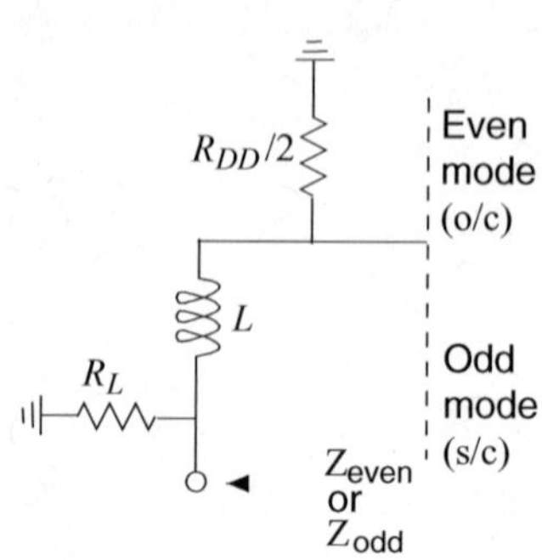

Figure 6-21 Modal subcircuits of the inductor-based biasing circuit of Figure 6-19 including single-ended load resistance R_L.

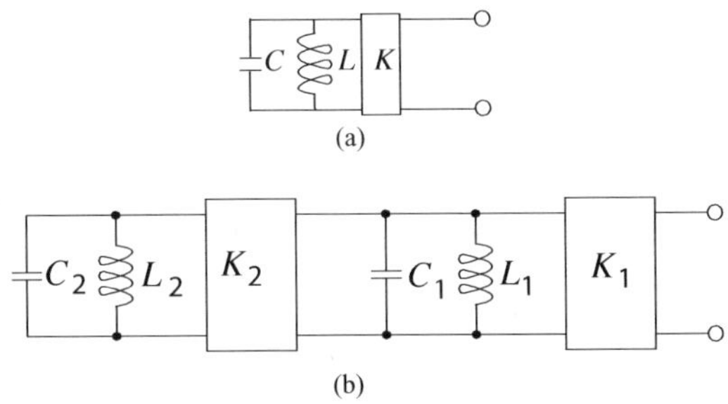

Figure 6-22 Coupled lumped element bandpass resonators: (a) single resonator coupled into by an inverter; and (b) a pair of resonators coupled by inverters.

to the even-mode impedance. Thus low even-mode impedance of the biasing network is desirable as this suppresses common-mode noise. This defines a vital design objective of the new biasing circuit.

6.4.2 Design Based on Analogy to Coupled Resonators

Coupled resonator structures are appropriate topology choices that present different impedances for the common- and differential-mode signals. In essence, bias is also a common-mode signal and must be accommodated in the chosen topology. A class of appropriate topologies is based on lumped bandpass resonator(s) coupled into by an inverter K shown in Figure 6-22(a). (Note that an inverter and unitary element (UE) are synonymous.) The inverter element is an idealized component that is independent of frequency and can be approximated by quarter wavelength long transmission line as well as by various other circuit configurations [397]. At resonance the bandpass resonator presents an open circuit and, following impedance inversion, the input impedance of the network is a short circuit. This then, is the kind of network required to be presented to the common-mode signals. The second part of the network synthesis problem is creation of the high impedance condition for differential-mode signals. These issues will be considered separately. Figure 6-22(b) illustrates a pair of bandpass resonators coupled by inverters. At resonance the input impedance of the network is infinite. This is because the high impedance of Resonator 2 transforms to a short circuit after the second inverter, K_2, and this subsequently appears as an open circuit at the input of the network. So by adding an extra resonator another topology suitable for differential-mode operation is derived. If a load is now connected in parallel with this network then at resonance this will be the total impedance presented to the output transistors of the pseudo-differential amplifier. In summary, presenting different impedances for common- and differential-mode signals is desired. It is shown in the next section that the Marchand balun structure has the requisite properties.

6.4.3 Distributed Biasing

Consider now the topology of the Marchand balun of Figure 6-20 with the balanced Ports 3 and 4 connected to the output of the pseudo-differential circuit. The aim of this subsection is to show that this topology has the requisite even- and odd-mode impedances. Performing modal analysis at the balanced ports, the modal subnetworks of Figure 6-21 are obtained. In analyzing this structure use is made of the network model of a pair of coupled lines, Figure 6-1. Then by direct application of the network model and application of the relevant port conditions dictated by the modal subnetworks of the biasing circuit at Point X, simplifies the equivalent subnetworks of Figure 6-23 to those of Figure 6-24. Application of the relevant Kuroda transformation [397]

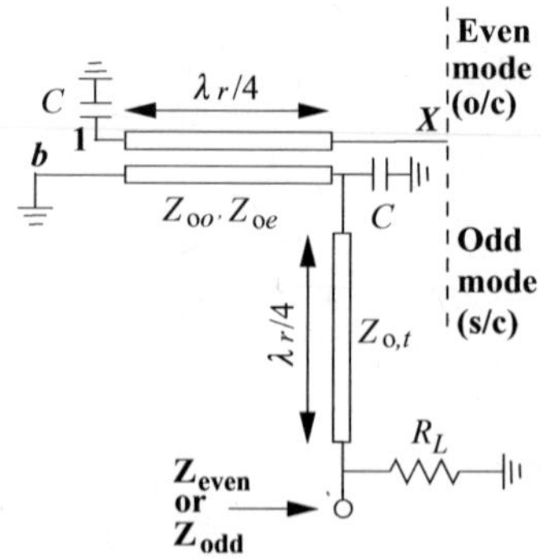

Figure 6-23 Modal subnetworks of the balun-like biasing circuit of Figure 6-20 with single-ended load resistance R_L (assuming negligible power supply source impedance, R_{DD}).

to distribute the inductance in Figure 6-24(c) results in the transformed subcircuits of Figure 6-24(d). Scrutiny of Figures 6-24(b) and 6-24(d) reveals that they are in fact analogous to the lumped-element coupled-resonator examples discussed earlier. The major differences are that transmission lines replace the inverters and the additional resistor R_c represents the loss associated with each capacitor. Also in the distributed domain each inductor or capacitor represents a short- or open-circuited stub, respectively, which is a quarter-wavelength long at the frequency f_q. The open-circuited stubs may, however, be approximated by lumped capacitors for miniaturization purposes. Thus over the operating band the performance of the even- and odd-mode subcircuits is very close to that of their lumped element counterparts. In summary, centered at a frequency f_0, the structure of the Marchand balun maximizes differential power delivery to the load while minimizing the gain of common-mode signals. DC bias may be easily applied through the short-circuited ports (labeled ***b*** in Figure 6-20) with the use of decoupling capacitors. The mathematical formulations for the total even- and odd-mode impedances with lumped capacitors are derived from Figure 6-24(b) and 6-24(c) as follows:

$$Z_{\text{even}}(S) = \frac{N_1 S^2 + N_2 S}{D_1 S^2 + D_2 S + D_3} \tag{6.47}$$

with coefficients:

$$N_1 = j\left(2Z_{0t}^2 Z_1 \pi f C R_L\right) \tag{6.48}$$

$$N_2 = Z_1 R_L Z_{0t} + Z_{0t}^2 R_L \tag{6.49}$$

$$D_1 = (Z_1 R_L) + j\left(2Z_{0t}^2 \pi f C Z_1\right) \tag{6.50}$$

$$D_2 = \left(Z_{0t}^2 + Z_{0t} Z_1\right) + j\left(2Z_{0t} R_L \pi f C Z_1\right) \tag{6.51}$$

$$D_3 = Z_{0t} R_L \tag{6.52}$$

and

$$Z_{\text{odd}}(S) = \frac{N_1 S^3 + N_2 S^2 + N_3 S}{D_1 S^3 + D_2 S^2 + D_3 S + D_4} \tag{6.53}$$

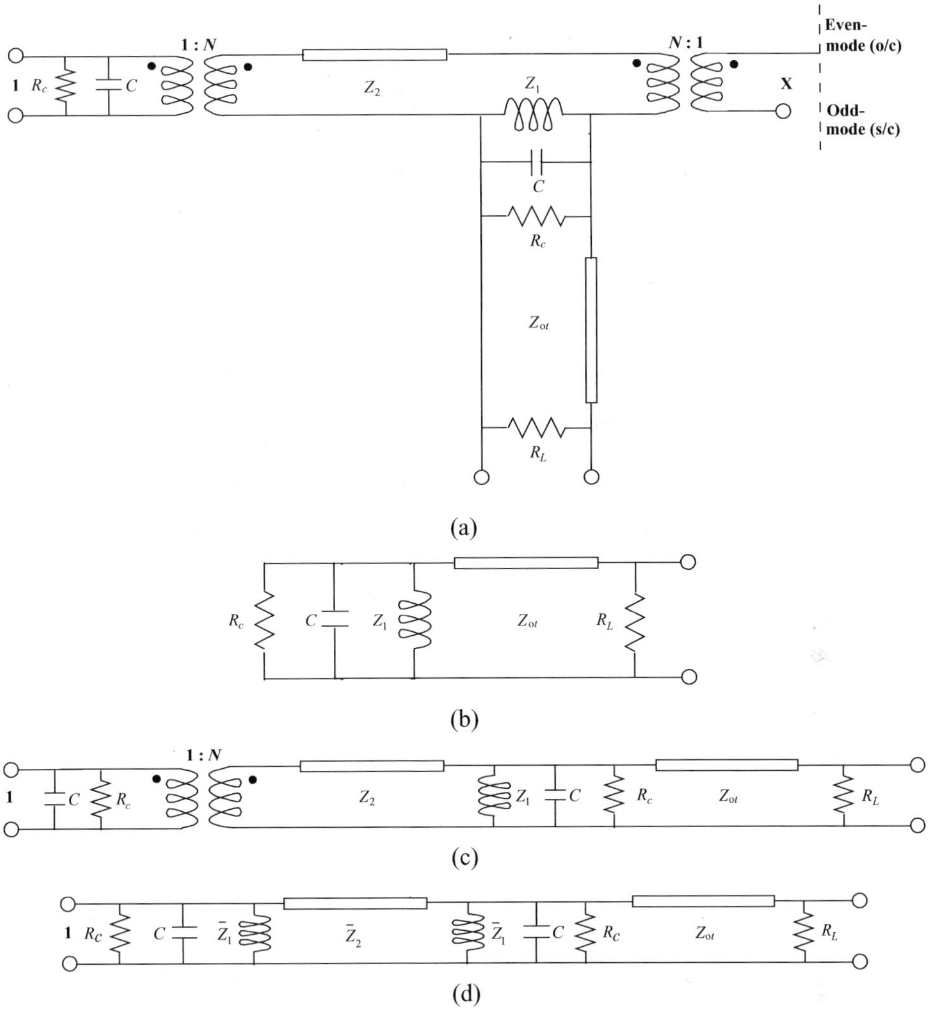

Figure 6-24 Equivalent subcircuits of Figure 6-23: (a) utilizing the network model; (b) even-mode subcircuit; (c) odd-mode subcircuit; and (d) odd-mode subcircuit after Kuroda transformation with $\bar{Z}_1 = 1/Zoe$, $\bar{Z}_2 = (Z_{0e} - Z_{0o})/(2Z_{0e}Z_{0o})$. $R_L = \infty$, $R_L = 100\Omega$, $R_L = 50\Omega$.

with coefficients:

$$N_1 = Z_{0t}^2 R_L Z_1 \tag{6.54}$$

$$N_2 = \jmath \left(2 Z_2 Z_{0t}^2 R_L \pi f C Z_1\right) \tag{6.55}$$

$$N_3 = Z_1 Z_2 Z_{0t} R_L + Z_2 Z_{0t}^2 R_L \tag{6.56}$$

$$D_1 = Z_{0t}^2 Z_1 \tag{6.57}$$

$$D_2 = (Z_1 Z_2 R_L + Z_1 Z_{0t} R_L) + \jmath \left(2 Z_2 Z_{0t}^2 \pi f C Z_1\right) \tag{6.58}$$

$$D_3 = \left(Z_2 Z_{0t} Z_1 + Z_2 Z_{0t}^2\right) + \jmath \left(2 Z_2 Z_{0t} \pi f C Z_1 R_L\right) \tag{6.59}$$

$$D_4 = Z_2 Z_{0t} R_L. \tag{6.60}$$

In Equations (6.47) and (6.53) S is the Richards transform [397] defined as $j\tan(\theta) = j\tan\left(\left(\pi/2\right)\left(f/f_q\right)\right)$ and f_q is the frequency at which the distributed lines of the balun are a quarter-wavelength long. In the above equations Z_{0t} is the characteristic impedance of the uncoupled transmission lines while

$$Z_1 = \frac{\sqrt{Z_{0e}Z_{0o}}}{\sqrt{1-\left(\frac{Z_{0e}-Z_{0o}}{Z_{0e}+Z_{0o}}\right)^2}} \tag{6.61}$$

and

$$Z_2 = \sqrt{Z_{0e}Z_{0o}}\sqrt{1-\left(\frac{Z_{0e}-Z_{0o}}{Z_{0e}+Z_{0o}}\right)^2}\left(\frac{Z_{0e}-Z_{0o}}{Z_{0e}+Z_{0o}}\right)^{-2} \tag{6.62}$$

with Z_{0e} and Z_{0o} being the even and odd-mode impedances of the coupled lines forming the balun. The total even- and odd-mode impedances presented to the pseudo-differential active circuit are in general complex, and thus may be written in the following notation:

$$|Z_{\text{even}}(S)| = \sqrt{\Re\{Z_{\text{even}}(S)\}^2 + \Im\{Z_{\text{even}}(S)\}^2} \tag{6.63}$$

$$|Z_{\text{odd}}(S)| = \sqrt{\Re\{Z_{\text{odd}}(S)\}^2 + \Im\{Z_{\text{odd}}(S)\}^2}. \tag{6.64}$$

Hence the procedure for determining the circuit parameters of the new biasing circuit is fairly simple. Suitable choices of the characteristic impedances Z_1 and Z_2 solve Equations (6.61) and (6.62) for the even- and odd-mode characteristic impedances Z_{0e} and Z_{0o} respectively. Selection of the characteristic line impedance Z_{0t} together with choice of capacitance value, C, fully determines Equations (6.47) and (6.53). The magnitudes of the total even- and odd-mode impedance functions may then be evaluated using Equations (6.63) and (6.64). The commensurate length of the transmission lines will be determined by the value of C. Zero capacitance results in the lines being a quarter-wavelength long at the center of the operating band otherwise miniaturization of the lines is feasible. An example that shows the overall characteristics of the magnitudes of the even- and odd-mode impedances without lumped capacitors is depicted in Figure 6-25 for a range of single-ended load resistances, R_L.

6.4.4 Discriminative Operating Bandwidth

The operating bandwidth of the balun is defined at the crossover frequencies where

$$|Z_{\text{even}}(S)| = |Z_{\text{odd}}(S)|. \tag{6.65}$$

At the crossover frequency points, see Figure 6-25, the CMRR of the pseudo-differential circuit is 1 and the balun has similar performance to that of the non-discriminative inductor-based biasing circuit. For a certain choice of circuit parameters the frequency points at which the even- and odd-mode impedances crossover may be altered facilitating adjustment of the discriminative bandwidth. This is demonstrated in Figure 6-26 through a set of examples using the design procedure described in the above section. It is worth emphasizing that the total even- or odd-mode impedance of Equations (6.47) or (6.53) is the impedance of the input of the balun in parallel with the single-ended load resistance R_L. This implies that in order for the pseudo-differential circuit to deliver

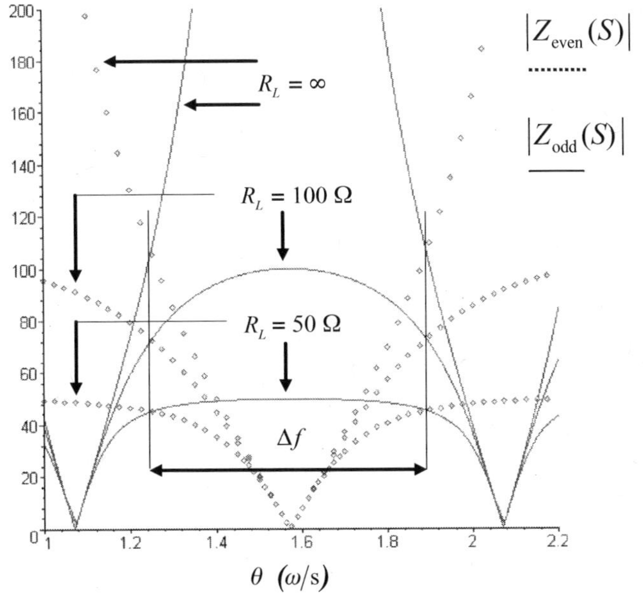

Figure 6-25 Calculated magnitudes of the total even- and odd-impedances for a range of single-ended load resistances with circuit parameters: $Z_{0e} = 208\ \Omega$, $Z_{0o} = 51\ \Omega$, $Z_{0t} = 137\ \Omega$, $C = 0$ pF.

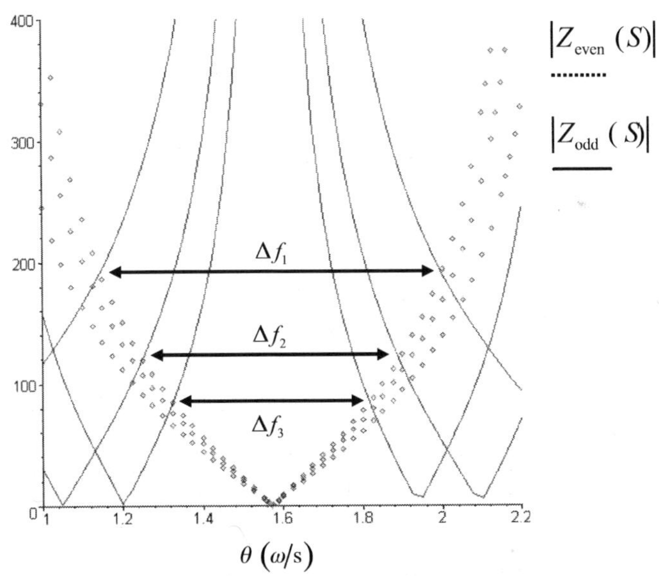

Figure 6-26 Calculated magnitudes of the total even- and odd-impedances for an infinite single-ended load resistance, R_L with the following circuit parameters: for Δf_1, $Z_{0e} = 157.32\ \Omega$, $Z_{0o} = 58.45\ \Omega$, $Z_{0,t} = 90\ \Omega$, $C = 0$ pF; for Δf_2, $Z_{0e} = 205.36\ \Omega$, $Z_{0o} = 47.82\ \Omega$, $Z_{0,t} = 100\ \Omega$, $C = 0$ pF; and for Δf_3, $Z_{0e} = 307.26\ \Omega$, $Z_{0o} = 22.24\ \Omega$, $Z_{0t} = 80\ \Omega$, $C = 0$ pF.

its power to the load, it is necessary that the output signal at each arm sees the correct impedance. Of course there will only be a perfect match at the center of the band (since the input balun impedance is then infinite) but not at the crossover frequency points. However a slight mismatch will only degrade the transfer of power slightly and is therefore acceptable. Figure 6-25 illustrates this argument for a 50 Ω system for a particular choice of balun circuit parameters. In this example the magnitudes of the total even- and odd-mode impedances at the crossover frequencies are approximately 43 Ω. However for a 100 Ω system it is obvious (see Figure 6-25) that the impedance levels are much lower than 100 Ω at the crossover frequencies causing a big mismatch that needs to be compensated for by adjusting the element values of the balun. Generally the higher the system impedance the higher the coupling required between the coupled lines of the balun.

6.4.5 Effect of Loss on Distributed Biasing

The remaining design aspect to be discussed is the effect of transmission line loss and also loss of the lumped capacitors if circuit miniaturization is desired. Modification of the total even- and odd-mode impedances to take into account the effect of substrate, conductor and capacitor shunt resistance, R_c, leads to

$$Z_{\text{even}}(S) = \frac{N_1 S^2 + N_2 S}{D_1 S^2 + D_2 S + D_3} \tag{6.66}$$

with coefficients:

$$N_1 = \left(Z_{0t}^2 Z_1 R_L\right) + \jmath\left(2Z_{0t}^2 Z_1 \pi f C R_L R_C\right) \tag{6.67}$$

$$N_2 = Z_1 R_L Z_{0t} R_C + Z_{0t}^2 R_L R_C \tag{6.68}$$

$$D_1 = \left(Z_1 R_L R_C + Z_{0t}^2 Z_1\right) + \jmath\left(2Z_{0t}^2 \pi f C Z_1 R_C\right) \tag{6.69}$$

$$D_2 = \left(Z_{0t}^2 R_C + Z_{0t} Z_1 R_C + Z_{0t} Z_1 R_L\right) + j\left(2Z_{0t} R_L \pi f C Z_1 R_C\right) \tag{6.70}$$

$$D_3 = Z_{0t} R_L R_C \tag{6.71}$$

and

$$Z_{\text{odd}}(S) = \frac{N_1 S^3 + N_2 S^2 + N_3 S}{D_1 S^3 + D_2 S^2 + D_3 S + D_4} \tag{6.72}$$

with coefficients:

$$N_1 = Z_{0t}^2 R_L Z_1 R_C \tag{6.73}$$

$$N_2 = \left(R_L Z_2 Z_{0t}^2 Z_1\right) + \jmath\left(2Z_2 Z_{0t}^2 R_L \pi f C Z_1 R_C\right) \tag{6.74}$$

$$N_3 = Z_1 Z_2 Z_{0t} R_L R_C + Z_2 Z_{0t}^2 R_L R_C \tag{6.75}$$

$$D_1 = Z_{0t}^2 Z_1 R_C \tag{6.76}$$

$$D_2 = \left(Z_1 Z_2 R_L R_C + Z_1 Z_{0t} R_L R_C + Z_{0t}^2 Z_1 Z_2\right) + \jmath\left(2Z_2 Z_{0t}^2 \pi f C Z_1 R_C\right) \tag{6.77}$$

$$D_3 = \left(Z_2 Z_{0t} Z_1 R_C + Z_2 Z_{0t}^2 R_C + Z_1 Z_2 Z_{0t} R_L\right) + \jmath\left(2Z_2 Z_{0t} \pi f C Z_1 R_L R_C\right) \tag{6.78}$$

$$D_4 = Z_2 Z_{0t} R_L R_C. \tag{6.79}$$

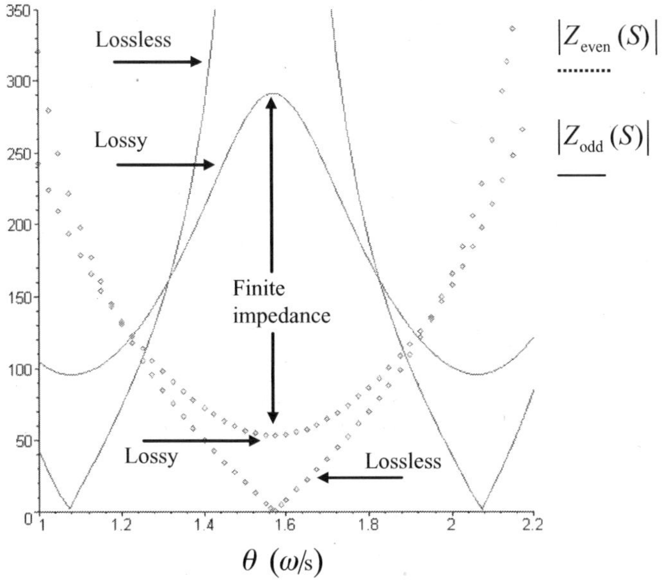

Figure 6-27 Calculated magnitudes of lossy even- and odd-impedances for an infinite single-ended load resistance, R_L with circuit parameters: $Z_{0e} = 208\ \Omega,\ \ Z_{0o} = 51\ \Omega,\ \ Z_{0t} = 137\ \Omega$, $C = 0$ pF, $\alpha = 0.2$ Nepers/meter.

Now S is the general Richards transformation [366] accounting for loss defined as

$$S = \tanh(\alpha + \jmath\theta) = \tanh\left(\alpha + \jmath\left(\frac{\pi}{2}\right)\left(\frac{f}{f_r}\right)\right) \tag{6.80}$$

and α is the attenuation constant in Nepers/meter. With no miniaturization capacitors, (i.e., $C = 0,\ R_c = \infty$) and a finite value of α, the same circuit parameters that generated the plots of Figure 6-25 are now fed back into the modified equations, Equations (6.66) and (6.72), from which their magnitudes are evaluated using Equations (6.63) and (6.64). Figure 6-27 highlights the effect of parasitic loss present. This plot assumes infinite single-ended load impedance, i.e., it represents the input impedances of the balun itself. The effect of loss manifests itself as a degradation of the magnitudes of the even- and odd-mode impedance levels. At resonance, the odd-mode impedance is very high but finite and likewise the opposite is true for the even-mode impedance. These impedances, in parallel with the single-ended load impedance directly impact the achievable CMRR of the system and increase insertion loss. However the plots of Figure 6-27 assumed no capacitors for miniaturization purposes. Normally lumped capacitors will possess finite Q that will further deteriorate the CMRR and reduce power delivery to the next stage. This point is demonstrated by measurements in the next section.

6.4.6 Experimental Results

A balanced power amplifier (HELA-10B) from Mini-Circuits [398] with a gain of 10 dB was selected for experimental proof of principal. The chip is powered by a single +12-V-DC power supply via a pair of external lumped inductors and consists of a pair of amplifiers in a pseudo-differential amplifier configuration. As the amplifiers are on the same chip, their gains and phase are well matched. The amplifier has an excellent second order intercept of 88 dBm operating from

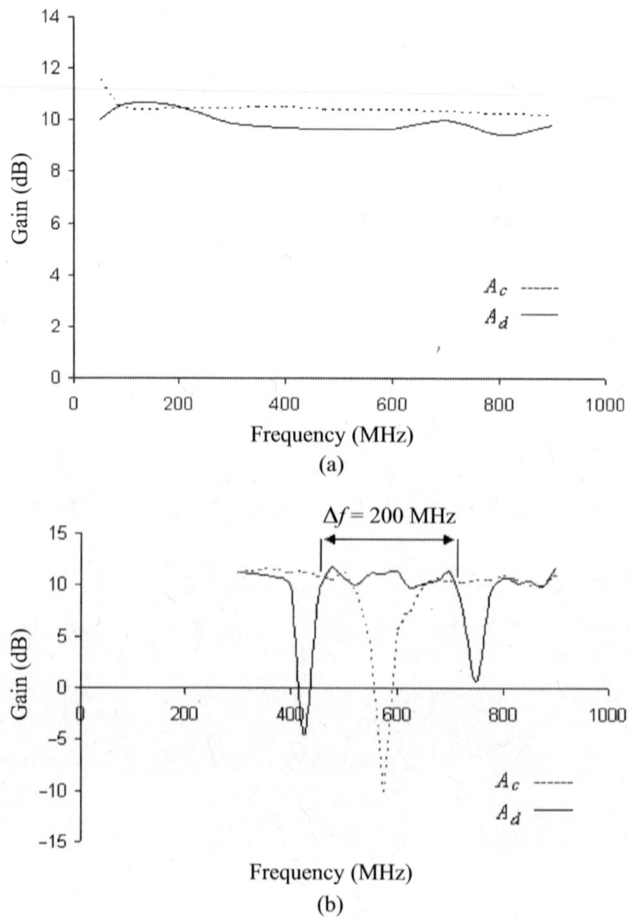

Figure 6-28 Measured common-mode gain, A_c and differential-mode gain A_d: (a) with inductor-based biasing circuit with lumped inductors of $0.75\mu H$; and (b) with balun-like biasing circuit without lumped capacitors with circuit parameters: $Z_{0e} = 208\ \Omega$, $Z_{0o} = 51\ \Omega$, $Z_{0t} = 137\ \Omega$, $\tan\delta = 0.016$.

50 MHz to 1 GHz in a 50 Ω system. A number of distributed biasing circuits were implemented and integrated with the pseudo-differential amplifier. The design objective was to achieve broadband bandwidth of 1.5:1, high odd-mode, and low even-mode RF bias circuit impedances.

6.4.6.1 Inductor-Based Biasing The amplifier was tested first with a conventional inductor-based biasing circuit. A pair of high Q, lumped inductors of value, 0.75 μH, was used and the circuit tested in both differential- and common-modes. The measured common- and differential-mode gains are 10 and 9.5 dB respectively as shown in Figure 6-28(a). This illustrates the non-discriminatory bandwidth produced using the conventional inductor-based biasing scheme.

6.4.6.2 Distributed Biasing To verify the properties of the differential biasing circuit the Marchand balun structure with a center frequency of 580 MHz was designed. The optimized balun parameters are $Z_{0e} = 208\ \Omega$, $Z_{0o} = 51\ \Omega$, $Z_{0t} = 137\ \Omega$, and $C = 0$ pF. The even- and odd-mode impedances were calculated based on a choice of $Z_1 = 129.5\ \ \Omega$, $Z_2 = 222.92\ \ \Omega$ from Equations (6.47)–(6.64). By referring to Figure 6-25 it is seen that when the single-ended load impedance is 50 Ω the magnitude of the total odd-mode impedance is desirably flat over the specific frequency band. This represents optimum power transfer to the load in a 50 Ω system. The HELA-10B chip was integrated with the balun on an FR4 PCB with a substrate thickness of 6.2 mils (0.157 mm), relative dielectric constant of 4.7 and loss tangent of 0.016. The above impedances were then converted into physical dimensions using a commercial CAD tool [115] and the final circuit optimized. The resulting coupled lines were found to be 7 mils (0.177 mm) wide, 7 mils apart, and 3031 mils long. The transmission lines connecting the output ports of the RF chip to the coupled lines were 7 mils wide, and 3031 mil long. The differential- and common-mode gains are shown in Figure 6-28(b). While the differential-mode gain is over 10 dB in the center of the band, the common-mode gain dips to −10 dB at the resonant frequency. Thus the device now discriminates between the common- and differential-mode signals. The measured bandwidth, defined as: $(\Delta f / f_0)$ is 34% of the resonant frequency 580 MHz, i.e., 1.4:1 bandwidth.

6.4.6.3 Distributed Biasing with Miniaturization Capacitors In another design operating at 580 MHz the transmission lines where chosen to be resonant at 3×580 = 1740 MHz. This yields a reduction in overall size by a factor of three with $C = 5.6$ pF. In Figure 6-29(a) the measured response (with high Q capacitors: $Q \approx 400$) is depicted. It is worth investigating the effect of loading the capacitors with resistors as this drops their effective Q. This was done with the objective of replacing the lumped capacitors by tunable components having moderate Qs to improve overall system functionality. For example the Q of a BST capacitor ranges between 30 and 60. Figure 6-29(b) illustrates the gains after loading the lumped capacitors with resistors to drop their Qs to about 30. It is obvious from Figure 6-29 that the loss degrades the performance of the circuit. In the case of high-Q capacitors the measured differential-mode gain is 10 dB. However the common-mode gain is hardly affected and notches at the center frequency to −10 dB. In the case of low-Q capacitors the common-mode gain only notches to 0 dB and the differential gain reduces from 10 to 9 dB. The measured bandwidth in the two cases is 31% of the resonant frequency 580 MHz, i.e., 1.36:1 bandwidth.

It is also observed from Figure 6-29 that the common-mode gain does not notch exactly at the center of the operating band, as was the case without lumped capacitors, see Figure 6-28(b). This implies that the frequency at which the odd-mode impedance of the balun peaks is different from the frequency at which its even-mode impedance is a minimum. This conflict arises due to the addition of lumped capacitors that cause asymmetry in the characteristics of the even- and odd-mode impedances. This may be resolved by fine-tuning of the lumped capacitors and optimization of the rest of the circuit elements of the equivalent even- and odd-mode subcircuits of the balun structure.

6.4.7 Summary

This section introduced a new distributed biasing circuit that replaces conventional biasing using inductors. The physical structure comprises transmission lines and a pair of identical coupled lines in the form of a Marchand balun. The circuit is inherently discriminatory yielding finite CMRR. The differential-mode gain is largely unaffected by loss in the biasing circuit.

6.5 FOUR-PORT NETWORKS WITH BROADBAND COMMON-MODE SUPPRESSION

The previous section introduced a distributed biasing structure that is further exploited here to integrate the biasing function with differential filtering. In this section two classes of four-port microwave networks are presented with discriminative even- and odd-mode transmission characteristics. Design of the networks is based on filter prototypes with predefined sets of transmission zeros. These network prototypes were specifically developed for active differential circuits.

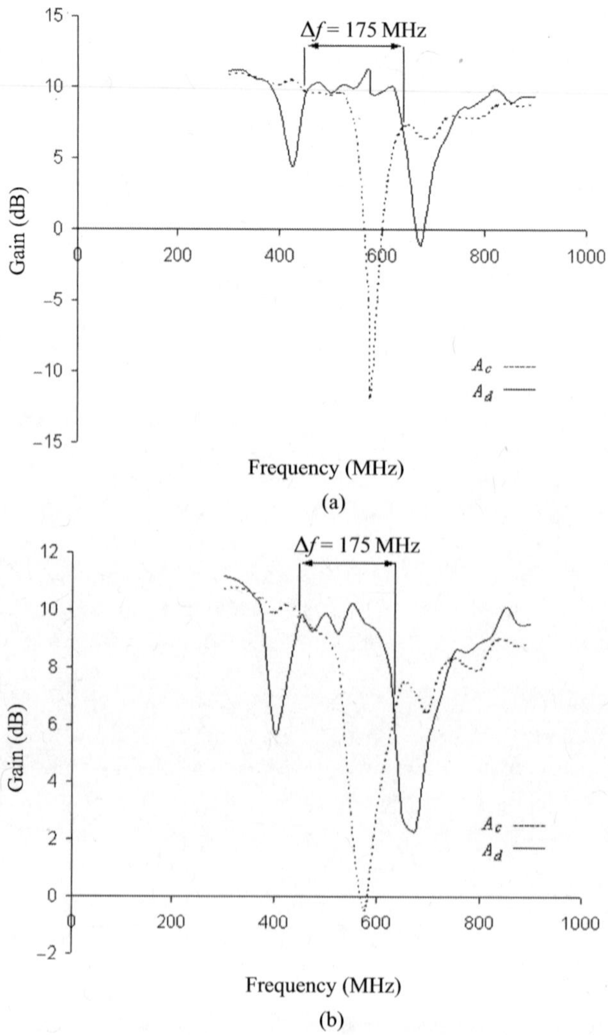

Figure 6-29 Measured common-mode gain, A_c and differential-mode gain A_d: (a) with miniaturized balun-like biasing circuit using high-Q lumped capacitors with circuit parameters: $Z_{0e} = 208\ \Omega$, $Z_{0o} = 51\ \Omega$, $Z_{0t} = 137\,\Omega$, $C = 5.6$ pF, $R_c = \infty$, $\tan\delta = 0.016$; and (b) with miniaturized balun-like biasing circuit using high Q lumped capacitors loaded with resistors with circuit parameters: $Z_{0e} = 208\ \Omega$, $Z_{0o} = 51\ \Omega$, $Z_{0t} = 137\,\Omega$, $C = 5.6$ pF, $R_c = 1.2\,\text{k}\Omega$, $\tan\delta = 0.016$.

Differential circuit design leads to stable, noise-tolerant monolithically integrated analog circuits and is compatible with current-mode biasing. Symmetrical differential circuits are generally less sensitive to component variations and when appropriately designed can reject common-mode signals, particularly coupled noise. However pseudo-differential circuits, as shown in the dotted box in Figure 6-30, typically have unit CMRR when biased by a pair of inductors. In this section two classes of four-port networks are introduced that are suited to biasing differential circuits and obtaining broadband common-mode signal suppression. DC biasing is achieved by appropriate selection of the set of transmission zeros of the odd-mode subnetwork of the four-port interstage circuit.

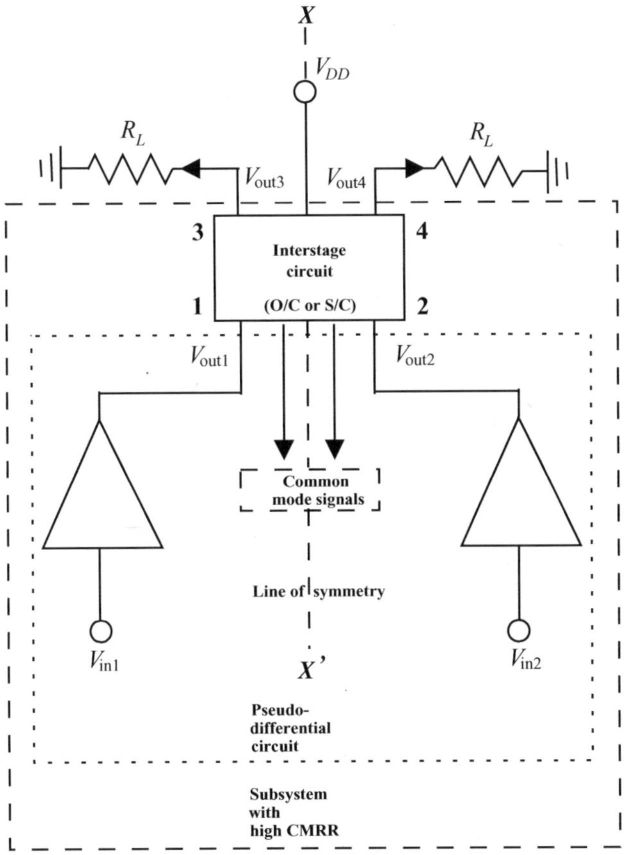

Figure 6-30 A pseudo-differential circuit commonly used in RFICs connected to a pair of single-ended loads R_L, separated by a four-port interstage network. Drain voltage, V_{DD}, is applied at relevant points within the interstage network to bias the active devices.

6.5.1 Class A Four-Port Interstage Network

The first class, denoted Class A, is shown in Figure 6-31(a) and comprises a pair of coupled-line sections connected to a pair of matching or filtering networks. The topology of the network may consist of coupled-line sections, transmission lines, and lumped components dependent on the choice of transmission zeros required to satisfy a particular electrical specification. Application of modal analysis to this structure results in the subnetworks of Figure 6-31(b) with equivalent even- and odd-mode prototypes as shown in Figures 6-31(c) and 6-31(d) respectively. These equivalent prototypes result because the plane of symmetry, $X - X'$, transforms to either an open or short circuit for common- or differential-mode signal excitations, respectively, at Ports 1 and 2. In the odd-mode effective network, differential-mode signals are passed to single-ended loads, R_L. Whereas in the even-mode equivalent circuit it is seen that common-mode signals do not pass through the network by virtue of the effective series open circuit present in the circuit. Thus the operation of Class A interstage networks in suppressing common-mode signals relies on the isolating property of the microwave section shown in the dashed box in Figure 6-31(c) [365,399]. Also Class A networks possess short-circuited points marked ***b*** in Figure 2(a) where DC bias of the active devices can be applied with the use of decoupling capacitors

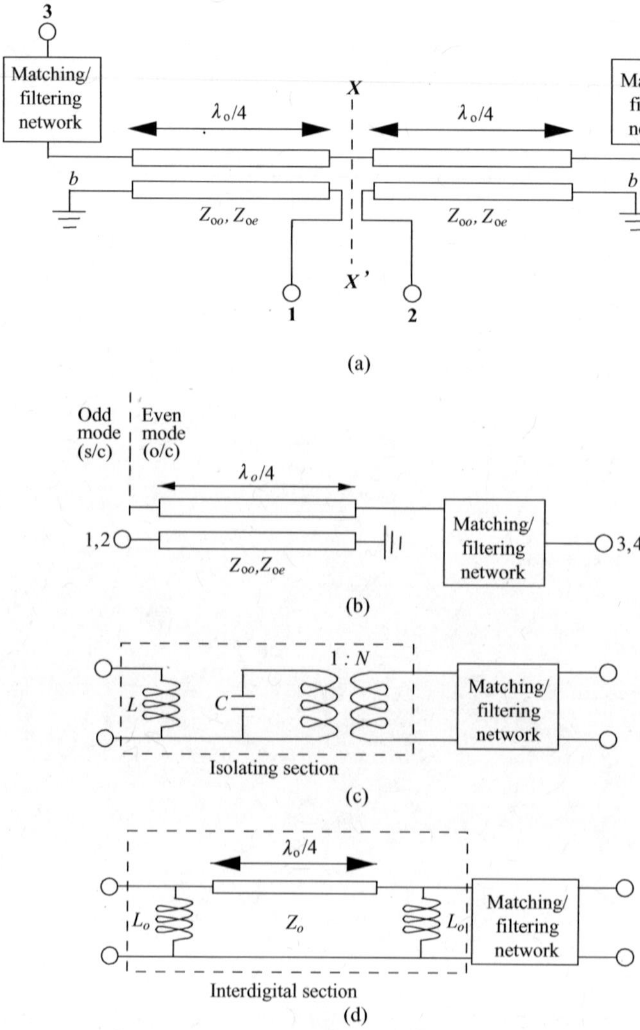

Figure 6-31 Generalized topology of Class A interstage network made of pairs of coupled lines and matching/filtering networks: (a) physical topology; (b) even- and odd-mode subnetworks; (c) even-mode prototype; and (d) odd-mode prototype with at least one zero at DC, and a single unit element (UE).

The square of the modulus of the transmission transfer function of the reciprocal odd-mode network of Figure 6-31(d) may now be defined in terms of its characteristic polynomial, $K(S)$, as

$$|S_T(S)|^2_{\text{odd}} = \frac{1}{1 + |K^2(S)|} \tag{6.81}$$

where $|S_T(S)|^2_{\text{odd}}$ replaces $|S_{31}(S)|^2_{\text{odd}}$ or $|S_{42}(S)|^2_{\text{odd}}$. With the even-mode network, ideally there is no transmission through the network and thus the design objective is

$$|S_T(S)|^2_{\text{even}} = 0 \tag{6.82}$$

where $|S_T(S)|^2_{\text{even}}$ replaces $|S_{31}(S)|^2_{\text{even}}$ or $|S_{42}(S)|^2_{\text{even}}$. In Equations (6.81) and (6.82) S is the Richards variable defined in Equation (6.8). Now the square of the modulus of the CMRR of the four-port interstage network may be defined as

$$|\text{CMRR}(S)|^2 = \frac{|S_T(S)|^2_{\text{odd}}}{|S_T(S)|^2_{\text{even}}} \tag{6.83}$$

and consequently the resulting CMRR of the subsystem configuration consisting of the pseudo-differential circuit in cascade with the four-port interstage network (the dashed box in Figure 6-30(a)) is

$$\text{CMRR}(S) = \frac{A_d(S) \times |S_T(S)|_{\text{odd}}}{A_c(S) \times |S_T(S)|_{\text{even}}} \tag{6.84}$$

where $A_c(S)$, and $A_d(S)$ are the single-ended common- and differential-mode gains respectively of the pseudo-differential circuit. For a pseudo-differential circuit, $A_c(S)$ and $A_d(S)$ are identical and thus the subsystem CMRR is dominated by the performance of the interstage network.

An example of a Class A interstage network is shown in Figure 6-32. It consists of a $2n$-th order network where n is the number of resonators in either the even- or odd-mode subnetworks. It is seen

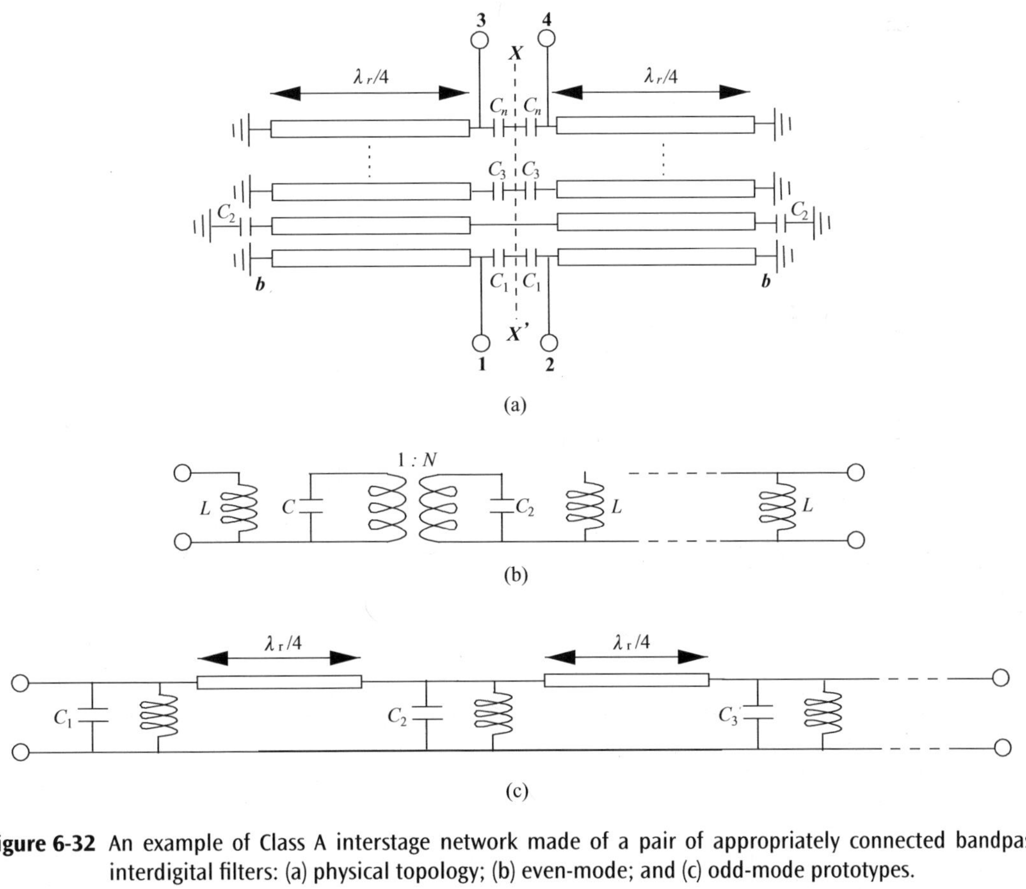

Figure 6-32 An example of Class A interstage network made of a pair of appropriately connected bandpass interdigital filters: (a) physical topology; (b) even-mode; and (c) odd-mode prototypes.

that the n-th resonators of the even-mode prototype, Figure 6-32(b), are completely de-coupled from each other, thus offering optimum isolation of common-mode signals. The odd-mode prototype, shown in Figure 6-32(c), consists of cascades of short-circuited interdigital coupled-line sections [365,366,384,395,400] loaded by open-circuited stubs. Thus the odd-mode network [395] possesses a zero at DC, n zeros at infinity, and $(n-1)$ zeros at $S=1$ leading to the existence of $(n-1)$ quarter-wave long transmission lines (or UEs).

The general design procedure for the interstage network starts with the synthesis of an n-th order S-plane odd-mode prototype corresponding to an f-plane response with a specific passband centered at f_0, with lower and upper bandedge frequencies f_1, and f_2 respectively, and a commensurate frequency f_r. A set of transmission zeros is selected to suit a particular electrical specification with a minimum of one zero at DC and a UE may be chosen to generate the appropriate characteristic polynomial [367, 394, 395] after which the square of the modulus of the reflection coefficient of the odd-mode prototype is constructed. Assuming a lossless system the square of the modulus of the odd-mode reflection transfer function is, from Equation (6.81),

$$|S_R(S)|^2_{\text{odd}} = 1 - |S_T(S)|^2_{\text{odd}} \tag{6.85}$$

which can be redefined in terms of $K(S)$ as

$$|S_R(S)|^2_{\text{odd}} = \frac{|K(S)|^2}{1+|K(S)|^2}. \tag{6.86}$$

$S_R(S)$ may then be found with the knowledge that

$$|S_R(S)|^2_{\text{odd}} = S_R(S)|_{\text{odd}} \cdot S_R(-S)|_{\text{odd}}. \tag{6.87}$$

From this the input impedance of the odd-mode prototype can then be derived in a 1 Ω system using

$$Z_{\text{in}}(S)|_{\text{odd}} = \frac{1+S_R(S)|_{\text{odd}}}{1-S_R(S)|_{\text{odd}}} \tag{6.88}$$

and synthesized using standard element extraction.

A possible source of CMRR degradation in Class A networks is the leakage of common-mode energy through the network. This is particularly a concern with inhomogeneous media such as microstrip or coplanar waveguide. To investigate this further consider the inhomogeneous transfer admittance of the isolating section (the dashed box in Figure 6-31(c)) [365]:

$$Y_{12} = -j\frac{\cos(\theta_e)-\cos(\theta_o)}{Z_{0e}\sin(\theta_o)\cos(\theta_e)+Z_{0e}\cos(\theta_o)\sin(\theta_e)} = Y_{21} \tag{6.89}$$

where Z_{0e} and Z_{0o} are the even- and odd-mode impedances of the coupled lines and θ_e and θ_o are the even- and odd-mode phase lengths respectively. It is seen from Equation (6.89) that the numerator is zero only when the even- and odd-mode phase lengths are equal resulting in the equivalence of Figure 6-31(c). Thus the deterioration in subsystem performance due to the inhomogeneous media will be manifested in lower achievable CMRR since the transmission of the even-mode network, $|S_T(S)|^2_{\text{even}}$ will have a finite value.

6.5.2 Class B Port Interstage Matching Network

Another four-port circuit with valuable characteristics is the quadrature coupler of Figure 6-33(a) with even- and odd-mode prototypes illustrated in the Figures 6-33(b) and 6-33(c). The

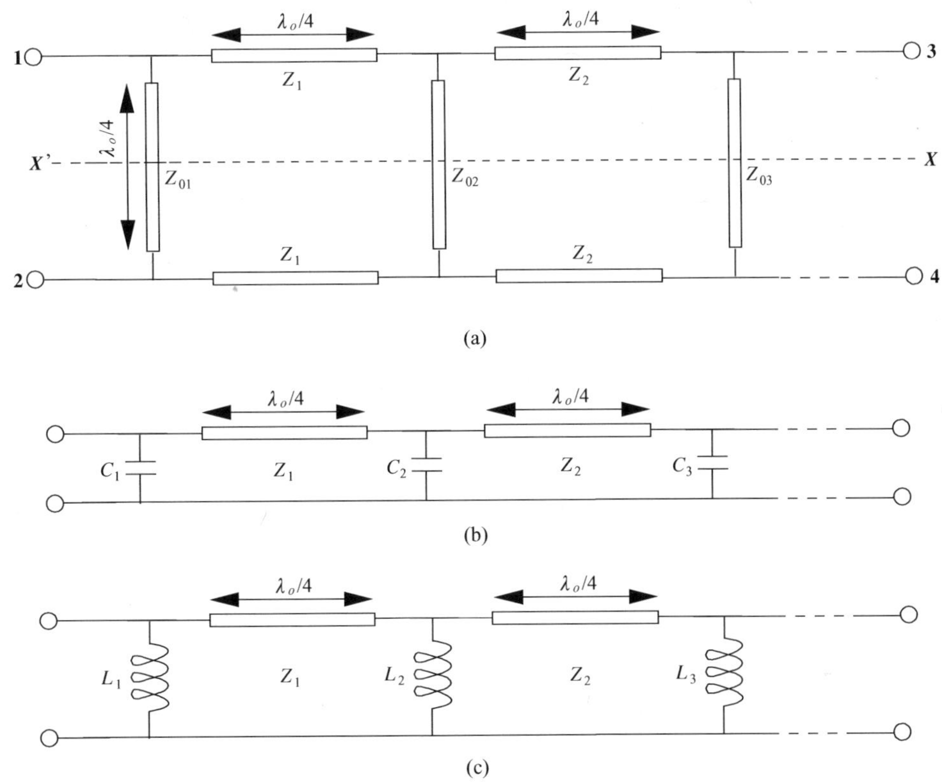

Figure 6-33 A quadrature hybrid coupler: (a) physical topology; (b) even-mode; and (c) odd-mode prototypes.

even- and odd-mode prototypes typically consist of sections of shunt eighth-wavelength long open- and short-circuited stubs respectively separated by quarter-wavelength long sections of transmission lines of uniform impedance, Z_i. Each quarter-wave transmission line section may then be split into two equal sections and the resulting ABCD matrix of a cascade pair of eighth-wavelength long transmission lines is

$$\begin{bmatrix} A & B \\ C & D \end{bmatrix} = \left\{ \frac{1}{\sqrt{1-S^2}} \begin{bmatrix} 1 & \frac{S}{Z_i} \\ Z_i S & 1 \end{bmatrix} \right\}^2 = \frac{1}{1-S^2} \begin{bmatrix} 1+S^2 & \frac{2S}{Z_i} \\ 2Z_i S & 1+S^2 \end{bmatrix}. \tag{6.90}$$

From Equation (6.90) it is seen that the resulting matrix is invariant with respect to the transformation

$$S \to \frac{1}{S} \tag{6.91}$$

except for a change of sign [401]. This leads to a very interesting property. The above implies that the odd-mode prototype of the coupler can be derived from its even-mode, and vice versa, by simply replacing the open-circuited stubs by short-circuited stubs [401]. Now in the S-plane the even- and odd-mode prototypes have lowpass and highpass responses respectively. In the f-plane the lowpass and highpass responses become periodic functions with bandstop and bandpass responses respectively at the center frequency f_0. Thus over a frequency band centered at f_0 the even-mode network suppresses common-mode signals while the odd-mode prototype passes differential-mode signals. So the CMRR of this network has the relationship already defined in Equation (6.83). In this case the square of modulus of the even-mode transmission transfer function, $|S_T(S)|^2_{\text{even}}$ only dips to zero at the center of the band, and thus the CMRR will have a varying characteristic with enhanced performance in the middle of the operating band.

The quadrature coupler utilized as an interstage network has a maximum bandwidth ratio ($f_2/f_1 : 1$ where f_1 and f_2 are the passband edge frequencies) of 3:1 over which there is discrimination of common- and differential-mode signals. This condition is satisfied only if the cutoff frequency, f_1, of the even-mode prototype is chosen to be

$$f_1 = f_0/2 \tag{6.92}$$

leading to an upper cutoff frequency of

$$f_2 = 3f_1. \tag{6.93}$$

So in principle the quadrature coupler could be used as an interstage network but in contrast to a Class A network it is a lowpass structure. This structure can be modified for the purpose of biasing active devices with the inclusion of a pair of wide-band bias-Ts [402] at the input arms. However, by proper selection of the relevant set of transmission zeros a coupler-like network that supports DC bias can be designed. This leads to the Class B interstage network shown in Figure 6-34. The generalized odd-mode S-plane prototype of a Class B circuit is shown in Figure 6-34(b) and consists of a minimum of five zeros at DC and i UEs, where i is an even number to maintain physical symmetry [367]. This has the benefit of providing DC bias to other parts of the system if required. It is now intuitive to conclude that the even- and odd-mode subnetworks of the coupler-like network will provide the necessary attenuation and matching of common- and differential-mode signals. This is illustrated in Figure 6-35 for a set of synthesized Class B prototypes with a 3:1 passband centered at 500 MHz, for various filter degrees. It is seen that the even-mode and odd-mode responses are considerably different from each other over the frequency range of interest. The main advantage of a Class B network is simplicity in design, control of the even-mode attenuation level, and inherent broadband operation. It is worth pointing out that the effective electrical lengths of the transmission lines forming the branches of the coupler-like section of Class B networks are a half wavelength long, whereas the branches of the true quadrature coupler are a quarter wavelength long. In general Class B networks are particularly suited for systems operating at very high frequency as electrical miniaturization of a Class B structure is not possible as is the case with Class A networks.

6.5.3 Implementation of a Class A Interstage Network

A pseudo-differential amplifier (HELA-10B) [398] was selected and integrated with a Class A network. The amplifier has a gain of 11 dB and broadband operation from 50 MHz to 1 GHz in a 50-Ω system. The chip requires a single +12V DC power supply. An odd-mode S-plane bandpass filter prototype was designed corresponding to an f-plane response with a center frequency, f_0, of 500 MHz, bandedge frequencies (f_1 and f_2) of 250 and 750 MHz, a commensurate frequency, f_r, of 2 GHz, and a return loss of 16 dB. The set of transmission zeros chosen for this filter are three zeros at DC, three zeros at infinity, and two non-redundant UEs. Based on this the characteristic polynomial, $K(S)$, of the filter prototype is constructed using a classic procedure [394, 395].

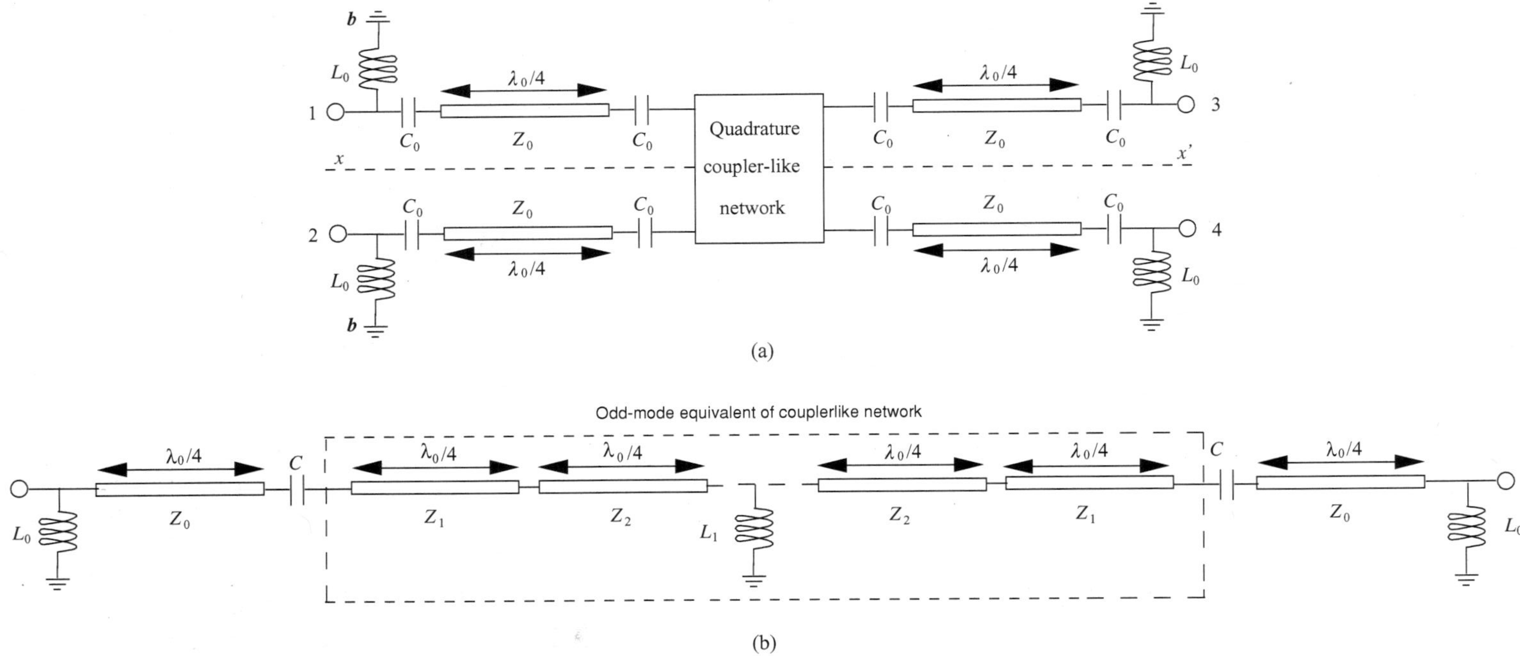

Figure 6-34 Generalized topology of Class B interstage network: (a) physical topology; and (b) odd-mode prototype with five zeros at DC, and i UEs ($i = 4, 6, 8, 10, ...$).

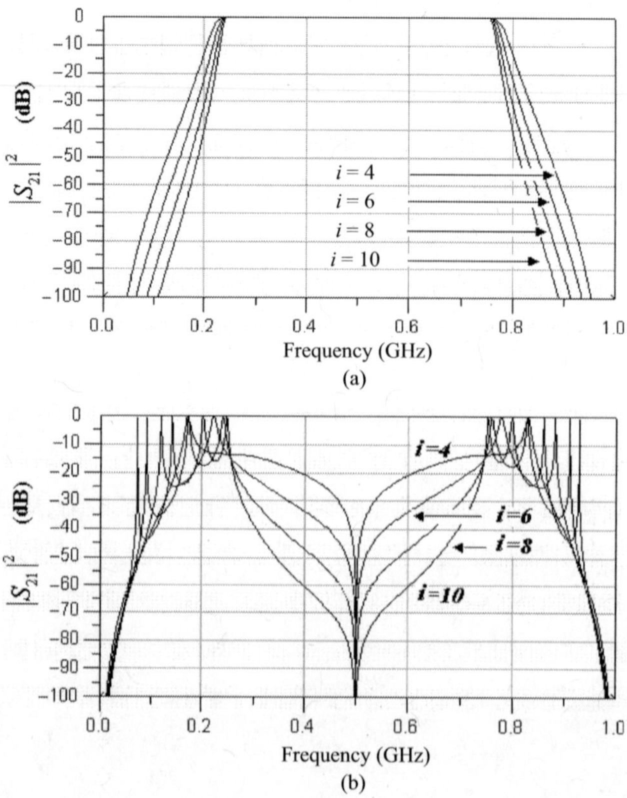

Figure 6-35 Simulated transmission characteristics of Class B interstage network for different degrees: (a) odd-mode; and (b) even-mode responses centered at 500 MHz with a 3:1 bandwidth.

This gives

$$K(S) = \frac{\left(2\,S\sqrt{1-S^2}\right)}{(-56.51122588\,S^4 + 0.01613466\,S^3 - 13.39620723\,S^2 + 0.0001404614\,S - 0.629533102)} \tag{6.94}$$

from which the input reflection coefficient is evaluated using Equations (6.86) and (6.87) leading to

$$S_R(S)|_{\text{odd}} = \frac{-56.51122588\,S^4 + 0.01613466\,S^3 - 13.39620723\,S^2 + 0.0001404614\,S - 0.629533102}{(56.51122588\,S^4 + 23.2890949\,S^3 + 18.19509163\,S^2 + 3.1689416\,S + 0.629533102)}. \tag{6.95}$$

The input impedance, $Z_{in}(S)|_{odd}$, was then evaluated using Equation (6.88). The synthesized prototype is shown in Figure 6-36(a) in a 50 Ω system. Construction of the four-port interstage network is now possible after transformation of parts of the odd-mode prototype to a structure made of coupled lines and lumped capacitors. Figure 6-36(b) shows the final layout.

Implementation on an FR4 PCB with a substrate thickness of 62 mil (1.57 mm), relative dielectric constant of 4.7 and loss tangent of 0.016, translate coupled line even- and odd-mode impedances of 276.589 Ω and 48.69 Ω to a pair of PCLs that are 10 mil (0.254 mm) wide and 7 mil (0.177 mm) apart. The coupled lines are 910 mil (23.1 mm) long. These dimensions were obtained

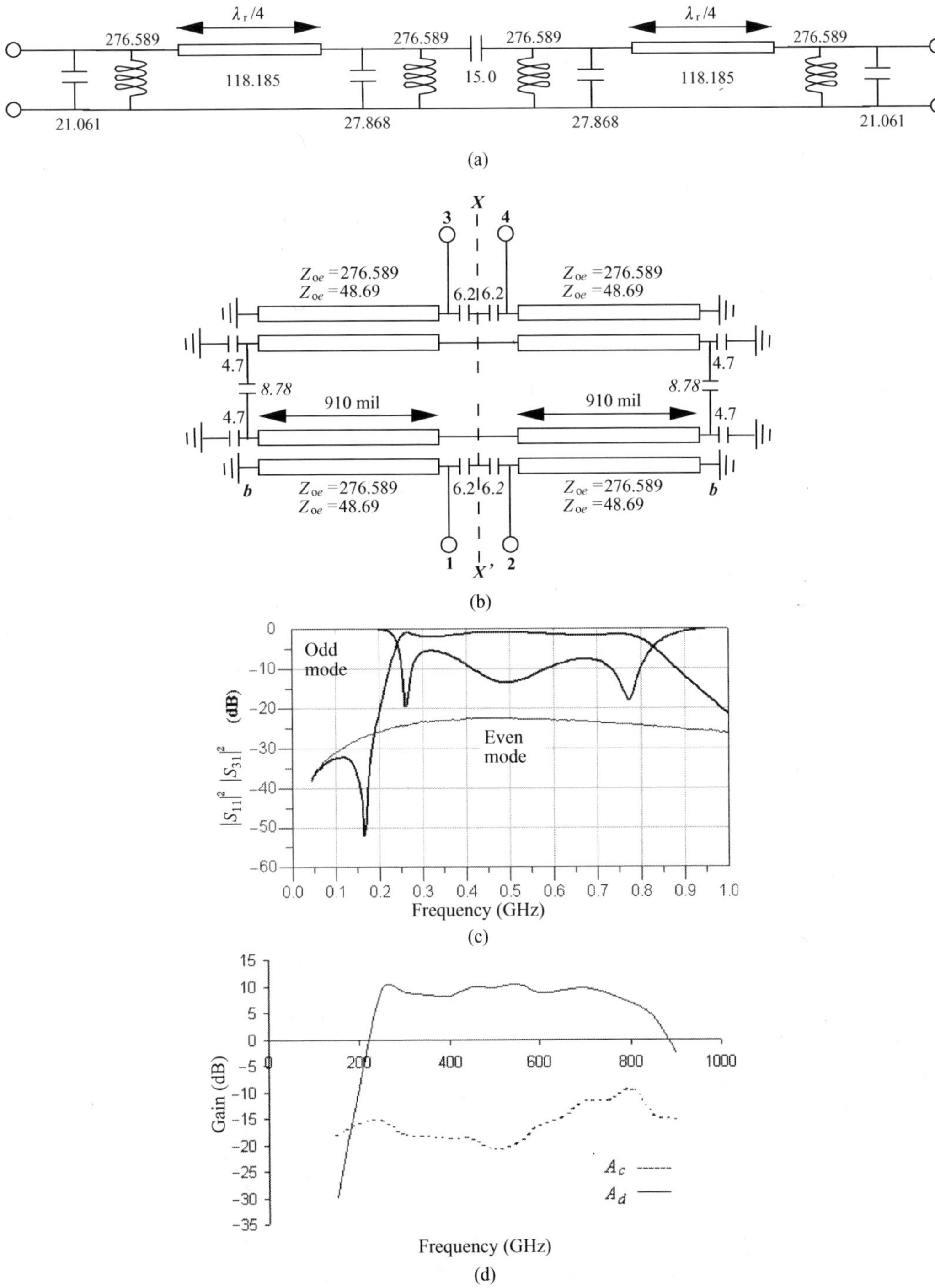

Figure 6-36 A Class A interstage network in a 50-Ω system: (a) synthesized odd-mode prototype: (b) physical layout with capacitors in picofarads; (c) measured even- and odd-mode transmission characteristics of (b); and (d) measured common- and differential-mode gains of the pseudo-differential amplifier in cascade with the interstage network.

using ADS [115]. The even- and odd-mode transmission characteristics of the physical layout of Figure 6-36(b) were measured, and are presented in Figure 6-36(c). In the odd-mode, a bandpass response with 3:1 bandwidth ratio is observed when the even-mode isolation is at least 22 dB. The finite even-mode isolation level is primarily due to the difference between the even- and odd-mode phase velocities of the microstrip media leading to the transmission of common-mode signals through the isolating sections. Another source of energy leakage contributing to this finite isolation level is the parasitic coupling beyond nearest neighboring resonators leading to unwanted couplings throughout the even-mode networks of the interstage network. This is particularly true since there is weak coupling between closely-spaced pairs of resonators of isolating sections giving rise to crosscoupling routes.

At this point the pseudo-differential amplifier and the interstage network were integrated and the overall characteristics measured. The result is shown in Figure 6-35(d). Very low common-mode gain, A_c, of less than -12 dB was recorded over the operating bandwidth while a 3:1 bandwidth ratio was achieved with a differential-mode gain, A_d, of about 10 dB. The excess 1 dB of loss is attributed to loss in the interstage network. Also it is observed that the differential-mode gain ripples over the measured passband and rolls off rapidly towards the upper band edge frequency, i.e., 750 MHz. This is primarily due to the impedance level mismatches since the measured return loss of the interstage circuit is only 6 dB, see Figure 6-36(c). The deteriorated return loss was due to the restriction imposed by the FR4 board manufacturing process that only allowed a minimum of 7 mil spacing between the coupled lines. To achieve a better return loss, for example 12 dB, a 5 mil spacing between the lines is required.

6.5.3.1 Implementation of a Class B Interstage Network A Class B interstage network was also designed on an FR4 board with the same specification to that used in the above subsection. In this case the objective is to highlight the characteristics of the even- and odd-mode networks of a Class B circuit without integration with the pseudo-differential amplifier. The odd-mode prototype was designed for an f-plane response identical to that defined in the above subsection but with a return loss (RL) of 36 dB. Here the commensurate frequency, f_r, of the f-plane prototype coincides with the center frequency, f_0, since the prototype corresponds to an S-plane highpass. The set of transmission zeros of the odd-mode prototype of the coupler-like network are chosen to be five zeros at DC and four UEs. The characteristic polynomial, $K(S)$, was directly generated using [367] to give

$$K(S) = \frac{\begin{pmatrix} -.3234410760S^{12} - 4.831961462S^{10} - 12.43391743S^{8} \\ +6.369465650S^{6} + 29.01088612S^{4} - 1.294045780S^{2} - 16.49698604 \end{pmatrix}}{2S^{5}\left(S^{2}-1\right)}. \tag{6.96}$$

In a manner similar to that used previously, $S_R(S)|_{\text{odd}}$ and $Z_{\text{in}}(S)|_{\text{odd}}$ are constructed using Equations (6.86)–(6.88).

Direct synthesis of the above transfer function leads to the prototype of Figure 6-37(a) in a 50 Ω system. Application of the appropriate Kuroda transformations to the circuit of Figure 6-37(a) results in the circuit of Figure 6-37(b). It is seen from Figure 6-37(b) that the characteristic impedances of the inner pair of shunt short-circuited stubs are 33.27 Ω. This pair of stubs was forced to have a value of 50 Ω by application of circuit optimization leading to a degradation of the return loss from 36 dB to about 15 dB with a slight increase in bandwidth. The prototype is shown in Figure 6-37(c). At this point the layout of the interstage network was constructed from the odd-mode prototype. However the input and output sections (series open-circuited stubs separated by a UE) are difficult to realize using microstrip technology and equivalence to coupled-lines must be

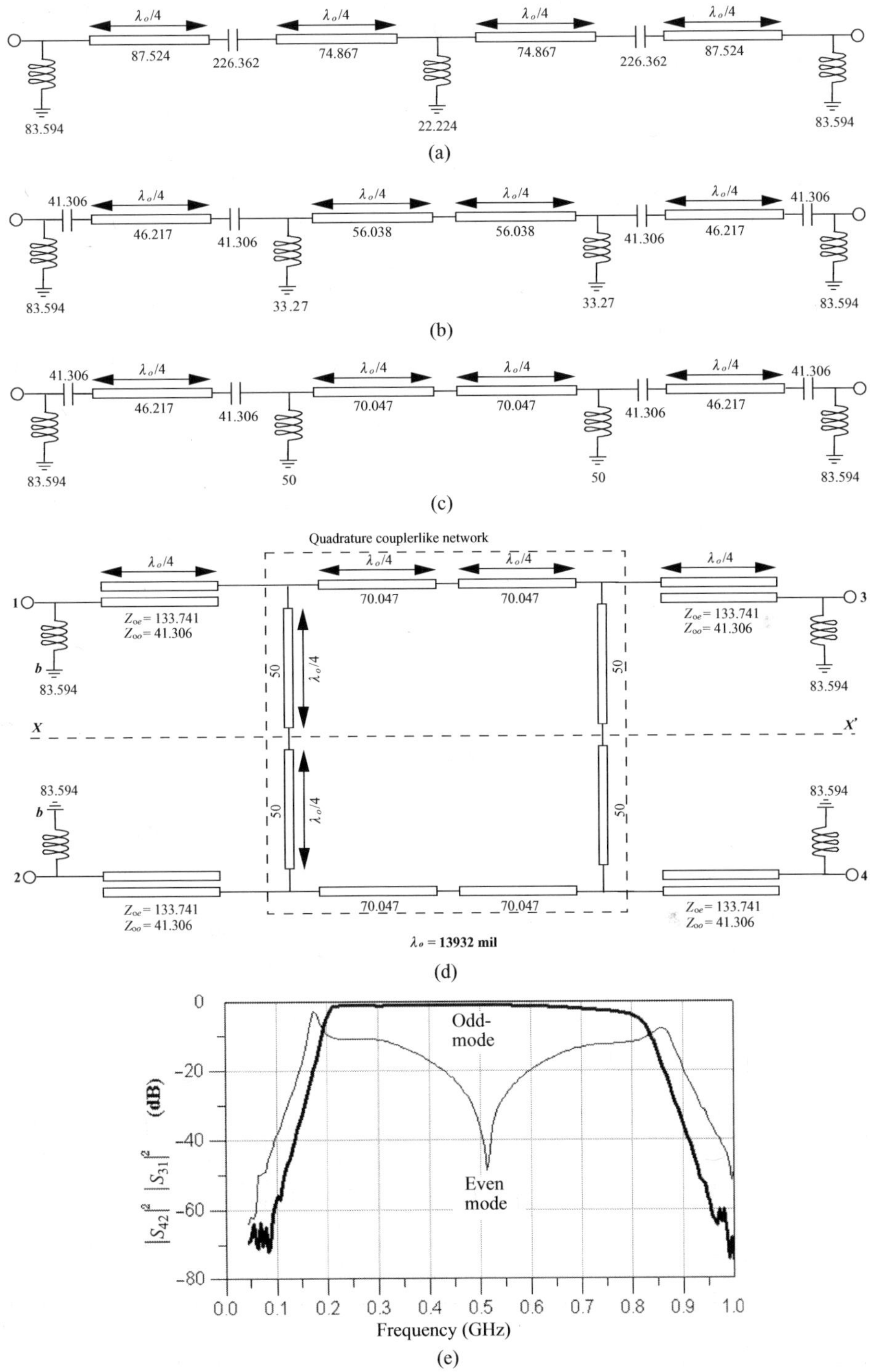

Figure 6-37 A Class B interstage network: (a) odd-mode prototype: (b) after application of Kuroda transformations; (c) after circuit optimization; (d) physical layout; and (e) measured even- and odd-mode transmission characteristics of (d).

made leading to the final network of Figure 6-37(d). Now the even- and odd-mode impedances of values 133.741 and 41.306 Ω of each coupled-line section were translated into physical dimensions resulting in a width of 27 mil (0.685 mm), spacing of 7 mil (0.177 mm), and length of 3483 mil (88.5 mm). The width of the other transmission lines of 50 Ω, 70.047 Ω, and 83.594 Ω were 110 mil (2.794 mm), 57 mil (1.447 mm), and 38 mil (0.965 mm) respectively. Also the lengths of all the transmission lines were 3483 mil (88.5 mm).

The measured even- and odd-mode transmission characteristics of the interstage network are shown in Figure 6-37(e) showing significant difference as expected.

6.5.4 Summary

This section presented the derivation of and design procedures for two classes of four-port microwave interstage networks that bloke common-mode signals over a very broad frequency range. The design procedure presented for the new networks is based on the synthesis of an odd-mode prototype with broad bandwidth and out-of-band rejection levels. Subsequent integration of either class of these networks with pseudo-differential circuits leads to dramatic improvement of the subsystem CMRR over large bandwidth. Both classes of networks possess fundamentally different even- and odd-mode transmission characteristics and reflect undesired common-mode energy. Class B networks could be designed to posses a finite attenuation level to reject common-mode signals. This feature does not exist with Class A networks and basically the rejection level of common-mode signals will depend on how homogeneous the realization medium is. Choice of which class to use is dependent on the system application.

6.6 TAPPED MARCHAND BALUNS FOR STABLE MATCHING APPLICATIONS

Subsystems with active devices generally require matching networks which, as well as providing optimum transfer of power within an operating band, often determine the stability of the subsystem by their out-of-band characteristics. Essentially a matching network must extract maximum power from the source for transfer to the load, but must also present the source with a specific immittance profile both in and out of its operating band. For example, the impedance level of a matching network connected to the output of a power amplifier approximating a current source must drop to a low level in the out-of-band region to ensure device stability. These design criteria impose requirements on the matching network necessitating that its design be based on a singly terminated [403–406] rather than a doubly terminated network [388]. In the case of a singly terminated matching network the real part of its input immittance function becomes the design parameter and hence is tailored appropriately to suit the characteristics of the source.

This section presents structures that combine the functionality of baluns and matching networks. The structures are derived from several filter prototypes [388] that can be realized in the form of Marchand baluns. The baluns represent a solution to the problem of matching between terminals of extreme impedance (or admittance) values without using elements of extreme characteristic impedances in the balun structures. This is by virtue of tapping the balanced resonators of the baluns, thus allowing a wide range of load-to-source immittance ratios to be realized.

The presentation here begins in Sections 6.6.1 and 6.6.2 by highlighting the difference in design objectives between doubly and singly terminated networks. Then Section 6.6.4 proposes a method of scaling the balanced load of a balun to specific values. Finally Sections 6.6.6 and 6.6.7 present the synthesis and implementation of two tapped baluns. Size miniaturization and specific load-to-source immittance ratios are achieved.

6.6.1 Doubly Terminated Matching Network

The fundamental difference between doubly and singly terminated matching networks concerns the design parameter synthesized to satisfy an electrical specification. First the doubly-terminated matching network will be considered.

A doubly terminated network attempts to achieve a constant return-loss level at its terminals over a specified passband. Referring to Figure 6-4, the design focus is the input reflection coefficient

$$\Gamma_{\text{in}} = \frac{Z_{\text{in}} - R_S}{Z_{\text{in}} + R_S} \tag{6.97}$$

for an impedance-based network where R_S is the source impedance. The input reflection coefficient can also be expressed as

$$\Gamma_{\text{in}} = \frac{Y_S - Y_{\text{in}}}{Y_S + Y_{\text{in}}} \tag{6.98}$$

for an admittance-based network where Y_s $(= 1/R_s)$ is the source admittance. In a lossless system the transmission characteristic is

$$|S_{21}|^2 = 1 - |\Gamma_{\text{in}}|^2 . \tag{6.99}$$

6.6.2 Singly Terminated Matching Network

An ideal impedance-based singly-terminated network is driven by a current source and presents a constant-impedance level to its driving source over its operating band. It can be shown that the transmission coefficient of a singly terminated network is actually equal to the real part of its input impedance [405, 406]:

$$|S_{21}|^2 = \Re Z_{\text{in}}. \tag{6.100}$$

Also, for an admittance-based network driven by a voltage source the transmission characteristic is

$$|S_{21}|^2 = \Re Y_{\text{in}}. \tag{6.101}$$

It is evident from Equation (6.99) to (6.101) that the transfer of power from the source to load is different in the two cases.

6.6.3 A Balun Example

To demonstrate the above points the performance of a third-order Class A Marchand balun [388] designed from the doubly and singly terminated prototypes presented in Sections 6.6.1 and 6.6.2 is considered.

Figure 6-38(a) presents plots of the real part of the balun's input admittance function at the single-ended port. Here it is shown how the characteristic of the real part changes as the return-loss level is varied from 10 dB to 20 dB. As expected, the passband ripple decreases as the return loss improves. However it is also seen, from Figure 6-38(a), that the real part peaks at frequencies outside the passband (defined by the vertical dotted lines) as the return loss improves resulting in an uncontrolled increase in bandwidth. Thus, in order to improve the in-band ripple of the input admittance level, the skirt selectivity must be degraded. In almost every application the skirt selectivity of a matching network is equally as important as the quality of the in-band match and has direct influence on the overall system performance.

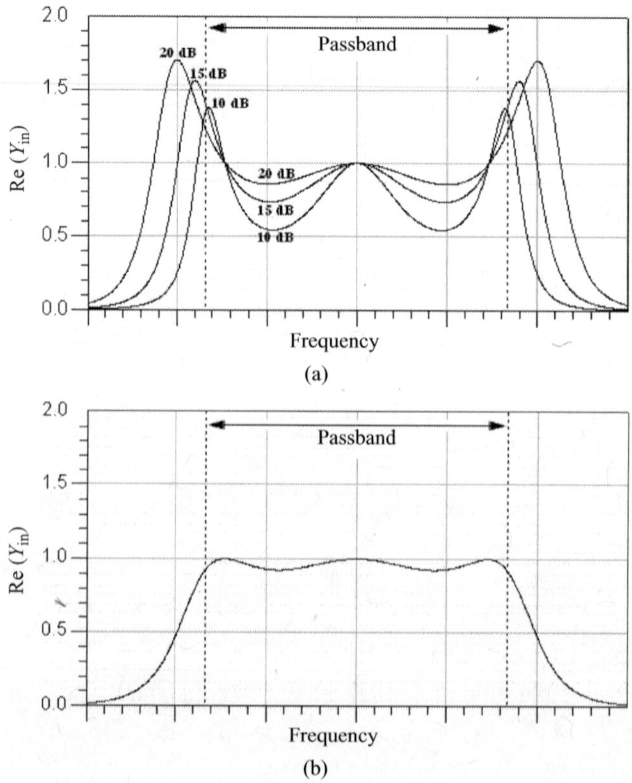

Figure 6-38 Input admittance of Class A Marchand balun normalized to a 1 Ω system: (a) real part of the input admittance of a balun designed from a doubly-terminated prototype for various single-ended return–loss levels; and (b) the same balun designed utilizing a singly terminated prototype with a specific passband ripple.

Now the same type of balun is synthesized on a singly terminated basis, see the characteristic in Figure 6-38(b). Here it is seen that the transition from passband to stopband is much faster than that of any of the functions presented in Figure 6-38(a). In addition there is a more controlled in-band ripple variation. The function depicted in Figure 6-38(b) is Chebyshev which is well known for its optimum performance in terms of trading off passband ripple and skirt selectivity [405].

Thus, based on the proceeding comparison, it is possible to conclude that a singly terminated balun integrated as a matching network in a subsystem configuration will extract maximum available power from its source and will possess the desired out-of-band characteristics for stability requirements of active circuits. Table 6-2 summarizes the type of parameter that is synthesized at either port with reference to the sets of transmission zeros used.

6.6.4 Tapping the Balanced Resonators of the Balun

This section shows that the balun of Figure 6-39(a) [388] can be designed for specific load-to-source immittance ratios. The equivalent prototype of this balun, Figure 6-39(b), can be synthesized for different commensurate frequencies, f_r. The commensurate frequency determines the electrical lengths of the transmission lines forming the balun and thus the larger the value of f_r the smaller the size of the resulting balun structure. In addition to fixing the overall size of the balun, the choice

Table 6-2 Type of parameter synthesized for each class of baluns. (S is the Richards transformation)

Prototype classes [9]	Distribution of zeroes	Type of parameter realized at single-ended port	Type of parameter realized at balanced ports
A	2 zeros at $S=0$ 2 zeros at $S=\infty$ 2 zeros at $S=1$	$\Re Y_{in}$	$\Re Z_{in}$
$C1$	2 zeros at $S=0$ 1 zero at $S=\infty$ 1 zero at $S=1$	$\Re Y_{in}$	$\Re Z_{in}$
$C2$	1 zero at $S=0$ 2 zeros at $S=\infty$ 1 zero at $S=1$	$\Re Z_{in}$	$\Re Z_{in}$
B	2 zeros at $S=0$ 2 zeros at $S=\infty$	$\Re Z_{in}$	$\Re Z_{in}$

of f_r also impacts the values of the immittances terminating the balun. In general the inherent values of the load and source immittances of a Marchand balun are functions of the passband ripple, operating bandwidth, and also f_r. An example of a synthesized balun with an octave bandwidth centered at 500 MHz is shown in Figure 6-40. It is seen that the inherent values of the single-ended and balanced impedances drop as f_r increases from 5 to 20 GHz. Thus miniaturizing such a structure has an impact on its terminal impedances. It is also observed that the inherent

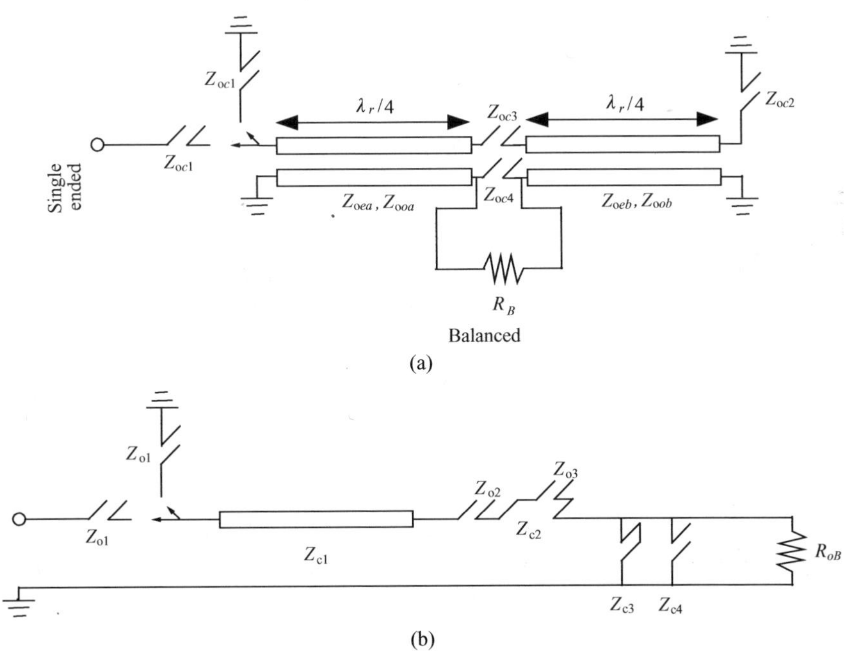

Figure 6-39 A bandpass Marchand balun with balanced load impedance R_B: (a) its physical layout based on Figure 6-6; and (b) its corresponding S-plane bandpass prototype (from Figure 6-8(b)).

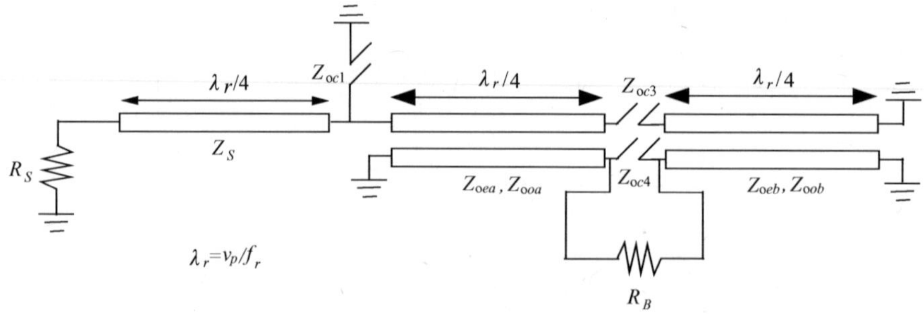

(a)

f_r (GHz)	Z_s (Ω)	Z_{0c1} (Ω)	Z_{0c3} (Ω)	Z_{0c4} (Ω)	$Z_{0ea} = Z_{0eb}$ (Ω)	$Z_{0oa} = Z_{0ob}$ (Ω)	R_s (Ω)	R_B (Ω)	R_B/R_s
5	47.55	1.56	3.117	4.744	164.946	58.253	6.166	59.988	9.728
10	48.028	0.385	0.787	1.184	166.842	59.421	3.127	30.567	9.775
20	48.148	0.095	0.197	0.294	167.502	59.827	1.569	15.373	9.797

(b)

f_r (GHz)	Z_s (Ω)	Z_{0c1} (Ω)	Z_{0c3} (Ω)	Z_{0c4} (Ω)	$Z_{0ea} = Z_{0eb}$ (Ω)	$Z_{0oa} = Z_{0ob}$ (Ω)	R_s (Ω)	R_B (Ω)	R_B/R_s
5	35.695	0.985	3.345	5.245	189.411	50.789	2.748	39.173	14.255
10	36.257	0.245	0.845	1.310	191.724	52.069	1.396	19.988	14.318
20	36.399	0.061	0.211	0.326	192.152	52.307	0.701	10.035	14.315

Figure 6-40 Class A balun: (a) the electrical layout with an octave bandwidth centered at 500 MHz; (b) element values for a prescribed impedance function at the balanced ports with a passband ripple of 0.5 dB; and (c) element values for a prescribed admittance function at the single-ended port with a passband ripple of 0.5 dB.

balanced-to-single-ended impedance ratios and the modal-impedances in the two cases remain practically unaffected with different values of f_r. Focus will now shift to tapping the balanced resonators of the balun to achieve a wide range of balanced-to-single-ended immittance ratios.

The detailed analysis of the generalized Marchand balun presented in Section 6.3 demonstrated that the balanced load is always much higher in value than the single-ended source impedance. It is common practice to scale down the balanced load impedance by using a pair of quarter-wavelength long transmission lines, e.g.,see Figure 6-15. While this is normally an acceptable technique, the pair of transmission lines drastically increased the overall size of the balun. Note that these matching lines resonated at the passband center frequency f_0 while the balun structure comprised transmission lines that resonated at the commensurate frequency f_r where $f_r > f_0$. This situation is undesirable in circumstances where size reduction is the primary objective. This drawback is overcome while simultaneously achieving a specific load-to-source immittance ratio by tapping the balanced resonators of the balun. It is common practice to alter the system impedance at either or both ports of a microwave filter [407] to yield a realizable circuit. Consequently tapping the resonators at the scaled node(s) of the filter matches it to the desired system impedance with the advantage of no added hardware. This approach is used here in designing Marchand baluns.

To begin the description of the tapped balun design, consider Figure 6-41(a) which illustrates the basic output subsection of a bandpass Marchand balun. The subsection consists of a short-circuited stub, representing the balanced resonators, and an open-circuited stub connected to the balanced

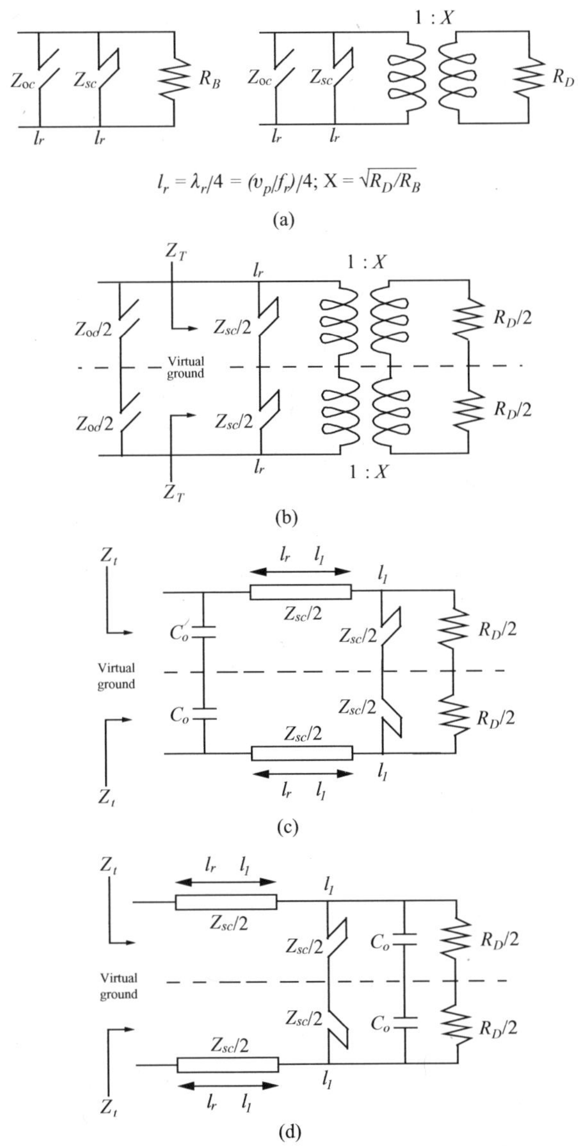

Figure 6-41 Balanced resonator tapping: (a) typical output section of a bandpass balun prototype after scaling its balanced load using an ideal transformer; (b) identification of the virtual ground plane; and (c) and (d) balanced resonators after tapping.

load, R_B. To scale the load impedance from R_B to R_D ($R_D < R_B$) an ideal $1:X$ transformer $\left(X = \sqrt{R_D/R_B}\right)$ is inserted at the output as shown in the Figure 6-41(a). If the balun is designed such that the virtual ground plane bisects the subcircuit in the center, then it is possible to exactly represent the subcircuit of Figure 6-41(a) by that of Figure 6-41(b). The subcircuit of Figure 6-41(b) (excluding the open-circuited stub) can then be approximated by that of Figure 6-41(c). Examination of the subcircuit of Figure 6-41(c) reveals that the total length of the balanced resonators is still equal to l_r but now the short-circuited stubs of length l_1 represent the parts of each resonator seen by each half of the scaled balanced load to the virtual ground. Also the existence of the lumped capacitor C_0 is required for fine-tuning purposes.

For a balun realizing a prescribed immittance function at the single-ended port the following approximation is valid for passband bandwidths up to an octave. At the center frequency f_0 the impedance Z_T in Figure 6-41(b) is

$$Z_T = \Re_T + j\Im_T \tag{6.102}$$

where

$$\Re_T = \frac{2R_D^2}{R_B\left(\frac{4R_D^2}{Z_{sc}^2(\tan(\beta l_r))^2} + \frac{4R_D^2}{R_B^2}\right)} \tag{6.103}$$

and

$$\Im_T = \frac{2R_D^2}{Z_{sc}\left(\tan\left(\beta l_r\right)\right)\left(\frac{4R_D^2}{Z_{sc}^2(\tan(\beta l_r))^2} + \frac{4R_D^2}{R_B^2}\right)}. \tag{6.104}$$

Now the ABCD matrix of half of the circuit of Figure 6-41(c) (without including the load impedance) can be derived based on the following matrix manipulations:

$$\begin{bmatrix} A & B \\ C & D \end{bmatrix} = \begin{bmatrix} 1 & 0 \\ j(2\pi f)C_0 & 1 \end{bmatrix} \begin{bmatrix} \cos\left(\beta\left(l_r - l_1\right)\right) & j\left(\frac{Z_{sc}}{2}\right)\sin\left(\beta\left(l_r - l_1\right)\right) \\ j\left(\frac{2}{Z_{sc}}\right)\sin\left(\beta\left(l_r - l_1\right)\right) & \cos\left(\beta\left(l_r - l_1\right)\right) \end{bmatrix}$$
$$\times \begin{bmatrix} 1 & 0 \\ j\left(\frac{Z_{sc}}{2}\right)\tan\left(\beta l_1\right) & 1 \end{bmatrix}. \tag{6.105}$$

Using Equation (6.105) the input impedance, Z_t, is evaluated at f_0 to give

$$Z_t = \frac{A\left(\frac{R_D}{2}\right) + B}{C\left(\frac{R_D}{2}\right) + D} = \Re_t + j\Im_t. \tag{6.106}$$

In Equations (6.103)–(6.105) $\beta\left(= (2\pi/\lambda) = (2\pi f/v_p)\right)$ is the propagation constant and v_p is the phase velocity in the transmission medium. The real and imaginary parts of Equations (6.102) and (6.106) may now be equated and solved simultaneously for the two unknowns l_1 and C_0. It should be appreciated, however, that there is a limit as to how much the balanced load impedance can be lowered from R_B to R_D using the tapping approach. For example, forcing R_D to be a very low value will likely degrade the ripple level of the real part of the input immittance at the single-ended port. In general the approximation is increasingly valid as R_D approaches $R_B/2$.

The above procedure can also be used to scale the balanced load of a balun realizing a prescribed immittance function at its balanced ports. In this case it is more appropriate to approximate the subcircuit of Figure 6-41(b) by that of Figure 6-41(d) in a similar fashion as was just described. The process of load scaling will be demonstrated as part of the examples to be discussed in the next section.

6.6.5 Design Example

The general form of the characteristic polynomial of an S-plane highpass network comprising m open- or short-circuited stubs, and n quarter-wavelength transmission lines is [367, 368]:

$$K(S) = \sqrt{\cos\left(m\cos^{-1}\left(\frac{S_c}{S}\right) + n\cos^{-1}\left(\sqrt{\frac{1-S_c^2}{1-S^2}}\right)\right)} \tag{6.107}$$

where S is the Richards variable defined in Equation (6.8). The prototypes to be synthesized in the next two subsections are chosen to have a pair of zeros at DC, ($m = 2$) and a single zero at $S = 1$ so ($n = 1$). For an octave bandwidth Equation (6.107) becomes

$$K(S) = \frac{-S^2\left(2S_c^2 + S_c^2\sqrt{1-S_c^2}\right) + \left(2\sqrt{1-S_c^2} + 2\right)}{S^3\left(S_c^3\sqrt{1-\frac{1}{S^2}}\right)} \tag{6.108}$$

with S_c equal to $j0.57735$. The FR4 PCB substrate utilized for the fabrication of the baluns had a thickness of 1.57 mm, relative dielectric constant of 4.7, and loss tangent of 0.016. The synthesis concept is implemented in the next two subsections for two types of baluns.

6.6.6 Balun with Prescribed Single-Ended Admittance

A Class A balun is required to have a prescribed immittance function at its single-ended port with a passband ripple, ε of 0.508 (i.e., $10\log_{10}\left[1/\left(1+\varepsilon^2\right)\right] = 1$ dB) centered at 500 MHz. According to Table 6-2, the Class A balun implements an admittance function at the single-ended port $\Re$. The balun is also required to transform a single-ended source impedance of 7 Ω to a balanced load of 100 Ω. A typical situation that requires this type of balun is in the matching stage between the output of a power amplifier having low output impedance and a balanced circuit such as an antenna.

Using Equation (6.108) the real part of the input admittance is derived from

$$\Re\left(Y_{\text{in}}(S)\right) = \frac{1}{1+\varepsilon^2\left|k(S)\right|^2} = \left|S_{21}(S)\right|^2. \tag{6.109}$$

Partial fraction expansion of Equation (6.109) is then applied after which the input admittance is extracted with the knowledge of the relationship [405, 406]

$$\Re\left(Y_{\text{in}}(S)\right) = 0.5 \times \left\{Y_{\text{in}}(S) + Y_{\text{in}}(-S)\right\}. \tag{6.110}$$

This gives the input admittance

$$Y_{\text{in}}(S) = \frac{(0.1662867861 + j0.3615083714)}{(S + 0.4719034609 - j1.681902406)} + \frac{(0.1662867847 - j0.361508366)}{(S + 0.4719034609 + j1.681902406)} - \frac{2.438018122}{(S + 3.734004591)} + 1. \tag{6.111}$$

Note that all of the poles of Equation (6.112) are located in the left-hand side of the S-plane to ensure that the input impedance is positive real. The above function is then synthesized to obtain an S-plane singly terminated prototype with element values as shown in Figure 6-42(a).

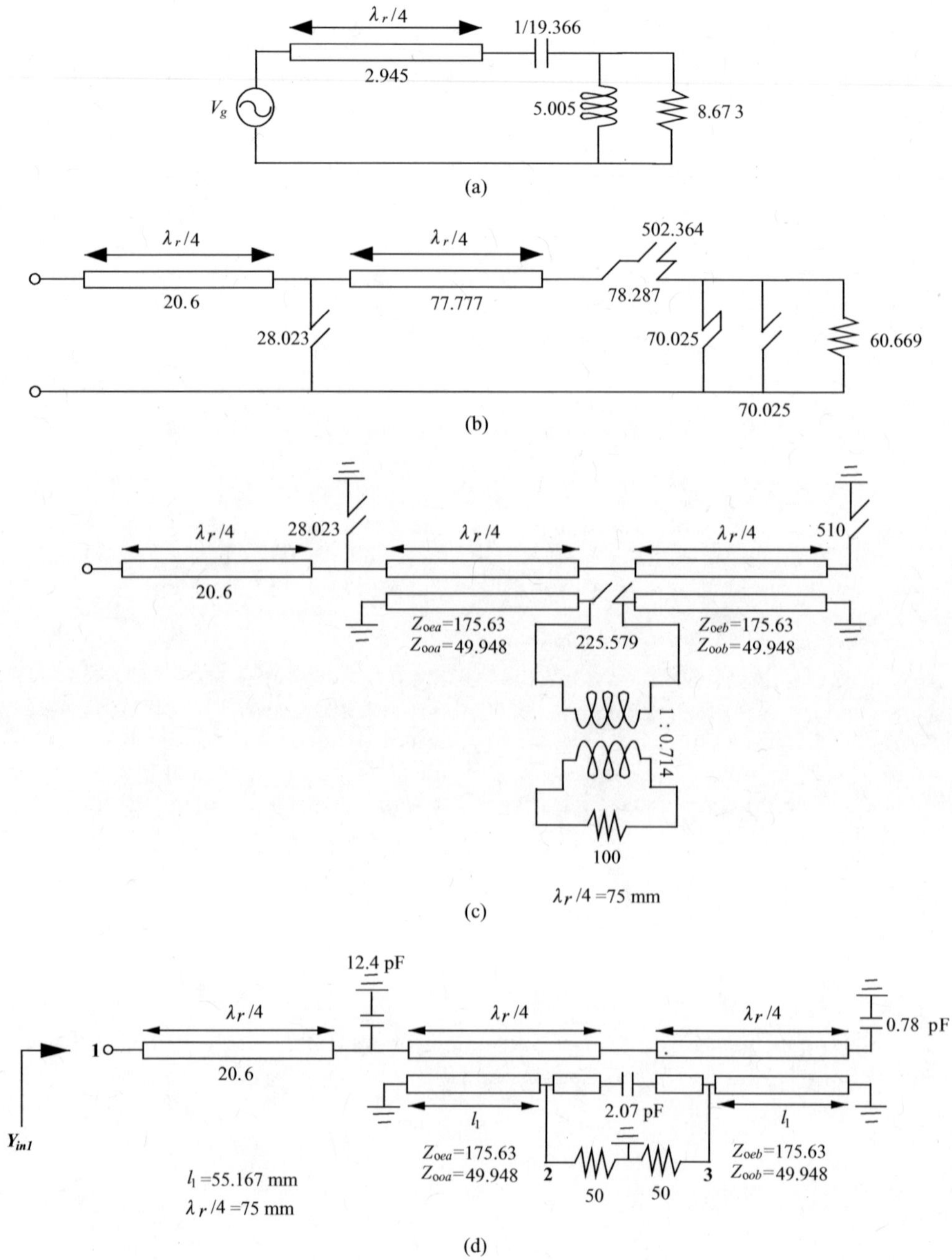

Figure 6-42 Class A balun with prescribed admittance function at the single-ended port: (a) S-plane singly terminated prototype; (b) admittance-scaled and circuit-transformed f−plane prototype; (c) layout; and (d) layout with tapped resonators.

It is worth mentioning that each S-plane capacitor and inductor of Figure 6-42(a) becomes an open- and a short-circuited stub respectively in the f-plane with characteristic impedances of $1/C$ and L respectively. The system impedance of the prototype is then raised from 1 to 7 Ω after which the Half-Angle [393] and relevant Kuroda transformations are applied resulting in a transformed bandpass prototype. This leads to the f-plane prototype of Figure 6-42(b). The element values of the circuit of Figure 6-39(b) may now be obtained from the circuit of Figure 6-42(b). They are as follows:

$$\begin{aligned} &Z_{01} = 28.023\ \Omega;\ Z_{c1} = 77.777\ \Omega;\ Z_{02} = 0;\ Z_{c2} = 78.287\ \Omega \\ &Z_{03} = 502.364\ \Omega;\ Z_{c3} = 70.025\ \Omega = Z_{c4};\ R_{oB} = 60.669\ \Omega. \end{aligned} \tag{6.112}$$

Substituting Equation (6.112) into Equations (6.12)–(6.14), the parameters of the balun of Figure 6-42(c) can be derived using Equations (6.15) to (6.20). Figure 6-42(c) depicts a load impedance of 100 Ω after the inclusion of a 1 : 0.714 transformer. This implies that the balun has an inherent balanced-to-single-ended impedance ratio of

$$\frac{R_D}{R_S} = \frac{195.879}{7} = 27. \tag{6.113}$$

From Figure 6-42(c) it is seen that the pairs of coupled-lines of the balun possess identical modal impedances. This is by virtue of the appropriate sequence of circuit transformations applied to the initial prototype of Figure 6-42(a). Due to this, the balanced load and the output shunt stubs are symmetrically bisected by a virtual ground plane right in the center. Thus the resonator tapping approach discussed in Section 6.6.4 can be implemented. For this balun the elements of the output subsection (corresponding to Figure 6-41(a)) are

$$\begin{aligned} &Z_{oc} = 225.579\ \Omega;\ Z_{sc} = 225.579\ \Omega;\ R_B = 195.879\ \Omega \\ &R_D = 100\ \Omega;\ l_r = 75\,\text{mm};\ X = 0.714 \end{aligned} \tag{6.114}$$

which upon solving Equations (6.102) and (6.106) yields the tapping height and the value of the capacitor as follows:

$$l_1 = 55.787\ \text{mm},\ C_0 = 1.313\ \text{pF}. \tag{6.115}$$

At this stage the open-circuited stubs in the balun structure of Figure 6-42(c) are approximated by lumped capacitors leading to the electrical layout of Figure 6-42(d). Subsequently the electrical layout of the balun is then converted into a physical layout using ADS [115] and fabricated as shown in Figure 6-43.

For practical purposes additional sections of 50 Ω lines interconnecting the SMA connectors to the actual balun ports were added as shown in Figure 6-43. The extra transmission lines at the balanced ports have no effect on the measured response at the single-ended port however the delay introduced between the SMA connector and the actual single-ended port of the balun must be de-embedded in order to accurately measure the input admittance. A pair of 50 Ω loads was connected at the balanced ports and the single-ended port of the balun connected to a digital sampling oscilloscope. The measured time delay due to the transmission line section at the single-ended port was 82.5 ps.

The measured performance of the balun is shown in Figure 6-44. It is observed that the center frequency of the passband has been offset by 6.6% and is now centered at about 467 MHz. This offset is due to the unaccounted parasitic inductances of the via holes and can be corrected by

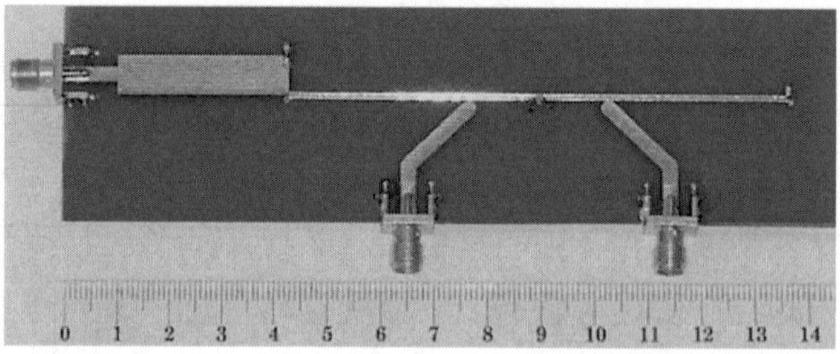

Figure 6-43 Class A balun fabricated from the layout shown in Figure 6-42(d).

slightly shortening the transmission line resonators of the balun. However since this implementation is only for demonstration purposes, the measured performance of the balun is compared with the simulated characteristic of the final circuit of Figure 6-42(d) but scaled down by 33 MHz. The correspondence is excellent. Next the amplitude and phase imbalances were characterized. This was done using a 50 Ω network analyzer notwithstanding the fact that the measured single-ended port impedance of the balun is 7 Ω. The measured amplitude and phase imbalances were found to be less than 0.2 dB and 1° respectively.

6.6.7 Balun with Prescribed Balanced Impedance

In this section a balun is synthesized with a prescribed immittance function at the balanced ports of a Marchand balun. The Class A balun of the previous subsection is required to transform a balanced source impedance of 100 Ω to a single-ended load of 50 Ω over the same frequency band, but with a ripple, ε of 0.349 (0.5 dB). According to Table I, the structure will realize an impedance function. A typical application that requires this type of balun is in the matching stage between the differential outputs of an RFIC feeding a single-ended circuit such as a filter.

The real part of the balanced impedance function was constructed in an identical fashion as done previously using Equation (6.108) to Equation(6.110) but on an impedance basis. The synthesized S-plane prototype is shown in Figure 6-45(a). The prototype is then scaled up in impedance such that a 50 Ω load results at the single-ended port resulting in the f-plane prototype of Figure 6-45(b). Going through identical steps as discussed in Section 6.6.4 the tapped balun of Figure 6-45(c) results with its simulated differential impedance depicted in Figure 6-45(d). The fabricated balun is shown in Figure 6-45(e).

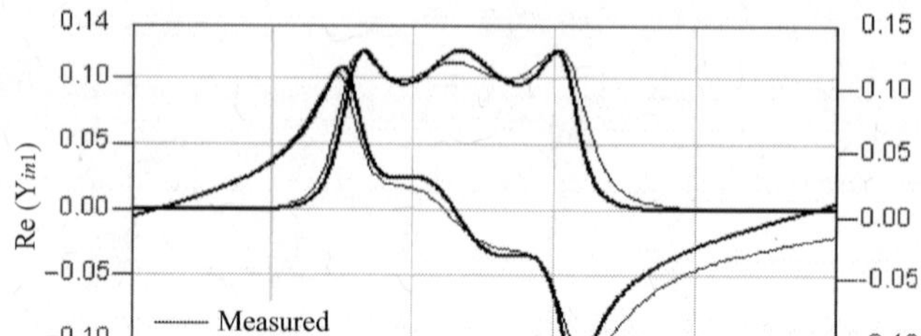

Figure 6-44 Measured characteristics of the fabricated balun of Figure 6-42(e).

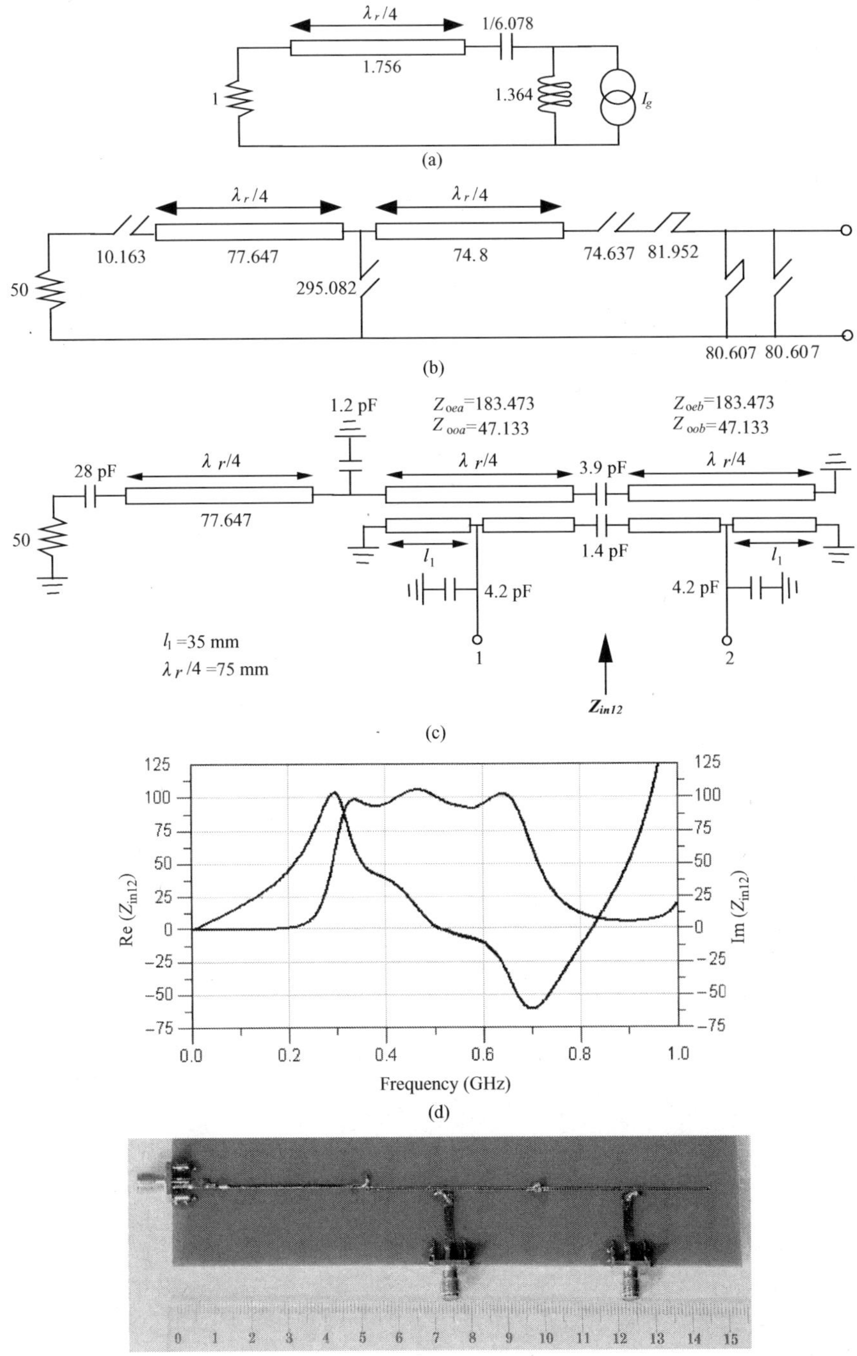

Figure 6-45 Class A balun with prescribed impedance function at the balanced ports: (a) S-plane singly terminated prototype; (b) impedance-scaled and circuit-transformed f−plane prototype; (c) electrical layout with tapped resonators depicting the modal impedances of the coupled-lines; (d) simulation of the differential impedance between ports 1 and 2; and (e) the fabricated balun.

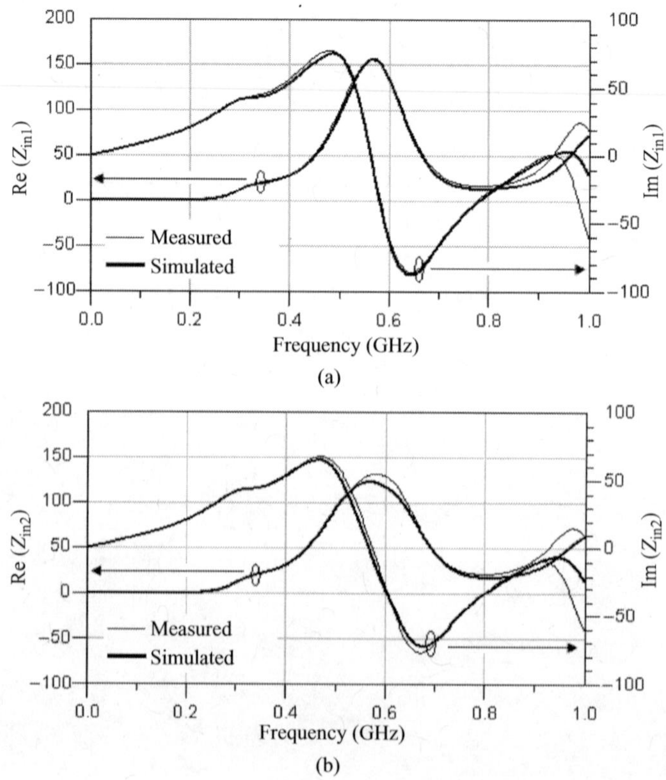

Figure 6-46 Measured characteristics of the balun of Figure 6-45(e): (a) and (b) the real and imaginary parts of the driving point impedances seen to ground at each balanced port with the other terminated in a 50 Ω load.

Balanced operation of the balun was tested by measuring the real and imaginary parts of the impedance function to ground at Port 1 of Figure 6-45(c) with Port 2 terminated to ground in a 50 Ω load. Similar measurement at Port 2 with Port 1 terminated with 50 Ω was also performed. The de-embedded measured characteristics are shown in Figure 6-46 together with the simulation. Very good agreement between the simulated and measured results is observed.

6.6.8 Summary

Singly terminated prototypes were used to derive new immittance-transforming Marchand baluns suitable for matching purposes. This led to miniaturized lumped-distributed baluns realizing specific immittance profiles at either their single-ended or balanced ports. Stability of active devices such as single-ended or differential amplifiers could be improved by controlling the immittance they see. This is achieved by utilizing the proposed baluns in those situations. For a specific operating bandwidth and passband ripple the commensurate frequency of the presented baluns can be varied resulting in a wide range of terminal immittance values. Additional tapping of the balanced resonators of the baluns leads to the realization of specific balanced-to-single-ended immittance ratios at no increase in overall size. Two baluns were designed using a synthesis procedure. Measured and simulated specifications were in close agreement.

6.7 FILTERS WITH ENHANCED STOPBAND PERFORMANCE

A new class of parallel-coupled line filters with broad stopband response is introduced in this section. The design is based on the synthesis of bandpass prototypes with predefined upper stopband characteristics. The new filters have uniform- and stepped-impedance resonators some of which are loaded by open-circuited stubs at their open-circuited ends.

Considerable efforts to improve the stopband performance of microstrip PCL filters have largely been directed at suppressing the harmonic spurious passband located at $2f_0$. This was done by compensating for the difference between the even- and odd-mode phase velocities of the inhomogeneous media in a variety of ways. In one approach the coupled lines were loaded by lumped capacitors [369] or meandered [370]. Other approaches introduced ground plane apertures underneath the coupled lines [371–373], or utilized suspended substrates with dielectric overlay [374, 375]. Other attempts to suppress the undesired spurious passbands introduced geometrical perturbations to the uniform-impedance resonators in the coupling regions [376–380], loaded them by split ring resonators [382], or alternatively utilized stepped-impedance resonators [383]. Using uniform- or stepped-impedance resonators is the most successful strategy for enhancing the stopband performance of PCL filters.

6.7.1 Uniform- and Stepped-Impedance PCL Filters

It was pointed out in Section 6.2 that filters using transmission lines have spurious passbands. To some extent it is possible to alter the positions of the spurious passbands of conventional PCL filters by using stepped-impedance resonators. An excellent account of this technique can be found in [383]. Figure 6-47(a) shows a Type 1 stepped-impedance resonator where $Z_s > Z_r$ [383]. The design approach using Type 1 resonators is briefly explained as follows. The uniform resonators of each PCL section in a conventional filter (see Table 6-1) are staggered by a defined amount. This effect is consequently compensated for by setting the characteristic impedances of the uncoupled line sections of the resonators to a fixed high value, Z_s. Consequently the length, l_s, of each un-coupled line in the filter is then optimized until the return loss of the fundamental passband is restored. A pair of stepped-impedance resonators is shown in Figure 6-47(b). Intuitively one could conclude that a change in the stopband performance of a filter will depend on the amount of stagger and the impedance ratio Z_s/Z_r. This process could be iterated on until a suitable stopband

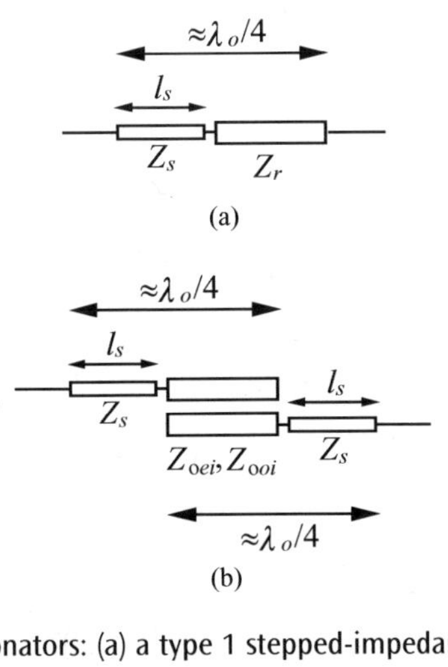

Figure 6-47 Stepped-impedance resonators: (a) a type 1 stepped-impedance resonator from [383]; (b) a pair of coupled stepped-impedance resonators.

performance is obtained. In general this technique offers limited stopband improvement and is best suited to applications where a small shift in the spurious passbands would block harmonics of the input signal from passing through the filter. Thus the spurious passbands of a filter are not eliminated but in effect the stopband characteristic is altered to suit a particular specification [383]. There is also a Type 2 stepped-impedance resonator with $Z_s < Z_r$ resulting in filters with slightly different stopband behavior to those utilizing Type 1 resonators [383]. In the following we investigate how the design of PCL filters based on S-plane bandpass prototypes would lead to better stopband performances.

6.7.2 Design of PCL Filters with Enhanced Stopband

More recently a PCL filter with uniform-impedance resonators loaded by open-circuited stubs at all its open-circuited ends demonstrated suppression of the second- and third-ordered harmonic passbands [408]. This approach was generalized in [381] and this development is described here. The new PCL filters are derived from synthesized bandpass prototypes with controlled upper stopband responses. This is done by systematic application of appropriate circuit transformations on the initial bandpass prototype resulting in a class of PCL filters comprising uniform- and stepped-impedance resonators some of which are loaded by open-circuited stubs at their open-circuited ends.

The prototype used here is shown in Figure 6-48(a). It is classified as an S-plane bandpass prototype with the Richards variable redefined as

$$S = j\Omega_{BP} = j\tan(\theta_{BP}) = j\tan\left(\frac{\pi}{2}\frac{f}{f_r}\right). \tag{6.116}$$

In Equation (6.116) f_r is the center frequency of the first upper stopband in the f-plane. The general design approach is to synthesize the prototype of Figure 6-48(a) in the S-plane by exact methods [368,394,395] and then transform it to the f-plane using Equation (6.116). In the f-plane the UEs, capacitors, and inductors of Figure 6-48(a) become transmission lines, open- and short-circuited stubs respectively and are all quarter-wave length long at f_r. It has been shown [405] that the f-plane response of an S-plane bandpass prototype is periodic with a fundamental passband centered at f_0 ($f_0 < f_r$) and the spurious passbands centered at

$$2if_r - f_0 \text{ and } 2if_r + f_0 \quad (i = 1, 2, 3, \ldots). \tag{6.117}$$

Thus synthesis based on an S-plane bandpass prototype has great flexibility in controlling the location of the first pair of spurious passbands in the f-plane. This is possible by selecting a large commensurate frequency, f_r to shift the upper spurious passbands further away from the fundamental passband.

A typical f-plane response of the proposed prototype is depicted in Figure 6-48(b) where it is seen that the series S-plane LC sections (i.e., f-plane short- and open-circuited stubs) contribute to multiple transmission zeros at if_r where $i = 1, 2, 3, \ldots$. Doubling the commensurate frequency (i.e., moving from f_r to $2f_r$) leads to the response shown in Figure 6-48(c) where the spurious responses are shifted leaving behind a broad stopband region.

Now attention is focused on the circuit structure of Figure 6-48(a). Through a number of circuit transformations this is converted into a form that can be implemented using distributed elements. Consider the S-plane subcircuit shown in Figure 6-49(a) which comprises a UE separating a pair of series capacitors followed by a series inductor. It is feasible to transform the series LC section across the UE by using relevant Kuroda transformations [368,405]. This leads to the subcircuit of

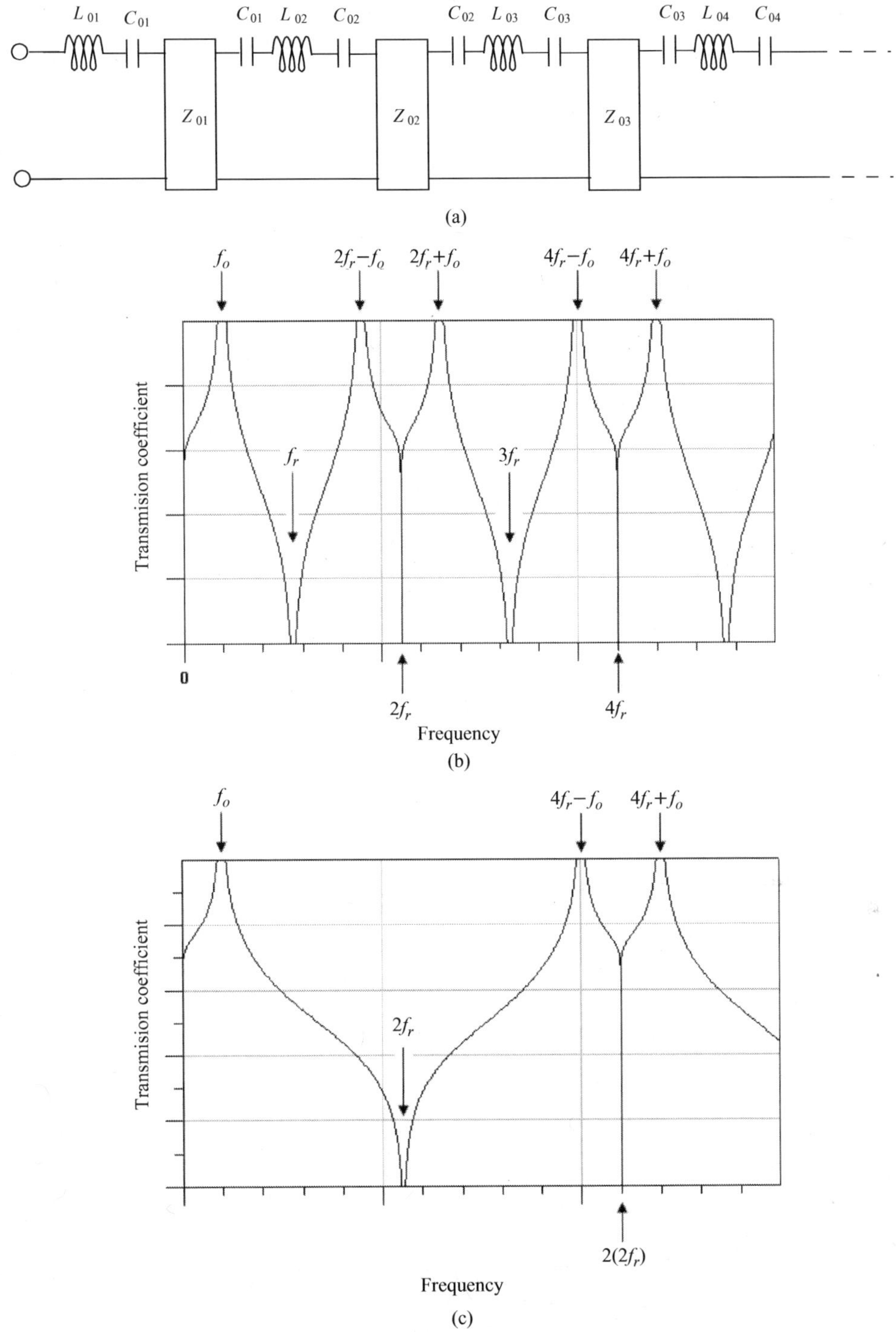

Figure 6-48 Bandpass PCL filter response: (a) *S*-plane bandpass prototype; (b) real frequency for one commensurate frequency f_r; and (c) response after doubling f_r.

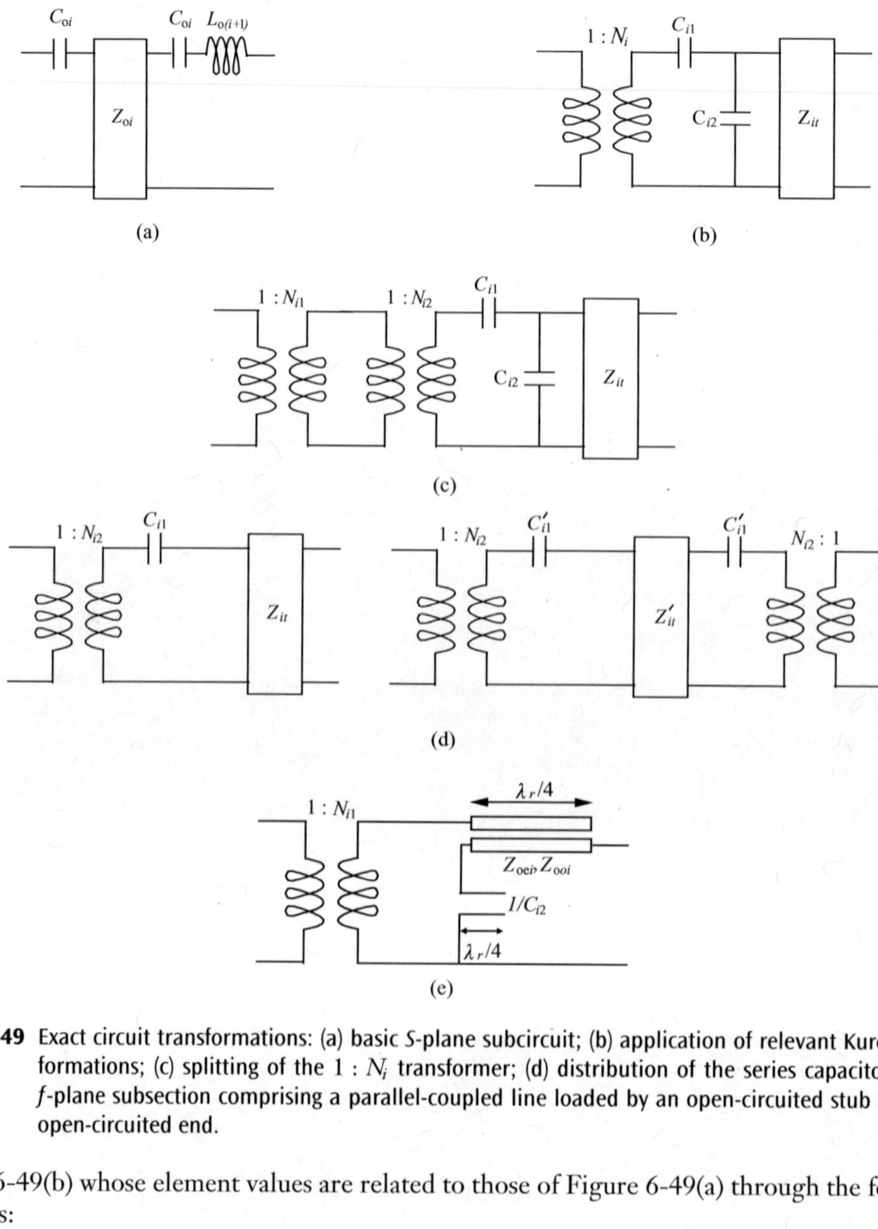

Figure 6-49 Exact circuit transformations: (a) basic *S*-plane subcircuit; (b) application of relevant Kuroda transformations; (c) splitting of the $1 : N_i$ transformer; (d) distribution of the series capacitor; and (e) *f*-plane subsection comprising a parallel-coupled line loaded by an open-circuited stub at the left open-circuited end.

Figure 6-49(b) whose element values are related to those of Figure 6-49(a) through the following relations:

$$N_i = 1 + \frac{1}{Z_{0i} C_{oi}}$$

$$C_{i1} = \frac{C_{oi}}{N_i (N_i - 1) C_{oi} Z_{0i} + N_i^2}$$

$$C_{i2} = \frac{L_{o(i+1)}}{N_i Z_{0i} \left(N_i Z_{0i} + L_{o(i+1)}\right)}$$

$$Z_{it} = N_i Z_{0i} + L_{o(i+1)}. \tag{6.118}$$

Now the series capacitor, C_{i1} and the UE of characteristic impedance Z_{it} form a single PCL section if a $1 : N_{i2}$ transformer is extracted from the existing $1 : N_i$ transformer. The value of the turns ratio of this transformer is

$$N_{i2} = \left(\sqrt{\frac{Z_{it}}{Z_{it} + \frac{1}{C_{i1}}}}\right)^{-1} \tag{6.119}$$

leaving a $1 : N_{i1}$ transformer of turns ratio

$$N_{i1} = \frac{N_i}{N_{i2}}. \tag{6.120}$$

This step leads to the circuit of Figure 6-49(c). The $1 : N_{i2}$ transformer together with the series capacitor and UE is the S-plane model of an open-circuited coupled-line section in homogenous media [368]. So for now ignore the shunt capacitor C_{i2}, and investigate the left-hand subcircuit of Figure 6-49(d). If a relevant Kuroda transformation is applied to spread the series capacitor, C_{i1}, then the following relations

$$\begin{aligned} C'_{i1} &= \frac{1}{N_{i2}\left(N_{i2} - 1\right) Z_{it}} \\ Z'_{it} &= N_{i2} Z_{it} \end{aligned} \tag{6.121}$$

hold for the circuits of Figure 6-49(d). Hence the modal impedances of the PCL section loaded by an open-circuited stub (of characteristic impedance $1/C_{i2}$) at the left open-circuited end are then evaluated [405] using Equation (6.121) to give

$$Z_{0ei} = \frac{\left(\frac{1}{C'_{i1}} + 2Z'_{it}\right)}{N_{i2}^2}, \quad Z_{0oi} = \frac{\left(\frac{1}{C'_{i1}}\right)}{N_{i2}^2} \tag{6.122}$$

This concludes the transformation of the subsection of Figure 6-49(a) into that of Figure 6-49(e) which is almost in the final desired form for implementation using transmission lines only. The ultimate transformation follows.

Figure 6-50(a) shows an S-plane bandpass prototype where the subcircuits enclosed in the dashed boxes can be systematically transformed as just described. The transformation results in the f-plane filter shown in Figure 6-50(b). This is only one among many that could have been obtained. In practical terms it might be desirable not to transform all the inductors across the UEs to form parallel-coupled lines loaded by open-circuited stubs. This is typically due to extreme element values that may be required that would lead to realization difficulties on a PCB.

To a great extent these constraints can be resolved dependent on how the initial S-plane bandpass prototype is transformed. For this reason consider application of the transformation steps to the subsections (enclosed in the dashed boxes) of the S-plane prototype shown in Figure 6-50(c). This prototype is identical to that of Figure 6-50(a) except that some of the inductors are now divided into pairs. Upon transformation the f-plane filter illustrated in Figure 6-50(d) results. As seen from Figure 6-50(d), the filter inevitably has some series short-circuited stubs in its main signal path. These can be approximated by sections of high impedance transmission lines as described below.

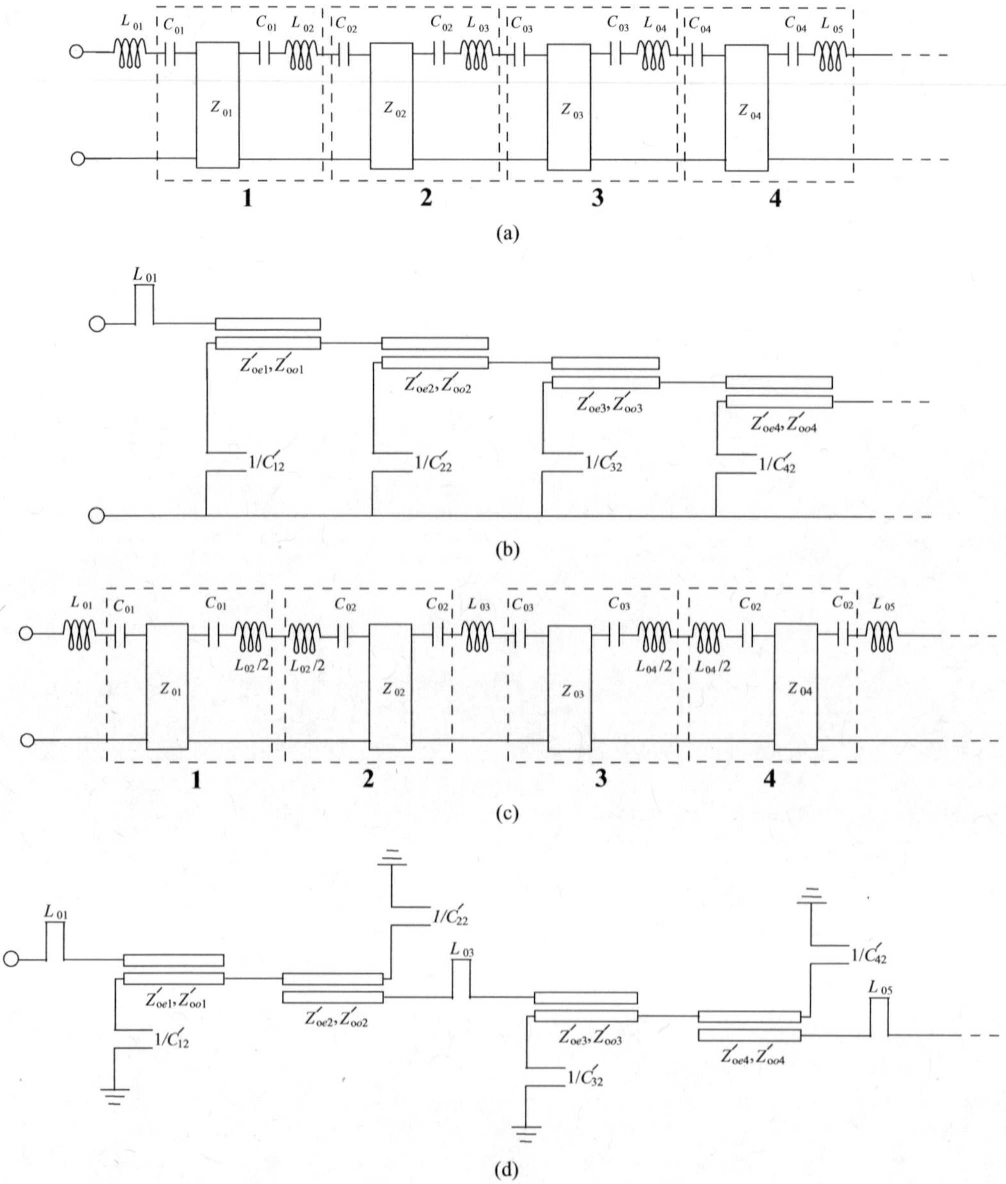

Figure 6-50 Filter 1: (a) *S*-plane bandpass prototype; and (b) its corresponding *f*-plane filter comprising PCLs loaded by open-circuited stubs at the left open-circuited ends. Filter 2: (c) *S*-plane bandpass prototype; and (d) its corresponding *f*-plane filter comprising PCLs loaded by open-circuited stubs at either the left or the right open-circuited ends. The filters in (b) and (d) are derived after application of the circuit transformation scheme described in Figure 6-49 on the subcircuits enclosed in the dashed boxes of the *S*-plane prototypes.

A section of transmission line of short electrical length at f_0 and of high characteristic impedance Z_0 has an input impedance approximately equal to

$$Z_{\text{in}} \simeq j\,(Z_0)\,(\theta') = j(Z_0)\left(\frac{\pi}{2}\frac{f_0}{f_r'}\right). \tag{6.123}$$

where f_r' is the quarter-wave length frequency of the transmission line. On the other hand, a series short-circuited stub of characteristic impedance L is resonant at f_r and has an input impedance at the center of the fundamental passband equal to

$$Z_{\text{in}} = j\,(L)\tan\left(\theta_{BP}\right) = j\,(L)\tan\left(\frac{\pi}{2}\frac{f_0}{f_r}\right). \tag{6.124}$$

The above equations are approximately equal for passband bandwidths up to an octave (i.e., 2:1 bandwidth ratio) and hence can be simultaneously solved for the resonant frequency required of the transmission lines yielding

$$f_r' = \frac{\pi}{2}\frac{Z_0 f_0}{L\tan\left(\frac{\pi}{2}\frac{f_0}{f_r}\right)}. \tag{6.125}$$

Subsequently the physical length of the transmission line can be evaluated for a given substrate specification. In general the higher the line impedance that can be realized the better the approximation. However, over half of the S-plane inductors in a prototype are transformed into open-circuited stubs and thus the shape of the f-plane transfer function of a prototype in the stopband region is maintained after the above approximation. This completes the filter design approach and now it is time for an example.

6.7.3 Example: PCL Filter with Enhanced Stopband

In this example a 7 th-order Chebyshev bandpass filter at 1 GHz with a broad stopband response is designed. The design begins with a seventh-order approximation function in the S-plane [368, 394, 395]. Derivation of the function from the outline procedure presented in the references sited is straight forward but lengthy. From the final function and using standard synthesis the physically symmetrical S-plane network results and one half of it is shown in Figure 6-51(a). In the f-plane the response of the network has a fundamental passband center frequency, f_0, of 1 GHz and a commensurate frequency, f_r, of 2.75 GHz which, according to Equation (6.117), leads to spurious passbands centered at 4.5 GHz and 6.5 GHz. The return loss of the fundamental passband was 16 dB and its bandwidth ratio was 1.22:1 (i.e., 20%) leading to lower and upper bandedge frequencies of 0.9 and 1.1 GHz respectively.

The next stage of design is transforming all of the inductors in the prototype into open-circuited stubs to yield a PCL filter similar in topology to that of Figure 6-50(b). From the first box (of the prototype of Figure 6-51(a)) the following parameters are obtained:

$$C_{01} = \frac{1}{119.025;} Z_{01} = 34.52;\; L_{02} = 713.332 \tag{6.126}$$

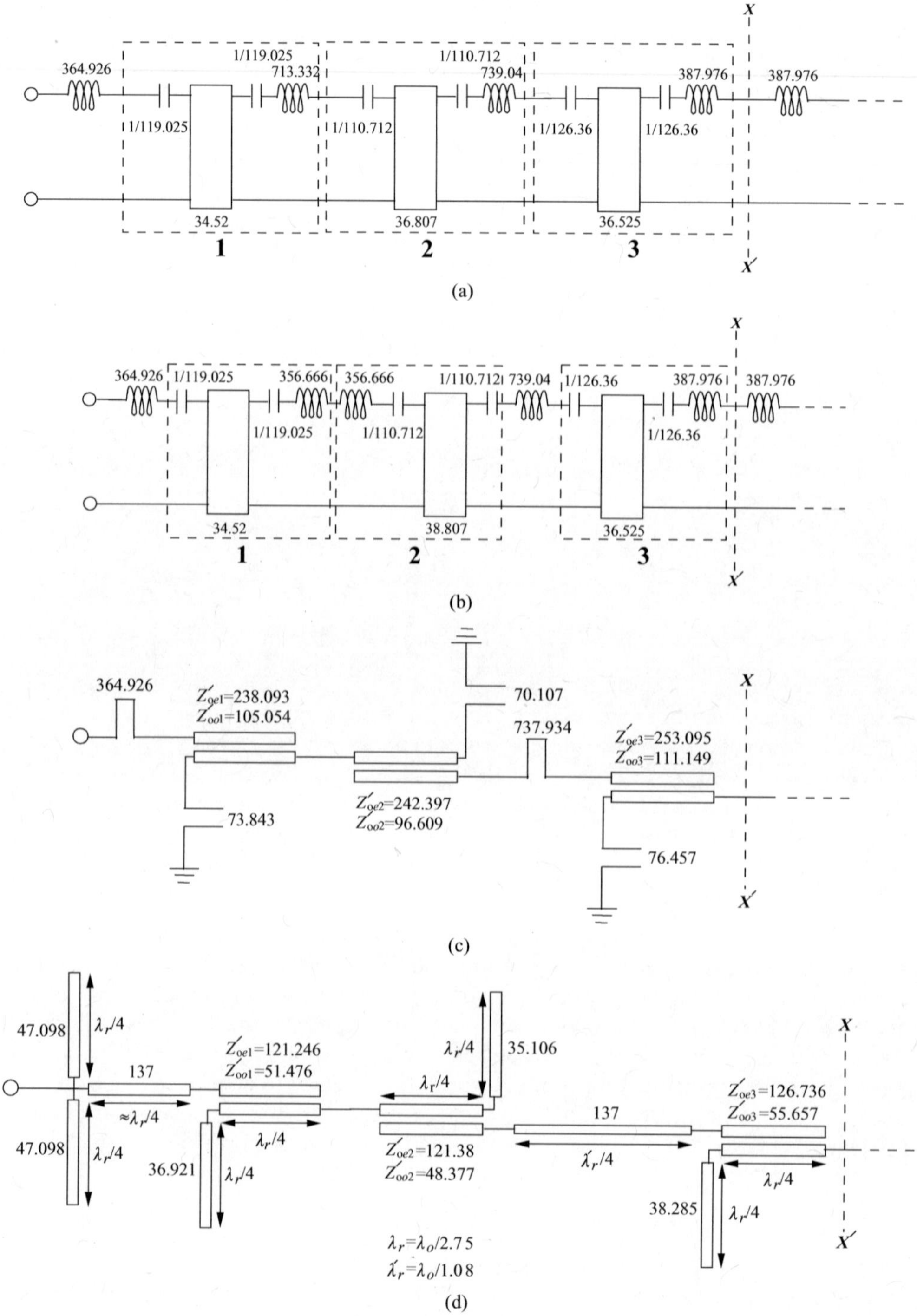

Figure 6-51 Development of a seventh-order parallel-coupled line filter: (a) the synthesized S-plane prototype in a 50 Ω system; (b) after splitting some of the inner inductors into pairs; (c) the topology of the f-plane filter depicting interresonator couplings; and (d) the filter's electrical layout. (The X–X'is the plane of symmetry).

corresponding to a subsection as shown in Figure 6-49(a). Now using Equation (6.118) a set of element values are evaluated to give

$$N_1 = 4.448;\ C_{11} = \frac{1}{2884.297}$$
$$C_{12} = \frac{1}{186.595};\ Z_{1t} = 866.876. \tag{6.127}$$

This transforms the subsection to that of Figure 6-49(b). A $1 : N_{12}$ transformer is then extracted from the $1 : N_1$ transformer using Equation (6.119). The value of the turns ratio of this transformer is

$$N_{12} = 2.08 \tag{6.128}$$

which according to Equation (6.120) leaves a $1 : N_{11}$ transformer of turns ratio

$$N_{11} = 2.138. \tag{6.129}$$

This transforms the subcircuit to the topology of Figure 6-49(c). The next step is to evaluate the element values of the subcircuit of Figure 6-49(d) using Equation (6.121) to give

$$C'_{11} = \frac{1}{1947.35;}\ Z'_{1t} = 1803.102. \tag{6.130}$$

Now according to Equation (6.122) the modal impedances of the first pair of parallel-coupled lines loaded by an open-circuited stub of characteristic impedance $1/C_{12} = 186.595\ \Omega$ at the left open-circuited end are calculated as

$$Z_{0e1} = 1283.643\ \Omega;\ Z_{0o1} = 450.108\ \Omega. \tag{6.131}$$

This concludes the transformation of the first subsection to a topology similar to that of Figure 6-49(e). In a similar fashion the subcircuits enclosed in the second, and third boxes are also transformed leading to the following element values

$$N_{21} = 2.14;\ \frac{1}{C_{22}} = 176.964\ \Omega$$
$$Z_{0e2} = 1351.136\ \Omega;\ Z_{0o2} = 403.706\ \Omega \tag{6.132}$$

and

$$N_{31} = 1.737;\ \frac{1}{C_{32}} = 231.266\ \Omega$$
$$Z_{0e3} = 765.535\ \Omega;\ Z_{0o3} = 336.182\ \Omega. \tag{6.133}$$

It must be appreciated that each transformed subsection has a $1 : N_{i1}$ transformer associated with it. Each of these transformers is eliminated by scaling down the modal impedances of the PCL sections together with the characteristic impedances of the stubs in the filter. This step leads to a symmetrical f-plane filter compatible with that of Figure 6-50(b) having the following element values:

$$L_{o1} = 364.926\ \Omega$$
$$Z'_{oe1} = 280.749\ \Omega;\ Z'_{oo1} = 98.454\ \Omega;\ \frac{1}{C'_{12}} = 40.805\ \Omega$$
$$Z'_{oe2} = 64.922\ \Omega;\ Z'_{oo2} = 19.72\ \Omega;\ \frac{1}{C'_{22}} = 8.442\ \Omega$$
$$Z'_{oe3} = 12.093\ \Omega;\ Z'_{oo3} = 5.31\ \Omega;\ \frac{1}{C'_{32}} = 3.653\ \Omega. \tag{6.134}$$

The design now enters a final phase taking into account fabrication constraints. Here the choice is to construct the filter on an FR4 Printed Circuit Board (PCB) with a substrate thickness of 62 mil (1.57 mm), relative dielectric constant of 4.7, and loss tangent of 0.016. Practical constraints impose a minimum track width and spacing between tracks of 7 mil (0.177 mm). From Equation (6.134) the coupling coefficients of the parallel-coupled lines were calculated and the *LinCal* component of Agilent's ADS simulator [115] was utilized to obtain a rough estimate of physical dimensions. Unfortunately the resulting dimensions are unrealizable on this type of board and also, as seen from Equation (6.134), some of the impedances of the stubs are too low. Therefore transformation of the initial prototype to the proposed network topology of Figure 6-50(d) was done. The steps involved are shown in Figures 6-51(b) and 6-51(c). It can now be seen from the circuit of Figure 6-51(c) that the modal impedances are actually high and would still be unrealizable on the PCB. Scaling down the impedance matrix representing this circuit is therefore necessary at this point. This step changed the system impedance of the filter from 50 to 50/2.5 Ω and imposes the placement of an impedance inverter of value 31.62 Ω at each port of the filter. Each inverter is then approximated by a transmission line and the scaled series short-circuited stubs (near the filter ports, see Figure 6-51(c)) of characteristic impedance $(364.926/2.5)$ Ω are subsequently transformed into open-circuited stubs using a Kuroda transformation.

For a seventh-order filter there is only a pair of series short-circuited stubs left in the main line of the filter that must be approximated by high impedance transmission lines. From Figure 6-51(c) the scaled characteristic impedance of one of these short-circuited stubs is $(737.934/2.5)$ Ω. The highest characteristic impedance, Z_o of a transmission that is realizable on the board is 137 Ω corresponding to a 7 mil (0.177 mm) wide track. Thus using Equation (6.125) the quarter-wave length frequency of the line is evaluated as follows:

$$f_r' = \frac{\pi}{2}\frac{137\left(1 \times 10^9\right)}{(737.934/2.5)\tan\left(\frac{\pi}{2}12.75\right)} = 1.134 \text{ GHz}. \qquad (6.135)$$

At this stage tuning of some of the element values of the filter is required to obtain a flat return loss over the passband. This leads to the electrical layout of the filter as shown in Figure 6-51(d). Examination of the interresonator couplings, and stub impedances show that they can now be easily realized on the PCB. The electrical layout of filter was then converted into physical dimensions using ADS [115] and the filter constructed, see Figure 6-52.

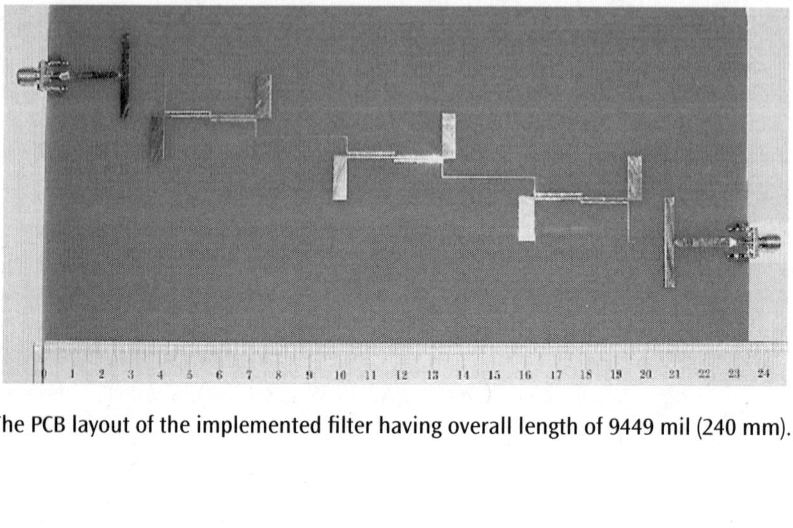

Figure 6-52 The PCB layout of the implemented filter having overall length of 9449 mil (240 mm).

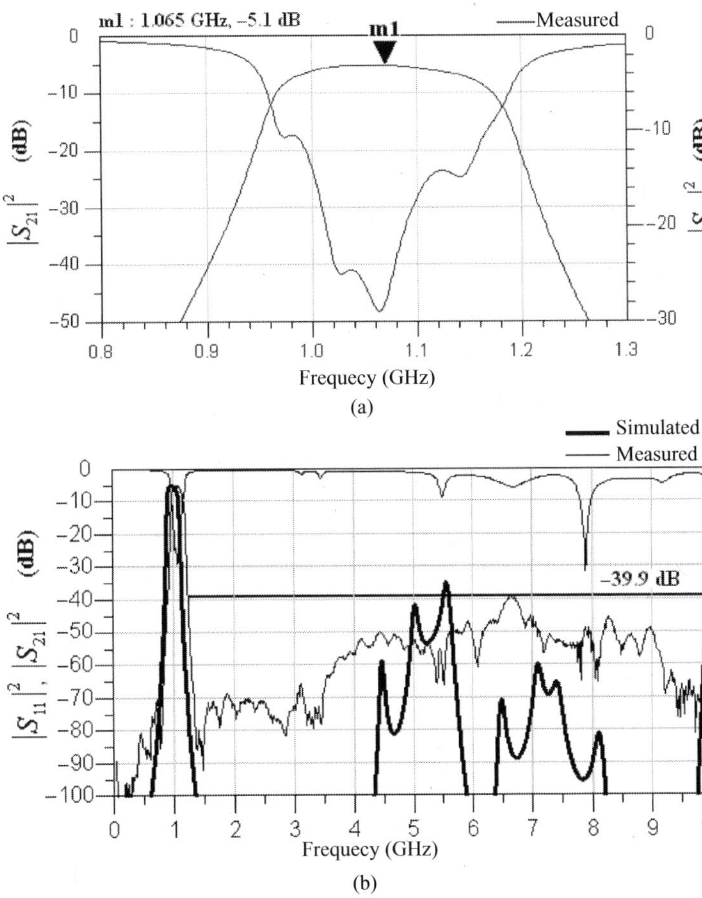

Figure 6-53 Measured and simulated performances of the implemented parallel-coupled line filter of Figure 6-52: (a) in-band frequency response; and (b) wide-band frequency response.

The measured in-band performance of the filter is shown in Figure 6-53. The center frequency of the passband is offset by 65 MHz (6.5%) from the design objective which could be compensated for by shortening the physical lengths of the transmission line resonators. The midband insertion loss was 5.1 dB and the return loss was greater than 10 dB as shown in Figure 6-53(a). The simulated and measured wide-band scattering parameters of the filter from DC up to 10 GHz are illustrated in Figure 6-53(b). This compares to the large number of spurious passbands that occur with a conventional parallel-coupled line filter, see Figure 6-3.

This filter was designed to have its first pair of spurious passbands centered at 4.5 and 6.5 GHz if it is realized on a homogenous media. However the simulated (dark) plot of Figure 6-53(b) predicts that the first two spurious passbands are approximately centered at 5.25 and 7.25 GHz instead. This difference is associated with the difference of the even- and odd-mode phase velocities of the microstrip media. The difference is also partially due to approximating the inner pair of short-circuited stubs in the f-plane filter by high impedance transmission lines. In summary the measured transmission performance of the filter followed the simulation and achieved an attenuation level exceeding 40 dB (i.e., an effective rejection level of about 35 dB) up to 10 GHz. As seen from Figure 6-53(b) the stopband of the filter has a tendency to show a spurious passband at the region where the theoretical spurious passbands would have existed. This is around 6 GHz as predicted by the simulation.

6.7.4 Summary

A circuit-oriented approach for the design of a new class of parallel-coupled line filters was presented in this section. The parallel-coupled line filters are driven from synthesized S-plane bandpass prototypes after appropriate circuit transformations. As a result the filters have uniform- and stepped-impedance resonators some of which are loaded by open-circuited stubs at their open-circuited ends. In comparison with the performances of conventional filters utilizing uniform- or stepped-impedance resonators the new filters have broader stopband characteristics. This was illustrated experimentally by the implementation of practical filters. It is believed that the new class of parallel-coupled line filters will find wide usage in advanced RF/microwave front-end transceivers.

6.8 DESIGN OF BANDSTOP FILTERS UTILIZING CIRCUIT PROTOTYPES

Conventionally-designed bandstop filters typically have a first spurious stopband at a frequency only three times that of the fundamental stopband center frequency, f_o [366, 409]. At the $3f_o$ frequency the electrical length of the transmission line components is $3/4\lambda$ and so they are the same electrically as at f_o where the line lengths are $1/4\lambda$ long. A solution to the spurious passband problem was recently proposed by Levy et al. [410] who introduced bandstop filter structures with extended upper passbands. Design of the filters in [410] begins with conventional bandstop networks and improvement of their upper passband performance is subsequently done by circuit optimization.

It is the purpose of this section to redesign the filter structures presented in [410] based on a systematic circuit-oriented approach. The method provides insight into the physical operation of these filters and requires minimal circuit optimization. Here a systematic circuit-oriented approach to the design of bandstop filters with predefined upper passband characteristics is presented. An experimental filter is built with a fundamental stopband centered at 1.085 GHz and a first spurious stopband centered at 5.4 GHz.

6.9 CIRCUIT-ORIENTED DESIGN

Figure 6-54(a) depicts a lumped element lowpass prototype that realizes finite-frequency transmission zeros that results from the series parallel-connected LC branches. This prototype is then transformed to that of Figure 6-54(b) (with impedance inverters) using the conventional bandstop transformation with specified stopband bandwidth. The shunt series-connected LC branches in Figure 6-54(b) are bandstop resonators with each the impedance function being

$$Z_i(s) = \frac{(L_i C_i)s^2 + 1}{(C_i)s} \quad (i = 1, 2, 3, \ldots). \tag{6.136}$$

where s is the Laplace variable. The lumped bandstop resonators result in transmission zeros in the fundamental stopband. From Equation (6.136) these zeroes are at

$$\omega_{oi} = \pm\frac{1}{\sqrt{L_i C_i}} \quad (i = 1, 2, 3, \ldots). \tag{6.137}$$

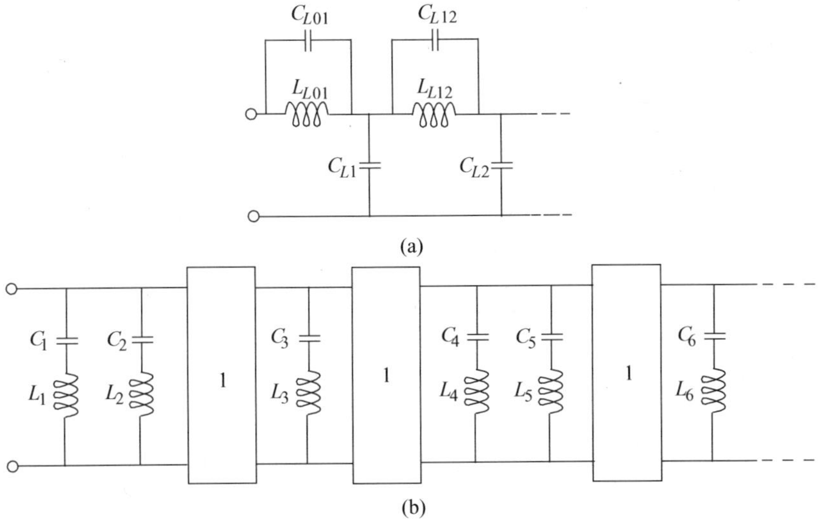

Figure 6-54 Lumped element prototypes: (a) initial lowpass prototype; and (b) corresponding bandstop prototype with unity impedance inverters.

Each lumped bandstop resonator can be transformed into a shunt section comprising series short- and open-circuited stubs of characteristic impedances Z_{SCi} and Z_{OCi} respectively. This leads to the distributed bandstop prototype shown in Figure 6-55. In the following the values of Z_{SCi} and Z_{OCi} will be bound to obtain exact equivalence at the frequencies of the zeroes. Now the impedance function of the i th distributed bandstop resonators can be expressed in term of the Richards' variable, S, as

$$Z_i(S) = \frac{(Z_{SCi})\,S^2 + Z_{OCi}}{S} \qquad (i = 1, 2, 3, \ldots). \tag{6.138}$$

Consequently the set of resonant frequencies of Equation (6.136) (the transmission zeroes) may now be rederived using Equation (6.138) as

$$\omega_{oi} = \pm\left(\frac{2\omega_r}{\pi}\right)\tan^{-1}\left(\frac{\sqrt{Z_{SCi}Z_{OCi}}}{Z_{SCi}}\right) \qquad (i = 1, 2, 3, \ldots). \tag{6.139}$$

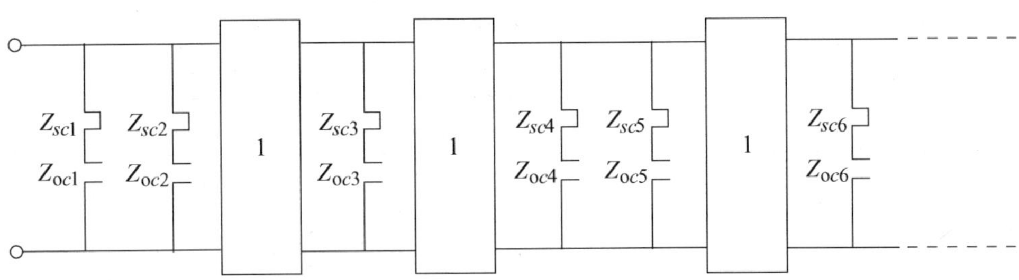

Figure 6-55 Distributed bandstop prototype with impedance inverters.

In Equations (6.138) and (6.139) S is equal to $j\tan\left((\pi/2)(\omega/\omega_r)\right)$ where ω_r is the angular frequency at which the stubs of the prototype are quarter-wavelength long. Thus to transform a lumped element bandstop resonator into its distributed counterpart, Equations (6.136) and (6.138) and their differentials must be equal at the resonant frequency, ω_{oi} for a selected commensurate frequency, ω_r. This leads to the following design expressions:

$$Z_{SCi} = \frac{\begin{pmatrix} \left(-\pi\omega_{oi}^3 L_i C_i + \pi\omega_{oi}\right) S^2 \\ -j\left(2\omega_r\omega_{oi}^2 L_i C_i + 2\omega_r\right) S \\ +\pi\omega_{oi}^3 L_i C_i - \pi\omega_{oi} \end{pmatrix}}{j2\pi\omega_{oi}^2 C_i \left(S^2 - 1\right) S} \quad (i = 1, 2, 3, \ldots) \tag{6.140}$$

$$Z_{OCi} = \frac{S\begin{pmatrix} \left(-\pi\omega_{oi}^3 L_i C_i + \pi\omega_{oi}\right) S^2 \\ +j\left(2\omega_r\omega_{oi}^2 L_i C_i + 2\omega_r\right) S \\ +\pi\omega_{oi}^3 L_i C_i - \pi\omega_{oi} \end{pmatrix}}{j2\pi\omega_{oi}^2 C_i \left(S^2 - 1\right)} \quad (i = 1, 2, 3, \ldots). \tag{6.141}$$

Examination of Equation (6.138) shows that the function approaches infinity at multiples of ω_r so that the first upper passband is centered at ω_r. Also, with the knowledge of Equation (6.139), it is seen that (Equation 6.138) also goes to zero at the following set of frequency points:

$$k.2\omega_r - \omega_{oi} \text{ and } k.2\omega_r + \omega_{oi} \quad (k = 1,\ 2,\ 3, \ldots). \tag{6.142}$$

The response of a bandstop filter with distributed resonators satisfying Equation (6.138) is shown in Figure 6-56. The interesting feature is that the location of the first pair of spurious stopbands is a function of the commensurate frequency, ω_r and thus, by specifying ω_r as large as possible, it is feasible to have a wide spurious-free upper passband. For example, by doubling the commensurate frequency (i.e., moving from ω_r to $2\omega_r$) results in a response where the spurious responses are shifted leaving behind a broad upper passband region.

The next step is the realization of the impedance inverters of the prototype. In an analogous fashion to the design approach of [366], each inverter in the distributed prototype of Figure 6-55

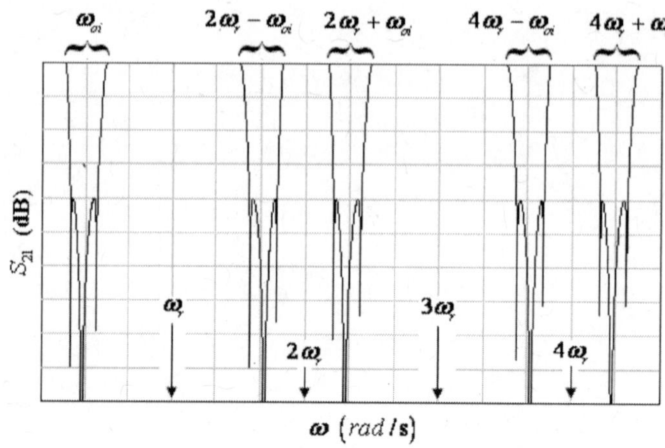

Figure 6-56 Frequency response of a distributed bandstop filter for a commensurate frequency ω_r.

must be approximated by a physical transmission line that is resonant at the fundamental stopband center frequency. This is a relatively broadband approximation to an inverter valid for filters with fundamental stopband bandwidths of up to 40%. The preceding step degrades the return–loss (R.L.) level of the passband and the level of degradation is directly proportional to the width of the fundamental stopband. To elaborate on this point, two filters were synthesized for the same center and commensurate frequencies but for different fundamental stopband bandwidths. The R.L. level in the passband was originally 20 dB. Upon approximation of the inverters by transmission lines the values of the characteristic impedances of the transmission lines were adjusted slightly to restore the R.L. level back to an acceptable level. The plots illustrated in Figure 6-56 show that the optimized R.L. level was better than 15 dB in the narrowband case but only better than 7 dB in the wideband case. Possible adjustment of the rest of the element values in the networks would eventually lead to better overall performance as is common practice in filter design based on prototype synthesis.

6.10 SYNTHESIS AND IMPLEMENTATION OF A BANDSTOP FILTER

In this section a network with the performance of Figure 6-57(a) is derived. Figure 6-58(a) shows the synthesized lumped element bandstop prototype. The center frequency, f_o, of the fundamental

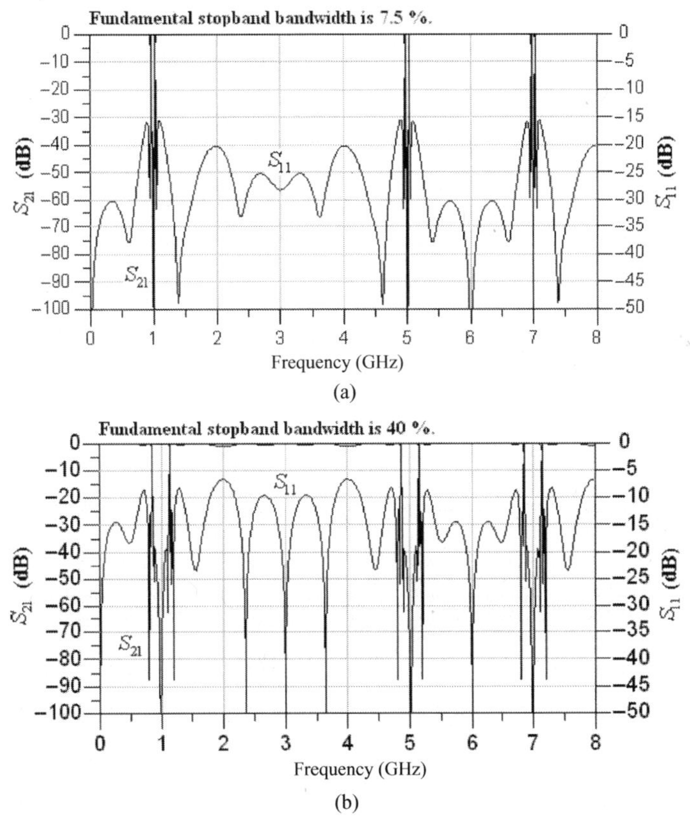

Figure 6-57 Simulated performances of bandstop filters after approximating the inverters in the prototypes by physical transmission lines. The fundamental stopband is centered at 1 GHz and the commensurate frequency is 3 GHz.

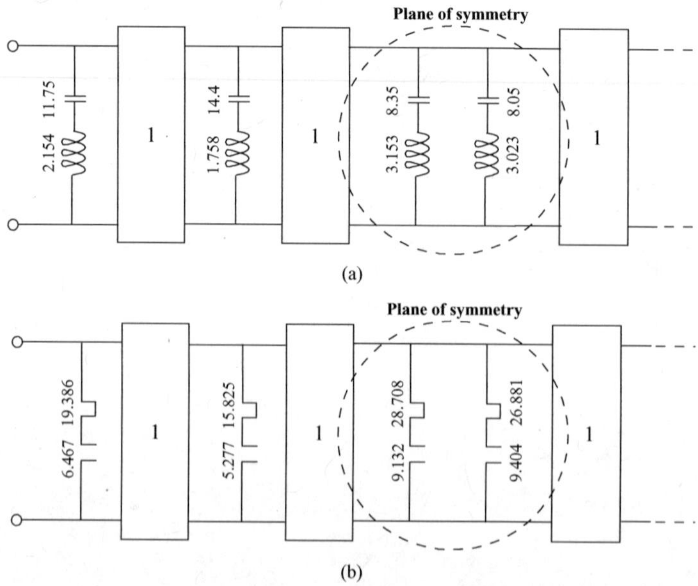

Figure 6-58 Bandstop filter design: (a) synthesized lumped bandstop prototype in 1 Ω system (inductors in nH and capacitors in pF); and (b) transformation to distributed prototype (all impedances are in Ω).

stopband is 1 GHz and the bandedge frequencies are 962.5 and 1037.5 MHz (i.e., 7.5% bandwidth). The R.L. level in the passband is 20 dB. The fundamental stopband attenuation level is 40 dB spanning from 977.5 to 1022 MHz. Transformation of this prototype into its distributed form using Equations (6.140) and (6.141) for a commensurate frequency, f_r, of 3 GHz leads to the circuit in Figure 6-58(b). Now according to Equation (6.142), the first pair of upper stopbands will theoretically be centered at frequencies of 5 and 7 GHz. Subsequently the unity inverters were approximated by quarter-wavelength long transmission lines resonant at f_o and with characteristic impedance of 1 Ω. At this stage the various transmission line sections have been optimized at different frequencies. Subsequently optimization in a circuit simulator is required to obtain the best trade off. The only adjustment required was fine-adjustment of the characteristic impedance of the first transmission line (on the left-hand side of the network) resulted in the characteristic impedance changing from 1 Ω to 1.032 Ω. Since f_r was selected as $3f_o$ the physical transmission lines were split into three sections as shown in Figure 6-59(a) (top left). Consequently in order to overcome realization difficulties the subsections inside the dashed boxes were transformed into parallel coupled lines (PCLs) using the circuit equivalences presented in [409–411]. The resulting electrical layout of the filter is illustrated in Figure 6-59(b) with small sections of transmission lines separating the PCLs.

A physical layout was then created using ADS [115] for construction on an FR4 printed circuit board (PCB) with a substrate thickness of 62 mil (1.57 mm), relative dielectric constant of 4.7 and loss tangent of 0.016. It is observed from Figure 6-59(b) that there is an open-circuited stub with a characteristic impedance of 141.659 Ω. This impedance level is the highest realizable with the minimum manufacturable line width of 6 mils. This implies that for the 7.5% fundamental stopband, the selected commensurate frequency of 3 GHz can be regarded as the optimum leading to the broadest spurious-free upper passband realizable on this PCB.

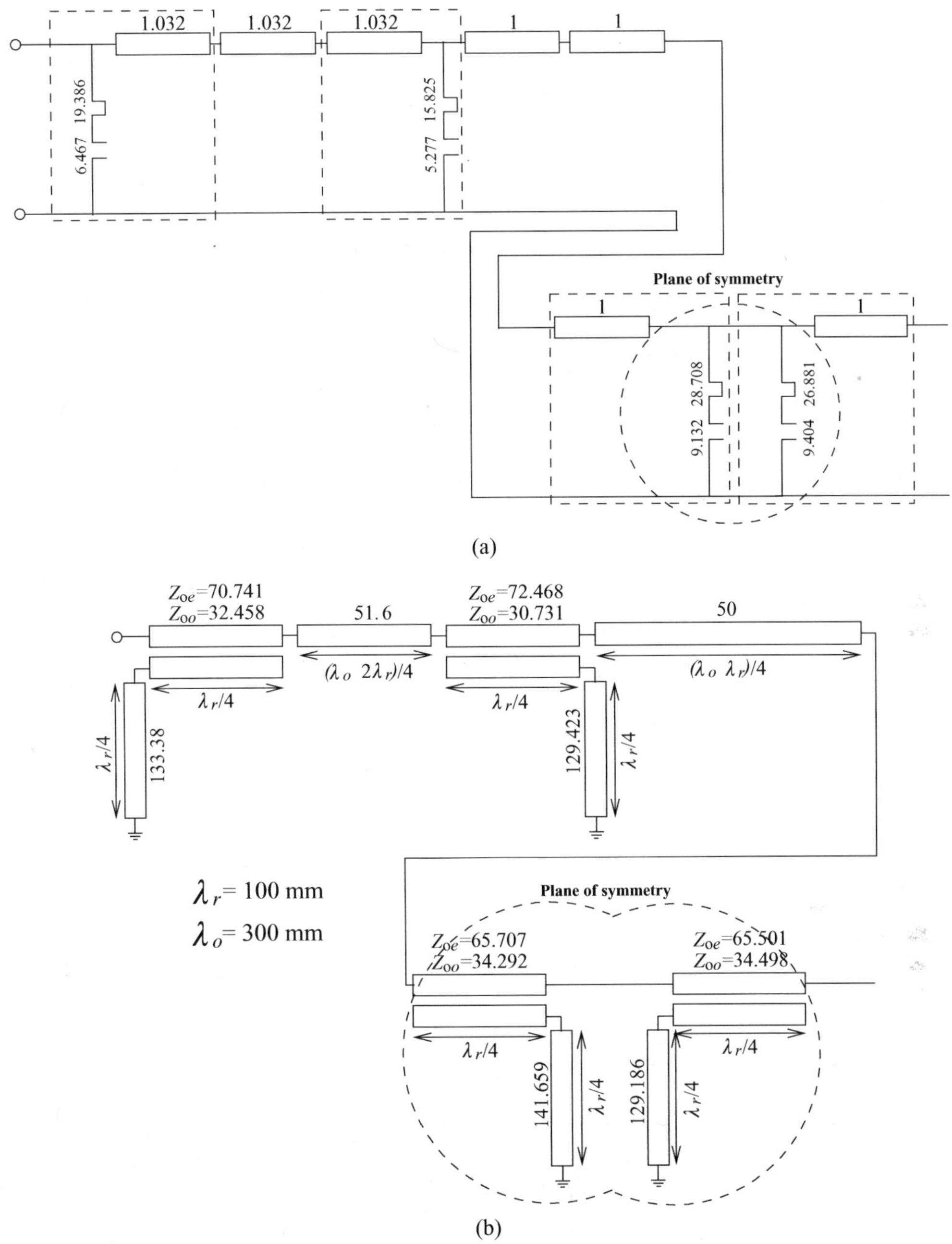

Figure 6-59 Bandstop filter design: (a) after approximating the inverters by transmission lines; and (b) the electrical filter layout in 50 Ω system.

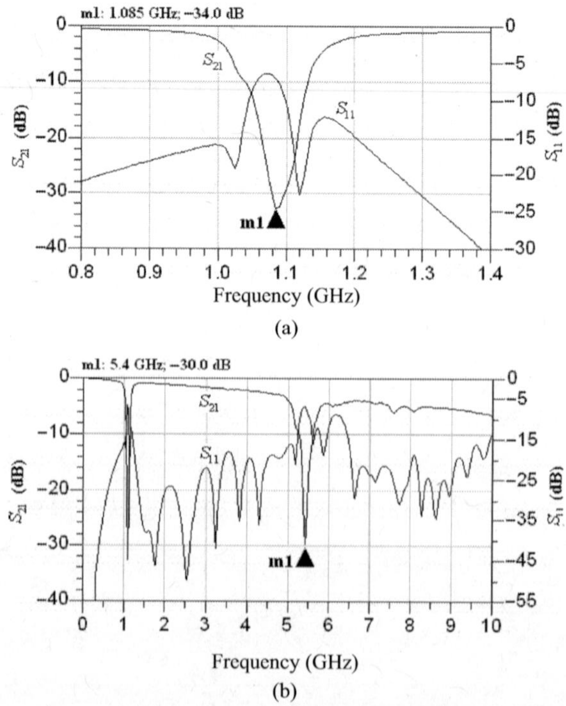

Figure 6-60 The measured performance of the constructed bandstop filter: (a) in-band; (b) wide-band frequency responses; and (c) comparison between measured and simulated wide-band performances.

The measured response of the filter is shown in Figure 6-60. It is observed from Figure 6-60(a) that the center frequency of the filter has shifted up by 8.5%. This can be easily compensated for by slightly increasing the lengths of the bandstop resonators. However, by examining the wide-band performance depicted in Figure 6-60(b) it is clear that the first upper stopband is centered at 5.4 GHz complying with theoretical expectations. It is also observed that the second spurious stopband (that was supposed to be centered at 7 GHz) has disappeared. This is primarily attributed to the parasitic losses in the filter as seen from the plot of Figure 6-60(c).

6.10.1 Summary

A step-by-step technique for the design of distributed bandstop filters with fundamental stopbands of up to 40 % is presented. Selection of the commensurate frequency of the distributed bandstop resonators of the filter determines the location of its first pair of upper spurious stopbands. The bigger the commensurate frequency, the smaller the overall size of the resulting filter and the broader its upper passband. The approach utilizes classical circuit synthesis as a starting point leading to filter structures that may require some optimization. A practical filter was implemented whose performance closely matches the simulations.

6.11 CONCLUSION

Multifunctional tunable and adaptive systems generally require narrow instantaneous responses that can be tuned over a broadband. This is in contrast to conventional RF and microwave circuits that are mostly designed for one specific purpose such as a operating in a particular frequency band and component design need only be broadband enough to accommodate signals the band of interest. With multifunctional systems many of the component subsystems must now be designed to be broadband and the tunable frequency selectivity restricted to just a few elements. In RF and microwave systems broadband component design must be approached as a network synthesis problem. This chapter applied network synthesis based on filter synthesis ideas to many of the subsystems encountered in multifunctional systems.

7

Tunable Filters

Wael M. Fathelbab and Michael B. Steer

7.1 INTRODUCTION

Adaptive radio, radar and sensors require small, highly selective and tunable filters with low insertion loss. Close assignment of bands requires that the transition from passband to stopband be rapid and nearly always zeroes are required above and below the passband so that adjacent signals, generally known as jamming signals, are sufficiently attenuated. In heterodyne architecture suppression at the image frequency[1] is particularly important. An asymmetric filter response is generally required for image channel suppression and also because channel spacing is linear whereas the pole-zero responses of electrical networks are logarithmic. An asymmetric response may also be required because of known jamming signals which may be higher on one side of the passband than on the other. This latter situation is also required with transmit/receive diplexers for mobile terminals [412, 413] and base stations [366, 414–416] of mobile communication systems. Filters with zeroes in their transmission response that can be tuned independently of the filter passband are also desirable. Sometimes the desired attributes can be satisfied by a filter that can be reconfigured to have two or more states. Many other times the filter must be continuously tunable [417, 418]. The design of both types of filters is addressed in this chapter with emphasis on the synthesis of classes of tunable filters.

Filter synthesis is the origin of a systematic approach to realizing circuits with desired frequency characteristics [366, 394, 395, 414–417, 419, 420]. A filter can also be thought of as a hardware implementation of a differential equation producing a specific impulse, step or frequency response. It is not surprising that a filter can be described this way as the lumped reactive elements, L and C, describe first-order differential equations and their interconnection describes higher order differential equations. The synthesis of a filter response begins by identifying the desired differential equation response. The differential equations are specified using the Laplace variable, s, so that in the frequency domain (with $s = \jmath\omega$) an n'th-order differential equation becomes an n'th-order polynomial expression in s. If possible, and this is usually the case, a lowpass version of a filter can be developed. The conversion to the final filter response, say a bandpass response, proceeds mathematically through a number of stages. Each stage has a circuit form and each stage is called a filter prototype

The second important aspect of filter design is the identification of particular structures that inherently have the desired frequency shaping attributes. Filter design is both science and art

1 The image frequency is the frequency which is at the same distance from a mixer's local oscillator as the input RF signal but on the opposite side. For example, if the local oscillator is 2 GHz and the RF is at 1.8 GHz, then the image is at 2.2 GHz.

matching synthesis with instinctive knowledge of physical and circuit structures. However, once a suitable filter prototype has been identified it can be optimized using optimization tools incorporated in circuit analysis packages.

Since the beginnings of circuit synthesis, particularly since the 1950's it has become common to present filter concepts using examples very often because the steps in realizing a filter maybe too difficult to specify algorithmically and there are many steps which require intuition. This procedure to present the thinking process in synthesizing a filter is followed here. This chapter presents designs for a reconfigurable bandpass filter and two designs for tunable bandpass filters. Design in all cases begins with the synthesis of prototypes from lumped element circuits. The prototypes are implemented in distributed form using parallel-coupled transmission lines in their combline configuration (discussed in Section 7.2.7 on Page 324 [421]). The significant aspect of these filters is the handling of zeroes as these must be tuned at the same rate as the tunable poles that establish the passband.

7.2 THE CHEBYSHEV APPROXIMATION

The reconfigurable and tunable filters developed in this chapter have a Chebyshev response and this section contains the necessary background for the filter synthesis described in Section 7.3 where the synthesis procedure is developed. Realization as a reconfigurable filter is described in Section 7.4 and as a tunable filter in Section 7.5. Most microwave filters have an in-band Chebyshev response. Chebyshev filters have optimal responses near the band edges with low insertion loss near the edge but at the cost of amplitude ripples in the passband. Chebyshev filters are referred to as equiripple all-pole lowpass filters. The primary advantage as a microwave filter is that Chebyshev filters provide good frequency selectivity. An example of the type of reflection and transmission responses are shown in Figure 7-1 for a seventh-order and a sixth-order Chebyshev lowpass filter.

Before proceeding terms commonly used with filters will be defined. RF and microwave filters are generally two-port networks and they provide maximum power transfer from a load to a source over a specified frequency range while rejecting the transmission of signals over other ranges. The network in Figure 7-2 is known as a doubly terminated network as both ports are resistively terminated and this corresponds to the type of network required at RF frequencies where loads and source impedances are finite. The established synthesis procedure for doubly terminated filter networks focuses on realizing the input reflection coefficient which is related to the transmission coefficient by the characteristic function. The input reflection coefficient of the doubly terminated network, Figure 7-2(a), is

$$\Gamma_1(s) = \frac{Z_{\text{in},1}(s) - R_S}{Z_{\text{in},1}(s) + R_S} \tag{7.1}$$

where the reference impedance is the source resistance R_S and s is the Laplace variable so the ideal response in the passband is when the reflection coefficient is approximately zero. There are several other parameters used with filters and these will be introduced now. The transducer power ratio (TPR) is defined as

$$\begin{aligned}\text{TPR} &= \frac{\text{Maximum power available from the source}}{\text{Power absorbed by the load}}\\ &= \frac{\dfrac{\frac{1}{2}(V_g(s)/2)^2}{R_S}}{\dfrac{\frac{1}{2}V_2^2(s)}{R_L}} = \left|\frac{1}{2}\sqrt{\frac{R_L}{R_S}}\frac{V_g(s)}{V_2(s)}\right|^2 .\end{aligned} \tag{7.2}$$

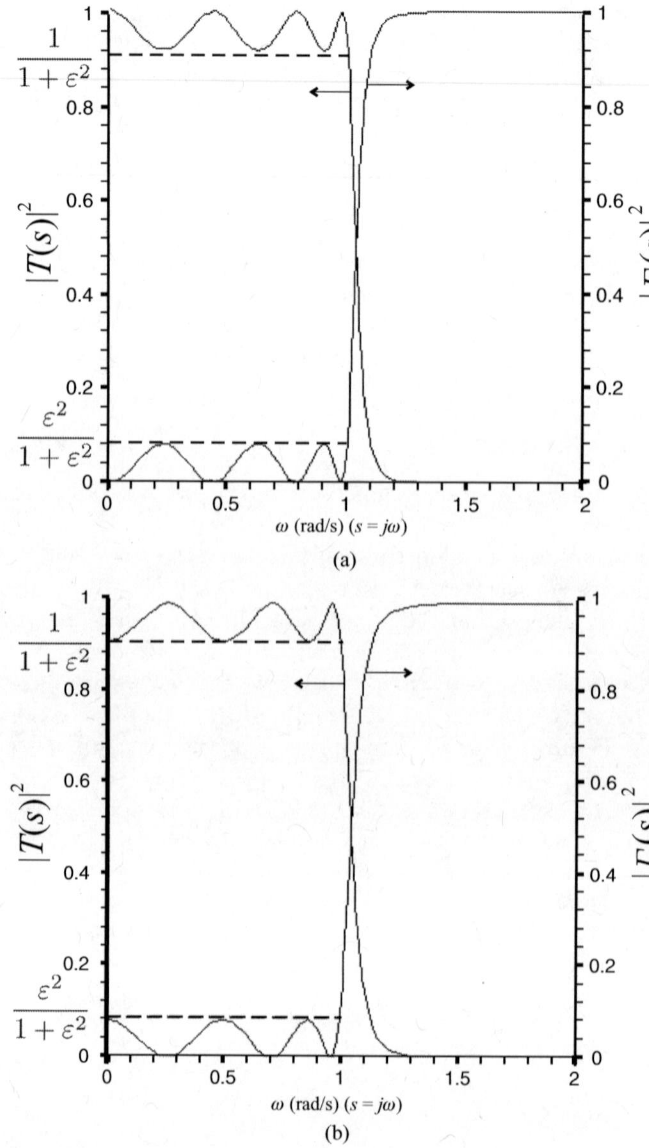

Figure 7-1 Chebyshev lowpass filter approximation: (a) seventh-order filter; and (b) sixth-order filter. $T(s)$ is the transmission response and $\Gamma(s)$ is the reflection response.

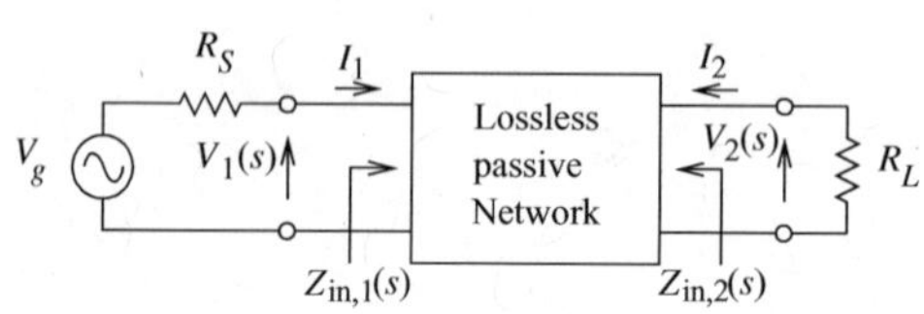

Figure 7-2 A doubly terminated network.

The transmission coefficient, $T(s)$, of the network is

$$T(s) = \frac{1}{\sqrt{\text{TPR}(s)}} = 2\sqrt{\frac{R_S}{R_L}\frac{V_2(s)}{V_g(s)}} \tag{7.3}$$

so that the insertion loss (IL) (or transducer function) is:

$$\text{IL}(s) = \text{TPR}(s) = |1/T(s)|^2 \ . \tag{7.4}$$

Note that this development assumes that the filter is lossless. The next step in development of the synthesis procedure is introduction of the characteristic function defined as the ratio of the reflection and transmission coefficients:

$$K(s) = \frac{\Gamma_1(s)}{T(s)} = \frac{N(s)}{D(s)} \tag{7.5}$$

where N is the numerator function and D is the denominator function. Ideally the filter network is lossless and so from the principle of energy conservation the sum of the magnitudes squared of the transmission and reflection coefficients must be unity:

$$|T(s)|^2 + |\Gamma_1(s)|^2 = 1 \tag{7.6}$$

or

$$|T(s)|^2 = 1 - |\Gamma_1(s)|^2 \ . \tag{7.7}$$

Dividing both sides of Equation (7.7) by $|T(s)|^2$ results in

$$1 = \frac{1}{|T(s)|^2} - \frac{|\Gamma_1(s)|^2}{|T(s)|^2} \ , \tag{7.8}$$

or

$$1 = |\text{IL}(s)|^2 - |K(s)|^2 \ . \tag{7.9}$$

Rearranging Equation (7.9) leads to

$$|\text{IL}(s)|^2 = 1 + |K(s)|^2 \tag{7.10}$$

and so

$$|T(s)|^2 = \frac{1}{1 + |K(s)|^2} \tag{7.11}$$

$$|\Gamma_1(s)|^2 = \frac{|K(s)|^2}{1 + |K(s)|^2}. \tag{7.12}$$

It is seen that both the reflection and transmission coefficients are governed by the characteristic function of the two-port network. A milestone has been reached. In an RF filter the frequency dependent insertion loss or transmission coefficient is of most importance as these are directly related to power flow. Equation (7.11) shows that this property can be expressed in terms of another function $K(s)$ which, from Equation (7.5) can be in-turn expressed as the ratio of two functions $N(s)$ and $D(s)$. For lumped element circuits $N(s)$ and $D(s)$ are polynomials. Very often design effort is directed at realizing $K(s)$ but sometimes it is better to focus design effort on realizing $\Gamma_1(s)$ but the quantities are interrelated.

7.2.1 Chebyshev Approximation

The general form of the Chebyshev transmission coefficient is

$$|T(s)|^2 = \frac{1}{1+\varepsilon^2|K(s)|^2} \tag{7.13}$$

where ε is an extra design parameter and defines the passband ripple (PBR):

$$\mathrm{PBR} = \frac{1}{1+\varepsilon^2} \quad \text{or} \quad \mathrm{PBR}\,|_{\mathrm{dB}} = 10\log_{10}\left(\frac{1}{1+\varepsilon^2}\right) \tag{7.14}$$

For the n'th-order Chebyshev approximation (to the ideal lowpass filter response) the square of the characteristic function is

$$|K_n(\omega)|^2 = \cos^2\left[n\cos^{-1}(\omega)\right] \tag{7.15}$$

which can be expressed as a polynomial. For example, with $n = 3$

$$K_3(\omega) = \cos\left[3\cos^{-1}(\omega)\right] = 4\omega^3 - 3\omega \tag{7.16}$$

and it is surprising that the trigonometric expression has such a simple polynomial equivalence. From Equation (7.11) the transmission coefficient is

$$|T(\omega)|^2 = \frac{1}{1+\varepsilon^2\cos^2\left[n\cos^{-1}(\omega)\right]} \tag{7.17}$$

and the reflection coefficient is

$$|\Gamma_1(\omega)|^2 = \frac{\varepsilon^2\cos^2\left[n\cos^{-1}(\omega)\right]}{1+\varepsilon^2\cos^2\left[n\cos^{-1}(\omega)\right]}. \tag{7.18}$$

Factorizing the denominator of either Equations (7.17) or (7.18) yields the following roots (of the denominator of $\Gamma_1(s)$ and $T(s)$):

$$\begin{aligned} s_i = {} & \sin\left[\frac{(2i-1)\pi}{2n}\right].\sinh\left[\frac{1}{n}\sinh^{-1}\left(\frac{1}{\varepsilon}\right)\right] \\ & + \jmath\cos\left[\frac{(2i-1)\pi}{2n}\right].\cosh\left[\frac{1}{n}\sinh^{-1}\left(\frac{1}{\varepsilon}\right)\right] \qquad i = 1, 2, 3, \ldots \end{aligned} \tag{7.19}$$

and the roots of the numerator of $\Gamma_1(s)$ in the s-plane are:

$$s_j = \jmath \cos\frac{(2j-1)\pi}{2n} \qquad j = 1, 2, \ldots, n \tag{7.20}$$

Equations (7.19) and (7.20) may be utilized to obtain the reflection and transmission coefficients directly in the s-domain.

It is possible to obtain the characteristic function of the Chebyshev approximation from the recursion formula:

$$K_n(\omega) = 2\omega K_{n-1}(\omega) - K_{n-2}(\omega) \tag{7.21}$$

with

$$K_1(\omega) = \omega; \qquad K_2(\omega) = 2\omega^2 - 1. \tag{7.22}$$

For example, with $n = 3$,

$$\begin{aligned} K_3(\omega) &= 2\omega K_{3-1}(\omega) - K_{3-2}(\omega) \\ &= 2\omega.\left(2\omega^2 - 1\right) - \omega = 4\omega^3 - 2\omega - \omega = 4\omega^3 - 3\omega. \end{aligned} \tag{7.23}$$

7.2.2 Chebyshev Prototypes

The n'th order lowpass filters constructed from the Chebyshev polynomials have the ladder circuit forms of either Figure 7-3(a) or Figure 7-3(b). Figure 7-3 uses several short hand notations commonly used with filters. First note that there are two prototype forms designated Type 1 and Type 2 and these are referred to as duals of each other. The two prototype forms have identical responses with the same numerical element values $g_1, \ldots, g_n$. Consider the Type 1 prototype of Figure 7-3(a). The right most element is the resistive load, which is also known as the $(n+1)$ th element. The next element to the left of this is either a shunt capacitor (of value g_n) if n is even, or a series inductor (of value g_n) if n is odd. So for the Type 1 prototype the shunt capacitor next to the load does not exist if n is odd. The same interpretation applies to the circuit in Figure 7-3(b).

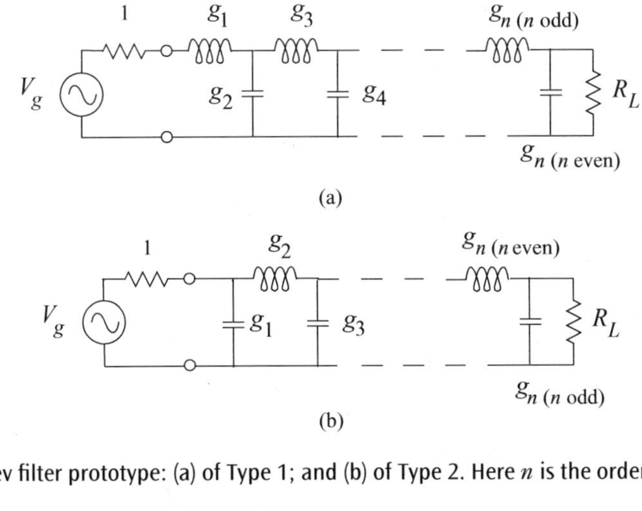

Figure 7-3 Chebyshev filter prototype: (a) of Type 1; and (b) of Type 2. Here n is the order of the filter.

For a Chebyshev response the element values of the prototype are found from the following recursive formula [400, 422]:

$$g_1 = \frac{2\sin\left(\frac{\pi}{2n}\right)}{\sinh\left[\left(\frac{1}{n}\right)\sinh^{-1}\left(\frac{1}{\varepsilon}\right)\right]} \tag{7.24}$$

$$g_r \cdot g_{r+1} = \frac{4\sin\left[(2r-1)\frac{\pi}{2n}\right].\sin\left[(2r+1)\frac{\pi}{2n}\right]}{\sinh^2\left[\left(\frac{1}{n}\right)\sinh^{-1}\left(\frac{1}{\varepsilon}\right)\right]+\sin^2\left(\frac{r\pi}{n}\right)}. \tag{7.25}$$

For an odd-degree function (n is odd) there is a perfect match at DC:

$$|T(0)|^2 = 1, \tag{7.26}$$

while for an even-degree function a mismatch exists of value

$$|T(0)|^2 = \frac{4R_L}{(R_L+1)^2} = \frac{1}{1+\varepsilon^2} \tag{7.27}$$

so that

$$R_L = g_{n+1} = \left[\varepsilon + \sqrt{(1+\varepsilon^2)}\right]^2. \tag{7.28}$$

7.2.3 Quarter-Wave Transformer

The structure known as a quarter-wave transformer is shown in Figure 7-4. This shows a section of transmission line whose length ℓ is $\lambda/4$ and hence the name quarter-wave transformer. The input impedance looking into the quarter wave transformer is given by

$$\begin{aligned} Z_{\text{in}} &= Z_0 \frac{Z_{01} + \jmath Z_0 \tan(\beta\ell)}{Z_0 + \jmath Z_{01}\tan(\beta\ell)} \\ &= Z_0 \frac{Z_{01} + \jmath Z_0 \infty}{Z_0 + \jmath Z_{01}\infty} \\ &= \frac{Z_0^2}{Z_{01}}. \end{aligned} \tag{7.29}$$

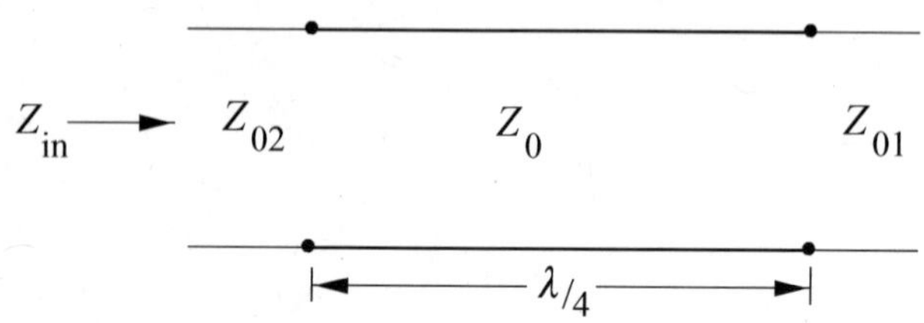

Figure 7-4 The quarter-wave transformer line.

Hence a section of the transmission line of length $\ell = \lambda/4 + n\lambda/2$ where $n = 0, 1, 2, \ldots$ can be used to match two different impedances Z_{01}, Z_{02} by constructing the line such that its characteristic impedance Z_0 is

$$Z_0 = \sqrt{Z_{01} Z_{02}}\ . \tag{7.30}$$

Note that for a design center frequency f_0, the electrical length of the matching section is $\lambda_0/4$, but at different frequencies the electrical length is different and a perfect match is no longer achieved. In general, perfect match is obtained only at the frequencies at which $\ell = \lambda/4 + n\lambda/4$.

There is an interesting property of a quarter-wave transformer that is widely used in filters. Examine the final result in Equation 7.29 which is repeated here:

$$Z_{\text{in}} = \frac{Z_0^2}{Z_{01}}\ . \tag{7.31}$$

This results applies to complex impedances as well. Equation 7.31 indicates that a quarter-wave long line is an impedance inverter, presenting, at Port 1, the inverse of the impedance presented at port 2, Z_{01}. This inversion is normalized by the characteristic impedance of the line. This inversion holds in the reverse direction as well.

7.2.4 The Inverter and the Quarter-Wave Transformer

An impedance inverter is a two-port element in the prototype network of many RF and microwave filters. The schematic representation of an impedance inverter is shown in Figure 7-5(a) and its constitutive property can be understood by considering the terminated inverter in 7-5(b). The input impedance of the impedance inverter is

$$Z_{\text{in}} = \frac{K^2}{Z_L}. \tag{7.32}$$

and so the inverter both inverts the load impedance and scales it. This does not fully define an inverter which is a two-port and this is done in the next subsection. An impedance inverter has the value K which is in Ω's and sometimes K is called the characteristic impedance of the inverter. Note that a $\lambda/4$-long line with a load has an input impedance which is the inverse of the load, normalized by the square of the characteristic impedance of the line. So an inverter can be realized at microwave frequencies using a quarter-wave long transmission line. An alternative schematic representation of an inverter as a line as shown in Figure 7-5(c) where it is understood from the context that a line is being used as an inverter. At lower frequencies an inverter can be realized using operational amplifiers. One of the principal advantage taken of inverters in developing networks

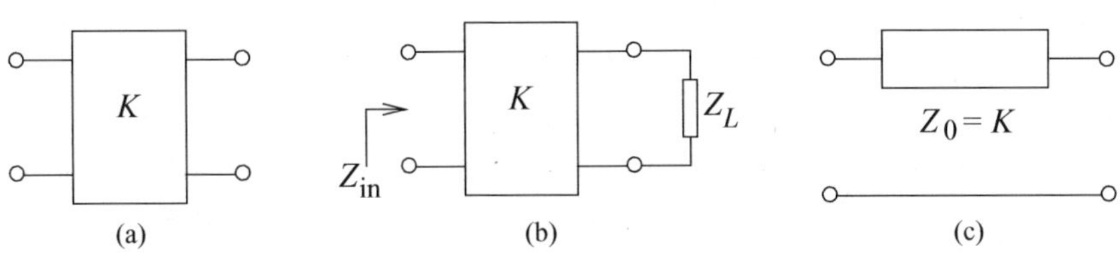

Figure 7-5 Impedance inverter (of impedance K): (a) represented as a two-port; (b) the two-port terminated in a load; and (c) the two-port implemented as quarter-wavelength long line of characteristic impedance Z_0.

is replacing series inductors and capacitors by shunt elements which are far more easily realized, consider for example the realization of shunt elements using shunt stubs.

At low frequencies it is necessary to use either transformers or active circuits based on operational amplifiers. In this section the impedance inverter can be treated as an abstraction. A two-port lossless, reciprocal and frequency-independent impedance inverter has the ABCD matrix:

$$\mathbf{T} = \begin{bmatrix} 0 & \jmath K \\ \dfrac{\jmath}{K} & 0 \end{bmatrix} \tag{7.33}$$

where K is called the characteristic impedance of the inverter. With a load impedance Z_L (at Port 2) the input impedance (at Port 1) is

$$Z_{\text{in}}(s) = \frac{A.Z_L + B}{C.Z_L + D} = \frac{\jmath K}{\left(\frac{\jmath}{K}\right) Z_L} = \frac{K^2}{Z_L} \tag{7.34}$$

Hence the impedance at the input port is proportional to the inverse of the load impedance. (Note that K has the units of Ω's.)

Now the ABCD matrix of the transmission line of Figure 7-5(c) is

$$\begin{bmatrix} \cos\theta & \jmath Z_0 \sin\theta \\ \dfrac{\jmath}{Z_0}\sin\theta & \cos\theta \end{bmatrix} \tag{7.35}$$

which is identical to Equation (7.33) when the electrical length $\theta = \pi/2$, i.e., when the line is $\lambda/4$ long. The inverter is shown in Figure 7-5(a) as a two-port and implemented as a quarter-wavelength long line in Figure 7-5(c). The bandwidth over which the line realizes an impedance inverter is limited however as it is an ideal inverter only at the frequency at which it is $\lambda/4$ long and when it is $(n\lambda/2 + \lambda/4)$ long.

7.2.5 The Admittance Inverter

The admittance inverter is functionally the same as the impedance inverter and the schematic is the same, see Figure 7-6(a). As will be shown the inverter can be realized using frequency-invariant lossless elements (that is elements whose reactance or susceptance do not vary with frequency) using the network of Figure 7-6(b). The distinction between the two types of inverters is that $\mathcal{J}$ is generally used with admittance inverters and K with impedance inverters. If not specified by the context, the inverter (with value specified by a number) defaults to being an impedance inverter. Alternatively units can be used to indicate which type of inverter is being used. The function of the inverter is the same in any case, both can be realized by quarter-wavelength-long lines

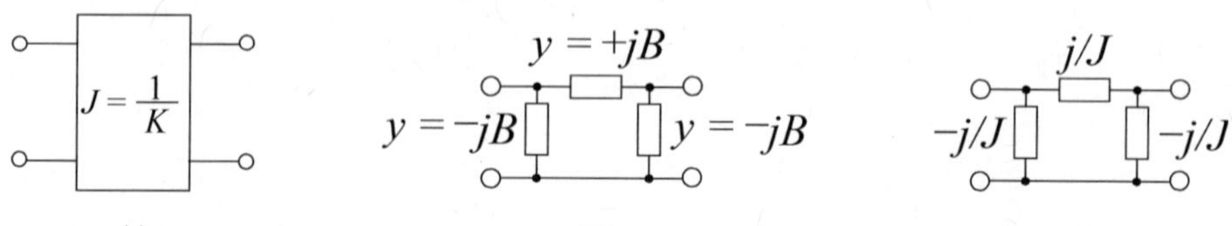

Figure 7-6 Admittance inverter: (a) as a two-port; (b) generalized inverter realized using lumped elements (identical if $B = -\mathcal{J}$); and (c) lumped equivalent circuit of the admittance inverter of (a) (the element values here are impedances).

Figure 7-7 Impedance inverter: (a) as a two-port; and (b) its lumped equivalent circuit (the element values here are impedances).

for example. For the remainder of this chapter it will be more convenient to use the admittance inverter as many calculations will be in terms of admittances since it is much easier to realize shunt elements in distributed form (for example using stubs) than to realize series elements.

Now it will be shown that the lumped-element network of Figure 7-6(b) realizes an inverter. To do this it must be shown that the inverter and the lumped-element network have the same two-port parameters. First, the ABCD matrix of an inverter of characteristic admittance $\mathcal{J}$ is

$$\mathbf{T}_{\mathcal{J}} = \begin{bmatrix} 0 & \dfrac{\jmath}{\mathcal{J}} \\ \jmath\mathcal{J} & 0 \end{bmatrix}. \tag{7.36}$$

Now, referring to Table 7-1, the circuit of Figure 7-6(b) has the ABCD matrix

$$T = \begin{bmatrix} 1 & 0 \\ -\jmath B & 1 \end{bmatrix} \begin{bmatrix} 1 & \dfrac{1}{(\jmath B)} \\ 0 & 1 \end{bmatrix} \begin{bmatrix} 1 & 0 \\ -\jmath B & 1 \end{bmatrix} \tag{7.37}$$

$$= \begin{bmatrix} 1 & \dfrac{1}{(\jmath B)} \\ -\jmath B & 0 \end{bmatrix} \begin{bmatrix} 1 & 0 \\ -\jmath B & 1 \end{bmatrix} = \begin{bmatrix} 0 & \dfrac{-\jmath}{B} \\ -\jmath B & 0 \end{bmatrix} \tag{7.38}$$

where B is the susceptance of the frequency-invariant elements.

Equation (7.38) is identical to that of Equation (7.36) if $B = -\mathcal{J}$. More practical equivalents of the circuit of Figure 7-6(b) can be derived as shown later on. For completeness the lumped element equivalent of the impedance inverter is shown in Figure 7-7. This was derived from Figure 7-6 using $\mathcal{J} = 1/K$.

7.2.6 Richards' Transformation

In this section Richard's transformation is used to realize a lumped element filter in distributed form. The design example begins by considering a Chebyshev lowpass filter with the transmission characteristic

$$|T(s)|^2 = \frac{1}{1 + \varepsilon^2 |K(s)|^2}. \tag{7.39}$$

With the Richards' transformation

$$s \rightarrow \jmath\omega \rightarrow \jmath\alpha \tan(\theta) \tag{7.40}$$

the response becomes

$$|T(\jmath\alpha \tan\theta)|^2 = \frac{1}{1 + \varepsilon^2 |K(\jmath\alpha \tan\theta))|^2}. \tag{7.41}$$

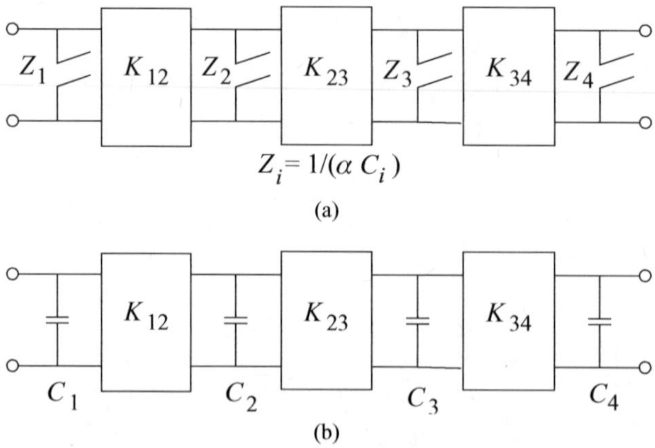

Figure 7-8 Lowpass prototypes: (a) lowpass prototype as a ladder filter with inverters; and (b) lowpass distributed prototype with open-circuited stubs (the impedance looking into the stub is indicated).

Thus the passband edge at $\omega = 1$ is mapped to $\omega = \theta_1$ as:

$$1 \to \alpha \tan(\theta_1) \tag{7.42}$$

$$\alpha = \frac{1}{\tan(\theta_1)}. \tag{7.43}$$

Since a capacitor is transformed into an open-circuited stub, Richards' transformation applied to the lowpass filter prototype results in a filter with transmission line elements only as shown Figure 7-8 provided that the inverters are realized using transmission lines.

The implementation of a lowpass filter in distributed form results in passbands and stopbands that repeat in frequency as shown in Figure 7-9. This occurs because transmission lines are used to realize lumped elements and the impedance of a transmission repeats every wavelength (or half-wavelength in some cases). For example, the input impedance of a stub is the same whether it is a half-wavelength long or a wavelength long.

7.2.7 COUPLED LINES

The lumped-element filters considered up to now were synthesized using a ladder-type element extraction procedure followed by a variety of transformations for impedance scaling, filter type and the Richards' Transformation to develop distributed forms of a filter. The other important aspect of RF and microwave filter design is identifying structures that inherently have responses that match the synthesized responses. The most important structure that that can be utilized to realize frequency selectivity is the coupled line. The aim in this subsection is to show how a coupled line of appropriate length relates to a bandpass filter configuration. A pair of coupled lines is shown in Figure 7-10(a) together with its capacitance model shown in cross-section in Figure 7-10(b). In Figure 7-10(b) the subscripts 1 and 2 refer to the line number and not the terminal indexes. The charge on each per unit length, Q_1 and Q_2, are related to the voltages V_1 and V_2 on the lines by the per-unit-length capacitances of the coupled line:

$$Q_1 = C_{11} V_1 + C_{12}(V_1 - V_2) = (C_{11} + C_{12}) V_1 - C_{12} V_2 \tag{7.44}$$

$$Q_2 = C_{22} V_2 + C_{21}(V_2 - V_1) = -C_{12} V_1 + (C_{22} + C_{12}) V_2 \tag{7.45}$$

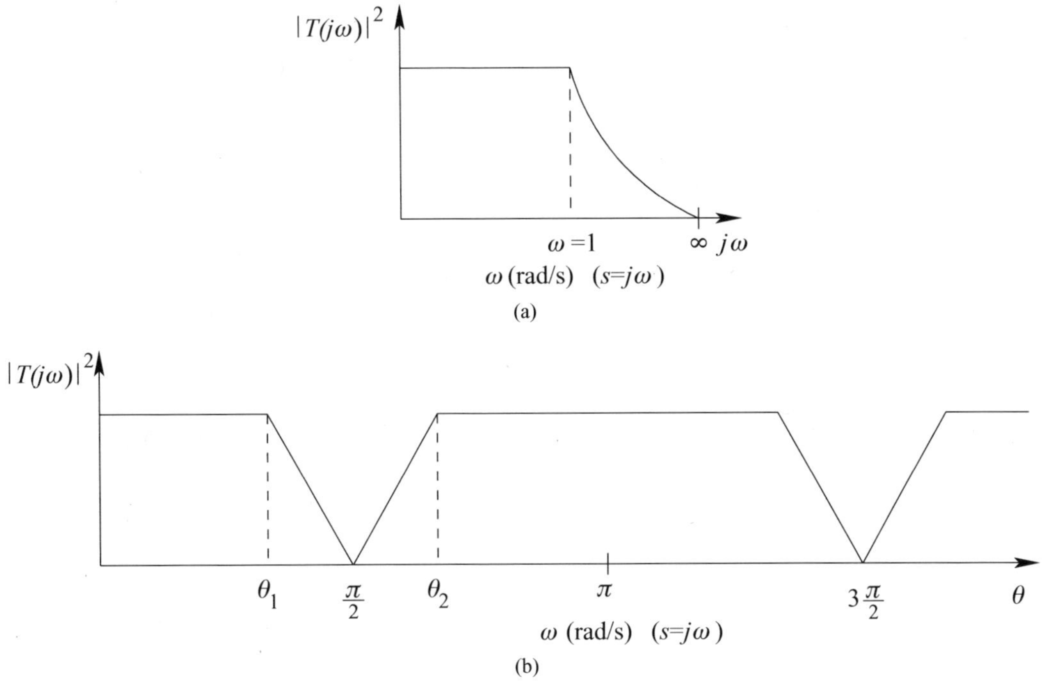

Figure 7-9 Lowpass to lowpass transformation: (a) lowpass filter response; and (b) distributed lowpass filter response.

The per-unit-length capacitance matric of the coupled lines is

$$C = \begin{bmatrix} C_{11} & C_{12} \\ C_{21} & C_{22} \end{bmatrix}. \tag{7.46}$$

The next step in modeling the coupled lines is to consider that the signals of the line can be realized as a combination of an even-mode and an odd-mode. To do this the capacitances of the lines are

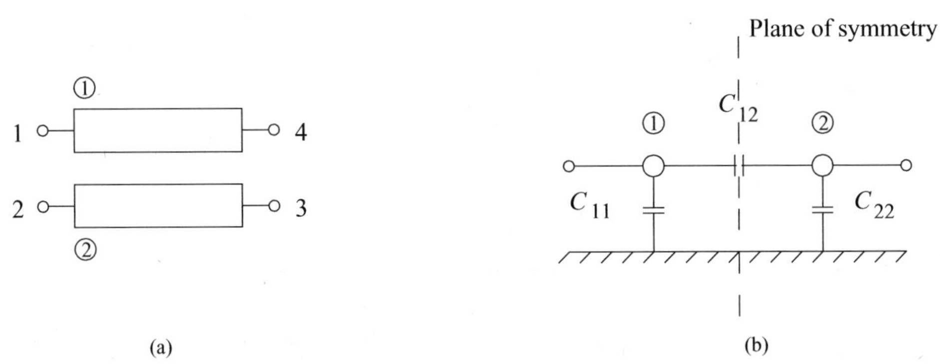

Figure 7-10 Coupled lines: (a) top view of a pair of coupled microstrip lines ; and (b) capacitance model of the lines in cross-section.

rephrased in-terms of even and odd modes. The even mode is defined as when the voltages on the two lines are the same, $V_1 = V_2 = V_e$, so that the even-mode charge is

$$Q_e = V_e(C_{11} + C_{22})/2. \tag{7.47}$$

This leads to the even mode per unit-length capacitance

$$C_e = \frac{Q_e}{V_e} = \frac{(C_{11} + C_{22})}{2}. \tag{7.48}$$

Similarly the odd-mode charge is when $V_1 = -V_2$ with the odd-mode voltage is $V_1 - V_2$:

$$Q_o = (C_{11} + C_{22} + 2C_{12})\,V_o = C_o V_o \tag{7.49}$$

and the odd-mode capacitance is

$$C_o = \frac{Q_o}{V_o} = \frac{(C_{11} + C_{22} + 2C_{12})}{2}. \tag{7.50}$$

The characteristic impedances of the odd and even modes can be calculated form the even-mode phase velocity, v_{pe}, and the odd-mode phase velocity, v_{po}, as

$$Z_{oe} = \frac{1}{v_{pe} C_e}, \qquad Z_{oo} = \frac{1}{v_{po} C_o} \tag{7.51}$$

It is now possible to represent a pair of coupled lines by the network model shown in Figure 7-11. The parameters of the network model are related to the model impedances as follows:

$$n = \frac{1}{K} = \frac{Z_{oe} - Z_{oo}}{Z_{oe} + Z_{oo}} \tag{7.52}$$

$$Z = \sqrt{(Z_{oe} Z_{oo})} \qquad Z_1 = \frac{Z}{\sqrt{1 - K^2}} \qquad Z_2 = Z\frac{\sqrt{1 - K^2}}{K^2} \tag{7.53}$$

Various terminating arrangements of the coupled lines result in several useful filter elements. The arrangement relevant here is what is called the combline configuration shown in Figure 7-12. This network can be used in realizing bandpass filters derived from a lowpass prototype. With appropriate planar microstrip coupled lines can be used to easily achieve narrowband low-loss

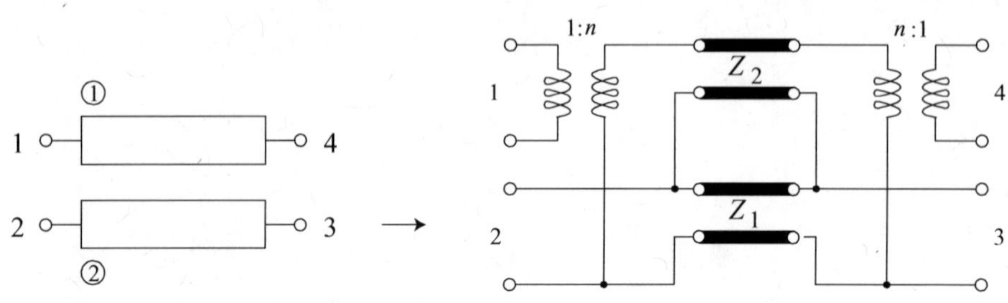

Figure 7-11 Network model of a pair of coupled lines.

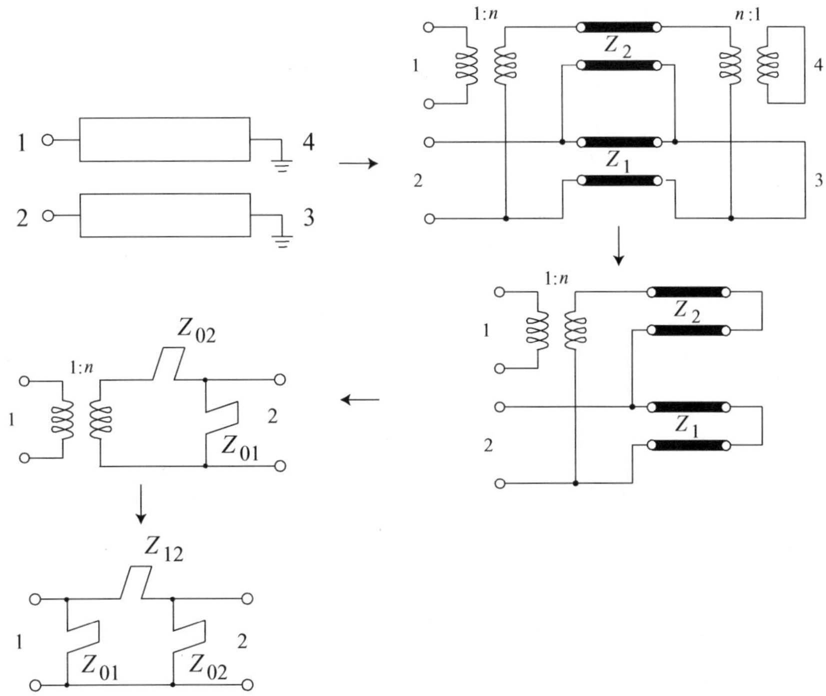

Figure 7-12 Combline section and network model. The final network model has two shorted transmission lines in shunt with characteristic impedances Z_{01} and Z_{02} separated by a series connection of a shorted transmission line with characteristic impedance Z_{12}. The transmission lines are a quarter-wavelength long at the commensurate frequency

bandpass filters. For tunable and reconfigurable filters it must be possible to incorporate tunable capacitors and switches into the filter topology. This is conveniently achieved with a combline filter.

7.2.8 Inverter Network Scaling

A filter that uses transmission lines is nearly always synthesized from a filter prototype that contains inverters. This is because an inverter can be approximated as a quarter-wavelength long line. The synthesis of a filter begins with a normalized prototype that is transformed to the desired frequency, impedance, and type. It is the purpose of this section to enable such scaling when inverters are present in the prototype. In particular it is shown that the scaling the admittance of the network of Figure 7-13(a), a common inverter subnetwork, by a factor x results in the network in Figure 7-13(b).

The nodal admittance matrix of the network in Figure 7-13(a) is

$$Y = \begin{bmatrix} 0 & \jmath J & 0 \\ \jmath J & y & \jmath J \\ 0 & \jmath J & 0 \end{bmatrix}. \tag{7.54}$$

Assigning nodal voltages to Terminals 1, 2 and 3, the nodal admittance matrix equation is

$$Y = \begin{bmatrix} v_1 \\ v_2 \\ v_3 \end{bmatrix} = \begin{bmatrix} 0 & \jmath J & 0 \\ \jmath J & y & \jmath J \\ 0 & \jmath J & 0 \end{bmatrix} \begin{bmatrix} v_1 \\ v_2 \\ v_3 \end{bmatrix} = \begin{bmatrix} J_1 \\ 0 \\ J_3 \end{bmatrix}. \tag{7.55}$$

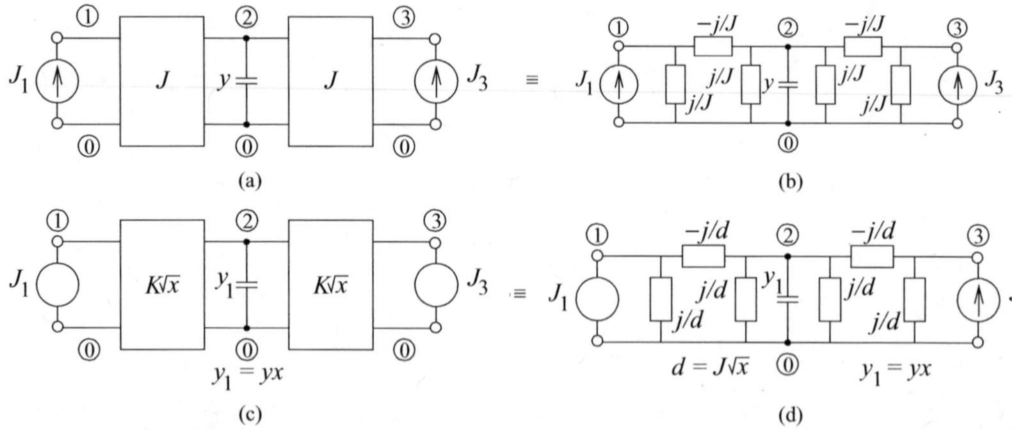

Figure 7-13 Inverter network: (a) network with two identical admittance inverters with an inserted shunt element of admittance y_1; (b) equivalent network using equivalence shown in Figure 7-6; (c) scaled original network; and (d) scaled equivalent network.

Eliminating Terminal 2 this reduces to

$$\begin{bmatrix} \dfrac{\mathcal{J}^2}{y} & \dfrac{\mathcal{J}^2}{y} \\ \dfrac{\mathcal{J}^2}{y} & \dfrac{\mathcal{J}^2}{y} \end{bmatrix} \begin{bmatrix} v_1 \\ v_3 \end{bmatrix} = \begin{bmatrix} \mathcal{J}_1 \\ \mathcal{J}_3 \end{bmatrix}. \tag{7.56}$$

and this describes the external characteristics of the subnetwork in Figure 7-13(a). Now the intent is to show that the network of Figure 7-13(c) which is a scaled version of the original network of Figure 7-13(a) has an identical external characteristic. That is, it has the same reduced nodal admittance matrix. The nodal admittance matrix of the scaled network in Figure 7-13(c) is

$$Y' = \begin{bmatrix} 0 & \jmath\mathcal{J}\sqrt{x} & 0 \\ \jmath\mathcal{J}\sqrt{x} & yx & \jmath\mathcal{J}\sqrt{x} \\ 0 & \jmath\mathcal{J}\sqrt{x} & 0 \end{bmatrix}. \tag{7.57}$$

Assigning nodal voltages to Terminals 1, 2 and 3 the nodal admittance matrix equation is

$$Y = \begin{bmatrix} v_1 \\ v_2 \\ v_3 \end{bmatrix} = \begin{bmatrix} 0 & \jmath\mathcal{J}\sqrt{x} & 0 \\ \jmath\mathcal{J}\sqrt{x} & yx & \jmath\mathcal{J}\sqrt{x} \\ 0 & \jmath\mathcal{J}\sqrt{x} & 0 \end{bmatrix} \begin{bmatrix} v_1 \\ v_2 \\ v_3 \end{bmatrix} = \begin{bmatrix} \mathcal{J}_1 \\ 0 \\ \mathcal{J}_3 \end{bmatrix} \tag{7.58}$$

and by eliminating Terminal 2 this reduces to

$$\begin{bmatrix} \dfrac{(\mathcal{J}^2\sqrt{x}\sqrt{x})}{(yx)} & \dfrac{(\mathcal{J}^2\sqrt{x}\sqrt{x})}{(yx)} \\ \dfrac{(\mathcal{J}^2\sqrt{x}\sqrt{x})}{(yx)} & \dfrac{(\mathcal{J}^2\sqrt{x}\sqrt{x})}{(yx)} \end{bmatrix} \begin{bmatrix} v_1 \\ v_3 \end{bmatrix} = \begin{bmatrix} \mathcal{J}^2/y & \mathcal{J}^2/y \\ \mathcal{J}^2/y & \mathcal{J}^2/y \end{bmatrix} \begin{bmatrix} v_1 \\ v_3 \end{bmatrix}$$

$$= \begin{bmatrix} \mathcal{J}_1 \\ \mathcal{J}_3 \end{bmatrix}. \tag{7.59}$$

Thus the original network shown in Figure 7-13(a) has identical external electrical characteristics as the scaled network of Figure 7-13(c) with the characteristic admittance of the inverters scaled by $\sqrt{x}$ and the shunt admittance scaled by x.

A generalization of this result (that is useful when there are additional connections between nodes 1 and 3, and at node 2) is that multiplying a row and a column of the nodal admittance matrix by a factor results in identical external characteristics. Note that the element sharing a row and column is multiplied twice. Nodal matrix reduction as used here for a 3×3 nodal admittance matrix can be used to prove this general result.

7.3 TUNABLE AND RECONFIGURABLE BANDPASS FILTER SYNTHESIS

In this section a tunable combline filter design is synthesized based on a conventional combline filter design with tunable lumped element capacitors. The combline filter is one of the most commonly used transmission line-based filters and is based on the combline section shown in Figure 7-12. Much of the development follows a conventional design except for the provision of lumped capacitors. The filter to be considered is a third-order combline filter and so consists of two coupled line pairs with one of the transmission lines of each pair shared. A suitable placement of capacitors that will be tuned is shown in Figure 7-14. Optimum tunability is obtained when the electrical length of the transmission line is 45°. That is L_1 is $\lambda/8$. Coupled with this the total capacitance at the end of the lines provides the equivalent electrical length of 45°. Furthermore the filter is synthesized so that the capacitances have the same value so that when they are tuned the same tuning voltage is applied across each BST varactor.

Each coupled transmission line pair is modeled by the network model of a combline section as shown in Figure 7-12 but with lumped capacitors at Terminals 1 and 2 to ground. The complete network model of the microstrip circuit of Figure 7-14 is shown in Figure 7-15. As is the general practice with developments in filter synthesis the combline filter design is be presented by way of an example.

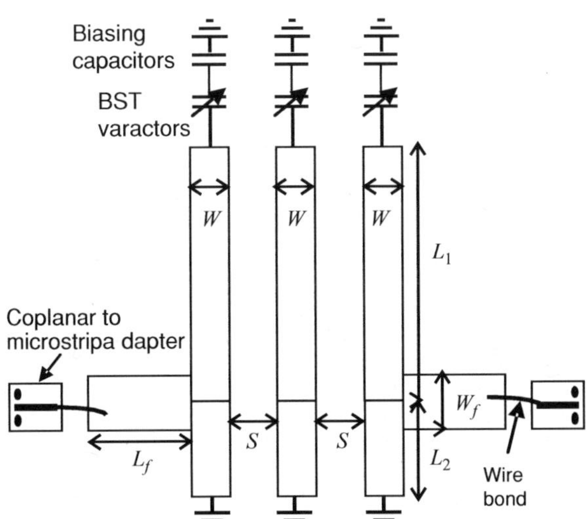

Figure 7-14 Layout of a combline bandpass filter with lumped element capacitors that can be replaced by tunable capacitors for a tunable filter.

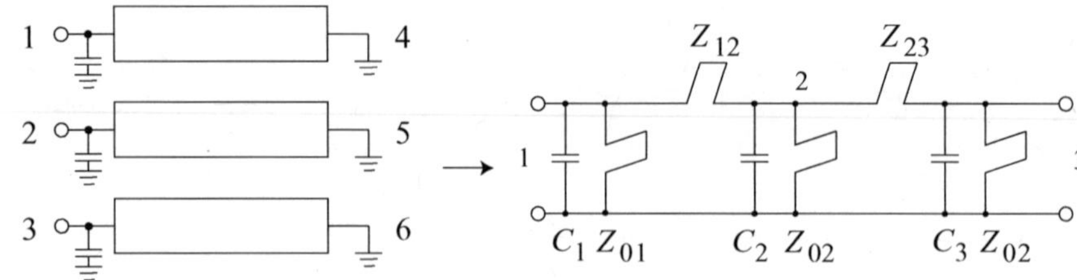

Figure 7-15 Network model of two coupled transmission line pairs as shown in Figure 7-14 without the impedance matching taps (the transmission lines identified by their length of L_f).

7.3.1 Scaled Prototype with Inverters

This subsection presents an example of how the inverter network scaling of Section 7.2.8 can be used with combline filters. A lowpass prototype with inverters, suitable for combline filters, is shown in Figure 7-16(a). There are three shunt capacitors in the prototype and, as will be made clearer in the combline filter design that follows, it is desirable that the shunt elements that realize these be identical. To achieve this the central capacitor must be scaled by a factor of 1/2. This can be achieved by scaling the impedance of the inverters by $1/\sqrt{2}$. So the network of Figure 7-16 can be replaced by a network where the shunt elements have the desired property of being identical.

7.3.2 Third-Order Chebyshev Combline Filter

The design of a third-order Chebyshev combline bandpass filter is presented here as an example of the filter design procedure based on synthesis. The choice of prototype is critical and it is an art to select a prototype the yields an implementable coupled line filter. As usual the design is presented here in steps. Design begins with the choice of a prototype that is expected to achieve the required specifications. If the choice is not suitable then another must be chosen. Here the third-order Chebyshev lowpass filter prototype shown in Figure 7-17 is considered with a passband ripple of $\varepsilon = 0.1$.

The next step in the transformation of the lowpass prototype to a bandpass filter is to replace the inductor by a shunt capacitor in cascade with admittance inverters as shown in Figure 7-18. Now the prototype can be transformed into a bandpass filter. Choosing a center frequency of approximately 1 GHz and 10% bandwidth corner frequencies are at $f_1 = 950$ and $f_2 = 1050$ MHz the bandpass prototype becomes that shown in Figure 7-19.

The bandpass resonator shown in Figure 7-19 will be transformed into a structure with lumped capacitors, transmission lines, and inverters that themselves can be realized using transmission lines. The transformation begins by transforming each lumped resonator (e.g., C_1 and L_1, in Figure 7-19) into a combined lumped-distributed resonator as shown in Figure 7-20. This is done by first deriving the input admittance of the lumped element resonator as follows. The input admittance of the lumped-element resonator is

$$Y_{\text{in}} = \frac{j\left(\omega^2 LC - 1\right)}{\omega L} \tag{7.60}$$

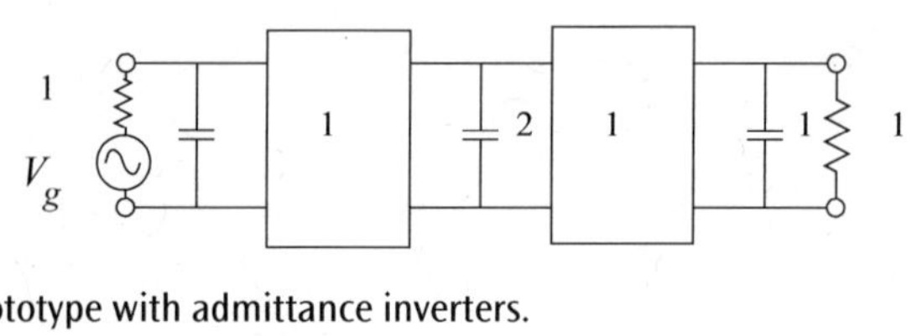

Figure 7-16 A lowpass prototype with admittance inverters.

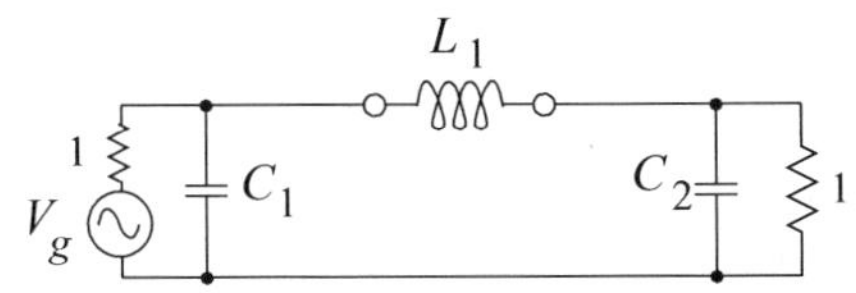

$$C_1 = 0.85239 \text{ F} = C_2 \quad L_1 = 1.10422 \text{ H}$$

Figure 7-17 Step 1. A third-order Chebyshev lowpass filter prototype.

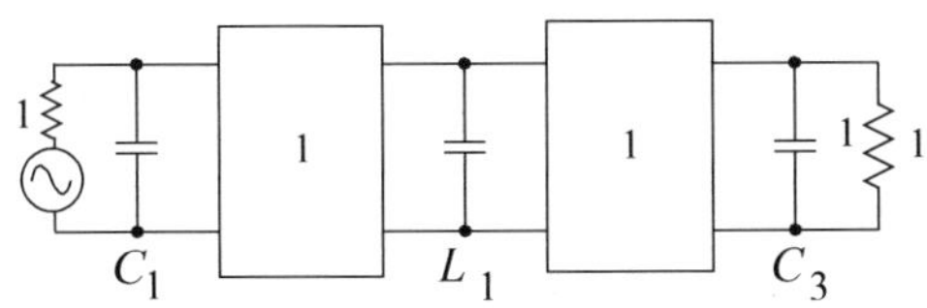

$$C_1 = 0.85239 \text{ F} = C3 \quad L_1 = 1.10422 \text{ F}$$

Figure 7-18 Step 2. Prototype filter with inductor replaced by a capacitor and admittance inverter combination. Note that L_1 has the units of farads.

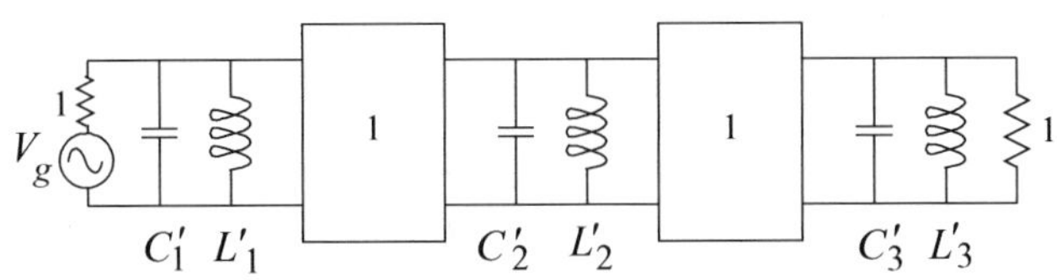

$$C_1' = 1356.6272 \text{ pF} = C_3' \quad L_1' = 0.018718 \text{ nH} = L_3'$$
$$C_2' = 1757.435 \text{ pF} \quad L_2' = 0.014444 \text{ nH}$$

Figure 7-19 Step 3. Bandpass filter prototype with inverters derived from the lowpass filter prototype of Figure 7-18.

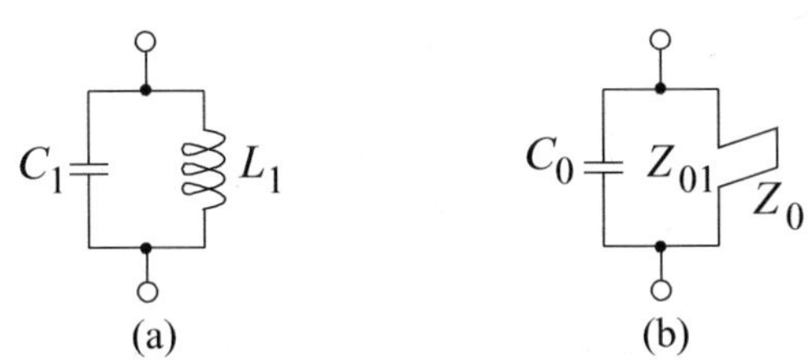

Figure 7-20 Resonator equivalence: (a) lumped resonator; and (b) mixed lumped-distributed resonator containing a resonant short-circuited stub of characteristic impedance Z_0.

with frequency derivative at resonance of

$$\frac{dY_{\text{in}}}{d\omega} = \frac{j\left(\omega^2 LC + 1\right)}{\omega^2 L} \, . \tag{7.61}$$

Similarly for the lumped-distributed element in Figure 7-20(b) the input admittance is

$$Y'_{\text{in}} = \frac{j\left(\omega C_o Z_o \tan\left(\frac{\pi}{2}\frac{\omega}{\omega_r}\right) - 1\right)}{Z_o \tan\left(\frac{\pi}{2}\frac{\omega}{\omega_r}\right)} \tag{7.62}$$

where ω_r is the resonant frequency. The frequency derivative at resonance is

$$\frac{dY'_{\text{in}}}{d\omega} = j\frac{1}{2}\frac{\left(2C_o Z_o \left(\tan\left(\frac{\pi}{2}\frac{\omega}{\omega_r}\right)\right)^2 \omega\omega_r + \pi + \pi\left(\tan\left(\frac{\pi}{2}\frac{\omega}{\omega_r}\right)\right)^2\right)}{Z_o \left(\tan\left(\frac{\pi}{2}\frac{\omega}{\omega_r}\right)\right)^2 \omega_r} \, . \tag{7.63}$$

For the lumped-element and distributed filters to be identical at the center frequency and approximating each other at nearby frequencies, Equations (7.60) and (7.62) are equated:

$$Y_{\text{in}} = Y'_{\text{in}} \tag{7.64}$$

and Equations (7.61) and (7.63) are also equated:

$$\frac{dY_{\text{in}}}{d\omega} = \frac{dY'_{\text{in}}}{d\omega} \tag{7.65}$$

Before performing this step the system impedance is scaled from 1 to 50 Ω leading to Figure 7-21.
Now solving Equations (7.64) and (7.65) for each lumped element resonator. Replacing each of the lumped resonators in Figure 7-21 by its lumped-distributed equivalent leads to the prototype shown in Figure 7-22.

The stubs in Figure 7-22 can be physically realized using transmission line segments. It is preferred that the impedances of the stubs (that is the impedances looking into the stubs) be identical as it is much easy to fabricte components that are identical. To achieve the impedance modification the method of nodal admittance matrix scaling described in Section 7.2.8 on Page 327 is used with the inverters scaled by $\sqrt{Z_{01}/Z_{02}}$ and the admittance by $C''_2//Z'_{02}$ scaled by Z_{01}/Z_{02}. This step scales each inverter impedance to 56.9102 Ω and also sets the impedance of all the shunt stubs to 7.558 Ω. The combline filter is now as shown in Figure 7-23.

Now the values of the approximate model of an inverter made of short-circuit stubs is obtained from the equivalence shown in Figure 7-24. The equivalence uses the ABCD parameters of the inverter, Equation 7.33, and the common two-ports shown in Table 7-1. Also the input impedance

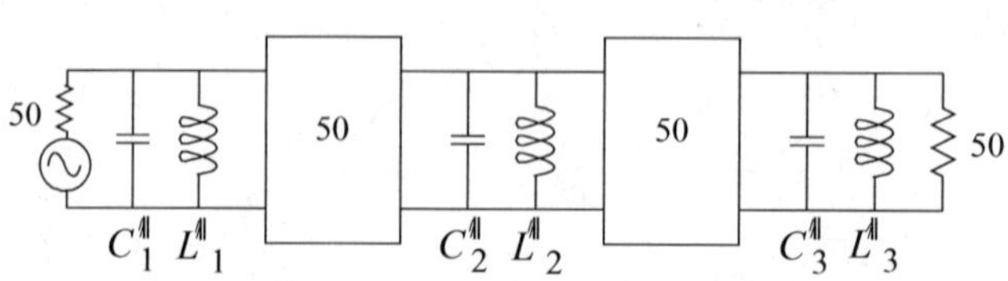

$C^{\sharp}_1 = 27.1325$ pF $= C^{\sharp}_3$ $L^{\sharp}_1 = 0.9359$ nH $= L^{\sharp}_3$
$C^{\sharp}_2 = 35.1487$ pF $L^{\sharp}_2 = 0.7224$ nH

Figure 7-21 Step 4. Bandpass filter approximation where the inverters are now impedance inverters.

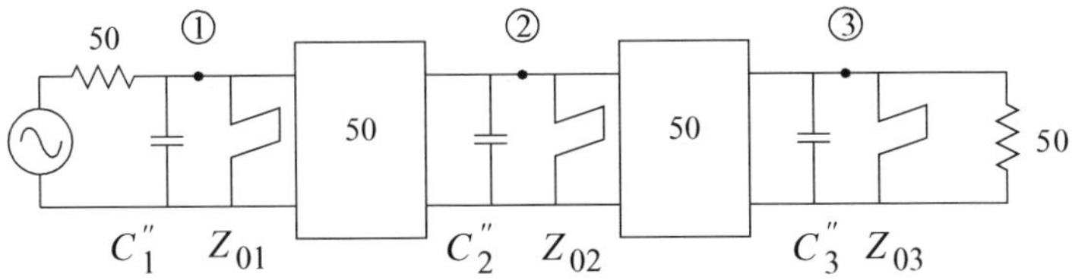

$$C_1'' = 21.123 \text{ pF} = C_3'' \quad C_2'' = 27.363 \text{ pF}$$
$$Z_{01} = 7.558\ \Omega = Z_{03} \quad Z_{02} = 5.834\ \Omega$$

Figure 7-22 Step 5. Bandpass combline filter with impedance inverters. The transmission line stubs presenting impedances Z_{01}, Z_{02}, and Z_{03}.

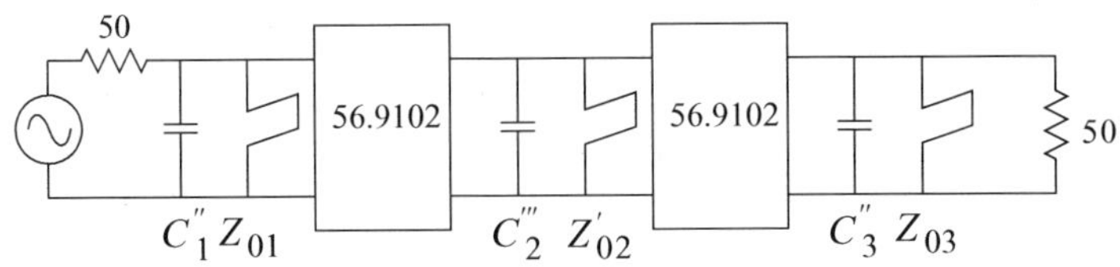

$$C_2''' = 21.1204 \text{ pF} \quad \text{and} \quad Z_{02}' = 7.558\ \Omega = Z_{01} = Z_{03}$$

Figure 7-23 Step 6. Bandpass combline filter with impedance inverters.

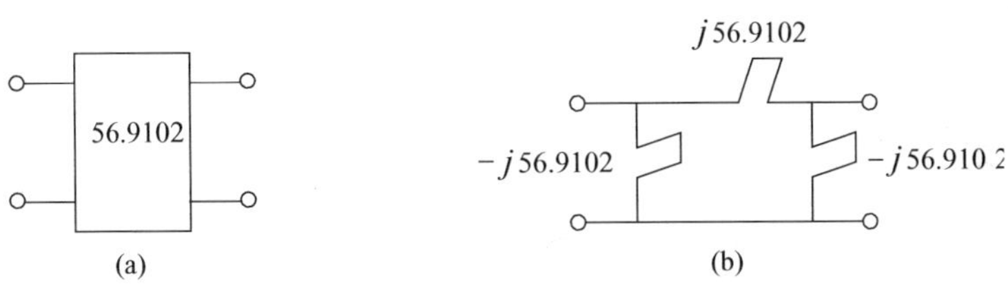

Figure 7-24 Inverter translation used with the combline filter. The $56.91\ \Omega$ impedance inverter in (a) is implemented in (b) using short-circuited stubs.

Table 7-1 ABCD parameters of several two-ports.

Series impedance, z	$A = 1 \quad B = z$ $C = 0 \quad D = 1$
Shunt admittance, y	$A = 1 \quad B = 0$ $C = y \quad D = 1$

of a short-circuited transmission line of electrical length θ is $Z_0\tan(\theta)$. So the equivalence shown in Figure 7-24 using ABCD parameters is mathematically as follows:

$$\begin{aligned} T &= \begin{bmatrix} 0 & \frac{J}{\mathcal{J}} \\ J\mathcal{J} & 0 \end{bmatrix} \\ &= \begin{bmatrix} 1 & 0 \\ \dfrac{-1}{[J Z_o \tan(\theta)]} & 1 \end{bmatrix} \begin{bmatrix} 1 & J Z_o \tan(\theta) \\ 0 & 1 \end{bmatrix} \begin{bmatrix} 1 & 0 \\ \dfrac{-1}{[J Z_o \tan(\theta)]} & 1 \end{bmatrix} \\ &= \begin{bmatrix} 1 & J Z_o \tan(\theta) \\ \dfrac{-1}{[J Z_o \tan(\theta)]} & 0 \end{bmatrix} \begin{bmatrix} 1 & 0 \\ \dfrac{-1}{[J Z_o \tan(\theta)]} & 1 \end{bmatrix} \\ &= \begin{bmatrix} 0 & J Z_o \tan(\theta) \\ \dfrac{J}{[Z_o \tan(\theta)]} & 0 \end{bmatrix}. \end{aligned} \tag{7.66}$$

Thus

$$\mathcal{J} = \frac{1}{(Z_0 \tan(\theta))} \tag{7.67}$$

and so the characteristic impedance Z_0 of each short-circuited stub is

$$Z_0 = \frac{1}{\mathcal{J}\tan(\theta)} = \frac{1}{\mathcal{J}\tan\left(\frac{\pi}{2}\frac{f}{f_r}\right)}. \tag{7.68}$$

By selecting the resonant frequency f_r of the resonator to be twice that of f_o then the the elements of the subnetwork in Figure 7-24(b) have the parameters:

$$\mathcal{J} = \frac{1}{K} = 56.9102\ \Omega = 17.5715\ \text{mS} \tag{7.69}$$

$$Z_0 = \frac{1}{\mathcal{J}\tan(\theta)} = \frac{K}{\tan(\theta)} = \left.\frac{K}{\tan\left(\frac{\pi}{2}\frac{f}{f_r}\right)}\right|_{f=f_0} = 56.9102\ \Omega \tag{7.70}$$

Finally, the prototype in Figure 7-25 is developed. Now this circuit will be scaled to a different impedance level to raise the characteristic impedances of the short-circuited stubs. We see that the middle short-circuited stub can be raised to 80 Ω if the system impedance is raised from 50 to 388.6681 Ω. This leads to the element values of the bandpass circuit of Figure 7-26.

For the filter to operate in a 50-Ω system, input and output impedance inverters are required to scale the source and load back to 50 Ω. The resulting circuit is shown in Figure 7-27 with input and output inverters of impedance $\sqrt{50 \times 388.61} = 139.404\ \Omega$. Note that inverters with an impedance of $-139.404\ \Omega$ would work just as well.

7.3.3 Realization of the Input/Output Inverters

At this stage the design 7-27 where the inverters at the input and output remain to be synthesized. Focusing on just the output inverter with a load R_L, as shown in Figure 7-28, it will be shown here

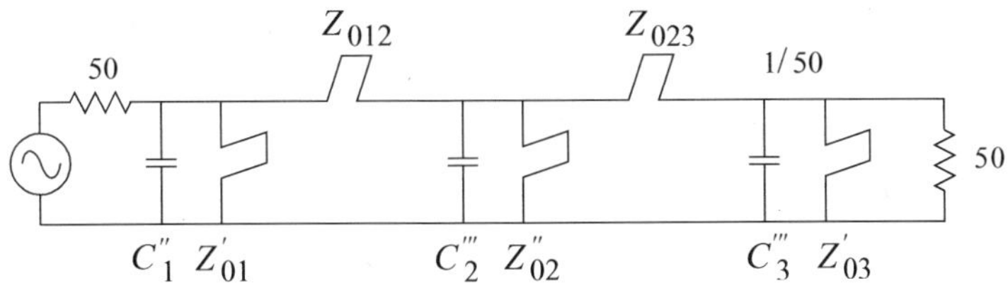

$$
\begin{aligned}
&C_1'' = 21.1215 \text{ pF} = C_3'' \quad C_2''' = 21.1204 \text{ pF} \quad Z_{02}'' = 10.2915\ \Omega \\
&Z_{01}' = 8.7161\ \Omega = Z_{03}' \quad Z_{012} = 56.9102\ \Omega = Z_{023}
\end{aligned}
\tag{7.71}
$$

Figure 7-25 Step 7: Bandpass filter approximation.

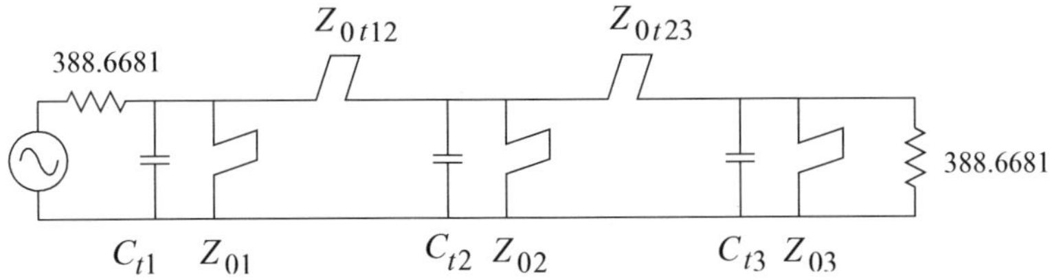

$$
\begin{aligned}
&C_{t1} = 2.7171 \text{ pF} = C_{t3} \quad C_{t2} = 2.7171 \text{ pF} \\
&Z_{0t1} = 67.7535\ \Omega = Z_{0t3} \quad Z_{0t2}'' = 80\ \Omega \quad Z_{0t12} = 442.3836\ \Omega = Z_{0t23}
\end{aligned}
$$

Figure 7-26 Step 8: Bandpass filter approximation.

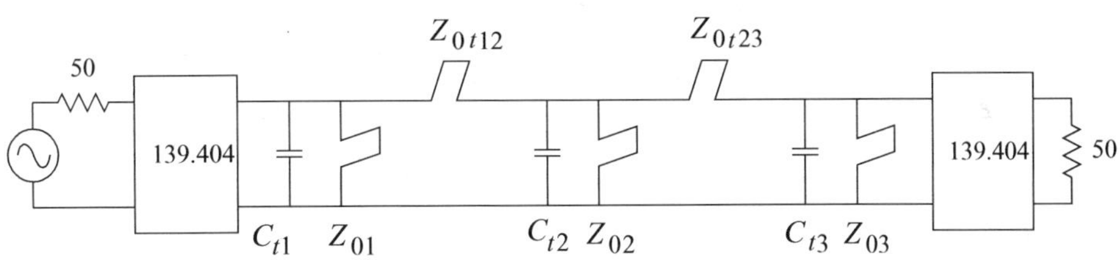

Figure 7-27 Bandpass filter approximation.

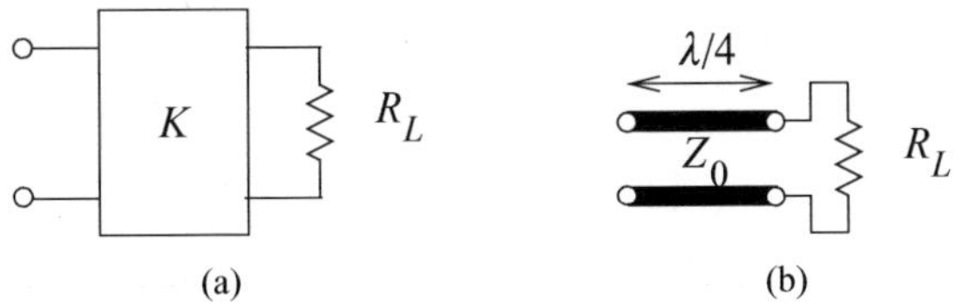

Figure 7-28 Inverter equivalence: (a) terminated impedance inverter; and (b) its equivalent transmission line form where $Z_0 = K$. The equivalence is exact when at the frequency where the transmission line is a quarter-wavelength long.

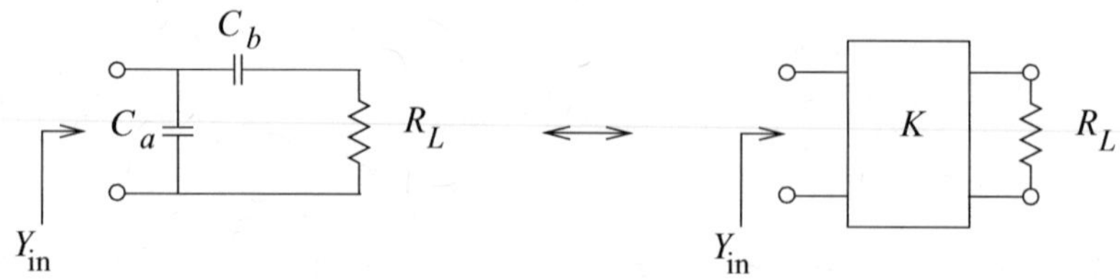

Figure 7-29 Equivalence of an inverter and a capacitor section. Note that one of the capacitors must be negative.

how this can be realized using the capacitive network of Figure 7-29. The input admittance of the inverter plus load resistor of Figure 7-28, is

$$Y_{in} = \frac{R_L}{K^2} \tag{7.72}$$

and this must be equated to the input admittance of the circuit in Figure 7-29 where $K = 139.404\,\Omega$. The input admittance of the circuit in Figure 7-29 is

$$Y_{in} = sC_a + \frac{1}{\frac{1}{sC_b} + R_L} \tag{7.73}$$

Thus, equating the real parts of Equations (7.72) and (7.73):

$$\Re(Y_{in}) = \frac{R_L\omega^2 C_b^2}{R_L^2\omega^2 C_b^2 + 1} = \frac{R_L}{(Z_0)^2} \tag{7.74}$$

and equating the imaginary parts yields

$$\Im(Y_{in}) = \frac{\omega\left(C_a\omega^2 C_b^2 R_L^2 + C_a + C_b\right)}{\omega^2 C_b^2 R_L^2 + 1} = 0. \tag{7.75}$$

Solving Equations (7.74) and (7.75) enables us to the inverters shown in Figure 7-27 to be replaced by the capacitive networks as shown in Figure 7-30. This capacitive network here is not a general replacement of an inverter. It only works when the inverter is terminated in a resistor. Note that C_a is negative and this is absorbed in the first and last resonators so that the final lumped element realization is as shown in Figure 7-31.

The prototype of Figure 7-31 can be directly translated into the combline bandpass filter, see Figures 7-14 and 7-15. The next step in physical realization including determining the lengths and spacing of the coupled lines begins by determining the even- and odd-mode impedances as

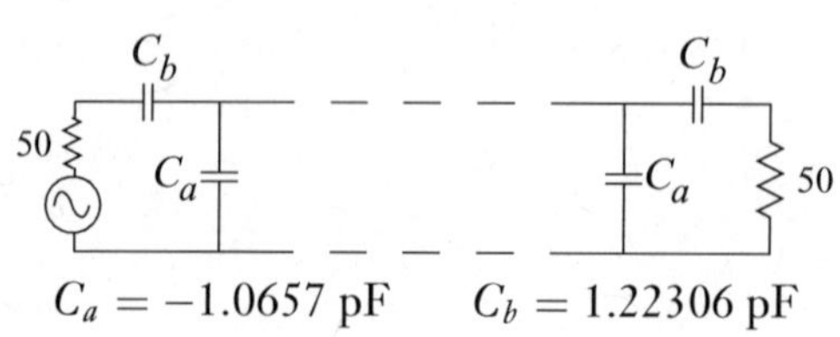

Figure 7-30 The external inverters of the prototype shown in Figure 7-27 replaced by capacitive networks.

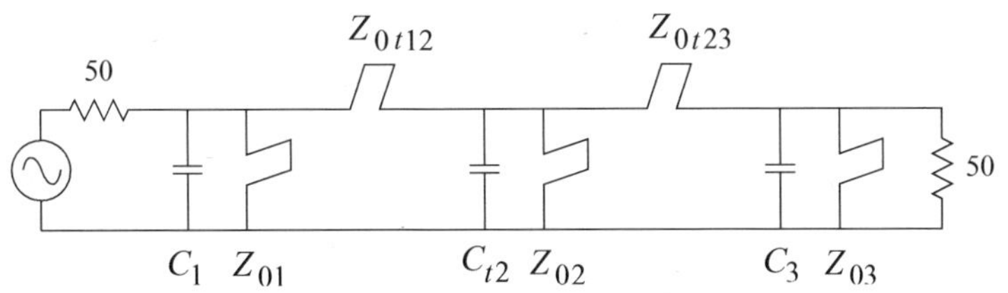

$C_1 = C_a + C_{t1} = 1.6515 \text{ pF} = C_3$ $C_{t2} = 2.7171 \text{ pF}$
$Z_{0t1} = 67.7535\ \Omega = Z_{0t3}$ $Z''_{0t2} = 80\ \Omega$ $Z_{0t12} = 442.3836\ \Omega = Z_{0t23}$

Figure 7-31 The bandpass filter approximation combining the capacitive equivalent of the inverters with the first and last resonators in the circuit of Figure 7-27.

shown in Section 7.2.7. Formulas can be used to convert these into physical dimensions [423] or microwave CAD programs can be used. Note that the lengths of the transmission lines is $\lambda/8$. The impedance matching at the input and output lines can take a variety of forms including the tapped lines shown in Figure 7-14. A final design using this structure is described in a paper by Nath *et al.* [418]. In the next section a tunable combline filter is described that uses closely coupled line for input and output impedance matching.

7.4 RECONFIGURABLE BANDPASS FILTER

This section focuses on the design of reconfigurable filters that alter their transition from passband to stopband by selectively moving finite transmission zero(s) from below or above the passband to the other side. By so doing the near-in rejection of the filter can be varied appropriately. This functionality is achieved here solely through the use of tunable capacitors. As an example, an RF multifunctional system utilizing a diplexer made of a pair of reconfigurable bandpass filters will be able to reconfigure its transmission zeros to control its transmit channel receive channel isolation level or to simply notch out-of-band interferers. Filters with asymmetric frequency responses are therefore favored to those with symmetric characteristics to minimize the number of resonator sections thus lowering the in-band insertion loss while maintaining a reasonable level of signal rejection. Reconfigurable filters provides additional functionality to systems that are required to adapt to suit their RF environment.

7.4.1 Candidate Bandpass Filter Prototypes

This section highlights some of the underlying difficulties encountered in implementing tunable and reconfigurable planar filters. Figure 7-32(a) shows the cascaded trisection (CT) [366,415,419,420] commonly used in filter prototypes to realize a single imaginary-axis transmission zero. If the cross-coupling inverter has an opposite sign to that of the main line impedance inverters, a transmission zero is placed in the lower stopband. Alternatively, if the inverters have the same sign, an upper stopband zero is realized. This implies that a tunable phase shifter in place of the cross-coupling inverter would result in a reconfigurable filter architecture although with an increase in overall design complexity. An alternative approximate equivalent to the CT suited for combline realization is shown in Figure 7-32(b). In a planar realization, the cross-coupling element may be directly realized in mixed lumped-distributed form [419] provided that a tolerable deterioration in stopband performance is accepted. Thus, another possible reconfigurable filter architecture is to use an active switch to interchange between the capacitive and inductive cross-couplings. Usually passband tunability is required in addition to reconfigurability and it is

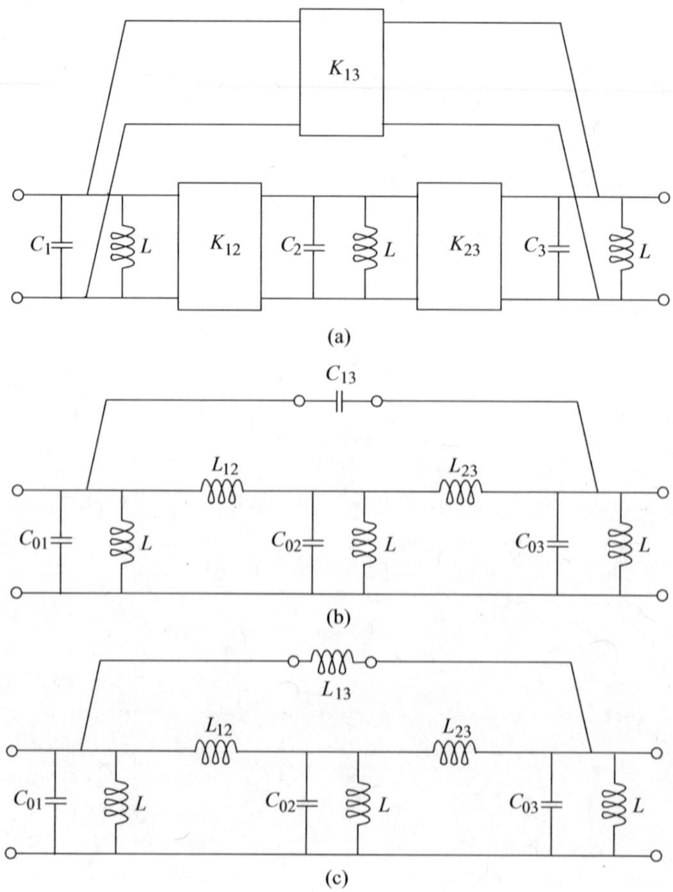

Figure 7-32 A bandpass cascaded trisection (CT) realizing a single transmission zero either in the lower or upper stopband: (a) with interresonator inverter couplings; (b) with inductive interresonator coupling and capacitive cross-coupling; and (c) with inductive interresonator coupling and inductive cross-coupling.

very difficult to design tunable inductors. Also the need for a switch is not desirable and the total number of switches required in a filter would be equal to the number of transmission zeros to be realized. A better approach is to achieve reconfigurability by solely using tunable capacitors and so an alternative network section is required. A suitable network is shown in Figure 7-33. This section has been implemented in the box configuration [414] for base station filters and is capable of realizing a single imaginary-axis transmission zero either in the lower or upper side of the passband. Upon examining the transmission transfer function of this network section, it can be shown that (with some return loss degradation) by interchanging the values of the second and fourth capacitors the location of the transmission zero can be reconfigured from the upper to lower stopband and vice-versa [414]. Consequently a bandpass filter made of this section has been proposed for complementary transmit/receive diplexers [414] where the bandwidths of the two channels are identical. The key point is that the filter can be reconfigured using only tunable capacitors [414]. The only disadvantage of the network section of Figure 7-33 is its complexity. It is of degree 4 and post-tuning and aligning of its response can be problematic

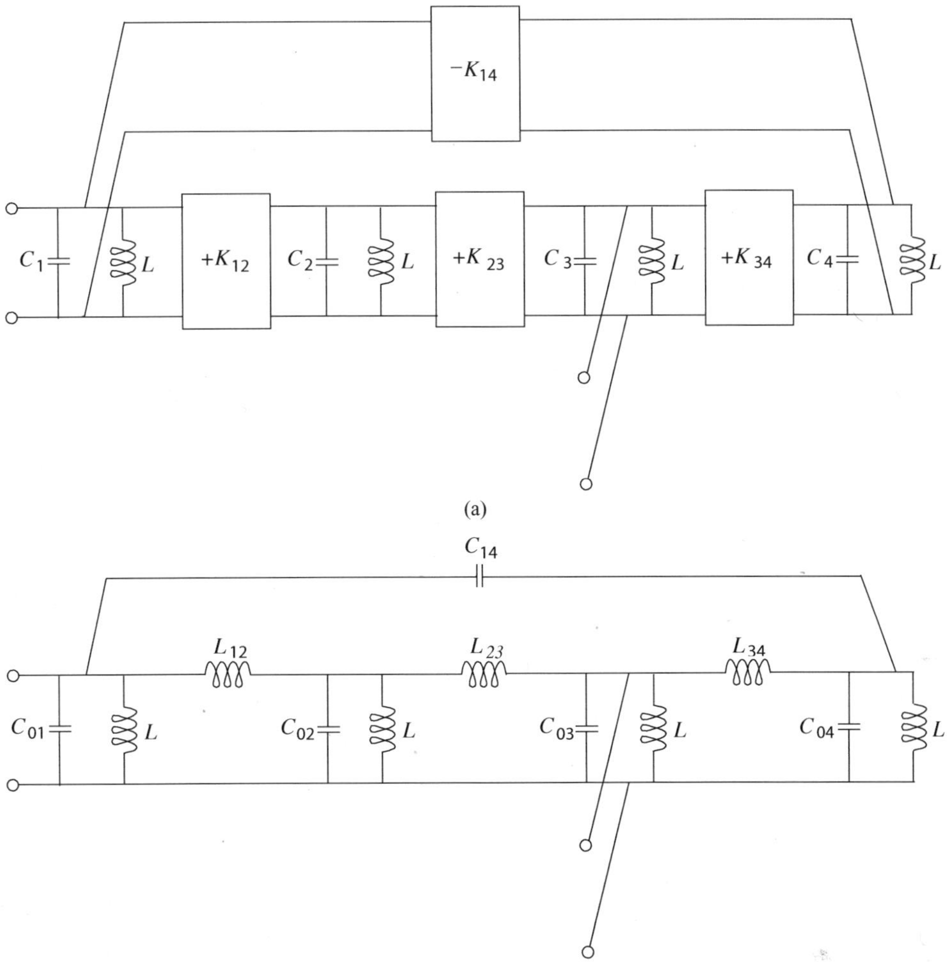

Figure 7-33 A bandpass fourth-order network section realizing a single transmission zero either in the lower or upper stopband: (a) with inverters; and (b) with inductive interresonator coupling and capacitive cross-coupling.

for planar realizations. In [414] the box-configuration was realized using coaxial resonators where tuning of the interresonator couplings was done manually. In the next section an optimum network section realizing a single imaginary-axis transmission zero is discussed. It will be shown that it is easily realizable in combline with a zero that may be re-positioned using tunable capacitors.

7.4.2 Reconfigurable Bandpass Filter Prototype

Consider a pair of resonators coupled by an ideal inverter as shown in Figure 7-34. An ideal impedance inverter has many approximate circuit equivalents [366] among which are the π models with simple elements having a pole at zero or infinity. A more general admittance inverter that has

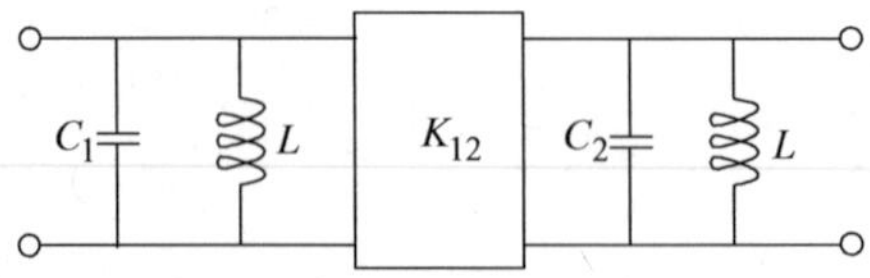

Figure 7-34 A pair of bandpass resonators coupled by an impedance inverter.

an attenuation pole at a finite frequency [420] rather than at zero or infinity is shown in Figure 7-35. This inverter has an attenuation pole at

$$f_p = \pm \frac{1}{2\pi\sqrt{L_p C_p}} \tag{7.76}$$

and the impedance of the inverter at mid-band is

$$K_{12} = \left(\frac{2\pi f_o L_p}{1 - 4\pi^2 f_o^2 L_p C_p} \right) \tag{7.77}$$

where f_o is the center frequency.

Substituting Equation (7.76) into (7.77) gives

$$K_{12} = \left(\frac{2\pi f_o L_p}{1 - \left(\frac{f_o}{f_p}\right)^2} \right) \tag{7.78}$$

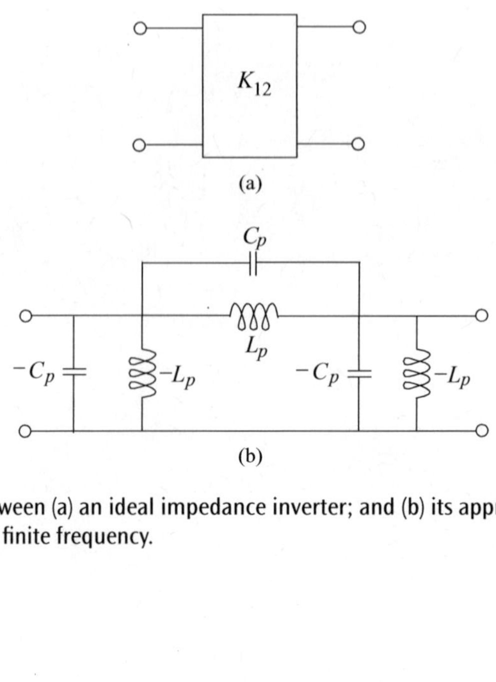

Figure 7-35 Equivalence between (a) an ideal impedance inverter; and (b) its approximate model with an attenuation pole at a finite frequency.

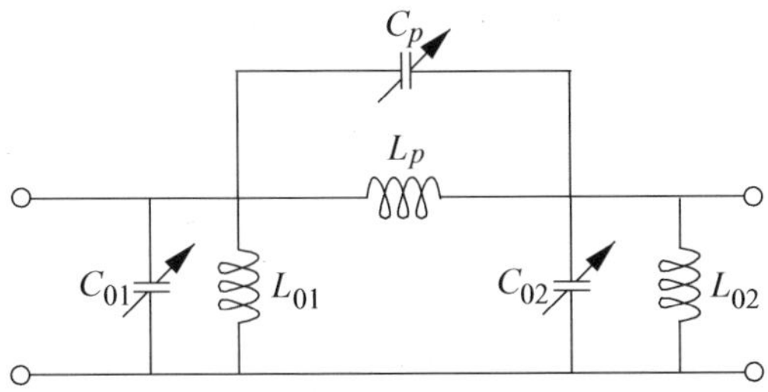

Figure 7-36 The pair of bandpass resonators of Figure 7-34 with mixed-type coupling.

or

$$K_{12} = \left(\frac{\dfrac{f_o}{(2\pi C_p f_o^2)}}{1 - \left(\dfrac{f_o}{f_p}\right)^2} \right). \tag{7.79}$$

It is seen from Equations (7.78) and (7.79) that if the inverter is positive then the pole lies above the passband and if negative, the pole is below the passband [420]. This property is very useful and is key to the design of reconfigurable filters. The negative shunt elements of the approximate inverter can be absorbed into their adjacent neighbors resulting in the mixed coupled circuit of Figure 7-36. This second-order circuit can be directly realized in combline form with a series coupling capacitor. If the series and shunt pair of capacitors were tunable then it would be possible to position the finite zero either in the lower or upper stopband. Thus the design of a reconfigurable filter commences with a prototype representing the filter with transmission zeros in the initial state, and then by tuning relevant capacitors specific zeros may be reconfigured accordingly. A simple design method [420] is to start from an n th-order prototype with resonators coupled together by ideal inverters and transform, using Equation (7.76) to (7.79), as many inverters as required to their corresponding approximates with attenuation poles at finite frequencies. Another more exact method is to use cascade synthesis in the transformed variable [394, 395] to synthesize a network prototype with a specific set of transmission zeros. The general form of a prototype with $(n-1)$ transmission zeros is shown in Figure 7-37 and had been used to realize elliptic function combline filters [421].

A key insight is that for a specific bandwidth and return loss level, the reconfigured transmission zeros are restricted to certain locations due to the interresonator mixed couplings of the filter that must be the same in both states. This dictates the achievable reconfigured locations of the

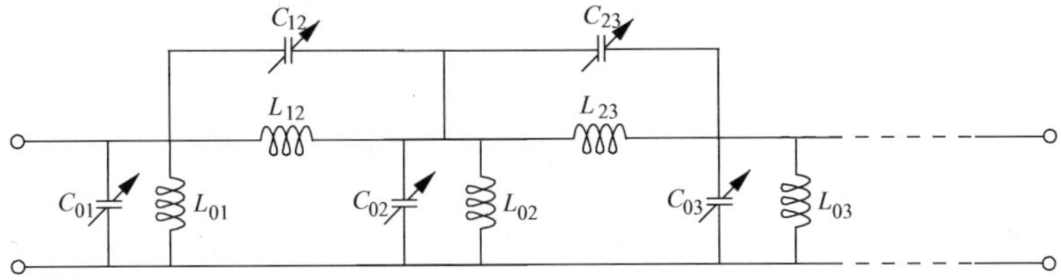

Figure 7-37 The n'th-order filter prototype realizing arbitrarily located transmission zeros with one zero at DC, one zero at infinity, and $(n-1)$ zeros at finite frequencies.

transmission zeros after perturbing the element values of the filter. This is a vital point that requires careful consideration in the design stage in order to satisfy a certain selectivity specification. In the next section a reconfigurable filter is implemented.

7.4.3 Filter Design and Implementation

A third-order narrowband receive filter is constructed on a printed circuit board (PCB) as an example of an approach to designing a reconfigurable filter. The PCB has a substrate thickness of 62 mils (0.157 mm), relative dielectric constant of 4.7, and loss tangent of 0.016. Using the procedure of [394], the characteristic polynomial, $K(S)$ of the filter is constructed with one zero at DC, three zeros at infinity, one finite frequency zero at 735.6 MHz, a center frequency of $f_o = 850$ MHz, bandwidth of 85 MHz, (10% of 850 MHz), and return loss (R.L.) of 20 dB. This gives

$$K(S) = \frac{\begin{pmatrix} -0.1048970136S^6 - 0.194888590S^4 \\ -0.1201173535S^2 - 0.02456206793 \end{pmatrix}}{(0.001006846557S^3 + 0.0004279173264S)}. \tag{7.80}$$

The input reflection coefficient is then evaluated from

$$S_{11}(S) = \frac{K^2(S)}{1 + K^2(S)} \tag{7.81}$$

leading to

$$S_{11}(S) = \frac{\begin{pmatrix} -0.1048970136S^6 - 0.194888590S^4 \\ -0.1201173535S^2 - 0.02456206793 \end{pmatrix}}{\begin{pmatrix} 0.1048970136S^6 + 0.02586220726S^5 + 0.1980767352S^4 \\ +0.03190556160S^3 + .1220119073S^2 + 0.009656678221S \\ +0.02456206793 \end{pmatrix}} \tag{7.82}$$

from which the input impedance is found in a 1-Ω system, using

$$Z_{\text{in}}(S) = \frac{1 + S_{11}(S)}{1 - S_{11}(S)}. \tag{7.83}$$

In Equation (7.80) S is the Richards transform variable defined as, $S = j\tan\left(\pi/2\left(f/f_q\right)\right)$ with f the real frequency variable, and f_q the frequency at which the commensurate lines are quarter wavelength long. For this design, f_q is chosen to be about 2000 MHz, i.e., the resonators are roughly 45° long at the resonant frequency, f_o.

A possible prototype obtained by direct synthesis of the input impedance is shown in Figure 7-38(a) where the short- and open-circuited stubs represent inductors and capacitors respectively at the commensurate frequency, f_q. Through application of Norton's transformation [422] the open-circuited stubs can be distributed throughout the network. Subsequently all the open-circuited stubs have been equated to lumped capacitors at the filter center frequency, f_o, and by appropriately scaling the network by a pair of impedance inverters, the transformed prototype of Figure 7-38(b) in a 50-Ω system results.

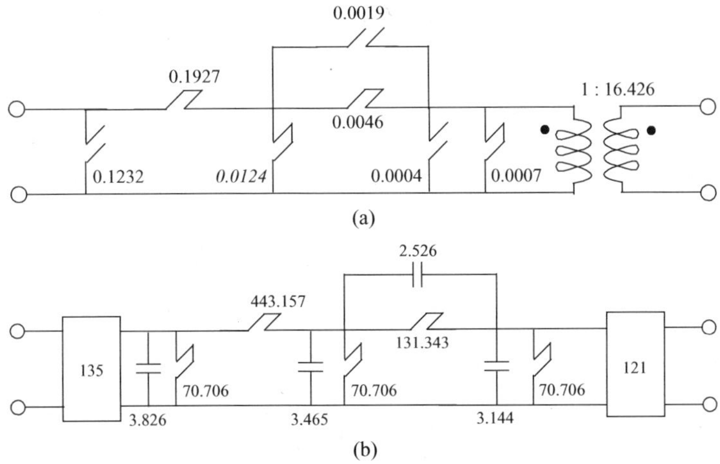

Figure 7-38 Distributed bandpass filter prototype with one zero at DC, three zeros at infinity, and one zero at 735.6 MHz: (a) after direct synthesis in a 1-Ω system; (b) after application of suitable circuit transformations and scaling in a 50-Ω system. (All capacitor values are in pF)

From Figure 7-38(b) the theoretical capacitor values required for the transmission zero located at 735.6 MHz are

$$\begin{aligned} C_{01} &= 3.826 \text{ pF}, \quad C_{02} = 3.465 \text{ pF} \\ C_{23} &= 2.526 \text{ pF}, \quad C_{03} = 3.144 \text{ pF}. \end{aligned} \tag{7.84}$$

The value of the impedance inverter with an attenuation pole at 735.6 MHz may now be evaluated from Equations (7.76) and (7.78). This is done after approximating the series short-circuited stub of impedance Z_o (of the mixed coupling branch) by a lumped inductor, L_p, at resonance using

$$jZ_o \tan\left(\frac{\pi}{2}\left(\frac{f_o}{f_q}\right)\right) = j2\pi f_o L_p. \tag{7.85}$$

This yields a value of L_p equal to 19.387 nH. However this step causes a slight shift in the location of the transmission zero which may be compensated for by fine-tuning the parallel capacitor, C_{23}, from a value of 2.526 pF to 2.414 pF. By referring to Figure 7-35(b) C_{23} becomes C_p. From Equation (7.76) f_p is 735.6 MHz and from Equation (7.78) K_{12} is −308.872 Ω. Now in order to reconfigure the zero to a location in the upper stopband, the absolute value of the inverter must remain the same but its phase must be altered from 90o to 270o, i.e., effectively changing sign. The new zero location, f_p is determined from Equation (7.78) as follows

$$308.872 = \left(\frac{2\pi\,(850\text{MHz})\,(19.387\text{nH})}{1 - \left(\dfrac{850\text{MHz}}{f_p}\right)^2}\right) \tag{7.86}$$

to give a value of 1042.517 MHz. This is essentially achieved by a new capacitance value of C_p (evaluated from Equation (7.76) at the new frequency, f_p in MHz) equal to 1.202 pF. With the knowledge of the new value of C_p the neighboring capacitors (C_{02}, C_{03}) may now be readily found.

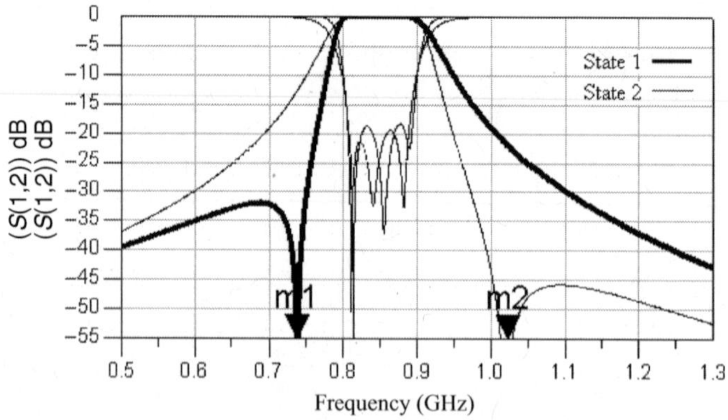

Figure 7-39 Simulated lossless responses of the prototype of Figure 7-38(b) in two states for the same interresonator couplings and return loss level. (m1: 735.6 MHz; m2: 1016.25 MHz)

With very little circuit optimization the new zero is finally located at 1016.25 MHz resulting in the following capacitor values

$$\begin{aligned} C_{01} &= 3.886 \text{ pF}, \quad C_{02} = 4.317 \text{ pF} \\ C_{23} &= 1.162 \text{ pF}, \quad C_{03} = 3.718 \text{ pF}. \end{aligned} \tag{7.87}$$

It is seen from Equation (7.87) that the value of C_{01} remained virtually unchanged implying that a total of three capacitors are only required to be tuned to reconfigure the transmission zero. The original (State 1) and reconfigured (State 2) responses are shown in Figure 7-39. At this point a pair of CT filters were synthesized in a 50-Ω system realizing the same transmission zeros as above (i.e., 735.6 and 1016.25 MHz). The element values corresponding to Figure 7-32(b) for a lower passband transmission zero are

$$\begin{aligned} &C_{01} = C_{03} = 32.931 \text{ pF}, \quad C_{02} = 34.332 \text{ pF} \\ &C_{13} = 1.84 \text{ pF}, L = 1.2172 \text{ nH}, \quad L_{12} = L_{23} = 10.609 \text{ nH} \end{aligned} \tag{7.88}$$

and those of upper passband transmission zero are

$$\begin{aligned} &C_{01} = C_{03} = 29.171 \text{ pF}, \quad C_{02} = 36.682 \text{ pF} \\ &L_{13} = 43.875 \text{ nH}, L = 1.2172 \text{ nH}, L_{12} = L_{23} = 10.609 \text{ nH}. \end{aligned} \tag{7.89}$$

Equations (7.88) and (7.89) suggest that all three shunt capacitors of a CT filter must be tuned. This is in addition to interchanging between capacitive and inductive cross-coupling elements to fully reconfigure the transmission zero highlighting the benefits of utilizing the filter prototype of Figure 7-38 with direct and mixed interresonator couplings.

The prototype of Figure 7-38(b) was then converted into a physical combline filter realization as shown in Figure 7-40. Using a commercial microwave CAD tool [115] the final realization shown in Figure 7-41 was developed. Figure 7-41 depicts the layout of the combline filter with even- and odd-mode impedances of each coupled-line section. All of the capacitors of the combline filter were then split into two and connected to high impedance quarter wavelength bias lines loaded by low impedance DC power sources. Initially high Q lumped capacitors ($Q_C = 65$) were used and the response was reconfigured by manually replacing the lumped capacitors with different values.

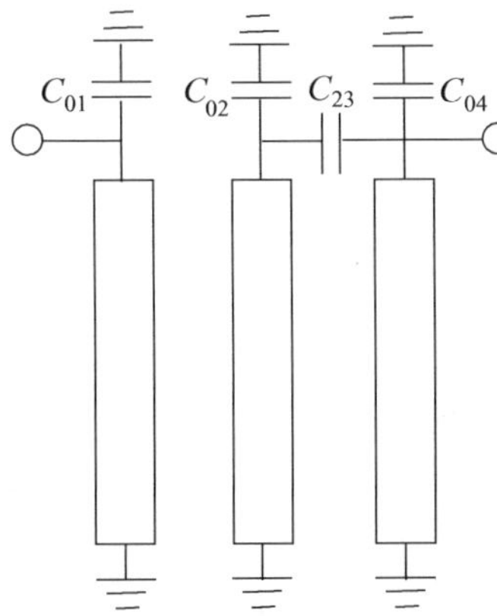

Figure 7-40 Layout of bandpass combline filter based on the prototype of Figure 7-38(b).

The practical capacitor values for the transmission zero located at 735.6 MHz are

$$\begin{aligned} C_{01} &= 4.3 \text{ pF}, \quad C_{02} = 3.3 \text{ pF} \\ C_{23} &= 3.3 \text{ pF}, \quad C_{03} = 3.9 \text{ pF} \end{aligned} \tag{7.90}$$

and for a re-positioned zero at 997 MHz (instead of 1016.25 MHz) are

$$\begin{aligned} C_{01} &= 4.4 \text{ pF}, \quad C_{02} = 4 \text{ pF} \\ C_{23} &= 1.4 \text{ pF}, \quad C_{03} = 4.4 \text{ pF}. \end{aligned} \tag{7.91}$$

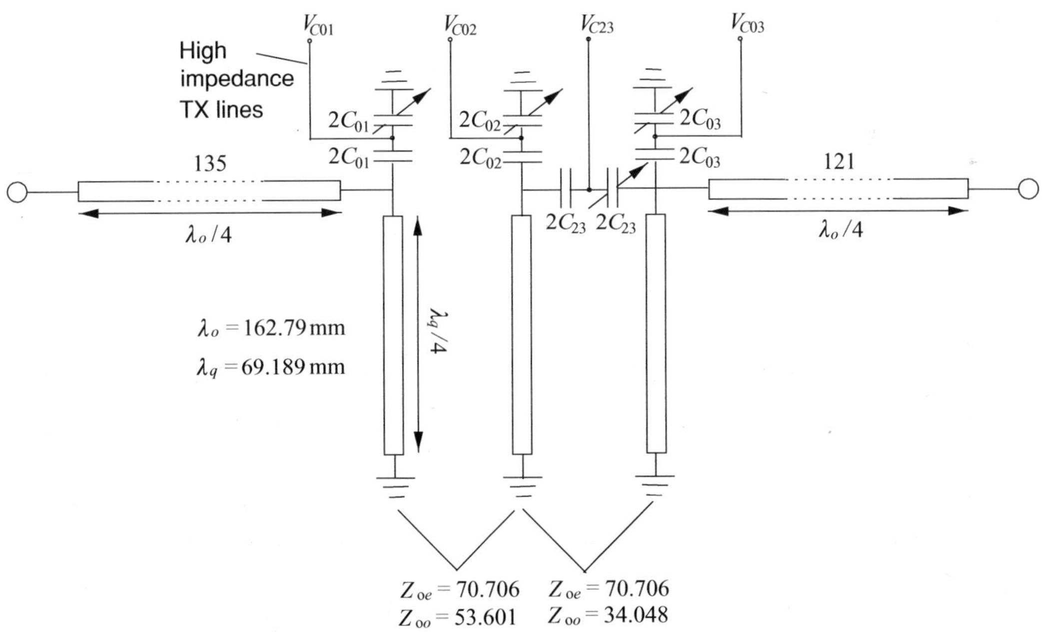

Figure 7-41 Layout of combline filter based on the prototype of Figure 7-38(b) with bias arrangement. (Half the capacitors are tunable).

These values differ slightly from their theoretical counterparts (Equations (7.84) and (7.87)) due to the microstrip vias on the PCB that were assumed ideal in the simulation and the slight offset in the re-positioned zero location in State 2. The capacitances values in Equations (7.90) and (7.91) suggest a maximum capacitance ratio for C_{23} of 2.3:1 $(3.3/1.4)$ to achieve reconfigurability.

Half of the lumped capacitors were then replaced by semiconductor varactor diodes with capacitance ratios of 10:1 [396] to fully construct the reconfigurable filter, see Figure 7-41. This was done to boost the effective Q of the series connection of capacitors. The varactor diodes offered a Q of about 12 at 850 MHz while the lumped capacitors had a Q of 65 at 850 MHz and thus the effective Q of the combination was approximately 20. Since half of the total number of capacitors were replaced by varactor diodes the maximum tuning ratio required by capacitor $2C_{23}$ calculated from Equations (7.90) and (7.91) is now 3.71:1. The measured response of the reconfigurable filter in the two states is shown in Figure 7-42. In both states the mid-band insertion loss level is below 6 dB and it is seen that the filter could be reconfigured to achieve two different selectivity specifications as required.

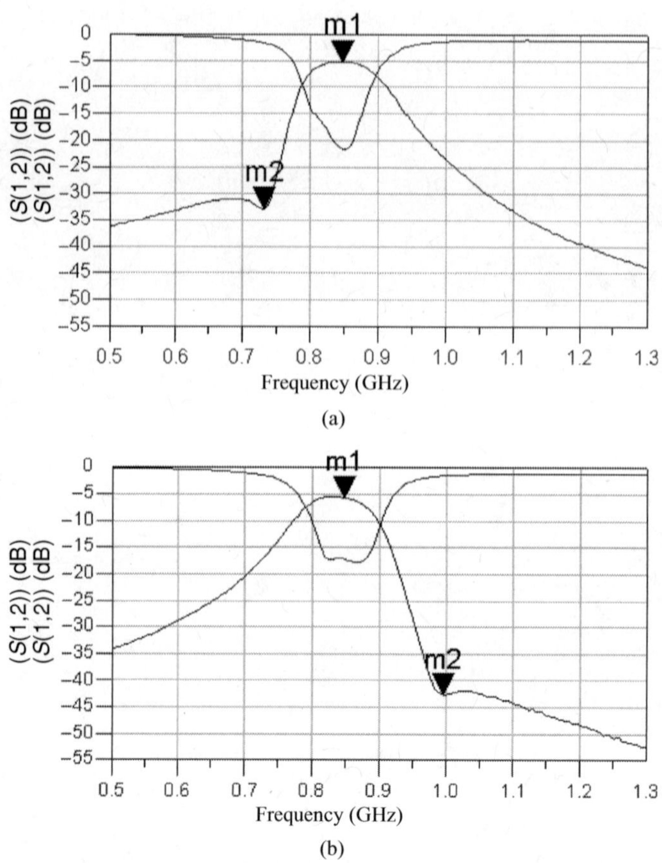

Figure 7-42 Measured scattering parameters of reconfigurable filter with a finite transmission zero: (a) in the lower stopband (m1: 850 MHz, −5.253 dB; m2: 732 MHz, −33.027 dB); and (b) in the upper stopband (m1: 850 MHz, −5.753 dB; m2: 996 MHz, −42.785 dB).

7.4.4 Summary

An asymmetric reconfigurable filter based on the traditional concepts of switching cross-coupling elements would require switching capacitive and inductive elements, or a tunable phase shifters. The inductive elements in particular must be avoided as reasonable tunable inductors are not available. This section presented the topology and design of a sub-class of reconfigurable filters that circumvent the above problem. The design theory is based on a network prototype realizing an asymmetric transfer function with possible mixed-coupling branches between adjacent resonators. Reconfiguring the transmission zeros of the filter enhances the steepness of its response on one side or the other of the passband to suit advanced filtering requirements for multifunctional systems. The synthesis approach used leads to a topology that can be realized using coupled lines in a combline configuration that supports the placement and biasing of variable capacitors. As an example a reconfigurable planar microstrip filter was developed.

7.5 TUNABLE COMBLINE FILTER REALIZATION

In this section a third-order tunable 950-MHz bandpass filter with a tunable zero is reported that tunes from 6.23 to 7.43GHz with the application of a bias voltage ranging from 0 to 40 V. The insertion loss varies from 8.4 dB to 6.3 dB and the return loss is better than 12 dB. The tunable zero above the passband that achieves enhanced stopband rejection. The filter is integrated on a ceramic substrate using resistive biasing. The bias voltage of interdigitated BST-based varactors, as used here, can be chosen to be much higher than the RF voltage so that nonlinear effects can be rendered insignificant [424]. The bandwidth requirement of tunable bandpass filters in many applications is around 500 MHZ to 1 GHz determined by the capabilities of analog to digital converters with sufficient bit-resolution (e.g., 10–12) for subsequent additional frequency selectivity through digital signal processing. The actual performance required also depends on the electromagnetic environment, in particular the level of interfering signals. Generally, however, it is important to achieve low insertion loss, high out-of-band rejection and steep transition between the passband and stop-bands. This latter requirement can require a tunable transmission zero.

7.5.1 DESIGN OF THE TUNABLE BANDPASS FILTER

The design of the filter is based on the more generic design presented in the Section 7.3 with coupled-line input and output matching. The filter layout adopted is shown in Figure 7-43. The three resonators and the coupling elements have an electrical length of 45°. This is an optimum choice of electrical length as it leads to high tuning range as well as yielding a miniaturized

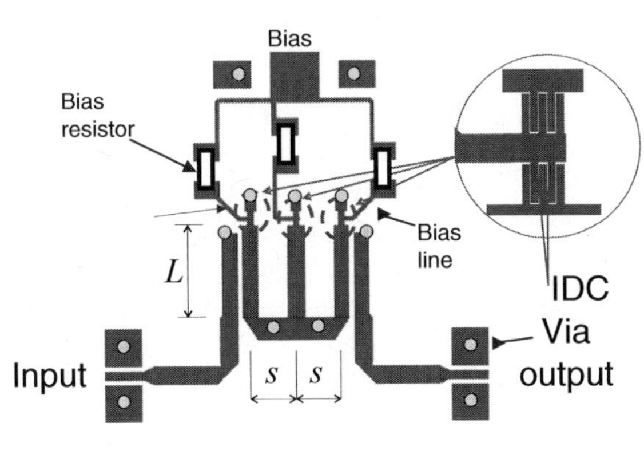

Figure 7-43 Layout of the tunable filter.

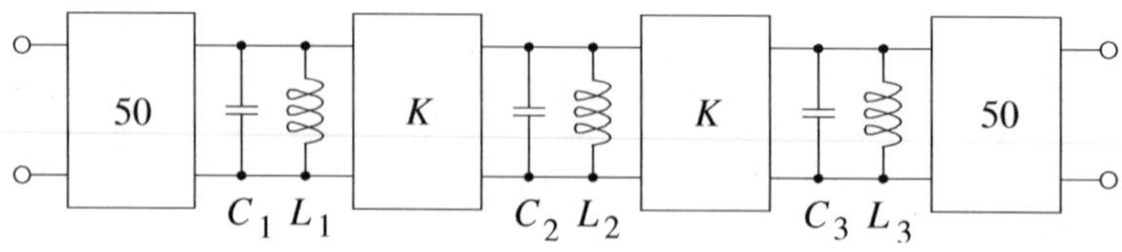

Figure 7-44 Lumped element bandpass prototype network in a 50-Ω system centered at 6.9 GHz with 10% fractional bandwidth; $C_1 = C_2 = C_3 = 4.056$ pF, $L_1 = L_2 = L_3 = 0.131$ nH, and K = 56.254 Ω.

overall circuit size. By altering the capacitance values of the loading capacitors frequency tuning is achieved. Furthermore at this electrical length the frequency tuning with capacitance change is the maximum. Each variable capacitor in the design is split into two series capacitors with the center point providing a convenient bias point at the common terminals of the pair.

The filter was designed based on the bandpass filter prototype shown in Figure 7-44. The prototype was synthesized for a center frequency of 6.9 GHz and a fractional bandwidth of 10%. From the prototype an initial estimate of the interresonator couplings was established. A polished 15 mil- (380 μm-) thick alumina substrate (from Coorstek, Golden, Colorado) with a 1 μm rms surface roughness, a dielectric constant of 9.9, and loss tangent of 0.0002 was chosen to construct the filter. One advantage of using alumina as a substrate is that its thermal expansion coefficient is closely matched to that of BST so that high temperature processing does not result in cracking of the BST thin film. Using the ADS circuit simulator [115] the electrical prototype was transformed to the physical layout shown in Figure 7-43 with resonator length $L = 1705$ μm, inter resonator spacing $S = 610$ μm, and 300μm diameter vias. Each tuning capacitor was 0.63 pF and each was realized as a pair of 1.2 pF IDC varactors in series. The IDC enabled a single photolithography step in the realization of the filter [425]. The narrow and thin (and hence high resistance) bias lines were connected through a surface-mounted 3 kΩ resistor to a bias supply. The bias connection at the top of the layout was capacitively coupled to the adjacent grounded pads by 20 pF capacitors. The inset in Figure 7-43 shows the detail of the series connected IDC with each finger being 20 μm wide and 100 μm long, with a gap between fingers of 3 μm.

7.5.2 Fabrication of the Tunable BST Filter

The first step in the fabrication process was forming the via openings which were subsequently filled with a proprietary gold-based compound compatible with subsequent thermal processing. The BST thin films were deposited on the alumina substrate using a radio frequency magnetron sputtering technique. A deposition time of 60 minutes resulted in a film thickness of 0.6 μm. Postdeposition anneal was done *ex-situ* in air at 900°C for 20 hours to crystallize and densify the BST films. Detailed deposition conditions are described elsewhere [426]. A hysteresis test was performed to confirm that the thin film was in the paraelectric phase. The crystalline structure of the film was investigated using a diffractometer with a CuK radiation source and showed a fully crystalline BST perovskite structure. Photoresist patterning was used to support subsequent metal lift-off. This proved successful in achieving the range of feature sizes ranging from 3 μm for the IDC gaps to the 5 mm overall size. A 0.3μm thick layer of silver was first deposited on the BST with deposition conditions optimized for adhesion to the BST and minimal deleterious oxidation effect at the metal/BST interface. After masking out the immediate vicinity of the IDCs this was subsequently plated up with silver and then copper to obtain good conductivity metal that was 3μm thick including the ground plane. The step-by-step process flow is as follows: (1) via drill; (2) via fill; (3) BST deposition; (4) BST annealing; (5) photographic patterning; (6) silver deposition; (7) metal lift-off; back; and (8) copper plating.

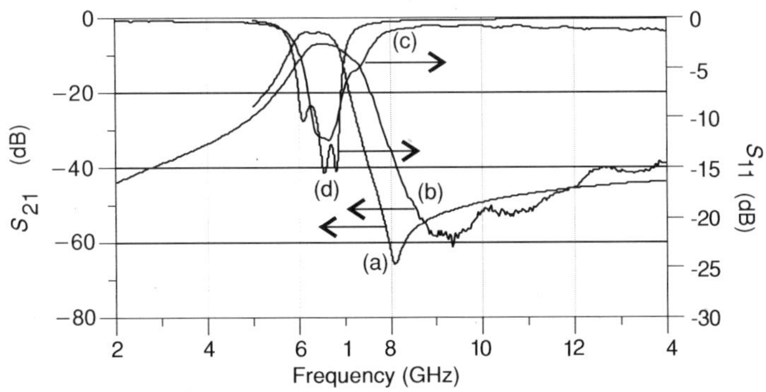

Figure 7-45 The measured S-parameters of the BST tunable filter without bias resistor and decoupling capacitors on: measured (b) and (c); simulated (a) and (d).

7.5.3 Measurement Results and Discussion

The measured capacitances of the BST varactors in the bandpass filter were 1.119, 1.152, and 1.151 pF which closely match the designed value of 1.2 pF. The corresponding loss tangent measured at 1 MHz were 0.0150, 0.0147 and 0.0144 respectively. The 2% variation of the IDC capacitance values is typical however the accuracy by which the capacitors are realized is much more variable due to the processing variability of the IDC gap width.

Microwave characterization of the filter used a vector network analyzer, 250μm pitch microcoax GSG probes from GGB Industries, and SOLT calibration. The measured and calculated response without the 3 kΩ bias resistors is shown in Figure 7-45. Here the simulated response used the measured capacitance and loss values above. The measured center frequency is 6.5 GHz with 3-dB bandwidth of 950 MHz and out-of-band rejection of 40 dB. The measured insertion loss is 6.2 dB and the measured return loss is around 10 dB. The simulated bandwidth is slightly smaller and this is attributed to the finite impedance of the bias as well as other to other parasitic effects not captured in circuit simulation.

The measured tuning response is shown in Figure 7-46 where the varactor bias is varied from 0 to 40 V in 10 V steps reducing the varactor capacitance by a factor of 1.4. The filter has a zero-bias insertion loss of 8 dB at 6.5 GHz. The center frequency of the filter was 7.4 GHz with 40 V bias providing a tunability of 19%. The insertion loss reduced to 6.3 dB at 40 V bias as the loss tangent of the varactor reduces at higher bias. The return loss at the upper end of the tuning range was better than 10 dB and the rejection at stop band is better than 40 dB for all bias voltages. Center frequencies and insertion losses at different bias voltages is given in Table 7-2.

Table 7-2 Summary of measured center frequencies and insertion loss.

Bias (V)	Center freq (GHz)	S_{21} (dB)
0	6.230	−8.457
10	6.500	−7.760
20	6.830	−7.187
30	7.100	−6.834
40	7.430	−6.316

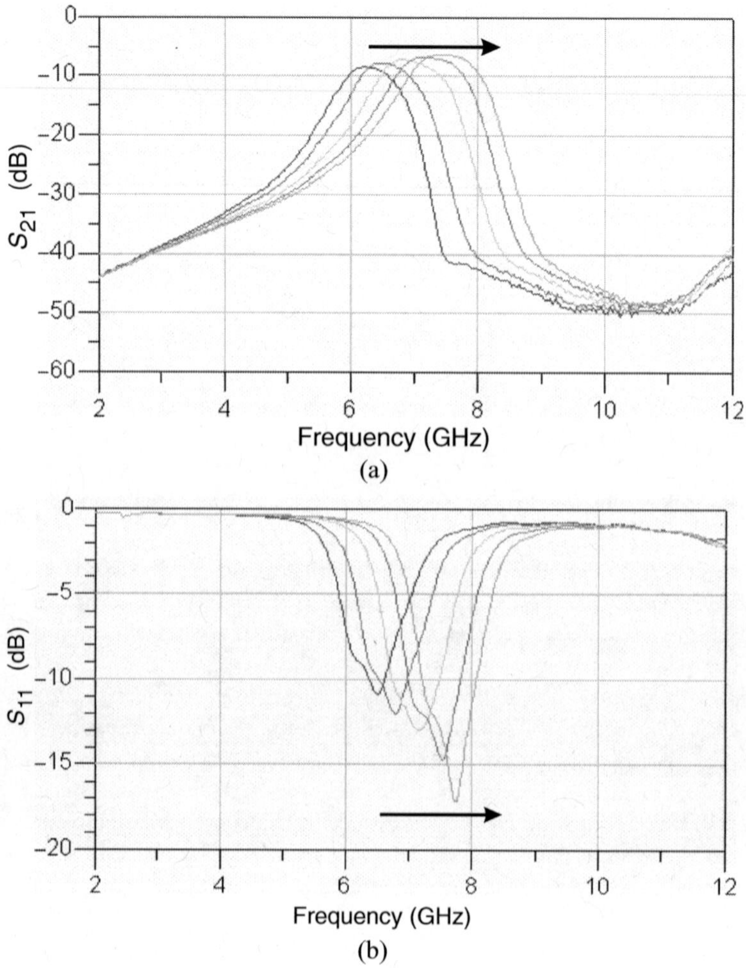

Figure 7-46 Measured S-parameters for bias from 0 to 40 V in 10 V steps: (a) S_{21}, (b) S_{11}.

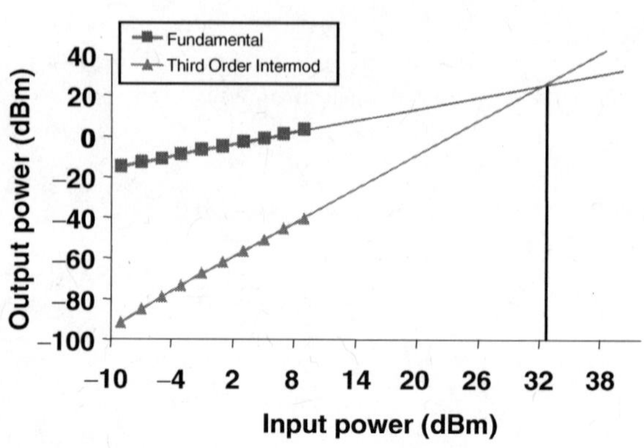

Figure 7-47 Adjacent channel power ratio (ACPR) of the filter for a CDMA 2000 signal at 0 V.

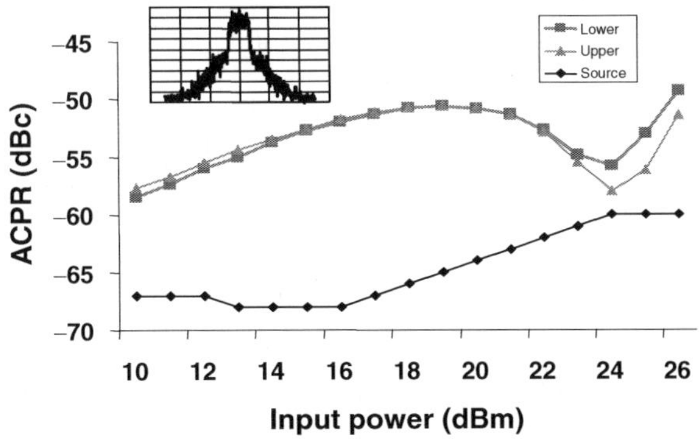

Figure 7-48 Two-tone intermodulation measurement of the filter at 0 V.

7.5.4 Nonlinear Characterization

The capacitance versus voltage characteristic of a BST varactor is nonlinear and so a tunable filter using BST varactors will also be nonlinear. In this section the nonlinear characterization of the tunable filter is done using a two-tone test and also a digitally modulated signal. The input power of the two tones was swept up to 8 dBm and the output power level of the tones and the intermodulation products was recorded. An IP3 of +32 dBm was found from extrapolation, see Figure 7-47. The filter was also characterized using a CDMA 2000 pilot signal with power levels up to 26 dBm. The ACPR was measured in a 30 kHz bandwidth at an offset of 750 kHz from the center frequency of 2.14 GHz. The ACPR was found to be better than −50 dBc, see Figure 7-48.

7.6 CONCLUSION

This chapter outlined an approach to developing electronically tunable and reconfigurable filters using ferroelectric varactors. The filters could just have easily been realized using MEMS varactors, and in the case of reconfigurable filters MEMS switches can be used. Yet again semiconductor varactors could also be used. The chapter was focused on ferroelectric varactors and showed what can be achieved and delved into the optimum system trade-offs. One of the first differences that comes apparent with tunable filters is the available Q is much lower than with filters using fixed capacitors or perhaps no capacitors at all in a totally distributed design. The limited Q, generally of 100 or less, does not enable filters of low fractional bandwidth to be developed. It was shown that filters with bandwidths of 500 MHz up to 18 GHz can be realized using a varactor with a Q of 60. This is an important bandwidth a analog-to-digital converters are available with analog bandwidths exceeding 500 MHz and 12 or more bits of resolution. This enables subsequent digital signal processing to be used to obtain the frequency selectivity required of modern radar, communication and sensor systems. The other interesting result of the limitation of the varactor Q is that the maximum RF voltage across the varactor is constrained. The voltage across the varactor in the considerations considered is proportional to the Q of the varactor so a very high Q would result in unacceptably large nonlinear effects. It was shown here that the linearity that can be obtained with ferroelectric-based tunable filters can be well within requirements. Tunable filter design technology is reasonably mature and justifies on-going research and development in tunable varactors.

Most of the work described in this chapter was performed at North Carolina State University, NCSU. Many people were involved including Professors Angus Kingon and Jon-Paul Maria. Dr. Zhiping Feng and Dr. Viera Vankova organization and all round expertise are appreciated. Many doctoral dissertations and masters theses were based on this work and the contributions of the people who wrote very full descriptions of their work and documented their inventiveness are thanked. The authors thank the contributions of Drs. Jayesh Nath, Brian Laughlin, Dipanker Ghosh. Others are working on their doctoral degrees at the time of writing continue to contribute. We thank Ms. Vrinda Haridasan, Mr. Peter Lam, James Kupferschmidt.

8

Retrodirective Array Technology

Darren S. Goshi, Kevin M. K. H. Leong, and Tatsuo Itoh

8.1 INTRODUCTION

This chapter will discuss retrodirective array (RDA) technology starting with a brief history, then covering a wide range of advanced architectures and applications. RDAs have the interesting property of automatically responding to an incoming signal's location without any prior knowledge of the direction of that source. This high-speed selftracking process makes an RDA useful for RFID (radio frequency identification) and point-to-point applications by offering an automatic maintenance of a high gain link between an interrogator and an RDA. Retrodirective systems can accelerate the detection of targets and reduce complexity of systems compared with conventional phased array-based smart antennas.

A retrodirective array is innately a transponder, as its main function is to wirelessly send a signal response upon being interrogated. Therefore RDAs are ideal for use as passive transponders for a wide range of applications. Battlefield IFF (identification friend or foe) transponders, and target or projectile surveillance detection radar for mobile vehicles, are some scenarios which would benefit from an RDA.

This chapter is organized as the following: The next section introduces the technology of RDAs and the mechanism that produces its functionality. The sections that follow are concerned with various implementations and demonstrations of RDAs with added functionality in terms of communication and sensing features.

8.2 RETRODIRECTIVE TECHNOLOGY

An RDA is a unique type of phased array where beam scanning is achieved automatically and instantaneously. It operates by responding back toward a source upon incident interrogation with a directivity based on array theory. In this section the concept and history of RDAs are discussed.

The RDA concept originates with the operation of a corner reflector [427], and the Van Atta array [428, 429] is the first circuit implementation. The Van Atta array is shown in Figure 8-1, and consists of pairs of antennas connected by equal lengths of transmission lines creating a symmetric array. As is well known, the received signal creates a phase gradient across the elements in the array based on the incoming signal's direction. The received signal at each element is then reradiated by its conjugate pair, effectively reversing the phase gradient in the transmitting beam, directing it back in the same direction as the originating source. Van Atta arrays can operate over a wide bandwidth but require that the incoming wavefront and the array itself be planar. Also, for large arrays, the design of the interweaving feed lines can become intricate. Although

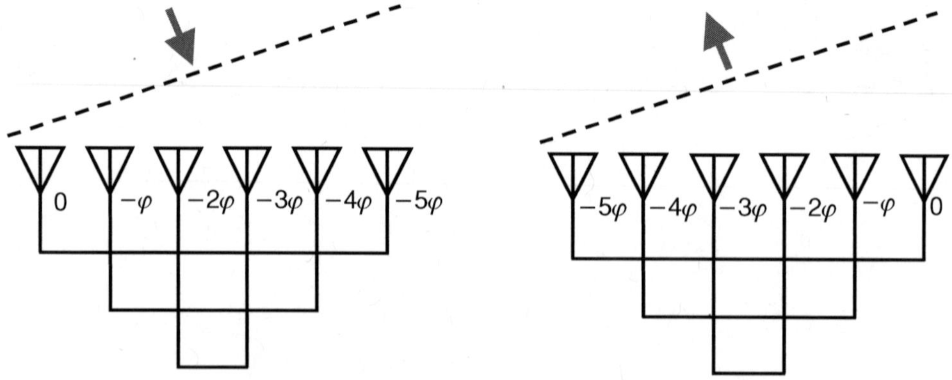

Figure 8-1 Van Atta array: (a) received signal showing the relative phases of the signal at each element of the array; and (b) transmitted signal traveling with required element phasing for transmission in the opposite direction to that received.

retrodirectivity can be accomplished using several methods, one of the most promising has been by using phase conjugation. This implementation, sometimes known as the heterodyne approach, achieves the phase reversal effect in an array at each element using phase-conjugating mixers [430, 431]. Figure 8-2 illustrates this conventional array architecture with a phase-conjugating mixing element at each antenna along with the mixing function being

$$\begin{aligned} V_{IF} &= V_{RF}\cos(\omega_{RF}t + \theta_n)\cdot V_{LO}\cos(\omega_{LO}t) \\ &= \frac{1}{2}V_{RF}V_{LO}\cos[(\omega_{LO} - \omega_{RF})t - \theta_n] + \cos[(\omega_{LO} + \omega_{RF})t + \theta_n]. \end{aligned} \tag{8.1}$$

The phase conjugating effect is demonstrated in Equation (8.1) where the lower sideband (or difference) component of the mixing product of an incoming RF (radio frequency) signal with an LO (local oscillator) signal consists of a signal with a conjugated phase based on the incoming signal. By mixing the incoming RF signal with a reference LO signal at twice that frequency, or $\omega_{LO} = 2\omega_{RF}$, the IF (intermediate frequency) of the lower side band product results in the original RF incoming frequency. Therefore using this relationship achieves the conjugate result at each element without changing the carrier frequency. The upper sideband signal and LO leakage

Figure 8-2 Phase-conjugating retrodirective array: (a) array; and (b) details of the frequency-mixing process.

can easily be filtered out since the frequency is far apart from the phase-conjugated signal at the original frequency.

8.2.1 Retrodirective Radar Cross Section Measurements

A figure of merit for an RDA can be evaluated by measuring the monostatic and bistatic patterns for an array. The monostatic radar cross section of an RDA is defined as:

$$\sigma_{\text{monostatic}} = \frac{\lambda^2}{4\pi} D_0(\theta_{\text{in}}) G_c D_e^2(\theta_{\text{in}}) \tag{8.2}$$

where G_c is the circuit gain, D_0 is the array directivity related to the array factor, and D_e is the element directivity. Note that the array factor term is not squared. This is because in the receiving process each element is separately processed without being combined. A monostatic measurement is taken by measuring the reradiated pattern over all angles while simultaneously sweeping an interrogating source. Therefore, the main beam of the RDA is always tracked since the incident angle (θ_{in}) is the same as the observation angle (θ).

The bistatic RCS of an RDA is given by

$$\sigma_{\text{bistatic}}(\theta, \theta_{\text{in}}) = \frac{\lambda^2}{4\pi} G_c D_0(\theta, \theta_{\text{in}}) D_e(\theta_{\text{in}}) D_e(\theta). \tag{8.3}$$

For a bistatic measurement, the interrogating source remains stationary, while a receiver or observation angle is swept. The resulting pattern then reveals the location of the interrogator since the retransmitting pattern is static. The test setups for these cases are shown in Figure 8-3.

8.2.2 Phase-Conjugating Mixers

Since the phase conjugator or mixing element is the key component of an RDA, the actual mixer is typically designed for the specific application and system specifications. Along these lines, specific system attributes can be incorporated in the mixing element by introducing components such as active mixers, or SOMs (selfoscillating mixers). For example, an active mixer design could eliminate the need for additional system amplifiers in the receive or transmit paths in the array by introducing signal gain directly into the phase-conjugating process. The addition of phase-locked SOMs would eliminate the requirement of potentially bulky and lossy LO distribution feed networks. Both of

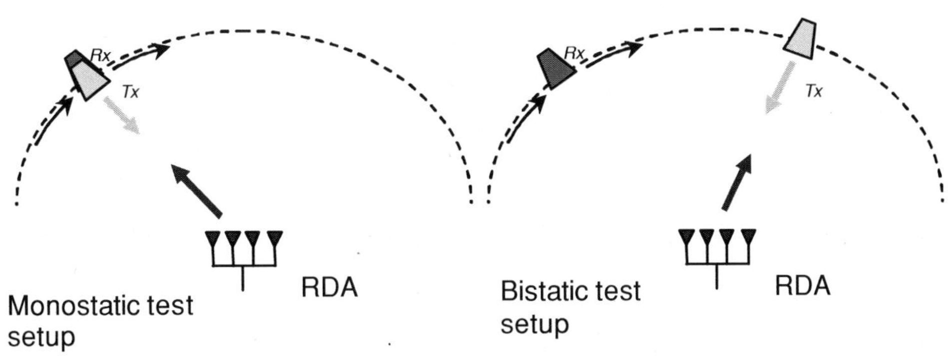

Figure 8-3 Monostatic and bistatic RCS measurement setups.

these integrated techniques would reduce the overall circuit size and possibly the cost of an RDA system.

The conventional approach to realizing phase-conjugation requires that the LO frequency be twice that of the RF carrier and can become taxing with higher carrier frequencies. This has driven the development of new phase-conjugating mixer designs. For example, an IF-based mixer using sub-harmonic mixing relaxes the requirement of having a high frequency LO. For second harmonic mixing, using the condition that $2\omega_{LO} > \omega_{RF}$, the IF result will be phase conjugated with respect to the RF signal. In addition, mixers can be designed to share the RF and IF ports permitting the use of a single antenna element for both transmit and receive.

In the following sections various retrodirective architectures are presented. Each architecture takes advantage of the innate selfsteering feature of an RDA and introduce novel communication or sensing features. Often enhanced array performance is obtained. Implementations of various phase-conjugating mixer designs are described throughout the chapter. Details regarding each design can be found in the references cited.

8.3 RECONFIGURABLE RETRODIRECTIVE ARRAY

The idea of reconfigurability can bring significant advantages to RF systems because it can reduce the total circuit size and cost by allowing the sharing of many front-end circuit components. In this section a reconfigurable array is presented which can function as a smart antenna for complex multiuser communication, or as an RDA for data extraction, or as a transponder [432]. The system can serve either as an RDA transponder or as a smart antenna receiver with the selection achieved by changing the frequency of the LO applied to the mixers. The system schematic for the reconfigurable digital wireless sensor is shown in Figure 8-4. The received signal is first amplified by a low noise amplifier and applied to passive FET mixers through a ratrace coupler. The reconfigurability is achieved though these mixers. Both modes are designed to operate at the same frequency. With an incoming signal at 5.8 GHz and an 11.6-GHz LO, the mixers operate as phase conjugators in the same manner as described by Equation (8.1) and thus the system functions as an RDA. When the LO is switched to 2.9 GHz, the mixers then operate as sub-harmonic direct conversion mixers and produce I and Q channel quadrature outputs which can be sent to a digital beamforming (DBF) receiver. This reconfigurability can be used to adapt the circuit to the communication environment. In this simple example it was shown that the retrodirective approach is flexible, and open to various integrated functions and added features.

8.4 FREQUENCY AUTONOMOUS RETRODIRECTIVE ARRAY

As discussed in the previous sections, when receiving a signal from an unspecified direction, RDAs are able to transmit a signal response in the reverse direction without any previous knowledge of the source location. This is accomplished without the use of phase shifters or signal processing,

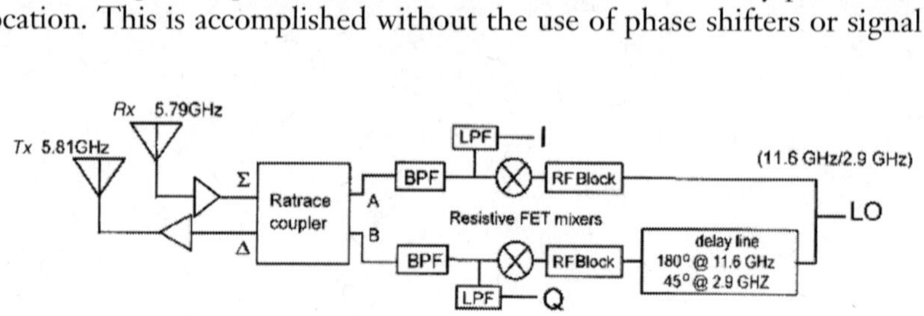

Figure 8-4 Reconfigurable smart antenna receiver/retrodirective array.

but instead exploiting the response of phase-conjugating mixers. The retrodirective functionality is achieved automatically, avoiding the digital circuit speed "bottleneck" of conventional smart antenna systems. Such unique features make the RDA an attractive candidate for advanced digital mobile communication systems where high-link gain and high-speed tracking is required.

In this section, an RDA with an added autonomous feature of not needing to know the exact source frequency is introduced [433]. This adds another dimension in system flexibility and increases the covert nature of the RDA. Frequency autonomousness is accomplished by using the received RF frequency power to generate LO power used by the phase-conjugating mixers. Therefore, the retrodirective array is now able to respond at the same frequency as the interrogation frequency automatically.

A schematic of the frequency autonomous RDA is shown in Figure 8-5. It comprise two main sub-circuits, namely an array of phase-conjugating mixers and the LO generator circuit. The signal received by the phase-conjugating mixers first passes through the circulator placed close to the antenna, then is fed to one of the ports of a 3-dB branch line coupler, via a second circulator. The signal is then spilt equally, feeding each Schottky diode mixer in quadrature phase. The RF signal is then phase-conjugated using heterodyne mixing. Because the LO frequency is exactly twice the RF frequency, the IF frequency will be the same as the RF frequency, therefore IF-RF isolation is achieved by using the relative phase relationship of the IF and RF leakage signals at each individual diode mixer. The phase-conjugated IF signal from one of the mixers combines at the same port where it entered while the non phase-conjugated RF leakage is dumped into the 50 Ω load terminating the opposite port of the branch line coupler. The phase conjugated IF signal passes back to the antenna through the circulators and is retransmitted. This circuit also allows for amplifiers to be included in either receive or transmit paths or both paths. The minimum mixer conversion loss was measured to be 5.1–9.1 dB over a 5–5.9 GHz frequency range with an optimum LO power level of 3 dBm per diode mixer. For this reason diode mixers are better suited for this system architecture than FET mixers because of the limited LO power. In addition, in hybrid-circuit fabrication, phase balance between a diode pair is more easily achieved.

A prototype 4-element RDA array based on the proposed phase-conjugating mixer was built on RT/Duroid of 25 mil thickness with $\varepsilon_r = 10.2$. To facilitate the relatively wide bandwidth of this system, the quasi-Yagi antenna [434], which is reported to have a wide operational bandwidth,

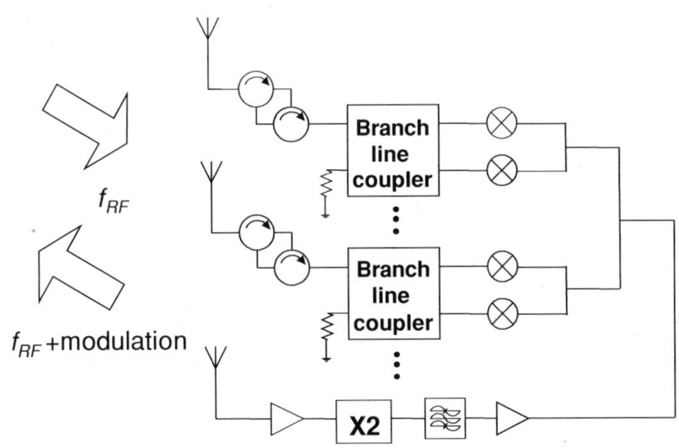

Figure 8-5 Schematic of frequency autonomous retrodirective array (RDA).

3.74–6.22 GHz (50%) was used. The antenna's radiation pattern is quite broad and it has low adjacent-element mutual coupling (−20 dB), making it suitable for use in antenna beam-scanning array applications. The antenna radiates in the end-fire direction so that the overall array circuit can be easily expanded into a 2-D array. The array spacing was chosen to be half-wavelength at 5.2 GHz. The small array spacing allows the array to avoid scan angle limitations due to grating lobes which become visible when the array spacing is large.

The frequency autonomous nature of this RDA comes from the fact that the incoming RF signal is used to generate the LO drive signal for the phase-conjugating mixers. This is most simply by first amplifying the received RF tone signal, and then feeding it into a frequency doubler. A filter is then used to reject any RF leakage. The doubled signal is further amplified and used to feed the phase-conjugating mixer array.

However, one of the main drawbacks of this LO generation method is that the LO power is dependent on the received RF power, which may change with different transmitter power or the location of the transmitter. Fluctuation of the LO power greatly impacts the conversion loss of the mixer array.

In the above system the optimum mixer conversion loss is dependent on maintaining a specific LO pump power. Therefore an alternative method to realize the LO generator circuit will now be considered. The circuit uses a phase-locked loop (PLL) to track the RF signal. The PLL output is at the same frequency as the incoming RF signal, however the power is kept constant with changing RF signal strength. The PLL circuit itself consists of two divide-by-8 circuits in series followed by a phase-detector. The output of the phase-detector is filtered by an active lowpass filter, leaving a DC control voltage. The control signal drives a voltage controlled oscillator (VCO). The VCO output power is fed back to another series of two divide-by-8 circuits, and back to the phase-detector. The dynamic range of the input level of the fabricated circuit was measured to be 30 dB, with a frequency range from 5.2–5.4 GHz. The output of the PLL is amplified and fed into a frequency doubler. After further amplification this doubled signal is used to drive the array of phase-conjugating mixers.

Retrodirectivity was measured by transmitting a single-tone interrogation signal at a fixed position and measuring the radiated response of the retrodirective array. Because of the system architecture, the return signal must be at exactly the same frequency as the interrogator's signal. In order to allow the receiver to distinguish between the array response and the interrogating signal, a 25 kHz sinusoidal modulation signal was mixed with the response signal transmitted by the retrodirective array. This demonstrates the arrays' functionality as an information transponder.

The array was first tested using the simpler LO generator sub-circuit consisting of an amplifier and a frequency doubler. The LO power level was manually controlled by adjusting the gain of the amplifier used to amplify the LO signal generated from doubling of the RF signal.

Figure 8-6 shows the measured bistatic RCS of the RDA with the interrogator source at broadside 0°, 15°, and −30° at 5.2 GHz. The same measurement is shown taken at 5.8 GHz in Figure 8-7. Retrodirectivity is clearly observed in all three cases at both frequencies, demonstrating the possible operating bandwidth and the frequency autonomy. Note that no grating lobes are observed due to the small array spacing. With the improved PLL circuit, the patterns were limited to the PLL bandwidth, or from 5.2 to 5.4 GHz. Scattering due to the measurement environment noticeably influences the measured radiation patterns. Other factors for pattern degradation include channel imbalances or mismatches within the circuitry.

In this section a novel frequency indiscriminate RDA was described. The array is able to respond to an interrogator at the same signal frequency with no a priori knowledge of its position or of its transmission frequency. This feature adds a sense of covertness by allowing automatic adjustment to frequency hopping.

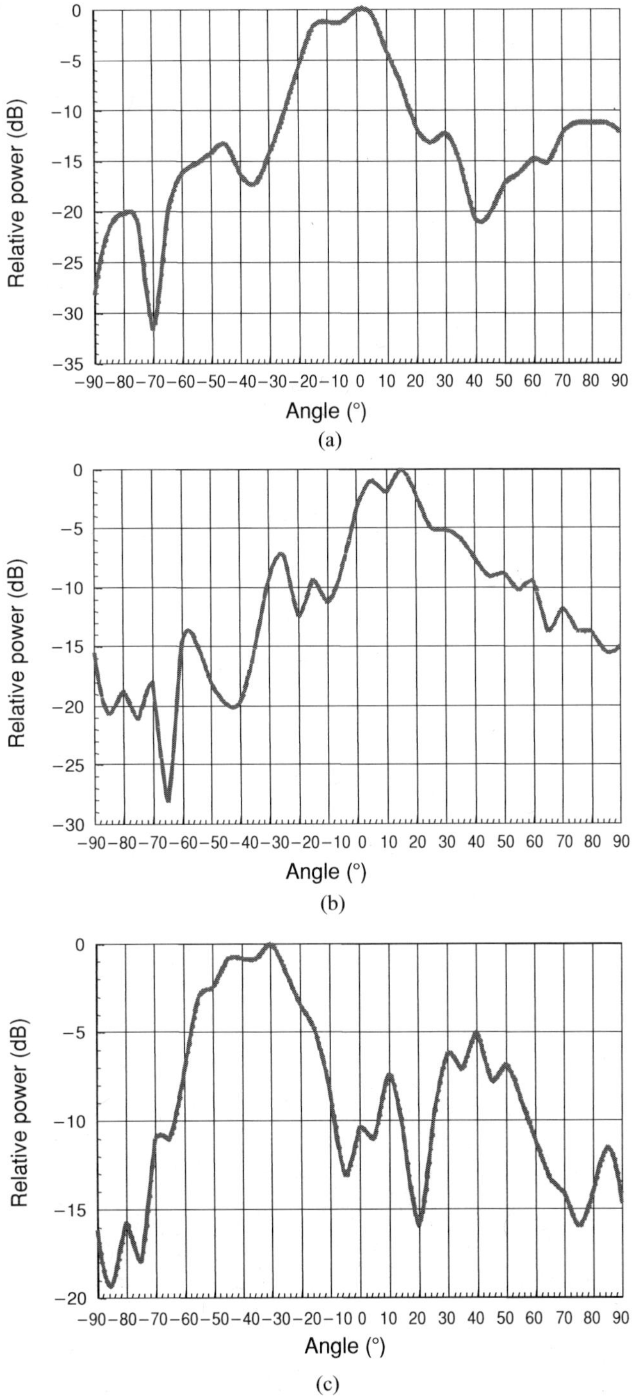

Figure 8-6 Measured RCS of the bistatic RDA at 5.2 GHz at various broadside angles: (a) 0°; (b) 15°; and (c) –30°.

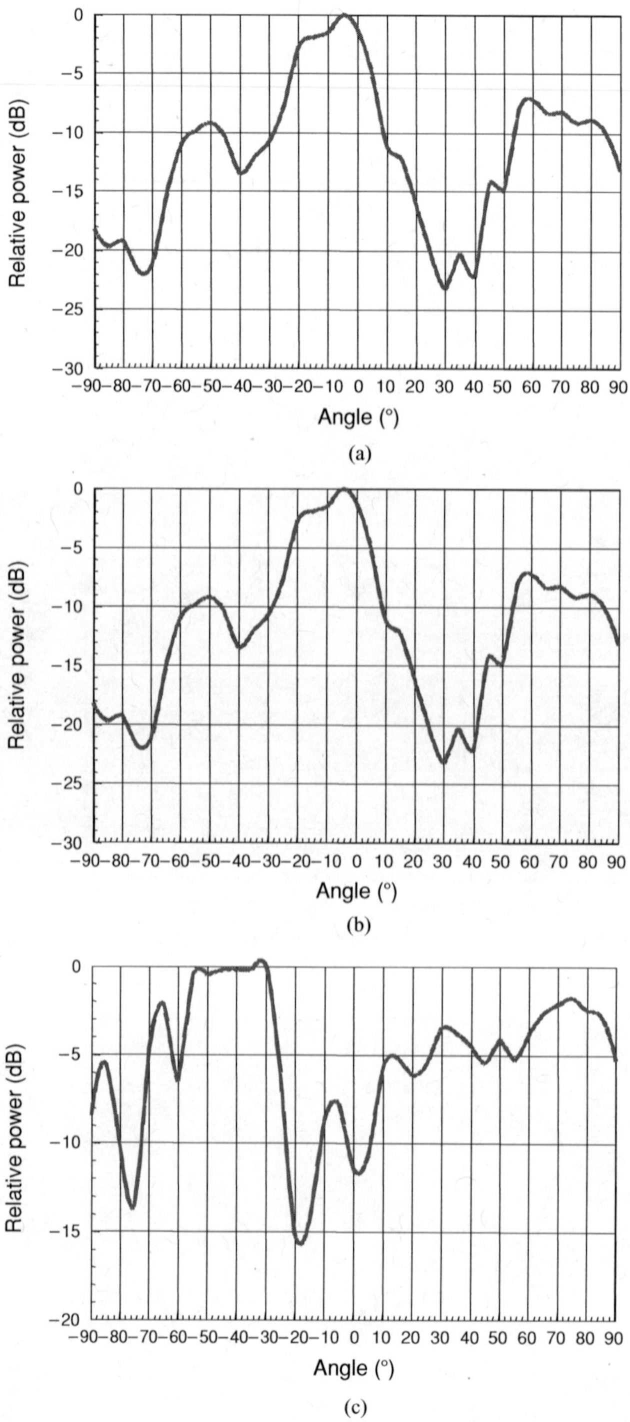

Figure 8-7 Measured RCS of the bistatic RDA at 5.8 GHz at various broadside angles: (a) 0°; (b) 15°; and (c) −30°.

8.5 FULL DUPLEX RETRODIRECTIVE ARRAY

In many earlier retrodirective works, including those described in Sections 8.3 and 8.4, receivers have been commonly integrated into RDAs enabling the recovery of data. Various modulation schemes have also been integrated into the transmission paths of RDAs enabling the retransmission of encoded data. However, these two functions are not typically performed simultaneously, since a clean CW signal is necessary for proper encoding of transmitted signals, thus limiting RDAs to one-way communication scenarios.

In this section, an RDA capable of full duplex communication is presented [435]. The system designed for full-duplex operation is illustrated in Figure 8-8. As shown, the incoming signal is split into two paths, one goes to a receiver and the other is for retransmission. The basic new feature is an added carrier recovery circuit introduced in the retransmission path. This circuit is used to eliminate the incoming modulation and allow remodulation onto a clean carrier. The system uses independent modulation schemes for transmit and receive. The complete system is based solely on analog circuits for all signal processing tasks including: target tracking; beam pointing; and carrier recovery. Analog processing maintains the high-speed functionality of the system. Two types of receiver architectures are used: AM (amplitude modulation) diode detectors for near-omni directional coverage; and digital beam forming receiver to enhance signal-to-interference ratio.

One of the main challenges in implementing this retrodirective transceiver system is preserving and processing the geometrical phase differences of the received signals from each antenna element, while removing the modulated data so that a new set of data can be re-modulated onto the carrier, and then transmitted back retrodirectively. In order to accomplish this task using the simplest means possible, the incoming signal is modulated using AM modulation, ensuring that the phase differences of the antenna element can be easily assessed. Since the data is contained

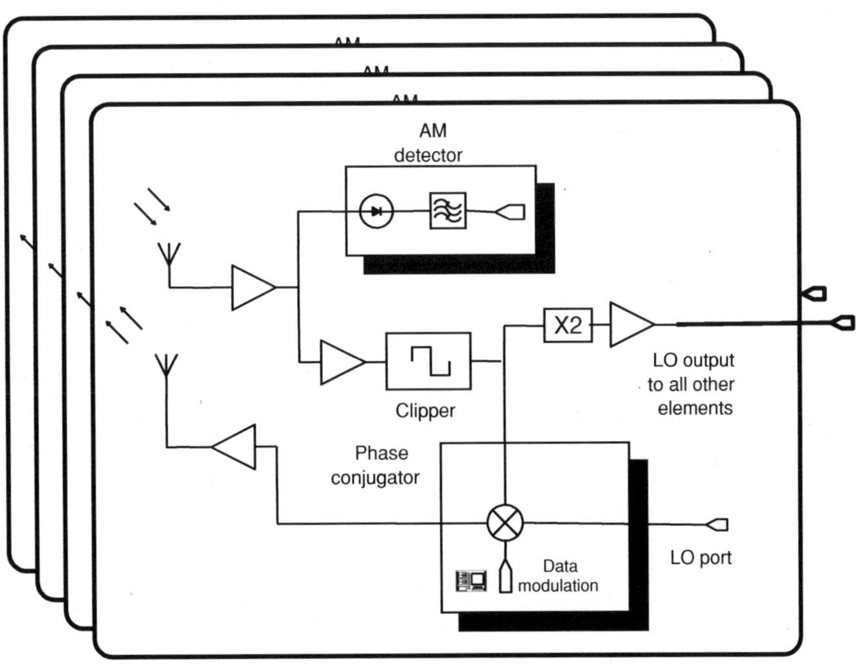

Figure 8-8 Schematic of a full-duplex RDA.

in the amplitude variation of the carrier signal only (i.e., the signal envelope), this data can be removed using a clipper circuit while still preserving the carrier phase. After the modulation is removed from the received waveform it is fed as an input to an array of phase conjugation mixers that also serve as modulators to re-modulate the signal with new data. A radiating beam is retrodirectively formed in the direction from which the received signal came. As depicted in Figure 8-8, this system also uses the same LO generation concept described in Section 8.4, thus maintaining a frequency autonomous architecture.

For retrodirective transmission, the original modulation is first removed using a clipper circuit. The clipper transforms received signal with a time-varying envelope to a signal of constant envelope. After filtering out the harmonics the carrier is recovered. A clipping circuit can be realized using a number of techniques, including using anti-parallel diodes and high gain saturated amplifiers.

A high gain amplifier with low output power was used for this application. In the frequency domain, the limiting effect can be seen from the reduction of modulation sidebands (for single tone modulation). The circuit was experimentally characterized by inputting a 5.8 GHz carrier signal AM modulated with a 100 kHz sinusoidal message signal. After passing through the clipper, the reduction of modulation sidebands was observed to be approximately 30 dB; from −35 to −65 dBc with an input power of 4 dBm. The second and third harmonics of the carrier were suppressed using a shorted-stub acting as a band stop filter along with a stepped-impedance lowpass filter so that only the carrier signal remains. This recovered carrier is then fed to a phase-conjugating mixer and remodulated for retrodirective transmission.

A balanced mixer design consisting of a pair of Schottky diodes is used for the phase conjugator. This phase conjugating circuit is shown in Figure 8-9 along with the complete fabricated system. The incoming signal is fed to two diodes 180° out of phase using a ratrace coupler. The LO drive signal is also fed 180° out of phase to the mixer pair using a Wilkinson power divider and a fixed delay line at 11.6 GHz (two times the RF frequency). The resulting mixed IF signals from the two diodes are then in-phase and therefore combine at the in-phase port of the ratrace coupler. However the RF leakage signal will be rejected at the in-phase output port since the leakage retains the out of phase relationship of the RF input. This scheme provides decent isolation between the phase-conjugated signal and RF leakage. In addition, both receive and transmit amplifiers can be included with this mixer increasing the transmission range of the array.

All of the circuitry, excluding LO amplifiers and doubler was built on three RT/Duroid boards with $\varepsilon_r = 10.2$ and 25 mil thickness and later connected using SMA connectors. The input amplifier, clipper circuit and phase conjugation mixers were etched on the central board. SMA connectors on the underside of the board connect to the output amplifiers placed on another Duroid board. Two receiver modules, namely the AM diode detectors and the direct down conversion mixers are interchangeably connected at the top most layer of the array. Two sets of $\lambda/2$ (at 5.8 GHz) spaced arrays of inset fed patch antennas, one for receiving and the other for transmitting were connected to the appropriate input/output ports of the array. LO power was generated by doubling the RF power of a single element using an external frequency doubler and amplified using an external high gain amplifier chain and then used to feed the entire RDA.

Using AM modulation results in many benefits for a full duplex retrodirective transceiver. First, because a strong carrier signal is transmitted along with the data it facilitates phase conjugation of the carrier tone for retrodirective transmission. Secondly, since the modulation spectral power can be kept low, potentially it can be compatible with conventional radar systems that simply transmit chirped or continuous-wave tones. Third, since AM detection does not require an LO, and the received signals are always directly down converted, this scheme is well suited to cope with a constantly-varying carrier frequency received by a radar target. Furthermore, AM demodulators

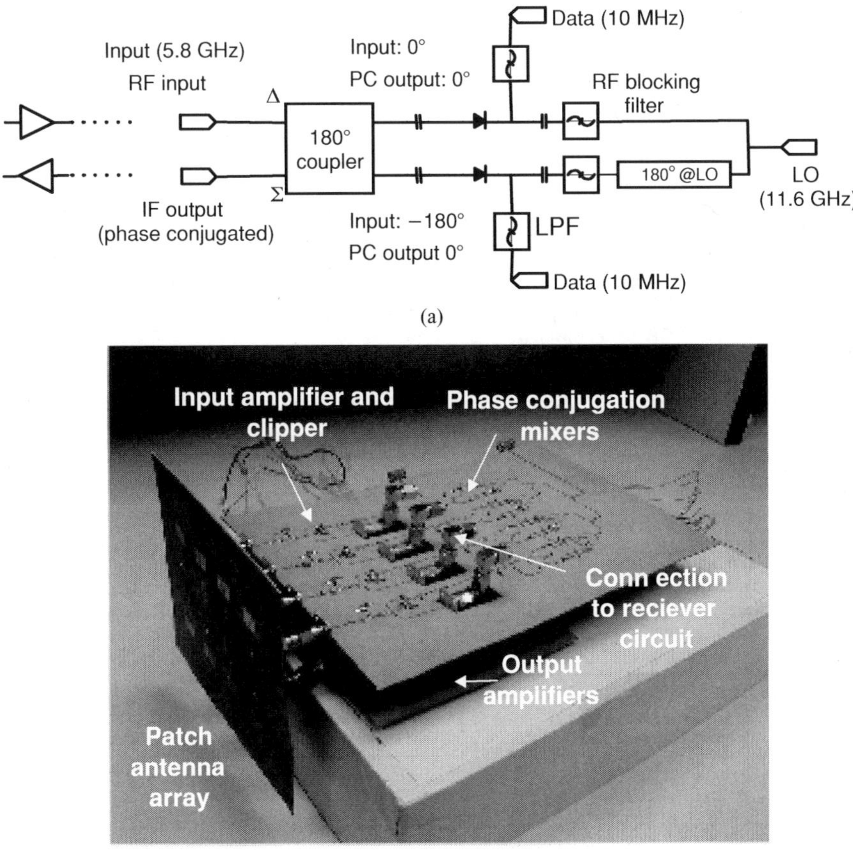

(a)

(b)

Figure 8-9 Phase conjugation mixer: (a) schematic; and (b) photograph of complete system.

can be made to be simple since they do not require frequency and phase synchronization. This is ideal for the case where the signal is to be broadcast from one source to a number of receivers, as is the situation for a SAR (synthetic aperture radar) platform and a retrodirective radar target transceiver.

One method of AM demodulation uses a Schottky diode detector. The diode detector was designed using a Schottky diode along with a lumped element lowpass filter as well as optimized microstrip sections for matching. In addition, DC bias provided forward biasing of the diode to improve conversion loss. To test this circuit, an RF carrier of 5.8 GHz was modulated with a 100 kHz signal with modulation indexes of 100%, 50%, 6%, and 3.6% respectively. The detector was measured to be linear with reasonable conversion loss. Several measurements were done to evaluate the system's performance. The full duplex capability of this system was tested using a single element of the antenna array first. Additionally, retrodirective re-transmission radiation patterns were measured. Receiving radiation patterns were also measured using both receiver topologies, i.e., Schottky diode detectors and coherent detection, to implement a digital beamformer.

The LO of the phase-conjugating mixer was supplied by first coupling a fraction of the signal outputted from the clipper. After filtering and frequency doubling, the 11.6-GHz signal was amplified to a level of 6 dBm and used to pump the phase conjugating mixer. The 'modulation-stripped signal' was also fed to the RF port of the phase conjugation mixer. Baseband data was applied to the circuit using an arbitrary waveform generator and used to re-modulate the phase conjugated signal. Simultaneously, the received AM signal was demodulated using the AM diode detector.

Several modulation combinations were experimented with. These combinations include receiving a sinusoidal signal while transmitting a digital signal, receiving a digital signal while transmitting another digital signal, and finally receiving a digital signal while transmitting a sinusoidal signal. The results for the last case are shown in Figure 8-10. The AM demodulated digital waveform follows the 10-Mbps pseudo-random reference signal quite well. Discrepancies in pulse width can be reduced by tuning the output resistances and capacitances of the output filter to compensate for the diode resistance and capacitance. Also, in Figure 8-10(b) the demodulated waveform at the output of the phase-conjugating mixer is a 10-MHz sinusoidal waveform as expected. This experiment obviously shows the validity of this system. Although, the input digital signal contained

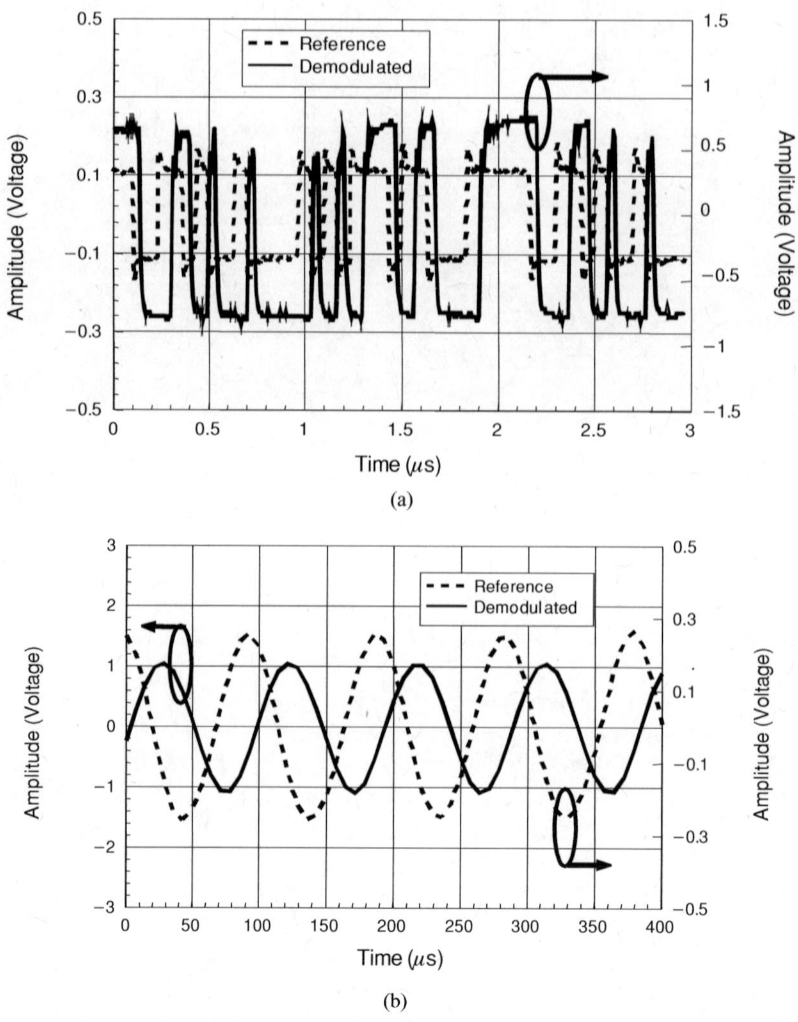

Figure 8-10 Demodulated waveforms: (a) digital waveform demodulated by an AM diode detector; and (b) 10 MHz waveform demodulated by interrogator receiver.

numerous frequency components the retransmitted signal contained only the 10-MHz signal. The waveform accurately follows the reference showing no interference from the original modulation that was removed by the clipper circuit.

The efficiency of the receiving and transmission functions are dependent on the modulation index of the received signal. If the modulation index is set higher, receiver conversion loss can be reduced; however this will have a detrimental affect on the BER (bit error rate) of the retrodirected signal. However, because of the resilience of digital signals, the modulation index can be best maximized if both received and transmitted modulation are in digital form.

Retrodirectivity was measured through the monostatic and bistatic patterns for this array. The results are shown in Figure 8-11 with the interrogator positioned at 0°, 30°, and −15°. Note that no grating lobes are observed due to the small array spacing. Since this experiment was not done in the controlled environment of an anechoic chamber, scattering due to the measurement environment noticeably influences the measured radiation patterns. Other discrepancies in the predicted and measured radiation patterns may also be attributed to nonidealities in phase and amplitude balance among the array elements. This receiver architecture is very simple and fits well with the concept of frequency autonomousness used in the whole system arrangement. In addition, since no directional phase information is retained after down conversion, the array pattern shows no null angles, which suggests that automatic near-omnidirectional signal reception can be achieved.

Digital beamforming capability can also be added to the receiver section of the transceiver, in place of the AM detectors. Digital beam-forming is a method of forming directional antenna beams using DSP and complex signal weighting. In this approach, the beam of the array may not only be directed to maximize the reception of the desired signal, but also can be tailored to suppress undesired interference signals. This process is usually done by first down converting a signal modulated on a carrier to baseband (or to a low IF) so that the signal can be directly digitized by an ADC (analog-to-digital converter).

In the proposed receiver hardware the input RF signal is fed into an array of down converting mixers. Each mixer output is first lowpass filtered and then sampled and digitized by a digital oscilloscope.

Proof of concept, digital processing of the sampled data was performed offline using a MATLAB code. After Fourier transforming the time-domain data, the relative phase and amplitude information was extracted. The circuit error was calibrated assuming that all elements receive the signal with equal amplitude and phase at broadside. The signals are weighted by multiplication of a signal with the required weight. The weighted signals are combined together to form a directional beam. It is also possible to place nulls in specified directions, thus increasing the system's Signal-to-Interference Ratio (SIR).

The DBF receiver was tested in two configurations. First, taking the more conventional approach, an onboard LO source was used to down-convert the incoming signal. An unmodulated RF carrier of 5.8 GHz was received, and mixed down to 1 MHz using an onboard LO at 5.799 GHz. DBF-formed beams at 0° and 15° are shown in Figure 8-12.

The second configuration that was tested uses a technique similar to that of the frequency autonomous RDA. The incoming RF signal was AM modulated by a 1-MHz sinusoid. As with the conventional configuration, these modulated signals were received and input to the mixer array. However in this case, the LO of the entire array of receiver mixers is supplied by a carrier-recovered, clipped signal output generated by one of the antenna elements rather than relying on an onboard LO. This allows the receiver to continuously track the carrier frequency of the interrogation signal while providing coherent demodulation. Again, the 1-MHz signal outputted at each mixer is digitized and processed. The beam patterns agree reasonably well with calculated results. Although, the beam tracking speed of this configuration is limited by the DSP speed, this

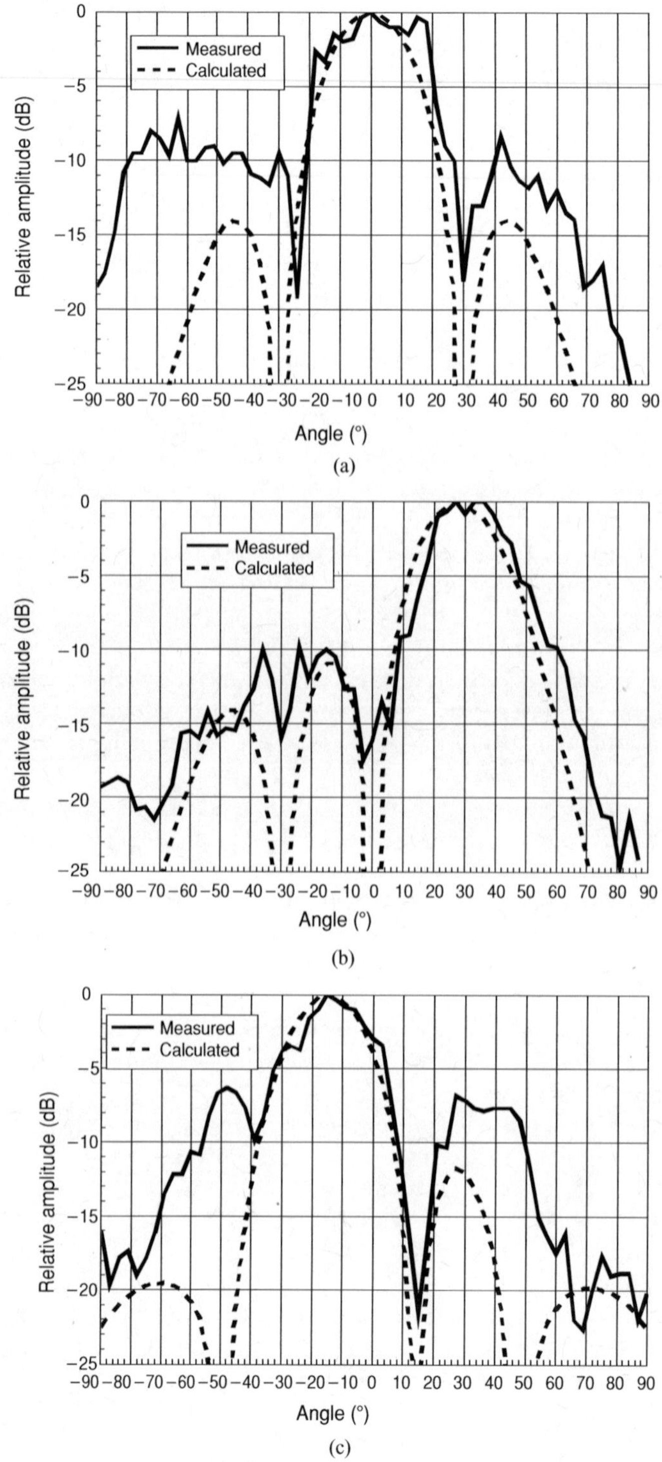

Figure 8-11 Bistatic RCS, interrogator at (a) 0°; (b) 30°; and (c) –15°.

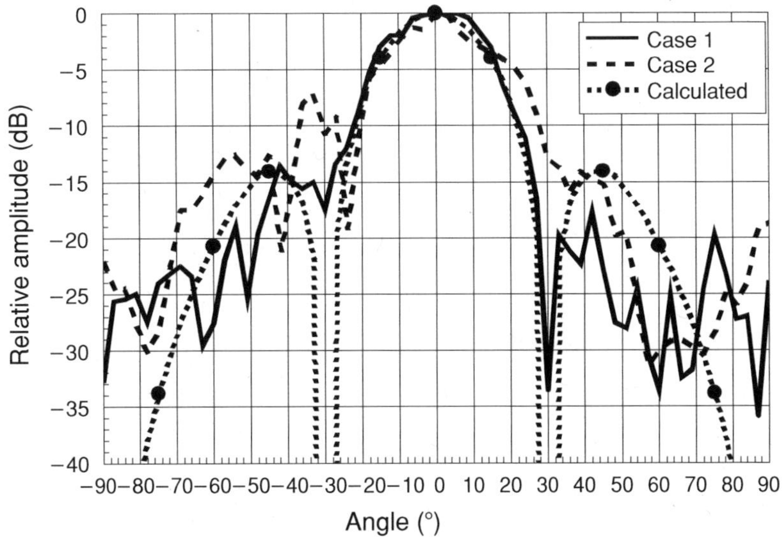

(a)

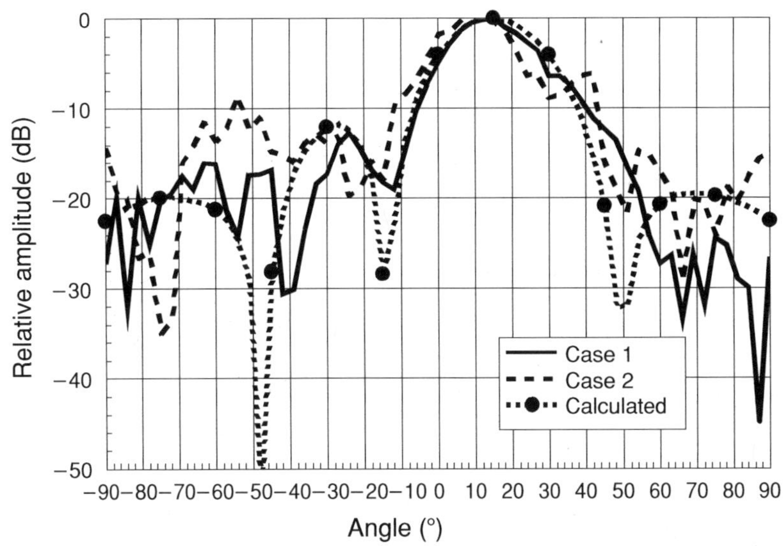

(b)

Figure 8-12 DBF beam patterns: (a) at 0°; and (b) at +15° (Case 1 on board LO, Case 2 frequency autonomous LO).

frequency autonomous DBF feature can be used to increase the SIR of the array when needed or used in slower moving or stationary communication scenarios.

The receiver portion of the array was implemented first using simple AM diode detectors that can provide near-omnidirectional receive coverage without any phase processing. This configuration complements the array's automatic omnidirectional transmitter property. Experiments using a DBF receiver as part of the retrodirective transceiver were also conducted. This configuration, offers the ability to form beams and nulls at any desired directions, potentially improving receiver SIR.

8.6 HYBRID SMART RETRODIRECTIVE ARRAY

When illuminated by an unknown interrogator, RDAs are able to automatically respond back toward the interrogator without any prior knowledge of the direction of the incoming signal. This is a truly interesting feature making these arrays a serious contender for use in communication systems where source tracking and high-link gain is desired. This process is completely analog and operates without phase shifters, offering a great advantage in terms of simplicity and speed compared with conventional phased arrays with phase shifters and digital processing. However, this generally beneficial feature could also pose a serious problem in a security sense. An RDA under multiple source interrogation will simultaneously respond back toward each source that interrogates it. This means that the RDA has essentially no control over which interrogator it would respond to, and in that context, a signal from the array could be stolen by simply interrogating the array.

In the scenario depicted in Figure 8-13, two sources are incident on an RDA, where one is a jamming signal. In this situation, the user (with onboard RDA) would firstly, not want to be affected by a jamming signal, and secondly would not want the interferer to 'eavesdrop' or intercept transmitted data.

The RDA presented in this section includes an additional feature of interference rejection through a hybrid smart antenna architecture. This adds a level of security in a retrodirective communication link by preventing others from "eavesdropping" and also increasing the quality of the link [436]. This architecture consists of a built-in receiver function, and therefore uses DSP initially and intermittently to detect the direction of an interfering source. The array continues to track the desired source through its autonomous high-speed retrodirective function thus having a computational advantage over conventional smart antennas, and a functionality additional to that of conventional RDAs. The system schematic is show in Figure 8-14. The key feature in this architecture is the integration of a two-element sub-array within each receiving channel, and serves as the antenna element for the RDA. In this manner the sub-array or element pattern—which can now be controlled—is essentially multiplied by the array factor, or retrodirective pattern of the system. Thus automatic beam tracking can be achieved along with the new feature of interference rejection. A single channel of the primary or main array consists of a two-element receiving sub-array, a sub-harmonic down-conversion phase-conjugating mixer, IF splitter, up-conversion mixer, input and output amplifiers, and a single transmitting antenna.

In conventional adaptive arrays, complex DSP algorithms are used for most or all implementations of source tracking, beamforming, and interference rejection. This time-consuming process

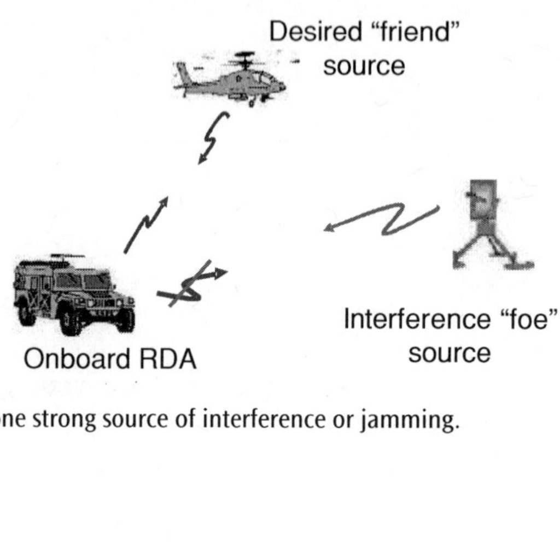

Figure 8-13 Scenario with one strong source of interference or jamming.

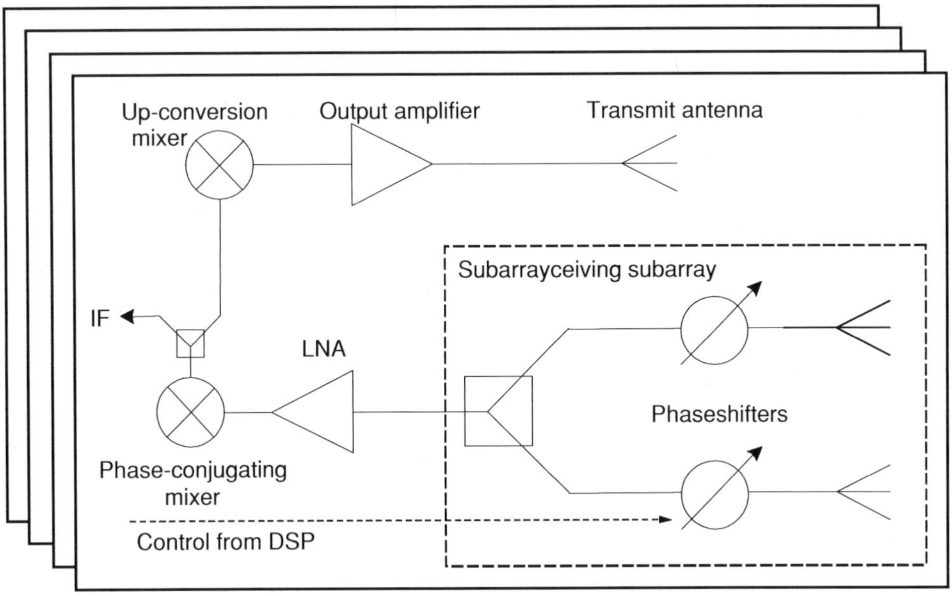

Figure 8-14 Schematic of the interference rejecting RDA.

of continuously tracking signals, computing weights, and creating beam patterns is known to be the "bottleneck" for many of these communication and radar systems. The "smart" retrodirective system considered here potentially will operate as a semi-adaptive array with minimal signal processing.

In this work, the principle of pattern multiplication is exploited. Equation (8.4) illustrates the basic concept of pattern multiplication where the overall array pattern, $F(\theta)$ is equal to the product of the element pattern $f(\theta)$ and the array factor:

$$F(\theta) = f(\theta) \cdot \sum_{n=0}^{N-1} I_n e^{jn\psi} \tag{8.4}$$

Notice here that the total radiated pattern is determined by the array factor as well as by the pattern of each antenna element in the array. Where as most adaptive antennas operate by manipulating the array factor by means of phase shifters or digital signal processing, this architecture demonstrates interference rejection by actually controlling the pattern of the individual elements in the RDA.

A demonstration of the basic operation of the hybrid smart RDA with interference rejection is illustrated in Figure 8-15. Figure 8-15(a) shows a hypothetical bistatic retrodirective pattern for two sources incident at $-30°$and $+25°$. The array pattern actually depicts two main beams, one directed back toward each source. For an RDA, this is an interesting natural feature, as this implies that multiple sources can be tracked without any additional complexity to the system. The baseline for this comparison is an adaptive array which would require additional computation time to account for a more complex scenario such as an added source. However, for an RDA, this can lead to a security issue as mentioned earlier, since another interrogator could steal a signal by simply interrogating the array. This scenario is investigated in the following where two sources are incident and one source is an interfering interrogator. Assuming the source at $+25°$ is determined to be the interference, all sub-arrays would be directed to set a null at that location and would have a pattern

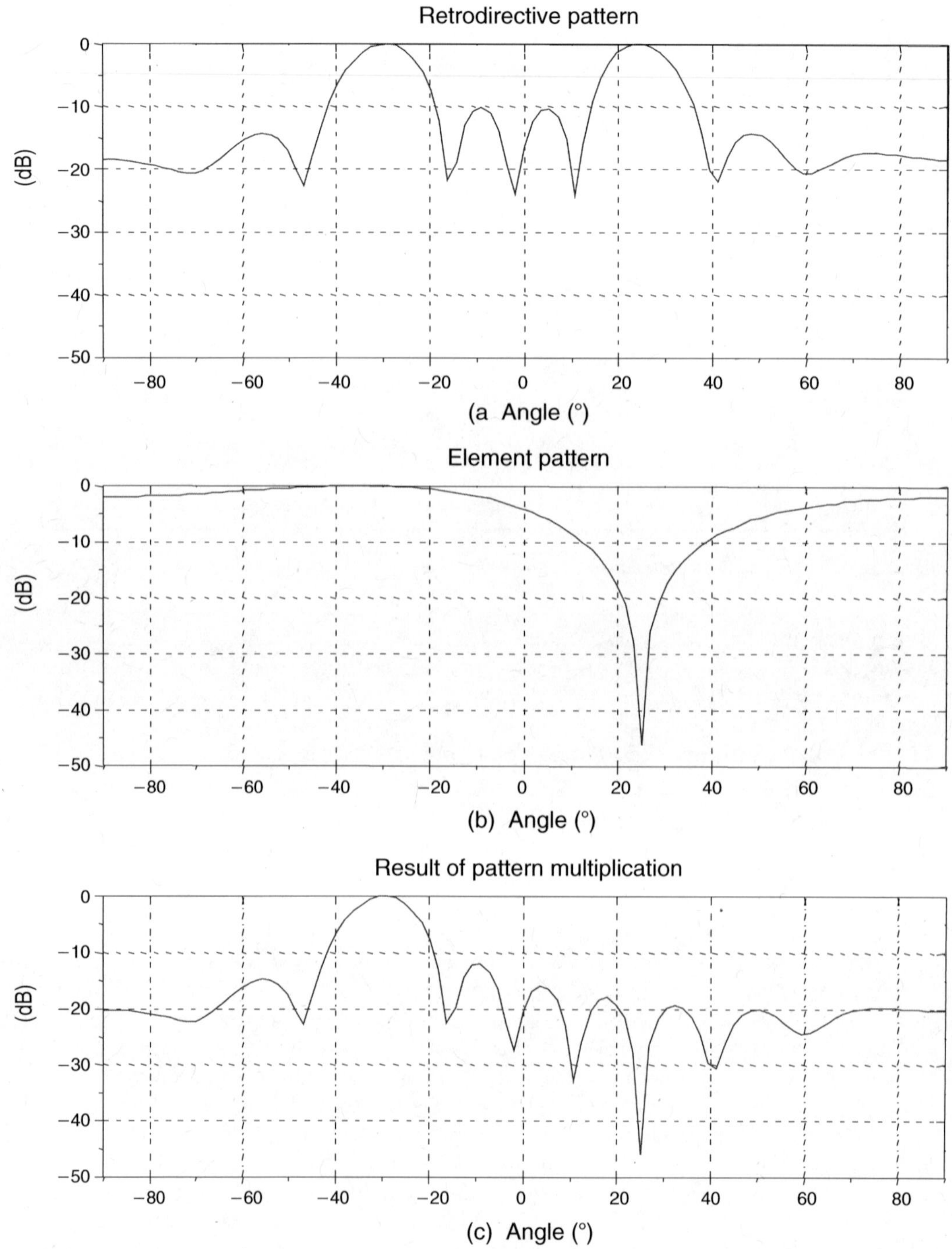

Figure 8-15 System concept demonstration: (a) two-source retrodirective pattern; (b) subarray, or element pattern with null; and (c) resulting pattern.

as in Figure 8-15(b). This pattern essentially would serve as the element pattern for the overall array. It is required that the subarray or element pattern have a deep null, for good interference rejection, and to have a flat, smooth pattern elsewhere so that signals in other directions can be seen and received. Figure 8-15(c) shows the results of pattern multiplication, if all subarray patterns are identically matched with nulls at the required locations. It is seen that the interfering signal has been eliminated and only a main lobe at the desired interrogator remains. This actually depicts the retransmitted pattern from the RDA, indicating that only the desired source location will receive a return signal. This is the basic principle underlying the operation of this system. The initial operation of this system occurs with the subarray in phase, operating as if no subarray existed. All incoming signals will be received, amplified, and mixed with an LO, downconverting and providing the required phase-conjugation. An IF splitter is then used to split the signal into two paths, one for a receiver and the other to be used in implementing retrodirectivity. The receiver is able to distinguish source and interference locations through a direction of arrival (DOA) estimation and simple identification test. The element pattern is then manipulated to null out reception from the direction of the interference, insuring that data can be securely transmitted from the RDA without fear of "eavesdropping." It is assumed that the interference source remains somewhat stationary so that DOA and null-direction adjustment need only to be periodically updated while a retrodirective beam is continuously tracking a "friendly" interrogator's movement. This task would be extremely computationally intensive for a conventional smart antenna to carry out, but the processing load is minimal for the proposed system. A comparison of the conventional smart antenna approach with the proposed "smart" retrodirective approach is shown in Figure 8-16. It is seen that the main difference is in the DSP load required to operate the system. A conventional smart antenna uses DSP to continuously locate all signals and to compute weights and form beam patterns for transmission and reception. The hybrid approach will use DSP to detect and distinguish the location of sources in a manner similar to that of the smart antenna approach;

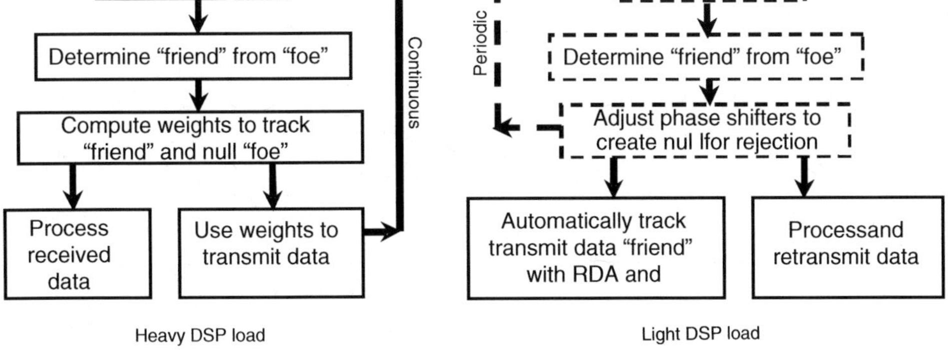

Figure 8-16 Comparison of computational load between the proposed system and the conventional smart antenna operation.

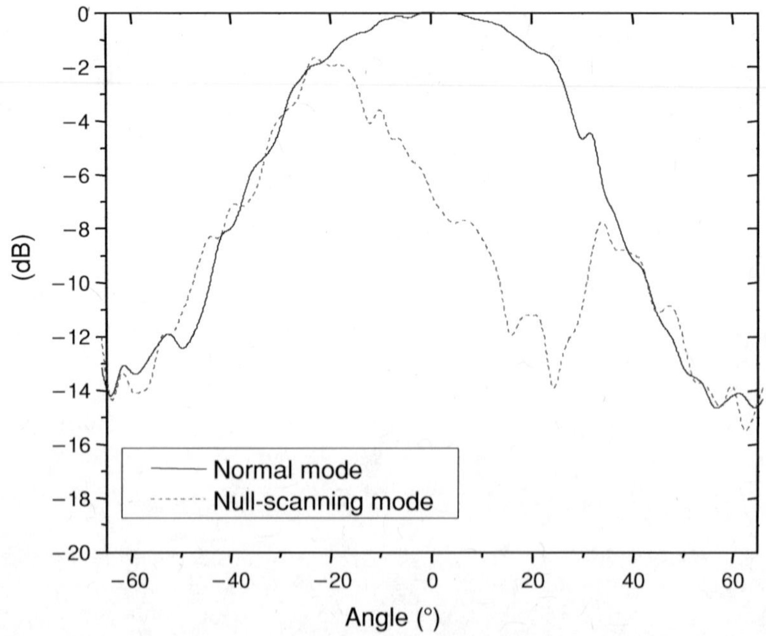

Figure 8-17 Monostatic pattern of the array in "normal" and "null-scanning" mode.

however, once this is established, the sub-arrays will be fixed to eliminate interference, while the innate, automatic retrodirective tracking function will continue to track the desired source. At this point, no digital processing for tracking or beam-forming is needed, thus resulting in a system with minimal processing. This procedure can be periodically applied to maintain an exclusive link and to take into account minor movement of the jamming signal.

The measured monostatic pattern demonstrates operation in two modes, the "normal" retrodirective mode, and the "null-scanning" mode. The patterns are shown in Figure 8-17. It is seen that the pattern of the normal mode has a smooth trend and when a null is formed at approximately $+20^\circ$, the monostatic pattern depicts a drop in that location. The measured bistatic pattern of the array further demonstrates the retrodirective function of the array and the implementation of the null-forming and null-scanning of the subarray. Figure 8-18 shows the measured bistatic pattern for the case where the RDA is operating in the 'normal' mode with one source present. Patterns are shown for a few different source locations, clearly illustrating that the array can function conventionally like other RDAs by automatically tracking and responding to an interrogator.

When multiple sources are incident upon an RDA, the RDA will respond by retransmitting a signal with main beams pointing toward each incident source location. Figure 8-19(a) shows measured bistatic patterns when two sources at slightly offset frequencies (f_1, f_2) are incident on the array. Each pattern has a main beam directed toward its respective source location. Assuming that the source transmitting at f_1 is the interference, Figure 8-19(b) shows the two respective patterns after a null is formed in the direction of the interference. It is seen that at f_1, the interference has been suppressed and the received power level at that location has dropped. Also at f_2 the main beam still exists and is at a higher power level than before the null is formed as shown in Figure 8-19(a). Therefore it is seen that the benefits of the null-scanning subarray is twofold: first the interference is suppressed from the desired location; and second, the link gain between the desired source increases.

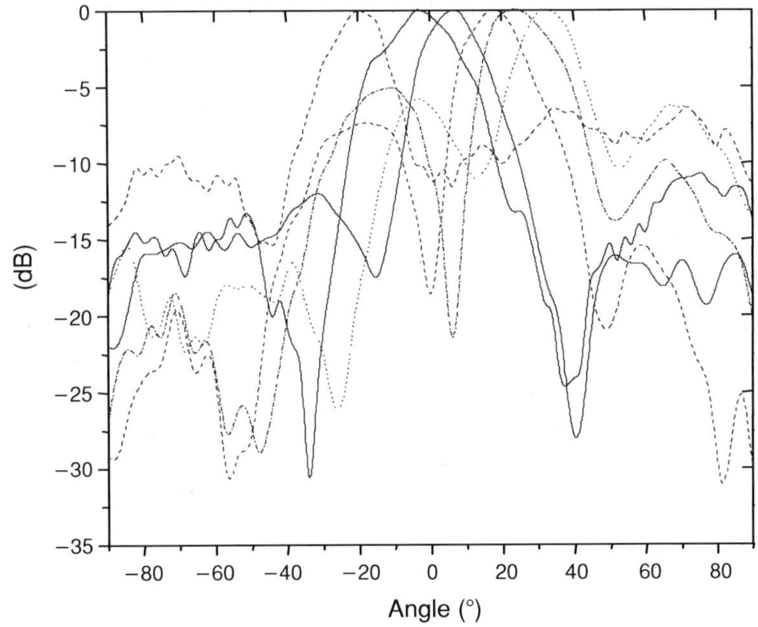

Figure 8-18 Bistatic pattern for various source locations.

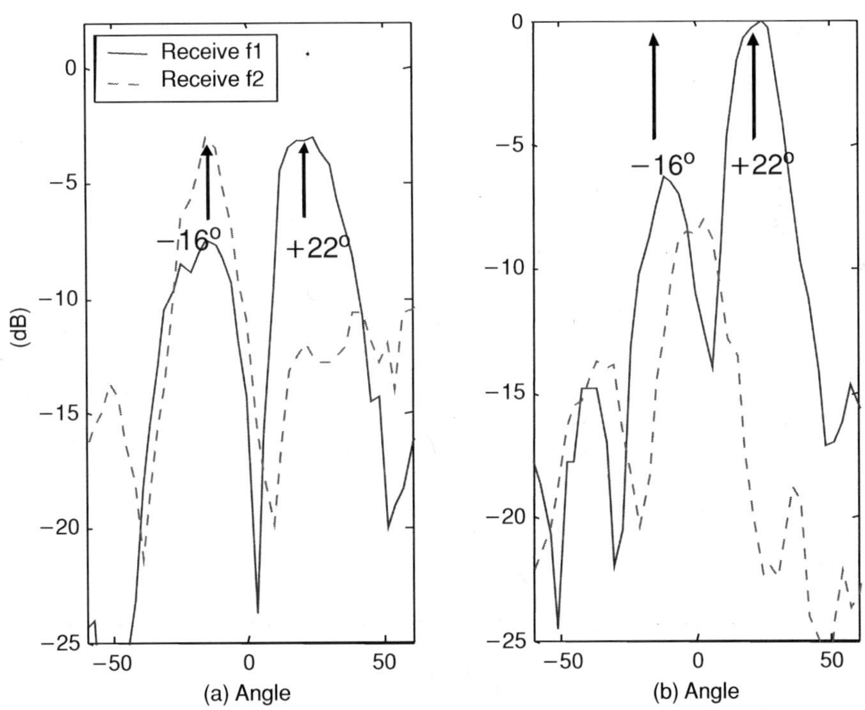

Figure 8-19 Antenna patterns: (a) before; and (b) after a null is formed in the direction of interference.

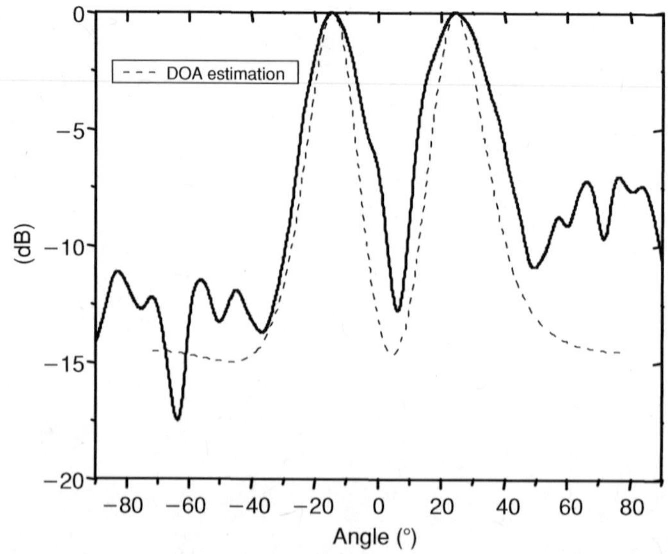

Figure 8-20 Bistatic pattern for two sources incident DOA estimation result also shown demonstrating proper detection of incident source location.

To better visualize this, two sources at precisely the same frequency are used. A single bistatic combined pattern is formed as shown in Figure 8-20. Observe that two main beams exist again implying that the retransmitted signal will be sent back to both interrogators. Also shown is the DOA estimation result after being computed in DSP. It is seen that the two incident locations are predicted accurately—this is important in determining the appropriate location to create a null. Once the source locations are appropriately found, the next step would be to determine which of the two sources is the interfering source, and this can be done by a simple identification test. A null will then be formed in each receiving sub-array at that location and only the desired source will be seen and communicated with. Figure 8-21(a) shows the RDA pattern for the case where two sources are present before and after a null is directed toward the interference. It is observed that only one main beam is seen and the interference location is suppressed. While the interference is suppressed by the sub-arrays, the desired source can be continuously tracked by the RDA and is shown in Figure 8-21(b). The amount of suppression will depend on several factors. First of all, the results shown in this study assume that the power levels from all sources are equal at the input to the RDA. Aside from the relative power levels of each source, the main factor in the determination of the degree of interference suppression is in the depth of the null created in each subarray, which corresponds to the calibration, consistency, and overall performance of each sub-array.

In this section a novel RDA with interference rejection capability was demonstrated. Under multiple source interrogation, this system can first distinguish a "friend" from a "foe" (i.e., it is an IFF system) then suppress the unwanted source by controlling the array element pattern, basically consisting of a two-element subarray, creating an exclusive link with the friendly source. Once this link is established, secure data communication to the target can proceed with automatic analog tracking. Bistatic pattern measurements of the array with one and multiple sources verified the retrodirective operation of the array as well as the effectiveness of the interference rejection mode. An interference link test proved a jamming signal to be detrimental to an RDA by possibly being a thief of information, or corruptor of the transmitted data in the communication link itself.

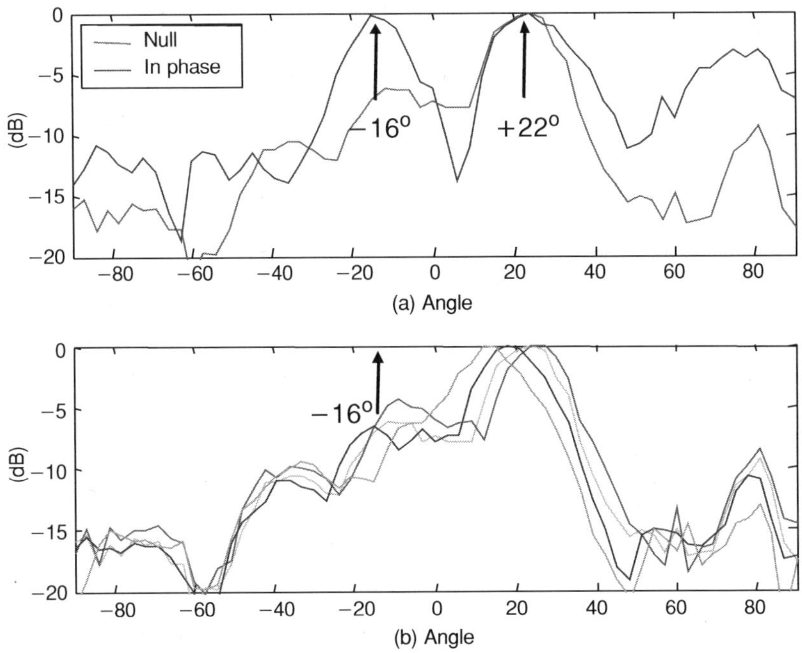

Figure 8-21 Retrodirective array: (a) bistatic pattern for two sources shown before and after a null is created; and (b) tracking of desired moving source after interference is suppressed.

8.7 HARDWARE REDUCING RETRODIRECTIVE ARRAYS

In any type of large phased array architecture, the extensive hardware requirements can boost the cost of such systems to the point that it may become impractical to fabricate. This is true in typically active antenna arrays or smart antennas where individual transceiver chains are required at each antenna element. This includes analog phase shifters, amplifiers, mixers, and ADCs. Therefore any reduction or relief in this area could lead to substantial savings in space, power and cost for a given system.

In this section two techniques employing RDAs are demonstrated that reduce hardware requirements. First a 4-element sparse configuration is demonstrated [437]. Then, an element switching scheme is introduced for hardware reduction [438]. Finally the two techniques are integrated into a single RDA architecture offering a maximum amount of hardware reduction [439].

8.7.1 Sparse Retrodirective Array

Sparse or thinned arrays have been applied to typically large arrays as a means to greatly reduce the overall number of elements, decreasing the amount of circuitry required, while maintaining certain array parameters such as beamwidth, beamshape, scan range, and sidelobe level [440,441]. Elements can typically be rearranged within a given aperture, thinned within that aperture, or even distributed over a larger aperture to represent a larger array with minimal array elements. This example demonstrates a 4-element sparse array, allowing for a hardware demonstration of the sparse concept in a retrodirective transponder architecture.

The antenna pattern of an optimized 4-element sparse array pattern with optimized element spacing $d = [0\ 0.46\ 1.71\ 2.61]\lambda_o$, is shown in Figure 8-22. (Sparse array theory and the rigors of

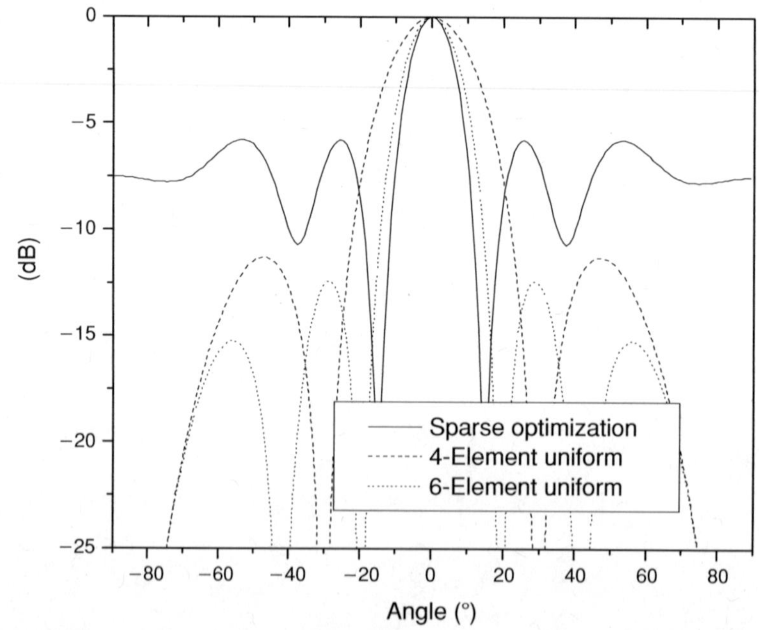

Figure 8-22 Antenna pattern of an optimized 4-element sparse array pattern within a 6-element aperture ("Sparse Optimization"). This is compared to the antenna pattern of a $\lambda_0/2$ uniform 4-element ("4-Element Uniform") and of a 6-element ("6-Element Uniform") arrays.

using an optimization algorithm to find the array element spacing are outside the scope of this chapter.)

It is seen that the antenna pattern of the 4-element sparse array has final configuration synthesized of approximately the same aperture as that of a 6-element, $\lambda_0/2$ spaced uniform array. Theoretically this offers more than a 33% reduction in RF channel circuitry. The optimized pattern has a flat sidelobe response. The achievable minimum peak sidelobe level of a thinned array actually depends on the number of elements, the aperture size being synthesized, and the desired scan range of operation [440,441]. Figure 8-22 also shows the antenna pattern of a $\lambda_0/2$ 4-element uniform array. The beamwidth of the sparse 4-element array closely matches that of the 6-element uniform array but not that of the the 4-element uniform array. Although a narrow beamwidth is achieved with the optimized sparse array it can easily be seen that the tradeoff is in the sidelobe level attainable. For a larger array, this sidelobe response can be significantly lowered by exploiting the added flexibility with a greater number of elements.

In the definition for the monostatic pattern from Equation (8.2) the incident and observation angle were stated as always coinciding leading to the main beam always being tracked. However the array factor or radiation pattern changes as the interrogator is swept and thus the array directivity, which is given by:

$$D_0(\theta) = \frac{|AF(\theta)|^2}{U_0(\theta)} = \frac{4\pi\ |AF(\theta)|^2}{\int |AF(\theta')|^2 \sin\theta' d\theta'} \tag{8.5}$$

also changes with scan angle. The variation of the array directivity mainly depends on the appearance of grating lobes, or in other words, it is related to the element spacing. Figure 8-23 shows the

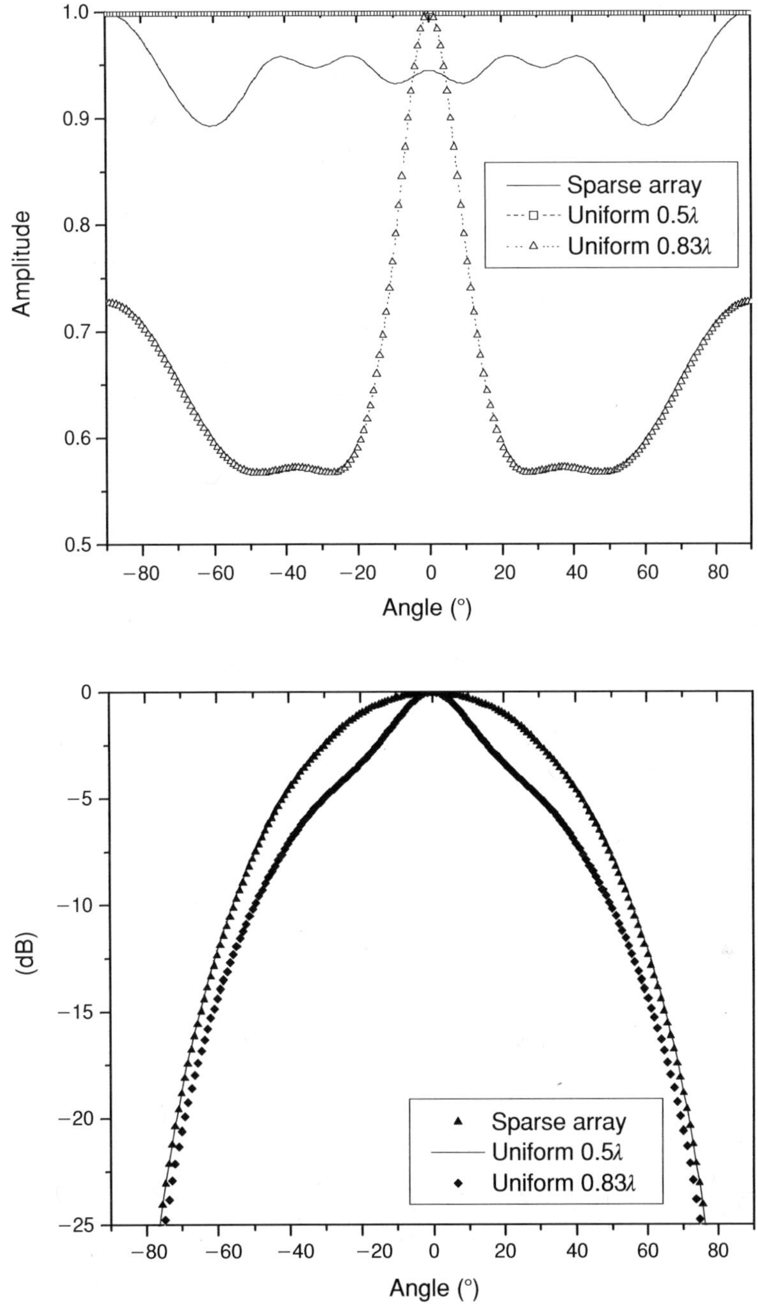

Figure 8-23 Monostatic retrodirective array: (a) Array directivity for the 4-element sparse array, a $0.5\lambda_0$ 6-element uniform array, and a $0.83\lambda_0$ 4-element uniform array and (b) monostatic patterns for same three cases.

normalized array directivity as a function of angle for the 4-element sparse array, a $0.5\lambda_0$ 6-element uniform array, and a $0.83\lambda_0$ 4-element uniform array, each of which have the same aperture size, and therefore another set of comparisons can be made. It is known that with the widely spaced uniform array, the scan range would be greatly limited due to grating lobes, resulting in the large and sharp drop-off in directivity as the array scans off broadside in the directivity plot. This is seen to be on the order of 50%. The $0.5\lambda_0$ uniform fully populated array does not suffer from grating lobes over a $\pm 90^\circ$ scan range, which is indicated in the directivity plot by a flat and basically constant response. The sparse array has the same number of elements and average element spacing as the scan-limited array, but due to the optimized design of element positions, has a much different directivity response, with only minor fluctuation over all angles. The monostatic patterns based on these plots and Equation (8.2) are also shown in Figure 8-23. As expected, the scan limited $0.83\lambda_0$ uniform array monostatic pattern is degraded, as the pattern shows a narrowing effect as the array would be scanned. This drop-off in power therefore affects the performance of the RDA. With the sparse array the element positions are optimally designed to suppress grating lobes over a given scan range and the directivity plot showed minimal variation over the entire range. As a result, the monostatic pattern is nearly identical to that of the ideal $\lambda_0/2$-spaced array, indicating that the sparse RDA is able to operate as well as the uniform array in terms of beam scanning.

The LO distribution network was designed to provide a common signal (in magnitude and phase) for each of the mixers by properly offsetting the transmission lines in each branch. The phase-conjugators for each element used a balanced mixer structure consisting of Agilent HSCH-5330 Schottky diodes fed through a 90° hybrid coupler. The RF input and IF phase-conjugated output of the circuit shared the same port, to simplify the design and to allow the use of a single antenna element for simultaneous receive and transmit. Using the $\omega_{LO} = 2\omega_{RF}$ relationship, the input (RF) and output (IF) frequencies will be the same, and therefore the hybrid coupler was necessary to achieve isolation between the incoming signal and phase-conjugated outgoing signal. More specifically, the RF leakage was steered away from the IF output port and dumped into a matched load at the opposite port of the coupler. The mixers produced less than 8 dB of conversion loss and achieved more than 15 dB of RF–IF isolation with the coupler. The circuit is shown in Figure 8-24.

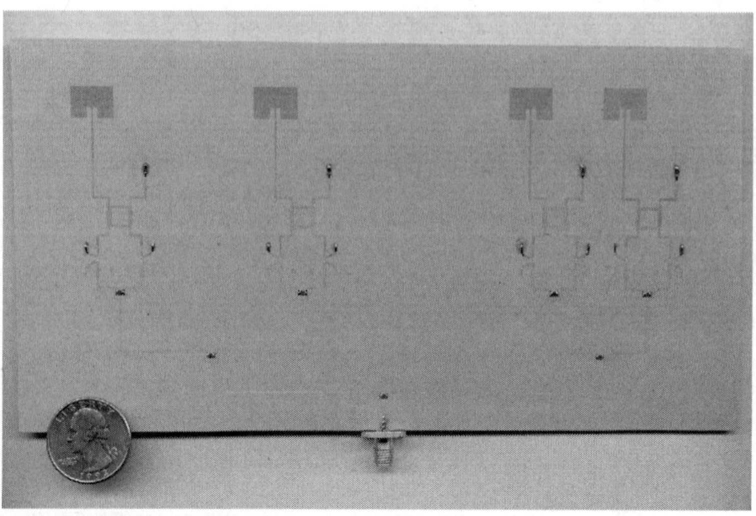

Figure 8-24 Photo of 4-element sparse RDA.

Figure 8-25 shows the bistatic pattern for the sparse RDA compared with sparse array theory for various scan angles demonstrating the self-tracking retrodirective function with the sparse architecture. The sparse design offers the synthesis of a narrow beamwidth and maintenance of proper array scanning equivalent to a full uniform array.

8.7.2 Element Switched Retrodirective Array

Typical RDAs operate by introducing either a phase-conjugator into each channel or one for each radiator. This allows for the received signal to be phase-conjugated individually at each element leading to a complete reversal of the phase gradient across the entire array. This directs the retransmitted signal back in the direction of the incident signal. In this section a time-sharing switching scheme is proposed that alleviates this requirement of phase-conjugating circuitry, without any loss of the retrodirective automatic beam-steering function.

The basic concept of this system is shown in Figure 8-26. By sequentially switching each antenna element connected to the feed network through the switching elements, a single phase-conjugating mixer can be shared by the signals received from all of the individual antennal elements. In this way, the amount of phase-conjugating circuitry/hardware required is reduced by a factor of N, for an N element array. Not only will the use of switches reduce the required RF array hardware to a single circuit, it will essentially eliminate channel imbalances that normally may occur due to the natural variation between active RF components.

The form of the phase-conjugated lower side-band product of the incoming RF and the reference LO signal was given in Equation (8.1), where the original incoming signal would be of the form $V_{RF}\cos(\omega_{RF}t + \theta_{RFn})$, the retransmitted (IF) frequency is $\omega_{IF}=\omega_{LO}-\omega_{RF}$, and $-\theta_{RFn}$ represents the conjugated phase. Each mixer is pumped by a common local oscillator (LO), which serves as the phase reference so that the relative phase between each channel (θ_{RFn}) is solely based on the element spacing and incident signal direction. In precisely the same manner, the single shared LO signal can be used as the common phase reference in the proposed switching scheme providing that the proper phase progression through each channel is based on the incoming RF signal's information. The only difference is the time sharing of the RF mixer by each of the antenna elements. This concept and its effect will be described in the following analysis.

In this scheme, a single-phase conjugator is sequentially time shared by all of the antenna elements. This is done by using a switching network which allows only one antenna element to access the phase conjugator at a time. This suggests that the shape of the output waveform of each antenna will be similar to an on/off or ASK (amplitude shift keying) type of modulated waveform. However, since the phase conjugation process occurs virtually instantaneously, the phase reversal at each antenna element needed for retrodirectivity is maintained.

The pulse switching waveform in the first channel for a 4-element array is given as:

$$x_1(t) = \begin{cases} 1 & 0 < t < \frac{T_s}{4} \\ 0 & \frac{T_s}{4} < t < T_s \end{cases} \tag{8.6}$$

with

$$x_n(t) = x_1[t - (n-1)t_o] \tag{8.7}$$

where the switching period, or time for all antennas to access the phase conjugator is T_s and the sequential time delay between each channel is t_o. These periodic signals can be represented as a Fourier series containing a DC term and harmonics based on the fundamental switching frequency $\omega_s = 2\pi/T_s$.

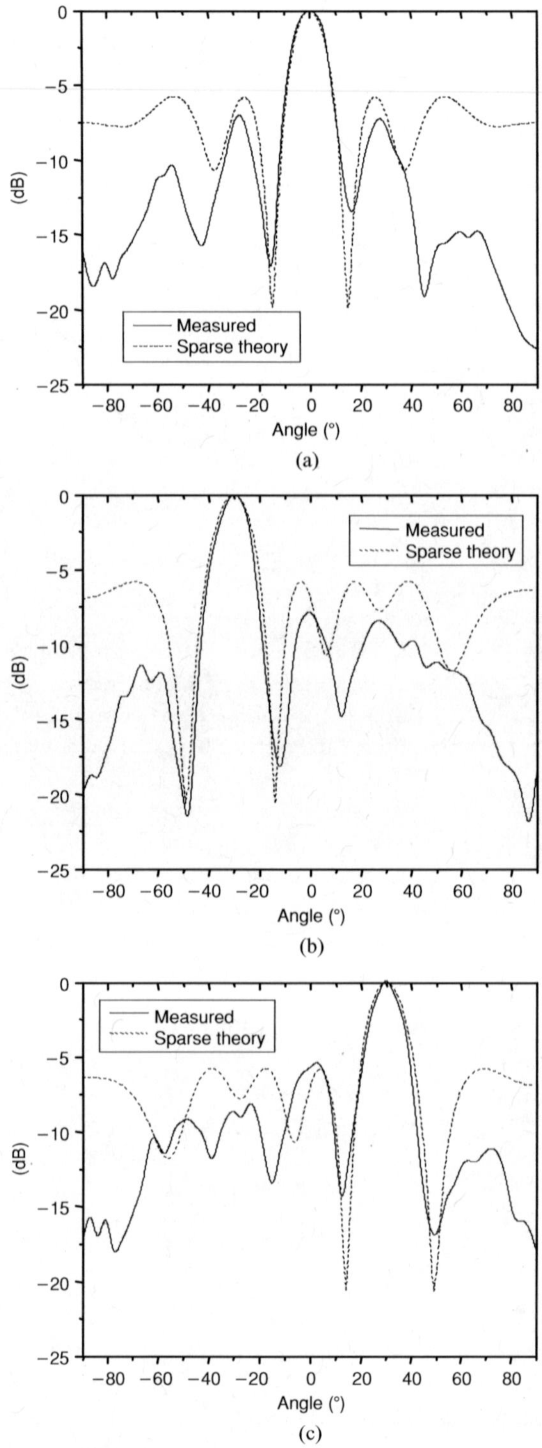

Figure 8-25 Bistatic pattern for the sparse RDA compared with sparse array theory for (a) broadside pattern; (b) -30° scan angle; and (c) $+30^\circ$ scan angle.

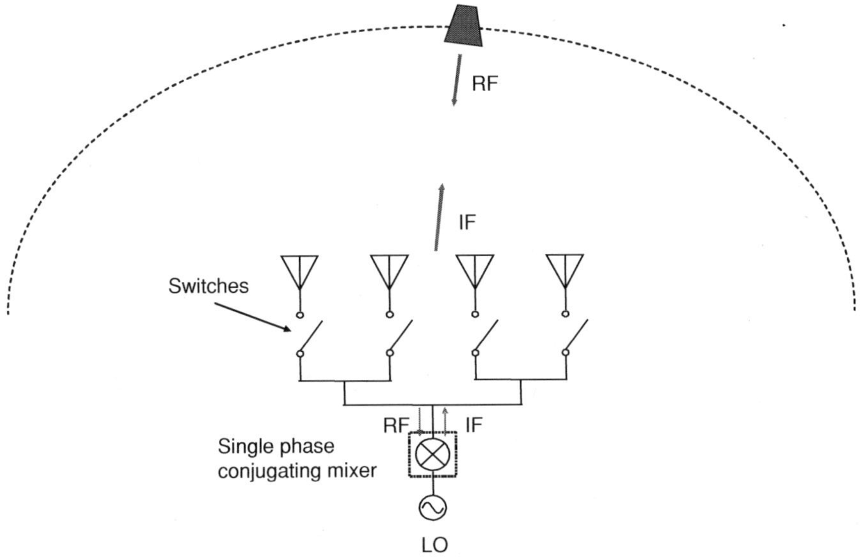

(a)

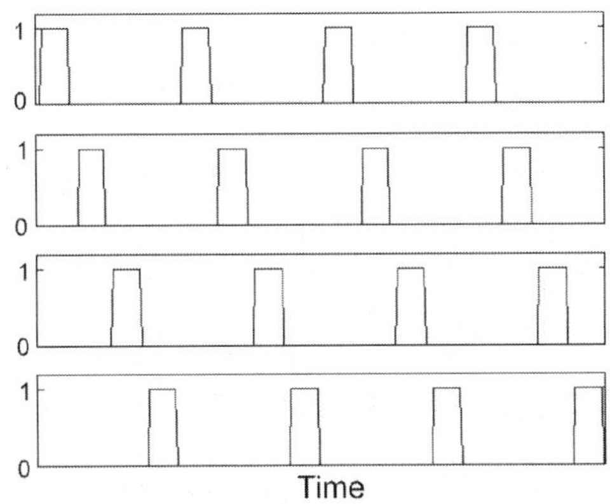

(b)

Figure 8-26 Schematic of the switched antenna RDA.

In a manner similar to the ASK format, when multiplied by the phase-conjugated signal from Equation (8.2), the periodic pulse train $x_n(t)$ acts as an amplitude modulator while keeping ω_{IF} and θ_{RFn} constant. The signal that is retransmitted from each channel is therefore:

$$S_n(t) = x_n(t) \cdot \frac{1}{2} V_{RF} \cdot V_{LO} \cdot [\cos(\omega_{IF} t - \theta_{RFn})] \tag{8.8}$$

with $x_n(t)$ simply introducing sidebands and harmonics to the signal based on the switching frequency. If these sidebands are filtered out, what would remain in each channel is the purely phase-conjugated carrier as in a conventional RDA from Equation (8.2). These retransmitted signals would then sum constructively and destructively in space based on their relative phases and form the radiated pattern.

The radiation pattern of the RDA with the time-shared phase conjugator is also formed by the summation of each element's transmitted signal, in precisely the same way as in traditional phased arrays. However, since only one antenna is active at a given instant in time, signal summation is not done instantaneously and should be considered over a complete switching period. In this way, each channel has some contribution at the carrier frequency and it is easier to visualize the actual signal that is retransmitted by the entire array. The complete retransmitted signal will therefore combine and be directed back toward the original source based on the array's retrodirective properties. In other words, even though only one channel is active and transmitting at a given instant, it is verified that the switching will have no influence on the beamforming and beam-steering of the array. To better illustrate this, Figure 8-27 depicts an example of the system in operation for a 4-element array, with an incident source at point $P1$. The original received signal creates a phase gradient across the array elements based on the direction of $P1$. The retransmitted signal from the array will be phase-conjugated and amplitude modulated, based on Equation (8.8) and is shown within the individual channels. Due to phase conjugation (phase gradient reversal) across the array, the received signal at $P1$ will be back in phase (seen in the individual retransmitted time domain signals), representing retransmission of a main beam by the array. The carrier frequency is reduced in the figure for better phase visualization and the waveforms are shown over two switching periods (i.e., $2T_s$). Again, the received signal at any location is essentially the summation in time of the signals from each channel:

$$S(t) \propto x_1(t) \cdot \cos(\omega_{IF}t + \varphi_1) + x_1(t - t_o) \cdot \cos(\omega_{IF}t + \varphi_2) + x_1(t - 2t_o) \cdot \cos(\omega_{IF}t + \varphi_3) + x_1(t - 3t_o) \cdot \cos(\omega_{IF}t + \varphi_4) \,. \tag{8.9}$$

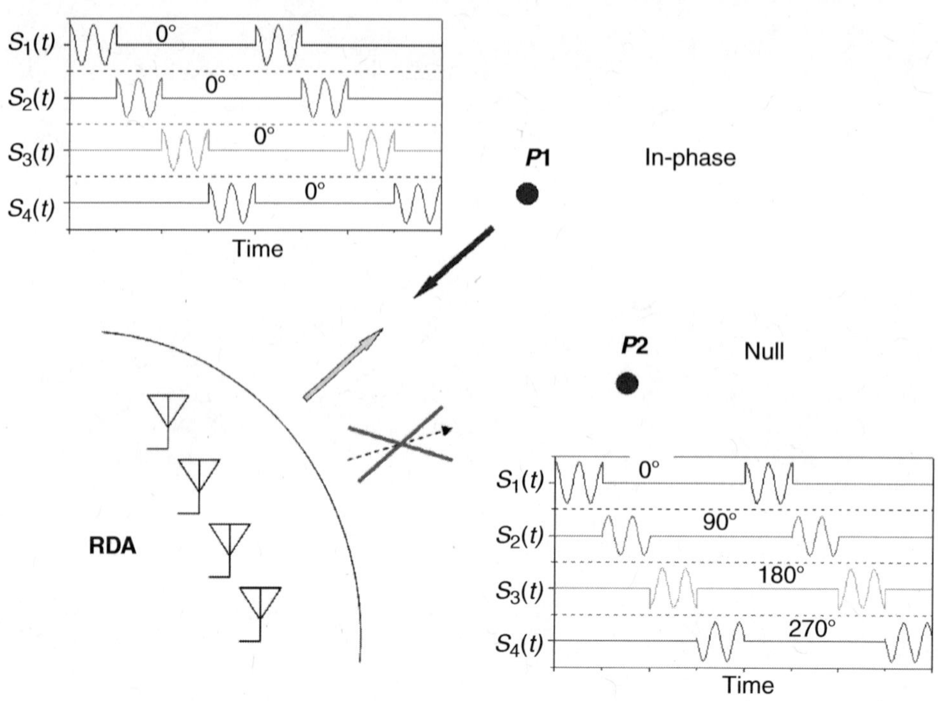

Figure 8-27 Communication scenario for switched-element network.

Here, ϕ_n represents the relative phase of each channel that is received at the observation point. A main-beam is therefore established when $\phi_1 = \phi_2 = \phi_3 = \phi_4$, or at the observation point where the phase-conjugated $-\theta_{RFn}$ is equivalently received from each channel.

The first null of the retransmitted pattern is shown to occur at $P2$, where the channel phase progression is 90°. At this point, the four signals add destructively based on the carrier phase. Based on the general array factor:

$$AF = 1 + e^{j\psi} + \cdots + e^{jn\psi} \tag{8.10}$$

where

$$\psi = \left(\frac{2\pi}{\lambda_o}\right) \cdot nd \sin(\theta) - \alpha \tag{8.11}$$

for a 4-element, $\lambda_0/2$ spaced array, this equates to a 30° angle off the main-beam direction in the array pattern, as expected. Therefore as the RDA with the time-shared switching scheme responds to interrogation, the retransmitted pattern will be based on the conventional case, without any switching. The signal strength at the carrier frequency will determine the radiated pattern, where the vector sum of N channels will vary depending on the relative phase between each channel, identical to a conventional phased array.

From both the previous example and Figure 8-27, it can be seen that although the retransmitted signal will be directive, with a beamwidth and pattern based on the array aperture, the obvious tradeoff is in the retransmitted power, which will be on the order of $1/N$ times lower due to the time-gating of each channel where only a single element would be active at a given instant, and the overall transmitted power is comparable to that of a single radiating element. Another side product of the switching would be the higher-order harmonics as mentioned previously. This can lead to some additional losses. Nonetheless, with this type of array architecture, a single active circulator circuit module could be integrated between the mixer and the feed network, allowing for the insertion of arbitrary gain blocks in both the receive and transmit paths.

For simple transponder-type applications the switching function can essentially double as a built-in modulator by making use of the on/off or ASK-type of waveforms generated from the switching. This would greatly simplify the overall system since the switching can be used for multiple purposes and no external modulator would be required for data transfer. In order to realize this effect, t_p, and t_o should be adjusted to force, or induce the existence of the switching frequency component into the spectrum of the transmitted signal. These parameters can be adjusted to increase and optimize the modulation index for the desired purpose. This effect is shown in Figure 8-28 for $t_p = t_o = T_s/5$. The phase conjugated carrier at 5.18 GHz is seen to contain its first spectral sideband component corresponding to the switching frequency. It is also illustrated that this frequency component will be shifted when the switching frequency is changed. This simple function can be used for basic binary data transfer within this proposed retrodirective transponder system by simply adjusting the modulation depth and varying the switching frequency.

Although in this demonstration it is proposed that the switching function be used as a built-in modulator, placing little constraint on the switching requirement (speed, etc.), it is also possible to transmit data through other forms of modulation schemes while using this architecture, adding to its versatility. If this were the case, the switching in each channel would be required to operate at or above the Nyquist rate to properly sample the signal bandwidth [442]. Of course this places a stricter requirement on switching and could therefore become a system limitation.

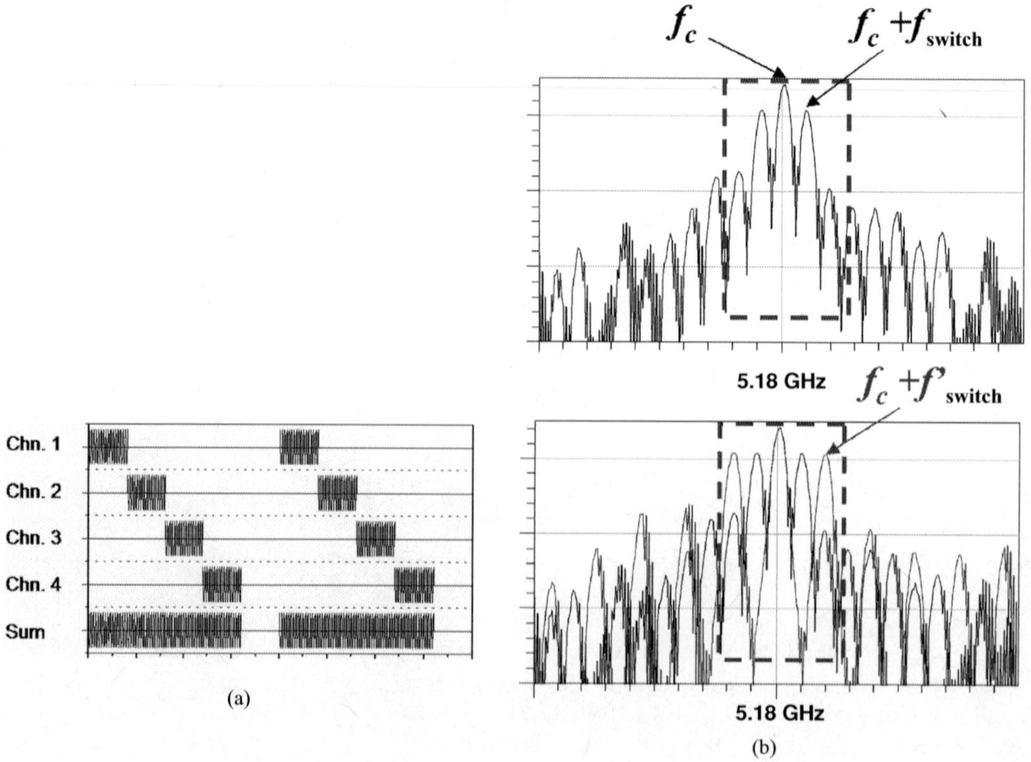

Figure 8-28 Introduction of ASK-type of modulation (a) and spectrum of retransmitted signal with switching component (b) top, and switching at second frequency (b) bottom.

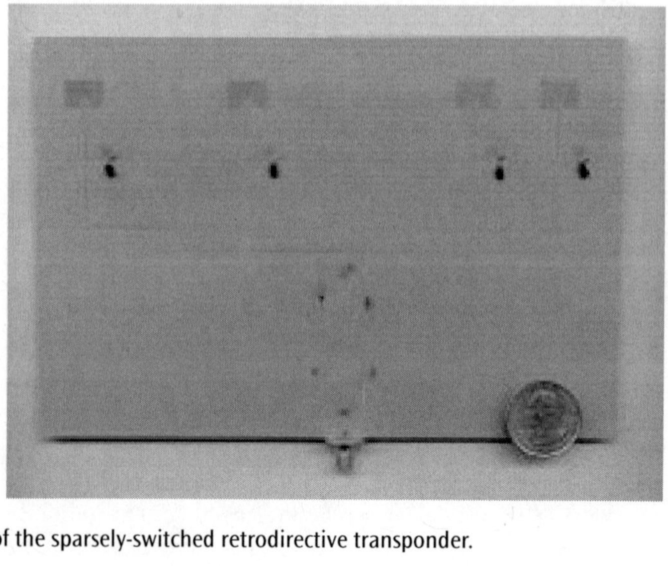

Figure 8-29 Photo of the sparsely-switched retrodirective transponder.

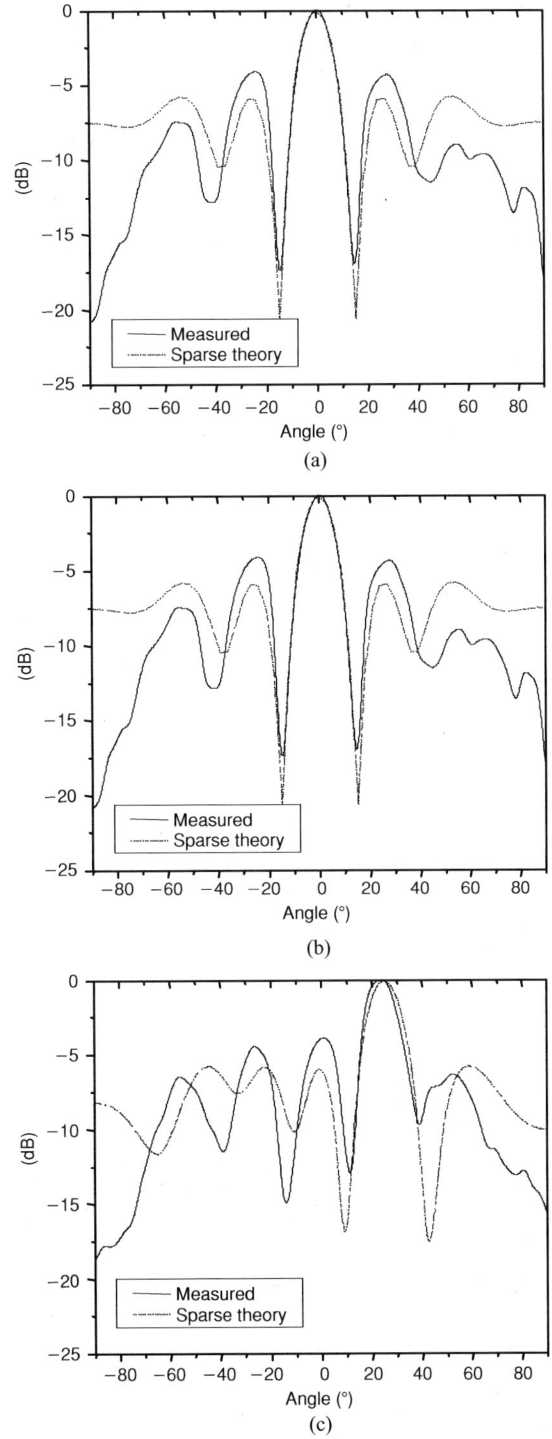

Figure 8-30 Bistatic pattern for the sparsely switched RDA compared with sparse array theory: (a) broadside pattern; (b) -30° scan angle; and (c) $+25^\circ$ scan angle.

8.7.3 Sparsely-Switched RDA

In this section, the two schemes from Sections 8.7.1 and 8.7.2 were combined to maximize the amount of hardware reduction. Figure 8-29 shows a photo of the completely redesigned switched array in a sparse configuration. The measured bistatic patterns are shown in Figure 8-30.

8.8 CONCLUSION

As demonstrated in this chapter, RDA applications have vastly advanced and expanded from the original simple self-tracking devices to more complex systems capable of wireless digital communication. Compared to DSP-based phased arrays, the retrodirective selftracking process is performed automatically without the use of phase shifters or digital circuitry, offering at least one key advantage over these other beam-steering architectures. Therefore RDAs can compete as a candidate for advanced radar, sensor digital mobile communication systems.

9

Multifunctional Radar, Communication and Radiometry Systems

Elliott R. Brown, Rahul S. Singh, George N. Saddik, and Austin C. Cotler

9.1 INTRODUCTION

In this chapter two new types of RF sensor systems are reported that were developed in simulation and experiment. One combines the radar and communications functions through one common aperture. It utilizes the technique of linear frequency modulated (chirped) waveforms and quasi-orthogonality of up- and down-chirps to combine the radar and communications pulses without the need for time-division multiplexing. It has been demonstrated at a center frequency of 750 MHz achieving an instantaneous bandwidth of 500 MHz, a 250 time-bandwidth product and a chirp rate of 1 MHz/ns. The system has a range resolution of 63 cm using targets having a Radar Cross Section (RCS) of 2.7 cm^2. With a radar and communication system operating simultaneously, the probabilities of detection (P_d) and false alarm (P_{fa}) were measured to be 99% and 7%, respectively. The measurement was conducted using a radar pulse width of 1.5 ns and pulse repetition frequency of 150 MHz, at a data rate of 1 Mbps. The bit error rate (BER) was also measured under the same above conditions and came out to be 2×10^{-3}.

The second radar system operates on the principle of retrodirectivity whereby a broadband noise waveform is transmitted by an array antenna and received by an identical but separate array. The received signal is fed back to the transmit array and re-transmitted, leading to retrodirective gain for targets lying inside a critical range. This system has been demonstrated at S-band (2.14 GHz) with two-element antennas and shown to have short detection time compared to conventional pencil-beam radar systems. Detection times less than 100 ns have been demonstrated out to ranges of about 10 m, and found to be limited primarily by the group delay through the transceiver electronics and a single round-trip time through free space.

Today, multifunctional systems are used in our daily life from PDAs to cell phones. The advantages of having a multifunctional system are low cost, reduced size, and simultaneous operation of different systems. The military sector has also taken advantage of multifunctional systems by developing broadband RF apertures that are capable of simultaneously operating communication, radar and electronic warfare [443, 444]. This chapter summarizes extensive efforts to develop two new multifunctional systems with radar as the primary function, and communications and radiometry, respectively, as the secondary functions.

In many modern RF systems combining the functionality of radar and communications would drive down cost and reduce size, especially through the elimination of multiple antennas. Phased arrays, for example, bring a prohibitive cost factor to high-performance, long-range systems, and multifunctionality would share this cost between different systems. In the present work, the quasi-orthogonality of the up- and down-chirp variants of linear-frequency-modulated (LFM) waveforms are utilized to co-locate and simultaneously operate communications and radar transmitter subsystems *without the constraint of time-division multiplexing*.

Recently an LFM-based communications systems demonstrated both the ability to use LFM waveforms in a wireless data application [445] as well as the mitigation of multi-path effects. This is advantageous in environments with many objects or RF reflectors. Besides enabling multipath mitigation, wideband chirped waveforms offer other advantages in multifunctional system performance well known in their earlier application in radar systems. For example, a chirp filter of given slope in a receiver approximates the ideal matched filter function when the transmitter has a chirp filter of identical characteristics but opposite slope. That may not be an important factor in communications but is generally important in radar with its much weaker signal detection requirements [446]. In addition, if the chirp filters are implemented as a passive surface acoustic wave (SAW) devices, as in the present work, they bring much greater simplicity in design and reduction in power compared to a chirped system based on electronically-generated LFM waveforms of comparable bandwidth. The second multifunctional system is aimed at improving RF sensing capability, including imaging, in close-range, cluttered environments such as an urban setting. Historically radar has usually been designed and implemented for mid- to long-range applications such as air defense, air traffic control, and collision avoidance. In these applications the decision times are long enough that the radar design need not be emphasized for fast detection and acquisition, but instead for high sensitivity and high resolution (i.e., for spatial and/or Doppler requirements). As such, when applied to short-range applications conventional radars must often be "cued" by another system, such as an ultrasonic sensor. There is a need in the radar field for RF sensors that can automatically detect and track a target without the need for a separate sensor to provide cuing. A need also exists for sensors that can detect very small targets.

Our approach to the radar challenges outlined above is based on retrodirective antenna arrays and noise-correlating signal processing. In rest mode the radar transmits broadband noise over a broad angle dictated by the antenna pattern of individual element an array. When a small target is present within a critical range, noise power is reflected back to the receiver where it is fully correlated between adjacent elements. This is good for target detection, and is also the basis for regenerative feedback to the target when the receive antenna array is interconnected to the transmit array retrodirectively. With each pass through the retrodirective feedback loop, the signal power grows exponentially, leading to very fast and confident detection.

9.2 COMMON-APERTURE RADAR AND COMMUNICATIONS SYSTEM

In this section an integrated radar and communications system is described with both of the subsystems sharing a common transmitting aperture. The system is ultra-wideband with simultaneous radar and communications operation achieved using quasi-orthogonality of the up- and down-chirp variants of the LFM. In 1962 Winkler first proposed the use of chirp signals for analog communications [447] and digital communication applications followed soon thereafter [448–452]. Chirped waveforms have been used in an indoor chirp spread-spectrum wireless communication system [453,454] that features chirped $\pi/4$-differential quadrature phase-shift keying

($\pi/4$-DQPSK) to achieve high data rates. This indoor system demonstrates a key advantage of wideband chirped waveforms in communications, which is mitigation of channel fading caused by multipath without using complicated signal processing [445].

Ultra wideband (UWB) signals date back to the first spark gap generators, with the first wireless demonstrations by Hertz and Marconi being examples of wide band transmission system [455,456]. In 1886 using a spark gap to generate RF radiation Hertz was able to detect oscillations at a distance of several meters in his lab. Later in 1895 Marconi was able to transmit and receive signals using a Hertz oscillator, antenna and a receiver [455]. In the late 1920's these types of signals were banned due to the inefficient use of the spectrum; however, the Federal Communication Commission (FCC) of the U.S.A. legalized the use of UWB for specific applications with well-defined restrictions. As part of the FCC standards signals are classified as being UWB if they have a fractional bandwidth greater than 25% [457], where fractional bandwidth is defined as [458]:

$$\Delta f = 2 * \left(\frac{f_{\text{upper}} - f_{\text{lower}}}{f_{\text{upper}} + f_{\text{lower}}} \right). \tag{9.1}$$

UWB signals are further classified based upon the specific waveform, such as a step function with a very short rise times, a very narrow pulse (i.e., impulse), and a single cycle sinusoid pulse [457]. Additionally the FCC designated frequency bands below 960 MHz for ground penetrating radar, and from 3.1 to 10.6 GHz for other ultra-wideband applications [456], including communications.

UWB signals have three propagation-related advantages over narrowband signals. First the propagation of UWB signals does not suffer from Rayleigh fading as is the case with narrowband signals. However their transmission is subject to multi-path dispersion effects [459]. Second, UWB is known for its penetration capabilities. Third, it can create images with less clutter. The second and third properties are independent of the system architecture [457]. For these reasons UWB signals have found application in military and commercial radar, and less commonly in communication systems. Narrowband radar is limited in the amount of information that can be collected from a target. In comparison UWB radar's short pulses can improve the radar range resolution and target recognition due to the increased information in the return pulse of the target as a whole and in the separate components. UWB can also reduce the passive interference from rain, mist aerosols, etc. due to the smaller scattering cross section of the passive interferer compared to the scattering cross section of the target [457]. In Table 9-1 the benefits of UWB radar and communications systems are summarized. The table emphasizes the differences of narrowband and UWB signals.

The integrated radar and communications system considered in the remainder of this section takes full advantage of the benefits of LFM waveforms. Along with the multipath mitigation, the use of chirp waveforms allow for a receiver that realizes matched filtering and is relatively inexpensive to build when SAW based filters are used. The system architecture is first presented, followed by simulation results and then experimental results for both the communications and radar operations.

9.2.1 Principles of Operation and System Architecture

Simultaneous operation is achieved using LFM signals of opposite slopes for the communications and radar transmitted pulses (see Figure 9-1) [460, 461]. It employs a single transmit antenna, the shared aperture, for both the radar and communications functions. The radar and communication receivers employ separate antennas for reception. The radar receive antenna is located adjacent to the radar transmitter antenna (this is the bistatic configuration), and the communication receiver antenna is located several meters down-range from the transmitter. The specific antenna placement is ultimately dictated by the application, with the present system configured

Table 9-1 Benefits of UWB signals in radar and communications applications.

Benefits of UWB radar	Benefits of UWB communications	Main differences between Narrowband and UWB signals
Increased range resolution.	Increased data rates.	Change of radar signal shape during target observation.
Reduction in radar "dead zone".	Reduction in Rayleigh fading.	Dependence of antenna parameters for signal shape and vice versa.
Recognition of target type and in forming an image.		Time variations of target RCS.
High radar immunity to passive interference.		Changes introduced in the form of the canonic radar range equation.
Increase in radar immunity to extraneous electromagnetic radiation and noises.		
Increase in target detection probability and reliable target tracking results in uniform RCS.		
Increase in target detection probability and target tracking reliability.		
Reliability of Target tracking at low elevation angles increases.		
Increase in radar operation security.		

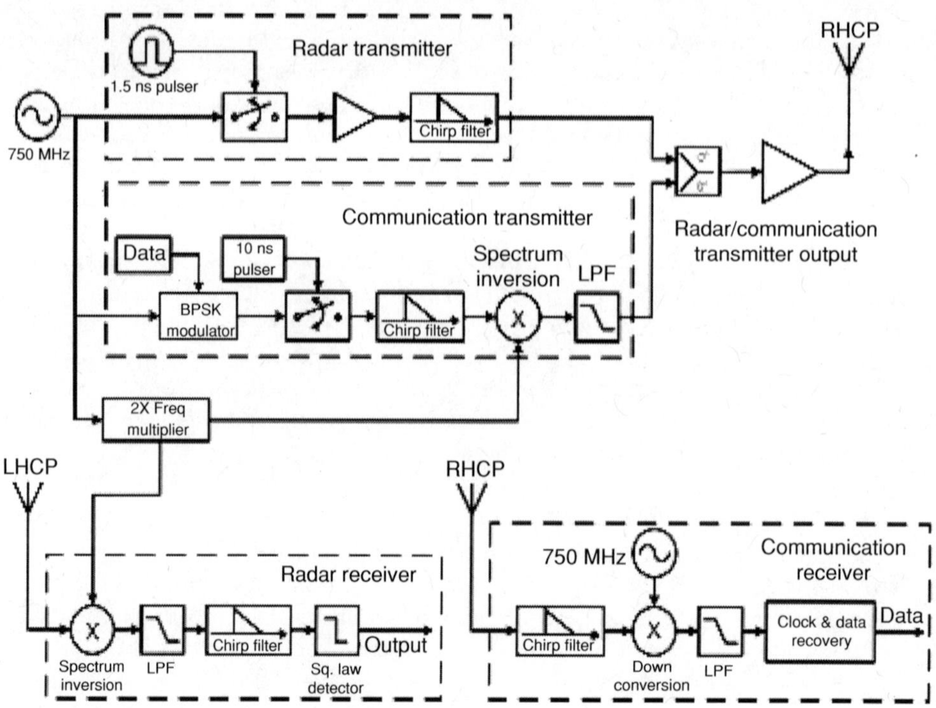

Figure 9-1 Block diagram of multifunctional communications and radar system.

only to demonstrate feasibility and characterize performance. Simulation studies of this system were performed previously [460, 461].

The quasi-orthogonality of opposite slope LFM signals is being exploited to allow for simultaneous communication and radar operation through a shared antenna aperture. Traditionally, LFM signals have been used for radar to improve range resolution while maintaining adequate average transmitting power [446].

The architecture of the multifunctional communications and radar system is presented in Figure 9-1. Operation and characteristics of the present system, Figure 9-1, are primarily defined by the expansive and compressive SAW chirp filters which are used in the transmitter and receiver, respectively. The communication and radar transmitters use LFM signals that are identical in frequency range but have opposite chirp slopes; positive (up-chirp) and negative (down-chirp), respectively. Because of the opposite chirp slopes, the communication and radar signals are quasi-orthogonal meaning that the cross-correlation in the time domain should be insignificant compared to the autocorrelation of either one. The correlation will be quantified later in the Section 9.2.2. The expander and compressor chirp filters both have down-chirp slopes; to achieve the up-chirp the spectrum must be inverted using a mixer with $f_{LO} = 2f_0 = 1500$ MHz either after the expander filter or before the compressor filter followed by a low pass Bessel filter to suppress the upper sideband [462]. The Bessel filter architecture was chosen for its linear phase response which is important in preserving the chirp waveforms. The center frequency of the chirp filters is 750 MHz with a bandwidth of 500 MHz, chirp duration of 500 ns, chirp rate of 1 GHz/μs, time bandwidth product of 250 and a fractional bandwidth of 0.66, calculated using the UWB definition in Equation (9.1).

The compressive filter provided has an advantage for the communication side of the system, it has a raised cosine window, which provides sidelobe suppression, and results in the reduction of intersymbol interference (ISI). However, this advantage comes at the expense of dispersion of the received pulse, which is a disadvantage to the radar side of the system since it will degrade the range resolution. Measurements of the expansive and compressive filters indicate an insertion loss of approximately 35 dB. Although this is a major disadvantage of the present SAW chirp filter, the high center frequency, the large bandwidth of 500 MHz, and the time-bandwidth product 250 (resulting in a processing gain of 24 dB) more than compensates for the insertion loss penalty. Finally the passive filters greatly reduce the complexity and power dissipation of the transmitter and receiver.

Three helical antennas were designed for the system; two were right-hand circularly polarized (RHCP) and the other was left-hand circularly polarized (LHCP). Oppositely polarized antennas were used for the bistatic radar because of the effect of reflection and antennas with the same polarization were used for the communication transmitter and receiver. The helical antenna was chosen for its high directivity, large fractional bandwidth, and circular polarization. The high directivity was desired as it increases target discrimination and reduces mutual coupling between the radar antennas. The large fractional bandwidth was needed to match the antenna to the chirp filters. Finally, circular polarization is beneficial because of its immunity to multiple reflections and multi-path effects. The system utilizes a single shared aperture transmit antenna for both the radar and communication. The radar and communication receivers employ separate antennas for reception. The radar receiver antenna is located adjacent to the radar transmitter antenna, and the communication receiver antenna is located several meters down-range from the transmitter. The bistatic configuration was chosen because of the inverting effect of reflection on helicity of polarization. This requires the receiver antenna to have opposite helicity as the transmitting helical antenna. Table 9-2 summarizes the designed and actual antenna parameters and dimensions.

Table 9-2 Designed and actual antenna parameters, and dimensions.

Parameters/Dimensions	Designed	Actual	Units
Helix diameter center to center	0.13	0.116	meters
Spacing between turns	0.1	0.089	meters
Ground plane diameter	0.3	0.460	meters
Conductor diameter	0.002 to 0.02	0.0017	meters
Pitch angle	14	13.7	°
Number of turns	4	4	turns
Impedance	140	127.5	Ω
HPBW	52	60.5	°
BWFN	115	133.9	°
Directivity	10.8	9.5	dBi
Axial Ratio	1.125	1.125	N/A

With the antennas and the system architecture chosen, a simple analysis was performed to evaluate the expected performance using Friis's equation (the classic radar equation). The maximum range was calculated to be 19.2 m assuming an SNR of 10 dB and using the radar equation

$$R_{\max} = \left(\frac{P_T G_T G_R \lambda_0^2 \sigma}{(4\pi)^3 S_{\min}} \right) \tag{9.2}$$

where P_T is the power transmitted, G_T and G_R are the gain of the transmitting and receiving antennas respectively, λ_0 is the free space wavelength, σ is the radar cross section (RCS), and S_{MIN} is the minimum detectable signal. The measured maximum range assuming an SNR of 10 dB at the output of the receiver is 15.2 m, which is within close agreement of the calculated maximum range with the same SNR requirement. The free space link loss was

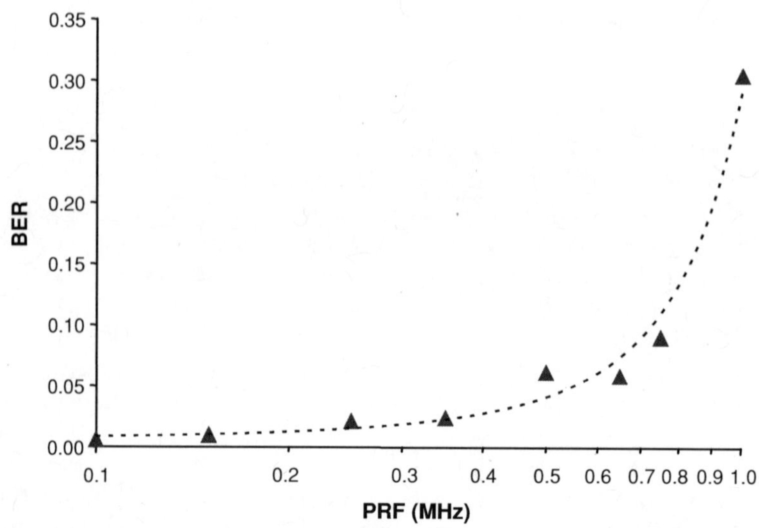

Figure 9-2 Bit Error Rate (BER) versus pulse repetition frequency (PRF) with data rate of 2.5 Mbps and radar pulse 1.5 ns.

calculated to be 87.9 dB using

$$L_{\text{freespace}} = \frac{\lambda^2 \sigma}{(4\pi)^3 R_{\text{MAX}}^4} \tag{9.3}$$

The receiver RF gain was calculated to be 44 dB, the LNA has a gain of 12 dB, the following gain stage has a gain of 20 dB, the mixer stage has a loss of 8 dB, the surface acoustic wave (SAW) chirp filter has a loss of 35 dB and a processing gain of 24 dB, and the next gain stage has a gain of 20 dB. The receiver bandwidth is 500 MHz which is limited by the SAW chirp filter. The receiver noise figure was calculated to be 2 dB, resulting in a calculated minimum detectable signal, MDS, of −75 dBm. Thus the output SNR, an SNR_{out} is 10 dB.

The communication transmitting subsystem uses a binary phase shift keying (BPSK) modulation scheme at a data rate of 1 Mbps. To generate the BPSK signals, a 750 MHz CW signal is fed into a 2 way 0°/180° splitter with its two outputs fed into a modulator controlled by the data. The modulated data signal is injected into a single-pole, single-throw (SPST) switch to reduce the bit time from 1 μs to 10 ns. For chirp compression and expansion, the input to the chirp filters should be no more than a few cycles and the 10 ns bit time was adequate to produce expanded and compressed waveforms. While the bit time could have been 1.3 ns (corresponding to a single cycle of 750 MHz center frequency), the added cycles allow for a simpler clock data recovery (CDR) architecture in the communications receiver. Having a bit width greater than the 2.6 ns limit reduces the ability of the chirp filter to expand and compress the bit. The modulated and bit-reduced data signal is fed into the expander down-chirp filter, which is followed by a mixer to invert the spectrum resulting in an up-chirp signal. The communication signal is combined with the radar signal through a Wilkinson combiner, amplified and transmitted through a single helical antenna.

The first stage of communication receiver is an LNA followed by a medium gain stage amplifier and a compressive chirp filter having the appropriate down-chirp slope. The output from the chirp filter is connected to an amplifier and that output is fed into a down-conversion mixer—the LO for the mixer is hard-wired from the transmitter's 750 MHz. The output of the mixer is the demodulated baseband signal. The down converted signal is then directed into a clock and data recovery circuit (CDR), which extracts the clock from the incoming non-return to zero (NRZ) data stream and uses this clock to retime the data. The basic idea behind the CDR is to use a PLL to extract the clock from the incoming data stream and use the recovered clock as the input to a D Flip-Flop to retime the data stream [463].

The radar signal has a single-cycle sinusoidal pulse waveform with a 750 MHz carrier, pulse duration of 1.5 ns, and PRF of 150 KHz. The CW pulse is amplified using a pulse amplifier and fed into the down-chirp filter followed by a gain stage. The two signals, radar and communication, now down-chirp and up-chirp signals respectively, are simply combined through a Wilkinson power combiner, amplified (+27dbm peak output power) and transmitted through the helical antenna.

The radar receiver also has an LNA and medium gain amplifier for the first stage which is followed by an inversion mixer to invert the spectrum from a down-chirp to an up-chirp. The output of the chirp filter is fed into a gain stage which is coupled to a square law detector filtered to remove the 750 MHz carrier frequency, amplifier and the output is displayed on an Agilent 54846 Infinium digital sampling oscilloscope.

9.2.2 System Simulations

To better understand the simultaneous operation of the communications and the radar systems, an analysis and simulation was performed on the communications receiver. The objective was to quantify the interference of the radar (down-chirp) signal on the communications (up-chirp) signal through the matched filter. The simulation was performed in MATLAB by filtering a communication and radar signal through the communication matched filter (i.e., the down-chirp compressor). The analysis was performed to verify the results and to provide a closed form expression for each signal through the communication's matched filter.

The communications and the radar waveforms are:

$$f_{\text{comms}}(t) = \text{rect}\left(\frac{t}{T_p}\right)\exp\left[j\left(2\pi f_c t + \pi k t^2\right)\right] \tag{9.4}$$

and

$$f_{\text{radar}}(t) = \text{rect}\left(\frac{t}{T_p}\right)\exp\left[j\left(2\pi f_c t - \pi k t^2\right)\right] \tag{9.5}$$

where k is the modulation rate (= 1 MHz/ns), f_c is the center frequency of operation (= 750 MHz), and T_p is the pulse width (= 500 ns). The communications matched filter (i.e., down-chirp compressor) has an impulse response function

$$h_{\text{comms}}(t) = \sqrt{4k}\,\text{rect}\left(\frac{t}{T_p}\right)\exp\left[j\left(2\pi f_c t - \pi k t^2\right)\right] \tag{9.6}$$

where t_{BW} is the time-bandwidth product (= 250). The time domain output signals through the matched filter were computed by the signals and the transfer function of the filter in the frequency domain (Fourier transformed) and then taking the inverse Fourier transform [464]. The resulting output matched filter signals for the communications and radar signals as inputs is then found to be:

$$y_{\text{comms}}(t) = \sqrt{t_{BW}}\,\text{sinc}\left(kT_p t\right)\exp\left[j\left(2\pi f_c t - \pi k t^2 + \tfrac{\pi}{4}\right)\right], \tag{9.7}$$

$$y_{\text{radar}}(t) = \frac{1}{2\sqrt{2}}a(t)\exp\left[j\left(2\pi f_c t - \pi\frac{k}{2}t^2 - \tfrac{\pi}{4}\right)\right] \tag{9.8}$$

where

$$a(t) = \text{erfz}\left[\frac{\sqrt{j2\pi k}}{2}\left(t + T_p\right)\right] - \text{erfz}\left[\frac{\sqrt{j2\pi k}}{2}\left(t - T_p\right)\right] \sim 2. \tag{9.9}$$

Here y_{comms} is the expected compression waveform of an up-chirp waveform, while the y_{radar} is the output of a down-chirp signal through a down-chirp compressor. Using MATLAB to numerically evaluate $a(t)$, it is found to be approximately 2 and is treated as a constant to evaluate the performance of the system. From Equation (9.8), it is observed that the output is not compressed but is rather a down-chirp signal that is twice as long as the input with the modulation rate halved, and the overall power halved. This leads to being able to state a peak power communication SNR for the system due to signal interference from the radar signal:

$$SNR_{\text{peak}} = 2 \cdot t_{BW} \cdot \frac{P_{\text{comms}}}{P_{\text{radar}}}. \tag{9.10}$$

The communications and radar signals, Equations (9.4) and (9.5), and the communication matched filter, Equation (9.6), were then simulated in MATLAB, with the waveform and spectra plotted in Figure 9-3 and the outputs plotted Figure 9-4. In the output plots the analytic results are overlaid atop the simulation showing the agreement between the simulation and Equation (9.6). Finally the output plots support the SNR expression above by showing the peak voltages as being equal 15.8 V for the communications signal and ½ V for the radar signal (the input signals were 1 V_{pp}).

9.2.3 Experimental Results

The radar experimental results were collected in a large open parking lot where the interference from other objects was minimized. The targets used in the experiment were two rectangular sheets of metal measuring 45.7 cm by 40.6 cm with a corresponding RCS of 2.7 m^2. From a single target the smallest return pulse width observed was approximately 3.6 ns. From this measurement the received radar pulse SNR was calculated to be 35 dB, and with the communication subsystem operating the SNR dropped to 29 dB. The criterion for minimum detectable range resolution is calculated to be ~43 cm using the relation $\Delta x = 0.5c\tau_P$, where c is the speed of light in air 3×10^8 m/s and τ_P is the measured pulse width.

As an example of the radar operation, the two targets were placed approximately 10 m away from the system and separated from each other in range by 63 cm. As shown in Figure 9-5, the experimentally determined range resolution is 63 cm which is in satisfactory agreement with the 43-cm prediction. The difference between the predicted and measured range resolution can be accounted for in the dispersion that the pulse experiences through the transmitter, and receiver. This dispersion is approximately 2.7 ns.

The probability of detection was measured using a logic analyzer to be 99% with the communication system off. With the communication subsystem operating at a data rate of 1 Mbps the probability of detection was 99% and the probability of false alarm was 7%.

The communication experimental results were obtained under the following conditions to achieve a benchmark results for the system. The output of the transmitter was connected directly to the input of the receiver with a 60 dB attenuator between them to protect the receiver from excessive power. The carrier signal was hardwired from the transmitter to the down conversion mixer in the receiver as the LO, and the data clock was hardwired to the receiver re-timing circuit. With these conditions a best-case eye diagram and BER measurement can be made. Figure 9-6(a) is an eye diagram of the communications waveform under the above conditions at 1 Mbps and Figure 9-6(b) is the eye diagram with the radar on simultaneously with the communication, the radar having a 1.5 ns pulse and a PRF of 150 kHz. As seen from the figures, the eye diagram closes in the vertical and horizontal directions by 6 dB and 4.7 dB, respectively, which are with in close agreement of the simulated results discussed above.

The BER measurement was conducted also under the above described conditions using a logic analyzer, and comparing the transmitted data to the received data—10^5 bits were collected and analyzed using MATLAB. The results of the analysis are as follows: the BER was less than 10^{-5} for the communication data at 1 Mbps and 0.002 for the communication data and radar simultaneously on with the data rate at 1 Mbps, and radar pulses of 1.5 ns, and PRF of 150 kHz. The BER was also measured while sweeping the PRF from 100 KHz to 1000 KHz, and as shown in Figure 9-2 an exponential increase in the BER as the PRF increases is apparent. These results show that in the best case scenario the radar has an insignificant effect on the communication. Table 9-3 summarizes some of the system performance parameters.

This system has demonstrated the feasibility of simultaneous operation of radar and communications using a shared transmit antenna aperture. The range resolution was demonstrated at

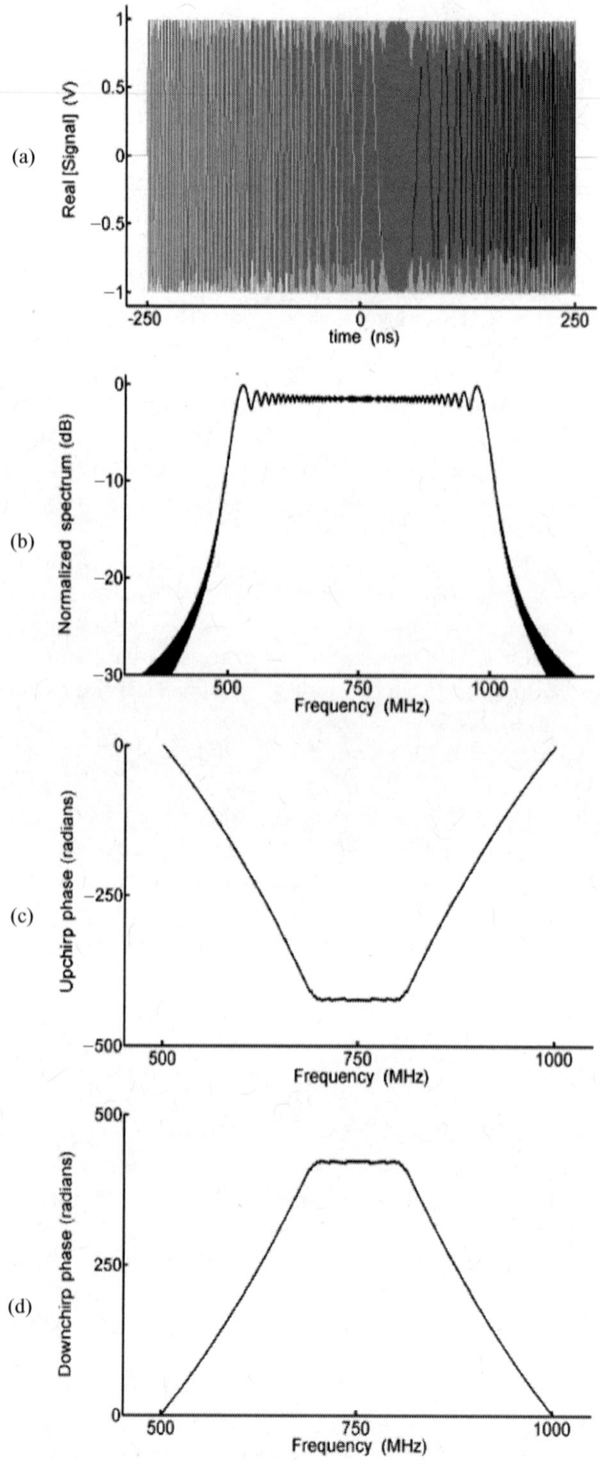

Figure 9-3 Simulation signals: (a) communications chirp (an up-chirp) in the time domain; (b) communications chirp in the frequency domain; (c) communications chirp phase in the frequency domain; and (d) radar chirp (down-chirp) phase in the frequency domain.

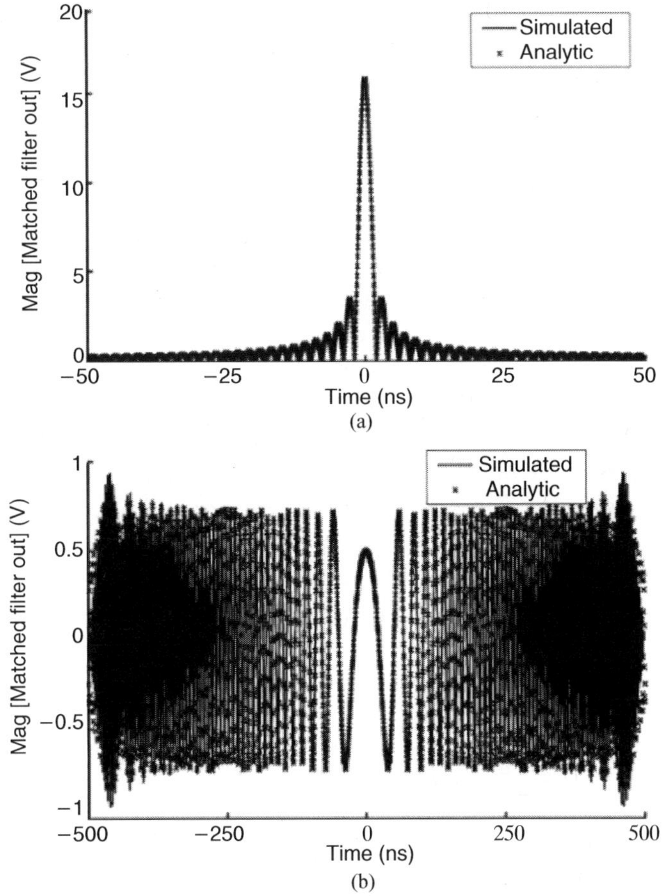

Figure 9-4 Simulated and analytically derived communications matched filter (down-chirp) output with: (a) communications (up-chirp) signal; and (b) radar (down-chirp) signal.

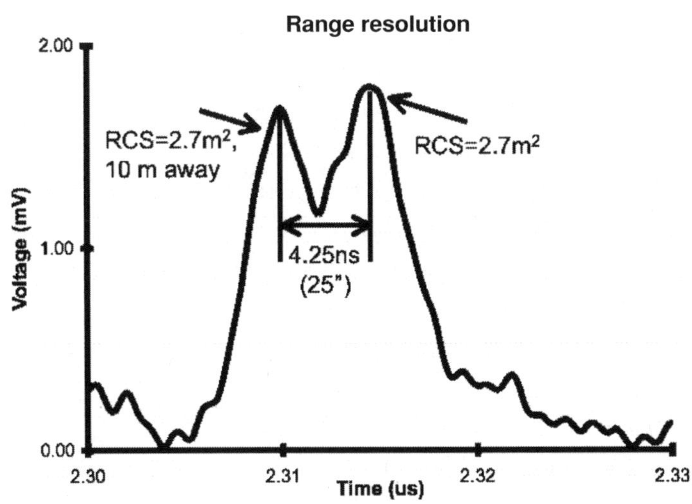

Figure 9-5 Range resolution at 10 meters away.

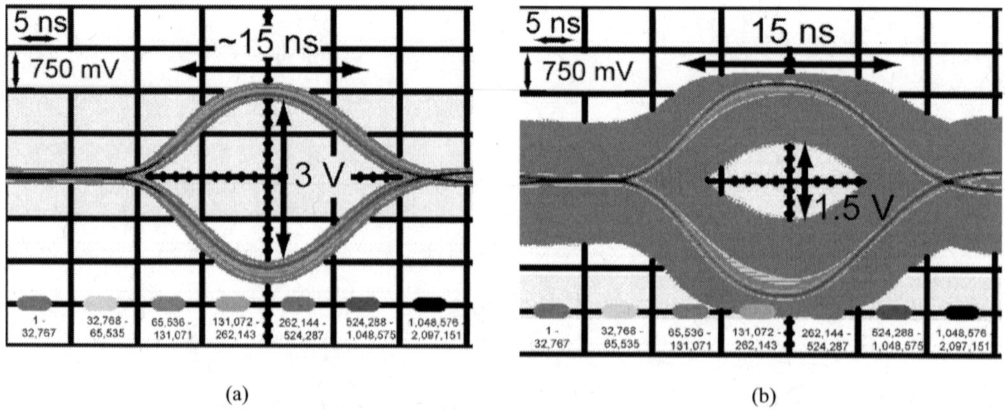

Figure 9-6 Communication eye diagram with: (a) radar off; and (b) radar on. The vertical resolution is 750 mV per division and the horizontal resolution is 5 ns per division.

10 m to be 63 cm and the probability of detection was 99%, and probability of false alarm 7%. The best case BER was measured to be 0.002 at a data rate of 1 Mbps, with the radar system on simultaneously and a radar pulse width and PRF of 1.5 ns and 150 kHz, respectively.

9.3 RETRODIRECTIVE NOISE CORRELATING RADAR

Retrodirective arrays (RDA) of the kind described in Chapter 8 are used here as the antenna array element in a noise radar system. The functionality exploited is the ability of an RDA to transmit a signal in the reverse direction to that of a received signal.

9.3.1 Principles of Operation and System Architecture

The retrodirective noise correlating (RNC) radar functionality is explained by the notional diagram in Figure 9-7. It starts in "rest mode" by radiating uncorrelated band-limited Gaussian noise from each antenna element in the transmitter (Tx) array, as represented by the first-pass output of Figure 9-7. The correlation between elements required for constructive interference, and thus array gain, is lacking. So the radiation is spread over the broad beam pattern of each element and the total power striking the target depends linearly on the number of elements N in the

Table 9-3 Measured radar performance with simultaneous communication at 1Mbps.

Parameter	Value
Measured maximum range	15.24 m
Measured minimum range resolution	63 cm
Pulse SNR (no communications)	35 dB
Pulse SNR (with communications)	29 dB
Calculated free space link	−87.9 dB
RF gain	40.5 dB
Receiver bandwidth	500 MHz
Receiver NF	2 dB
Minimum receiver level	−75 dBm

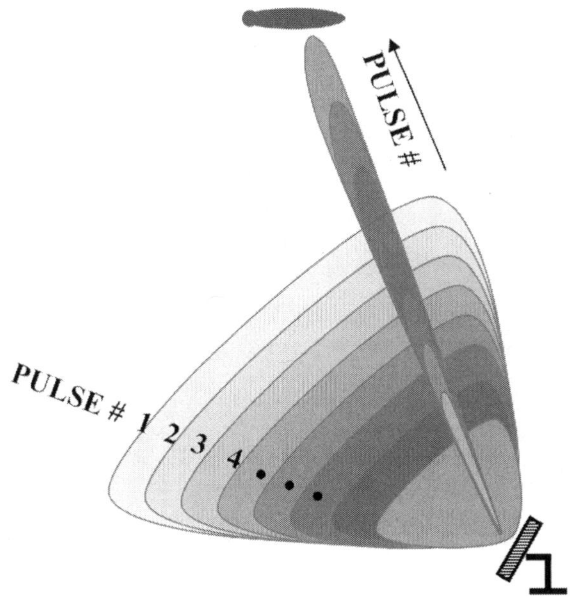

Figure 9-7 Behavior of an RNC radar in the presence of a small back-scattering target.

Tx array. If a small target suddenly presents itself to the RNC radar within the beam solid angle of each element, the target will intercept power from all the Tx antennas in proportion to its radar cross section, and scatters this power back to the receiver (Rx) array. A key feature of the RNC radar phenomenology is that this initial received radiation is correlated between adjacent elements of the Rx array. Although the scattered radiation remains stochastic in time, it is highly correlated between two points on a small target, as described by the Van-Cittert and Zernike theorem of statistical optics [465].

A second key feature of the RNC radar is that the transmit (Tx) and receive (Rx) antennas are *actively interconnected*, and in a retrodirective configuration [466], as shown schematically in Figure 9-8. In other words, the signal in each Rx element is amplified greatly within the RF band of the antennas, and connected to one and only one ("conjugate") Tx element such that the overall time delay is equal between all element pairs. Under this condition, the cross-correlated noise component is re-radiated into space from the Tx array with strong constructive interference and array gain, and is directed primarily toward the target, as shown for the 2nd and later transmit passes of Figure 9-8. The process repeats with a rapid growth of each subsequent transmitted pass, creating the "RNC feedback loop" and an inherent ability to detect and acquire targets very fast. The RNC radar has some commonality to recent retrodirective communications [467–469] and noise radar systems [470, 471]. But as a whole, it is a unique architecture particularly well suited to fast and automatic detection and acquisition of small and fast-moving targets.

9.4 EXPERIMENTAL DEMONSTRATION

9.4.1 Electronics

The RNC radar demonstrated in the present experiments consists of two identical transceiver chains as configured in Figure 9-8. The transmit and receive antennas are both Vivaldi antennas designed to have a broadside pattern with a directivity of approximately 6 dB. A photograph of

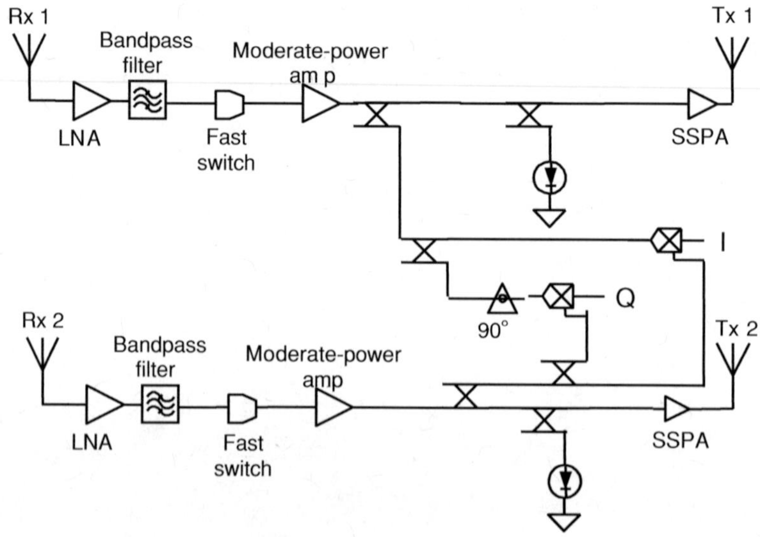

Figure 9-8 Schematic diagram of two-channel RNC radar.

these antennas is shown in Figure 9-9. They are both matched to 50 Ω and have a maximum return loss at the center frequency of 20 dB. Further details about the antennas and the transceiver design are discussed elsewhere [472]. Only a short summary is given here.

The front end of each chain consists of 4 LNAs each with a noise figure of 1.5 dB, an overall gain of 40 dB and a 1 dB compression of +10 dBm. The next component is the bandpass filter consisting of a SAW device having a center frequency of 2.14 GHz, a 3 dB bandwidth of 60 MHz, and an insertion loss of 2.6 dB. The next component is a fast semiconductor (pHEMT) switch having a risetime of 8 ns and an isolation of 32 dB. Following the switch are moderate-power RF amplifiers having 18 dB of gain across the passband and a 1 dB compression of +18 dBm. Finally, solid-state power amplifiers (SSPAs) drive the transmit Vivaldi antennas. The net electronic gain G_{amp} in each channel is 67 dB.

As shown in Figure 9-8, signal processing is carried out by sampling a small but equal portion of the power in each channel and doing in-phase (I) and quadrature (Q) cross correlation. The necessary electronics consists of 3-dB hybrids, a 90-degree phase shifter, and double-balanced mixers acting as analog multipliers. The RF and LO ports of each mixer were driven with equal portions of the adjacent-channel power, and the output was coupled from the IF port to either an integrating digital voltmeter or DVM (focusing on correlation strength) or a fast 50-Ω digital oscilloscope (focusing on detection time).

To do target detection the RF switch was turned off and on sequentially using a fast microcontroller in a burst of 8 pulses having 125 ns duration and 4 ns rise and fall times. The total duration of each burst was 4 μs and the off time between bursts was 5 μs. The reason for this rather complex switching waveform was to create a high enough duty cycle of the switch-on time to provide a measurable output on the integrating voltmeter, but a low enough duty cycle to prevent the amplifiers from saturating heavily or, worse yet, breaking into loop oscillations. Under these conditions, the criterion for target detection was defined as a cross-correlation SNR of approximately 10 as measured on either the DVM or the digital oscilloscope.

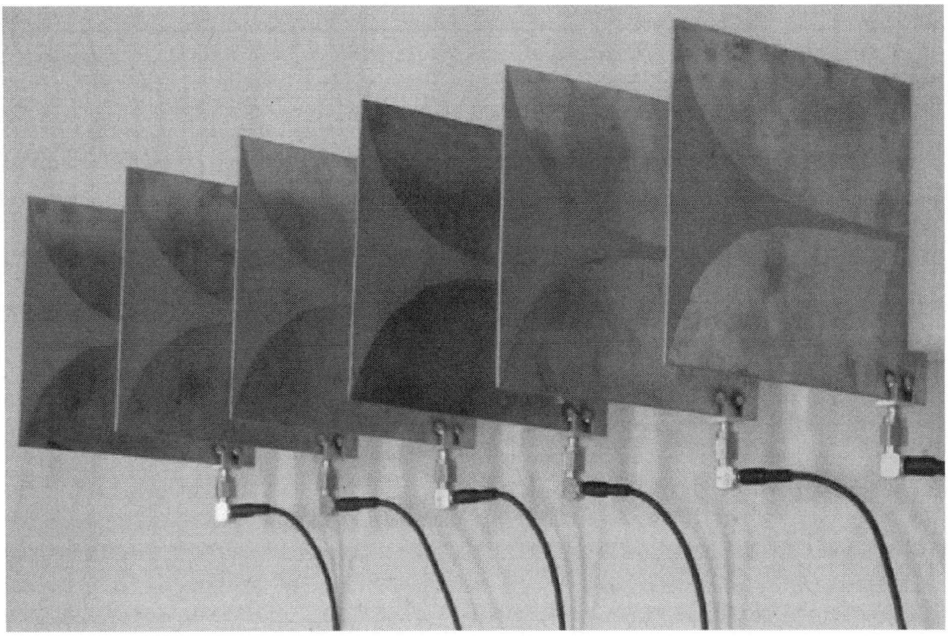

Figure 9-9 Vivaldi antennas used in S-band RNC radar demonstration.

9.4.2 Calibrated Range Measurements

To verify the behavior of the RNC radar, the first experiment was to check for target detection using an object of known RCS — an aluminized rubber sphere of radius $a = 11.2$ cm. From basic vector scattering theory, a metal sphere has an RCS σ best characterized as a normalized ratio $\sigma/(\pi a^2)$ versus the normalized frequency factor, a/λ [464]. For $a/\lambda << 1$, the RCS is proportional to $(a/\lambda)^4$ — the so-called long-wavelength or Rayleigh limit. For $a/\lambda >> 1$, $\sigma/(\pi a^2)$ approaches unity, consistent with geometrical optics. In our experiments, the center of the passband is 2.14 GHz where $\lambda = 14$ cm, $a/\lambda = 0.80$ and $\sigma \approx 1.2 \cdot \pi a^2 = 0.043 \text{m}^2$.

The aluminized sphere was placed on the end of a 7.6 cm diameter cardboard tube that could be positioned precisely at any range and orientation with respect to the 2-element RNC array. For the switching waveform described above, a strong I (in-phase) cross-correlation SNR of approximately 10 was measured by the integrating DVM out to a range R of 2.1 m. Removal of the sphere from the field-of-view of the radar caused the cross-correlation to drop into the noise floor. In comparison, a simple model discussed later, see Equation (9.13) predicts a range R of 2.9 m under the conditions, $G_{\text{AMP}} = 67$ dB, bandwidth $B = 60$ MHz, element gain $G_e = 4.0$, and number of elements $N_e = 2$. This is 38% above the measured range.

To confirm the retrodirective behavior of the two-element array, the sphere was moved in to 152 cm from the antennas and the I and Q signals were measured separately. The I signal was found to drop as the sphere was moved either way (left or right) from the broadside direction. The Q signal was found to become increasingly negative on one side and increasingly positive on the other, thus proving the essential nature of the RNC radar to transform uncorrelated AWGN (for additive white Gaussian noise) into strongly cross-correlated AWGN.

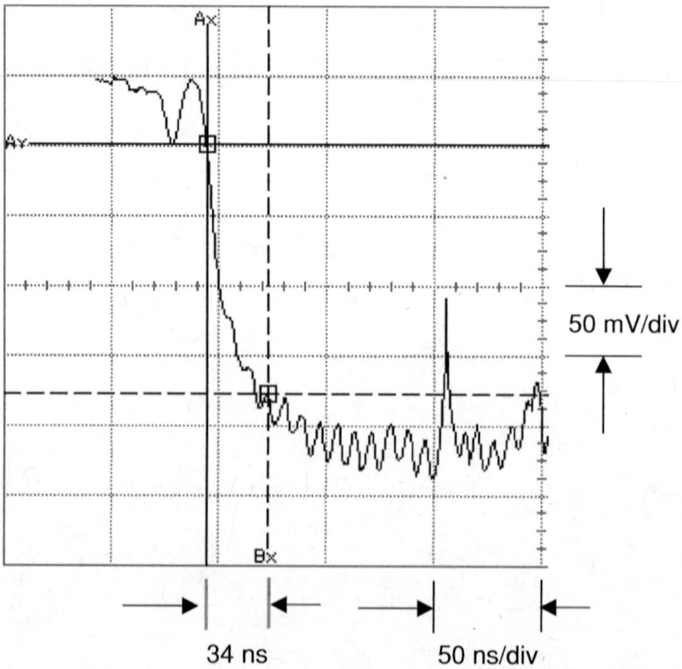

Figure 9-10 Fast oscilloscope I cross-correlation waveform showing acquisition of metallic spherical target with two-channel RNC radar. Note that the signal is negative going.

9.4.3 Measurement of Detection Time

Perhaps the most attractive aspect of the RNC radar is its inherently fast detection time. To verify fast detection it was necessary to first measure the electronic group delay, $t_g \approx 25$ ns, using a vector network analyzer. The first target detection experiments were then done with the spherical target very close (approximately 1 m) to the radar. The leading edge of the switching pulse was measured on the fast oscilloscope with the resulting I cross-correlation signal shown in Figure 9-10. The measured 10-90% risetime was 3 ns. This is remarkably close to the $t_{ACQ} = 32$ ns prediction of Equation (9.16) for $R = 1$ m and $m_{min} = 1$. In other words, the radar is acquiring the spherical target in one round-trip time through free space.

The accuracy of the measurement was improved by extending the range of the target so that the round-trip time in free space, $2R/c$, was at least equal to the total electronic group delay t_g. This required the use of a more specular target than the aluminized sphere and so a ~0.1-m^2 square of wire mesh was chosen. The 2.54 cm period of the mesh resulted in a high reflectivity in the radar passband. The mesh was aimed toward the radar to create a quasi-specular reflection, and target detection was achieved easily out to a range of at least 3.6 m as shown in the cross-correlation curve in Figure 9-11. Note also that the acquired signal strength varies with range approximately as $1/R^2$, confirming the quasi-specular nature of the target.

The plot of the I cross-correlation signal rise time as a function of range is shown in Figure 9-12. Note that t_{ACQ} varies almost linearly over this range—a fact that is at first surprising until one inspects the nature of Equation (9.16). At low values of R, the number of round trips to acquire the target is $m_{min} = 1$ and only the first two terms are important. Interestingly, it appears that for

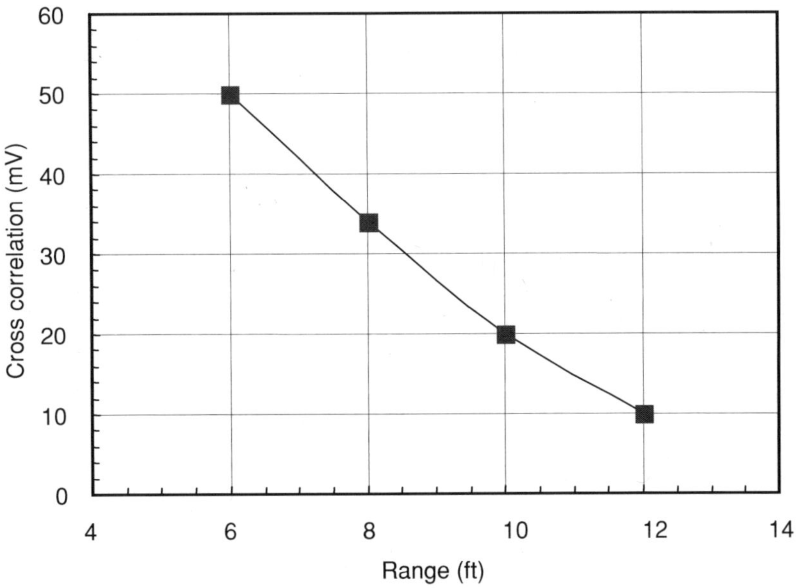

Figure 9-11 Time-averaged in-phase (I) cross-correlation signal as a function of range for a quasi-specular target.

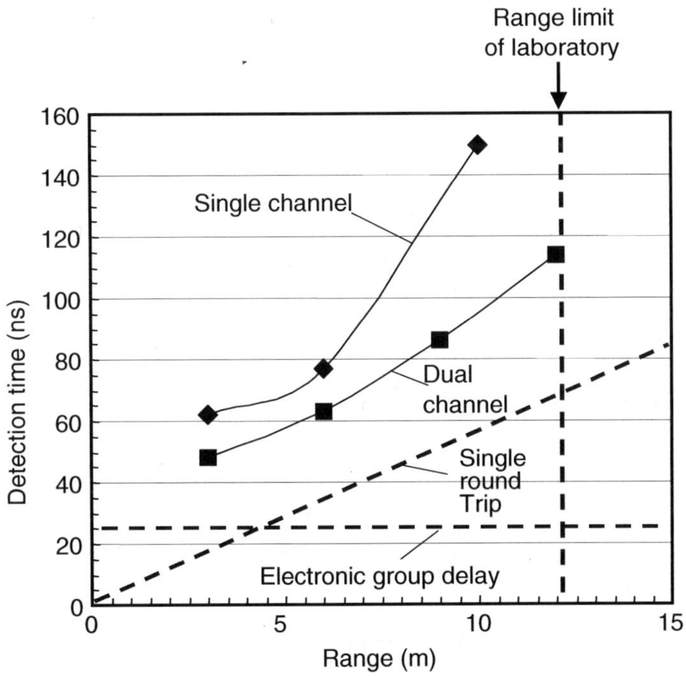

Figure 9-12 Cross-correlation (dual channel) signal acquisition time and autocorrelation (single channel) acquisition time measured for the RNC radar of Figure 9-1.

this quasi-specular target, $m_{\min}$ for the two-channel system remains at 1 even out to 12 m—the largest range possible in the laboratory.

9.5 ANALYSIS OF RANGE AND DETECTION TIME

9.5.1 Retrodirective-Gain Mode

The evaluation of the detection time in detail for a large target requires careful attention to electric field interference and distributed scattering effects for antennas radiating AWGN. Fortunately, the analysis becomes relatively simple for a small target in the far field and in the broadside direction where all elements of the transmit array have the same element gain, the array factor is unity, and each element of the receive array accepts the same amplitude (but different phase) of scattered radiation. Then if the target of cross section σ suddenly appears at a range R, it begins scattering power to each element that is fully correlated between the neighbors and after m scattering events (i.e., m passes through the transceivers) is given at each receiver by

$$P_{\text{out},m} = \left[G_{\text{link}} \cdot G_{\text{amp}} \cdot N_e^2\right]^m \frac{(F-1)k_B T_{300} B}{N} \tag{9.11}$$

where N_e is the number of elements in each array, G_{amp} is the net electronic gain of each transceiver channel, and B is the noise equivalent RF bandwidth of each channel. T_{300} is the ambient temperature nominally 300 K. The quantity G_{link} is the free-space link gain given by

$$G_{\text{link}} = \frac{\sigma}{4\pi R^2} \cdot \left(\frac{\lambda}{4\pi R}\right)^2 G_e^2 \cdot [F(\theta,\phi)]^2 \tag{9.12}$$

where G_e is the element gain and $F(\theta s \phi)$ is the element beam pattern, assumed to be maximum (unity) in the broadside direction.

Clearly, when $[G_{\text{link}} \cdot N_e^2 \cdot G_{\text{amp}}] > 1$ in Equation (9.11), this correlated 'signal' grows with each pass as does the signal-to-noise ratio. This inequality then defines the target range, which by rearrangement is found to be

$$R \approx \left[\frac{\sigma \cdot \lambda^2 \cdot G_e^2 \cdot G_{\text{amp}} \cdot N_e^2}{(4\pi)^3}\right]^{1/4}. \tag{9.13}$$

The greater the value of [G_{link} N^2 G_{amp}], the faster the detection will occur. The pre-detection signal-to-noise ratio after the m th pass is

$$\text{SNR}_m = \frac{P_{\text{out},m}}{(F-1)k_B T_{300} B} = \frac{\left[G_{\text{link}} \cdot G_{\text{amp}} \cdot N_e^2\right]^m}{N_e} \tag{9.14}$$

where F is the noise figure of the receiver. Given a minimally acceptable $\text{SNR}_{\min}$, this defines the minimum number of passes required for detection

$$m_{\min} = \frac{\log[N \cdot SNR_{\min}]}{\log[G_{\text{link}} \cdot G_{\text{amp}} \cdot N_e^2]}. \tag{9.15}$$

An approximation for the detection time t_d follows by assuming that the target is present and the radar is suddenly turned on by an RF switch (as described below). By properly accounting for the free-space and electronic delays then the acquisition time

$$t_{\mathrm{ACQ}} \approx \frac{2R}{c + t_g} + (m_{\min} - 1)\left[\frac{2R}{c + t_g}\right] \tag{9.16}$$

where t_g is the total electronic (antenna to antenna) group delay and c is the speed of light in air.

9.5.2 Noise Radar Mode

At long enough range or small enough target cross section (so that the stationary-to-cyclostationary transformation and associated mode-locking do not occur), the RNC radar as architected can still detect targets and determine their angular orientation in operation as a conventional bistatic noise radar. The dividing line between mode-locking operation and noise-radar operation is a function of the electronic gain, maximum transmit power, the noise dynamics and other effects, and is still being studied with the help of time-domain techniques. The noise-radar operation is much simpler to analyze and constitutes a "worst case" analysis of the radar range and target detection time. In noise-radar mode, the transmitter has no array gain so the power collected at each receive antenna element can be estimated as the incoherent sum of the maximum output power from each transmit antenna element, $P_{\max}$. For solid-state power amplifiers, the electronic gain is usually somewhat compressed in this region, but the nonlinearity associated with compression is not such a problem for noise radar as it can be for communications or coherent radar.

The peak power available at the output port of each receive antenna element is

$$P_{\mathrm{Rx}} = \frac{N_e G_e^2 P_{\max} \sigma \lambda^2}{(4\pi)^3 R^4}. \tag{9.17}$$

The average power will be somewhat less than this owing to the modulation required to discriminate target-scattered noise power from electronic noise. Assuming a simple amplitude modulation (i.e., fast chopping) with duty cycle D $(0 < D < 1)$, the pre-detection signal-to-noise ratio in each receive element is

$$\mathrm{SNR}_{\mathrm{pre}} = \frac{N_e G_e^2 (D P_{\max}) \sigma \lambda^2}{(4\pi)^3 R^4 k_B T_N \Delta\nu} \tag{9.18}$$

where k_B is Boltzmann's constant, T_N is the receiver noise temperature, and $\Delta\nu$ is the RF instantaneous bandwidth.

After sufficient RF gain in each channel to boost the signal well above the noise floor of RF mixers, a small portion of the power in each channel is cross-multiplied and then integrated (i.e., cross correlated) with all nearest neighbors[1]. Both I and Q cross-correlation are carried out to obtain target angle information. When the target is located in the broadside direction ($\theta = 0$), the post-detection signal-to-noise ratio at any I cross-correlator output will be

$$\mathrm{SNR}_{\mathrm{post}} \approx (\mathrm{SNR}_{\mathrm{pre}})^2 (2 \cdot \Delta\nu \cdot \tau) \tag{9.19}$$

where τ is the integration time. Finally, the (video) outputs of each I cross correlator are summed and demodulated. Because each I cross-correlator signal output is synchronized to the others,

1 Interferometric processing between channels (RF summation before multiplication) would lead to better range performance, but more difficult angle determination.

the SNR at the output of the summation circuit should improve as

$$\mathrm{SNR_{out}} \approx \mathrm{SNR}^2_{\mathrm{pre}} \cdot 2 \cdot C \cdot N_C \cdot \Delta\nu \cdot \tau \tag{9.20}$$

where C is a constant of order unity that depends on the exact modulation and synchronous de-modulation technique.

Our criterion for detection is that $\mathrm{SNR_{post}}$ have a minimal value $\mathrm{SNR_{min}}$ for both the total I and the total Q cross correlator summations. Equations (9.18) and (9.20) can then be solved for the range as a function of integration time, parameterized by the number of elements and cross correlations:

$$R \approx \left[\frac{N_e G_e^2 (DP_{\max})\sigma\lambda^2}{(4\pi)^3 k_B T_N \Delta\nu}\right]^{1/4} [2 \cdot C \cdot N_C \cdot \Delta\nu \cdot \tau]^{1/8}. \tag{9.21}$$

Equation (9.21) is plotted in Figure 9-13(a) versus integration time for $\mathrm{SNR_{min}} = 1$, four different array configurations ($N_e = 4, 8, 16, 32$), and N_C equal to the number of nearest neighbors in each configuration. Figure 9-13(b) is almost identical to Figure 9-13(a) except N_C is increased to the total number of unique pairs in each configuration, $N_C \approx (N_e)(N_e - 1)/2$.

The other key parameters in the curves are: $\sigma = 1\times10^{-3}$ cm^2, $\lambda = 14$ cm, $P_{\mathrm{MAX}} = 5$ W, $D = 0.5$ (e.g., PRF $= 500$ kHz, $t_p = 1$ μs), $T_N = 380$ K, $\Delta\nu = 60$ MHz, $G_e = 8.0$, and $C = 1.0$. As can be seen in Figure 9-13(a), for $N_e = 32$ and $N_c = 52$ (nearest neighbors only), the range passes 100 m at $\tau \approx 40$ μs. In Figure 9-13(b) with $N_e = 32$ and $N_c = 496$ (all unique cross-correlative pairs), the range passes 100 m at $\tau \approx 5$ μs. For reference, a Mach-1 (330 m/s) projectile travels approximately 13 mm in 40 μs and approximately 1.6 mm in 5 μs.

It is important to note that the cross section being simulated here, $\sigma = 1 \times 10^{-3}$ cm^2 (-30 dBsm), is quite small for a radar target (using HFSS, the high-frequency structure simulator). These statements are supported by X-band data for small objects indicating that even moderate-size flying insects (e.g., bees) have an RCS of approximately 16×10^{-4} cm^2 at 9 GHz [464].

9.6 NUMERICAL SIMULATIONS

9.6.1 Retrodirective Matrix and Loop Analysis with MATLAB

Like most electrical feedback effects, the RNC radar in the presence of a target must be analyzed with fields in space, and related current or voltage quantities in the transceiver channel. A flow diagram of the analytic technique used to model the RNC radar is shown in Figure 9-14 assuming four Tx and four Rx antenna elements arranged in the retrodirective architecture. Each antenna is modeled according to its far-field pattern, and the target is modeled as having a scalar radar cross section σ. Radiation reflected from a target can be received by any of the Rx elements, so the radiative path is highly cross-coupled as shown. The effect of the target is modeled as an electronic switch.

After reception, the radiation in each channel is amplified (with power gain G) and further randomized by additional AWGN from the receive electronics. The power in each channel is delivered without cross coupling to the transmit antenna where the process repeats. Hence the flow diagram in Figure 9-14 represents the RNC feedback loop (for 4 elements only). If the phase shift incurred by radiation around the entire loop is a multiple of 2π, then it will be constructively amplified on each pass, leading to rapid growth of the signal and, possibly, oscillation. If the phase shift incurred is some multiple of π, it will be destructively amplified and rapidly attenuate. Because of the natural linear phase progression of radiation with frequency, the interference effect

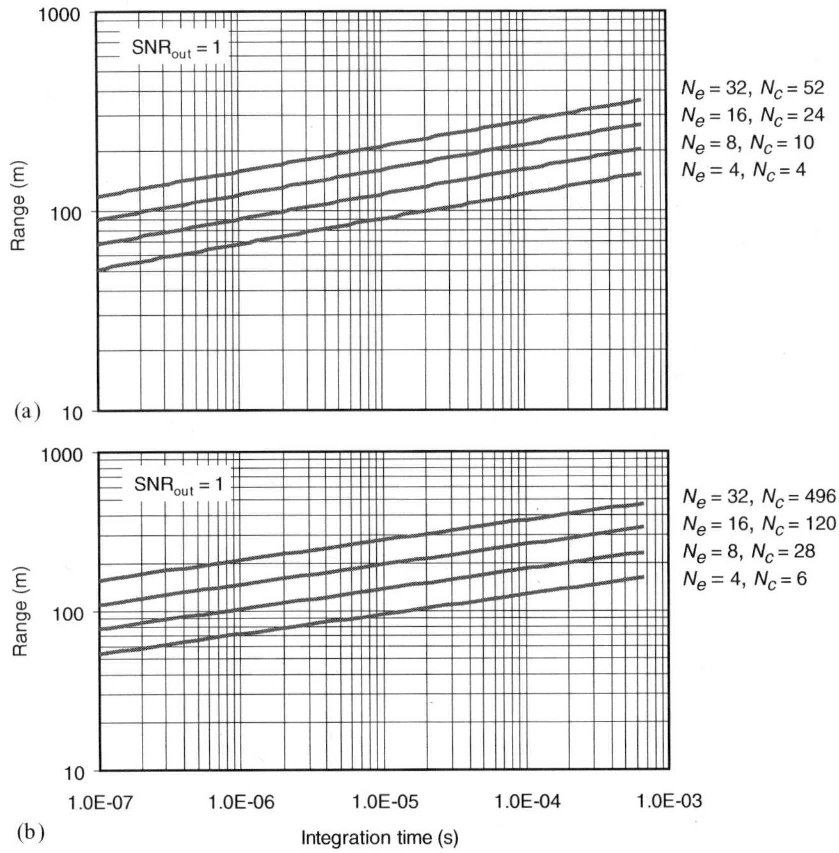

Figure 9-13 Calculated range of RNC radar operating in noise-radar mode versus integration time and parameterized versus number of elements and number of cross correlations. In (a) cross correlations are computed only between nearest neighbors. In (b) they are computed between all unique element pairs. Data is tabulated in Table 9-4.

will create a periodicity in the power spectrum, much as occurs during start up of a mode-locked laser. Each resonant frequency should satisfy a modulo-2π phase rule, corresponding to different longitudinal modes in the laser analogy.

A novel aspect of the analysis is a representation of the coupling through the free space array using matrix theory. The most general form of the matrix with an arbitrary number of

Table 9-4 Array correlations.

N_e	Array configuration	# Cross-correlations (nearest neighbors only)	# Cross-correlations (total unique pairs)
4	2 × 2	4	6
8	2 × 4	10	28
16	4 × 4	24	120
32	4 × 8	52	496

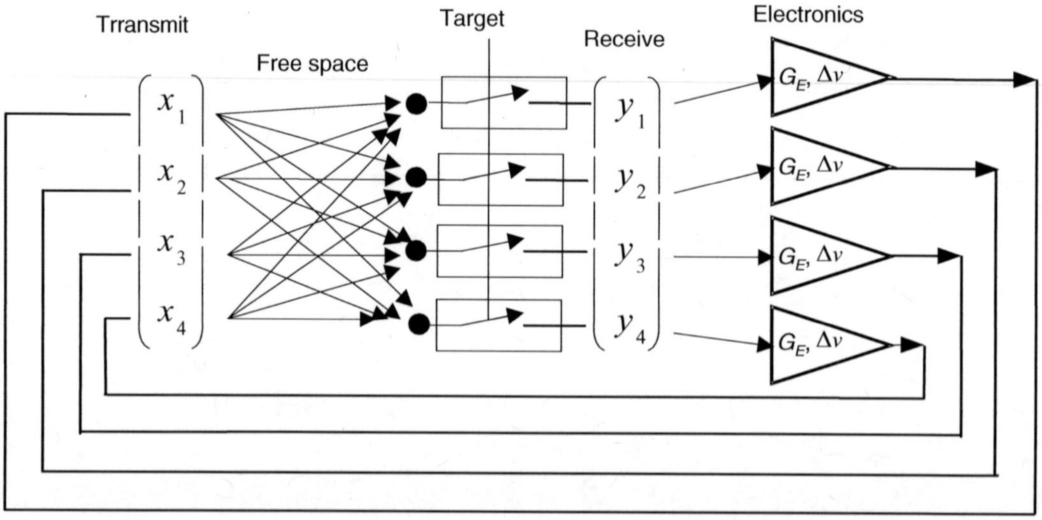

Figure 9-14 Flow chart of a representative (4-channel) RNC radar showing the cross coupling through free space and the band-limited feedback between the Rx and Tx arrays through power gain G.

elements is given by

$$\hat{X}_{Rm} = \sum_n g_{nm} \exp[-j\omega(R_{nt} + R_{tm})/c]\hat{X}_{tn} \equiv \sum_n M_{mn}\hat{X}_{Tn} \tag{9.22}$$

where $\hat{X}_{Rm}$ is the phasor of the electrical signal at the m th Rx element, $\hat{X}_{Tn}$ is the electrical signal at the n-th Tx element, g_{nm} is the free-space coupling gain between the n th Tx and m th Rx elements, and $M_{m,n}$ is the retrodirective matrix element. The quantities R_{nt} and R_{tm} are the range factors between the n th Tx element and the target, and between the target and the m th Rx element, respectively. From the trigonometry of Figure 9-15, $(R_{nt})^2 = (R_0\cos\theta_0)^2 + [R_0\sin\theta_0 - w - (n-1)d]^2$, and $(R_{tm})^2 = (R_0\cos\theta_0)^2 + [R_0\sin\theta_0 + w + (m-1)d]^2$, where R_0 is the range to the target from the geometric center of the arrays, w is half the gap width between the Tx and Rx arrays, and d is the inter-element spacing.

If the target is in the far-field of both arrays, the free-space coupling gain between elements can be estimated from the Friis transmission formula as

$$g_{nm}(\theta, v) \approx \left[\frac{G_n G_m F_n F_m \sigma \lambda^2}{(4\pi)^3 R_0^4}\right]^{\frac{1}{2}} \tag{9.23}$$

where F_n and F_m are the antenna pattern functions for the n th Tx element and the m th receive element. If the Tx and Rx elements are identical (so that $G_n = G_m = G_E$ and $F_n = F_m = F_E$ then the free space gain amplitude is

$$g_F = g_{nm} = \frac{G_E F_E(\theta)\lambda\sigma^{\frac{1}{2}}}{(4\pi)^{\frac{3}{2}} R_0^2}. \tag{9.24}$$

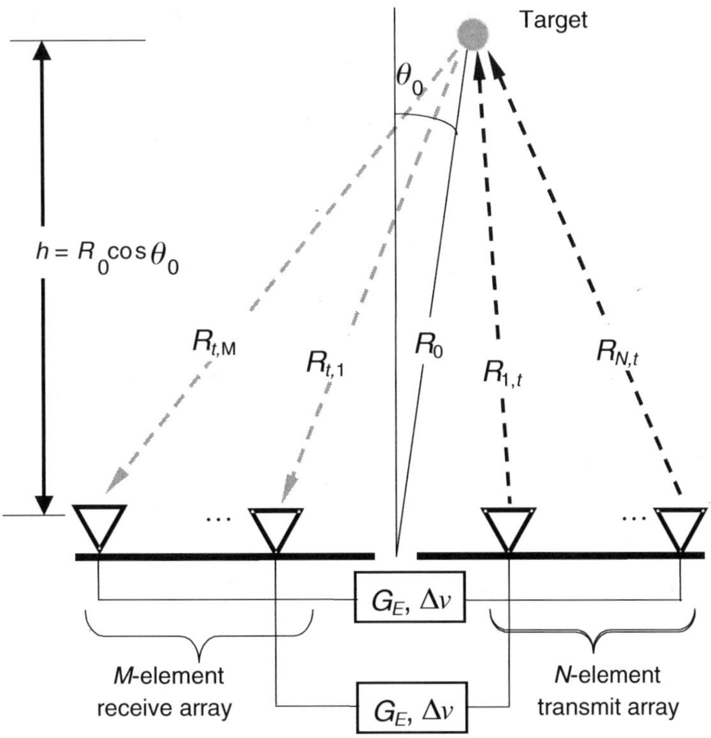

Figure 9-15 Geometrical layout of retrodirective antenna arrays relative to the target location.

Note that when the Tx and Rx arrays have the retrodirective architecture with uniform element spacing, the retroactive matrix (RM) $[g_{nm}]$ has special properties, including symmetry about the diagonal. The RNC loop is completed by connection the appropriate conjugate elements of the Rx and Tx arrays so that the phasor of the transmitted signal from the m th element is

$$\hat{X}_{Tm} = g_E \cdot \exp(-j\omega t_g)\hat{X}_{Rm} \tag{9.25}$$

where $g_E = (G_E)^{1/2}$ is the electronic gain and is assumed to be band-limited and identical in amplitude and phase in each channel.

The detection stage of the RNC radar operation was evaluated as follows. The initial signal from each Tx element X_{Tn} was assumed to be band limited, complex Gaussian noise, fully uncorrelated between the channels so that $X_{Tn} = N_{Tn}$ where N_{Tn} is the noise transmitted by the n th transmitter element. The target was assumed to appear suddenly at time $t = 0$, and the signal at each Tx port was computed after successive passes around the loop, each time injecting more AWGN. Thus the signal at the m th transmit element after p passes through the loop is

$$\hat{X}_{Tm}(p) = \sum_{k=1}^{p}\sum_{n} [(M_{mn})^k \hat{N}_{Tn}(1) + \hat{N}_{Tn}(k > 1)] \tag{9.26}$$

where M is the retrodirective matrix defined by Equation (9.22) and $\hat{N}_{Tn}(\mathrm{k} > 1)$ denotes new AWGN injected in each channel on the k th pass for $k > 1$.

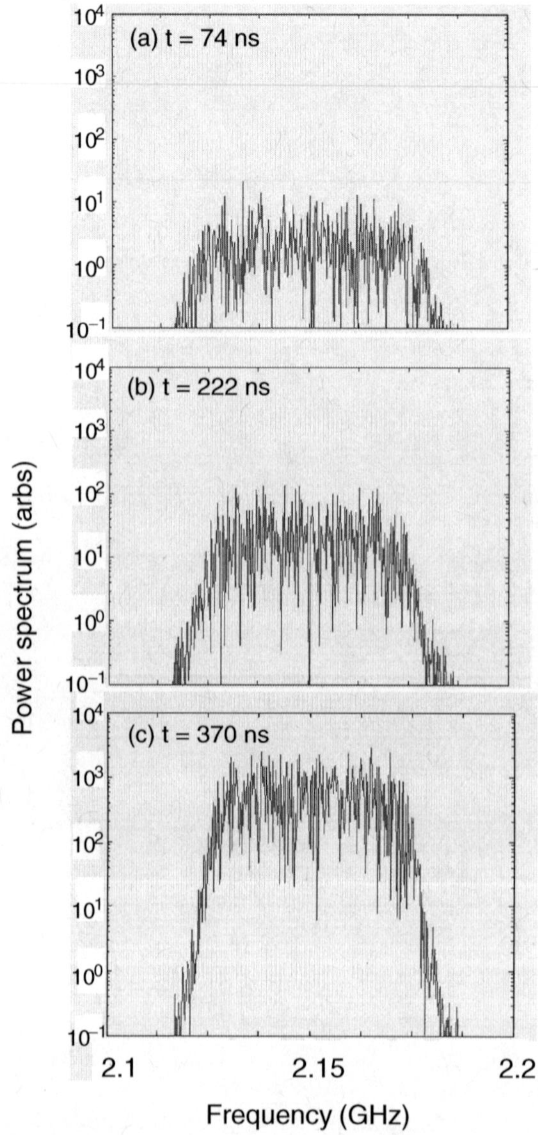

Figure 9-16 Sequence of power spectra for an one-dimensional, 16-element S-band array computed from the MATLAB model. The target is at $R = 7$ m and has an RCS$\sigma = 10$ cm^2. The time is an odd–integral multiple of the round trip time (49 ns for the through free space group delay plus 25 ns for the electronic group delay.)

To illustrate the RM method in retrodirective mode, the power spectrum of $X_{Tm}(p)$ is computed in Figure 9-16 at $t = 0$ and for successive passes around the loop of an RNC radar having 16 linear Tx and Rx elements, an electronic gain in each channel of 98 dB, and a band limit between 2.13 and 2.18 GHz. The target was assumed to be located at a range of 10 m and to have an isotropic RCS of σ_g=10 cm^2. With each successive pass through free space, the power spectrum grows across the

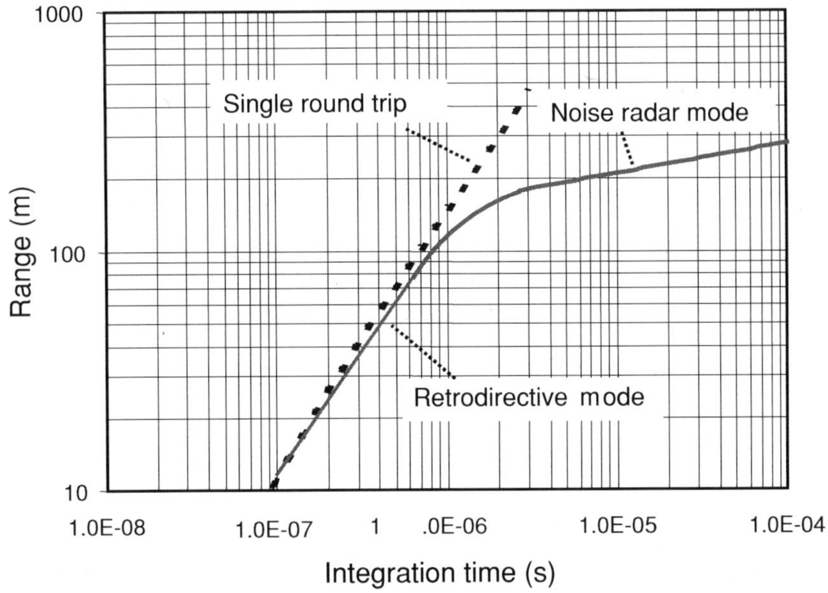

Figure 9-17 Predicted range versus detection time for S-Band 32-element RNC radar having cross-correlative signal processing between all nearest neighbors and assuming $\sigma = 10\text{cm}^2$. The retrodirect-mode range was obtained from the MATLAB simulations, and the noise-radar range was obtained analytically.

band but remains fully correlated between neighboring channels. This greatly accelerates target detection and also facilitates target angle determination by cross-correlative techniques.

9.6.2 RNC Radar Scaling

The RM method has also been used to predict the scaling of the RNC radar performance with range and element number. Figure 9-17 shows the predicted range versus detection time for a 32-element system having cross-correlative signal processing between nearest neighbors. The future shows two distinct regimes. When the target is close enough that the combination of electronic gain, antenna gain, and free-space link loss make the RNC 'loop gain' exceed unity, the power received from the target grows very quickly, and the detection time approaches a single-round-trip limit at short range of $2R/c + t_G$. When the target is outside of this critical range, it can still be detected by exactly the same (correlative) technique, but on a line of more gradual slope as shown in Figure 9-17. This is the more traditional noise radar mode.

9.6.3 ADS Modeling: Radar Dynamics

In this subsection the RNC radar is modeled in the Agilent ADS circuit simulator. First implementation of digital processing (DSP) will be described. Figure 9-18 shows the screen-shot of the top-level of the ADS simulation implementing the DSP in 5 boxes. The thick-lined boxes are time-domain sub-networks, and the thin-lined boxes do signal-conditioning and processing. On the far left-hand side, there is a data-based model for the antennas and the free-space coupling to the target. The model is detailed in Figure 9-19. Capturing free-space propagation (e.g., using Friis' transmission formula) from a radar is not straightforward in ADS, and here this accomplished using a transmission line equivalent circuit. The time delay and phase rotation that occurs when the signal travels a distance 2R from the receiver reference plane to the target and back are modeled

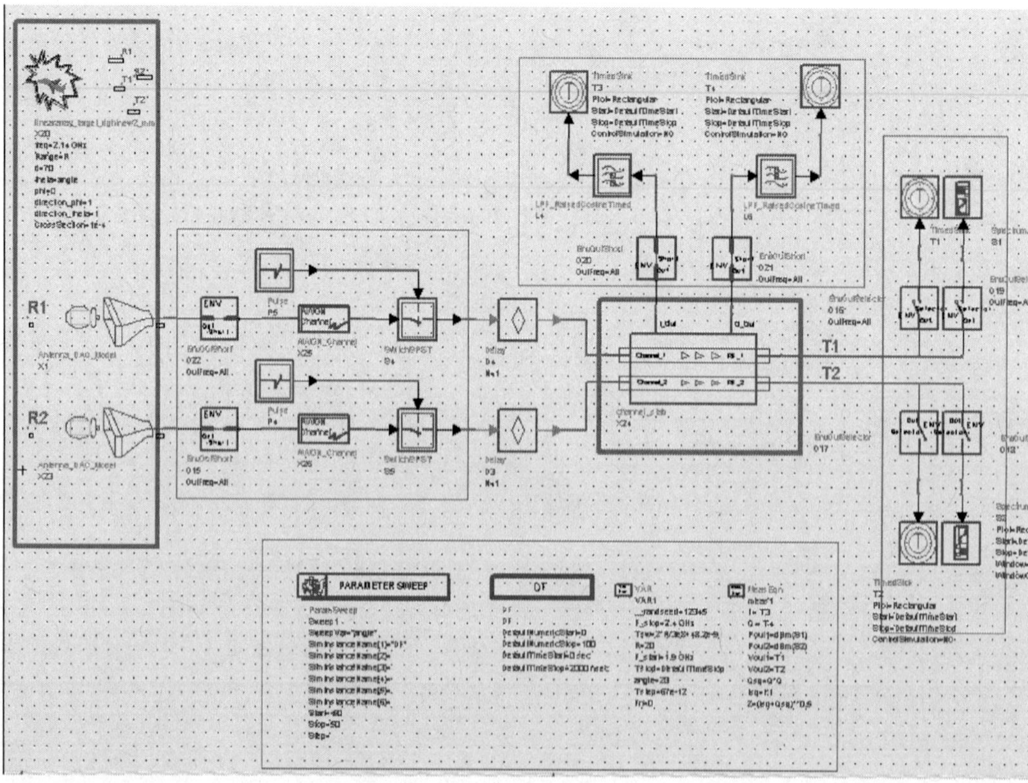

Figure 9-18 Top-level screen-shot of the ADS model of the RNC radar system.

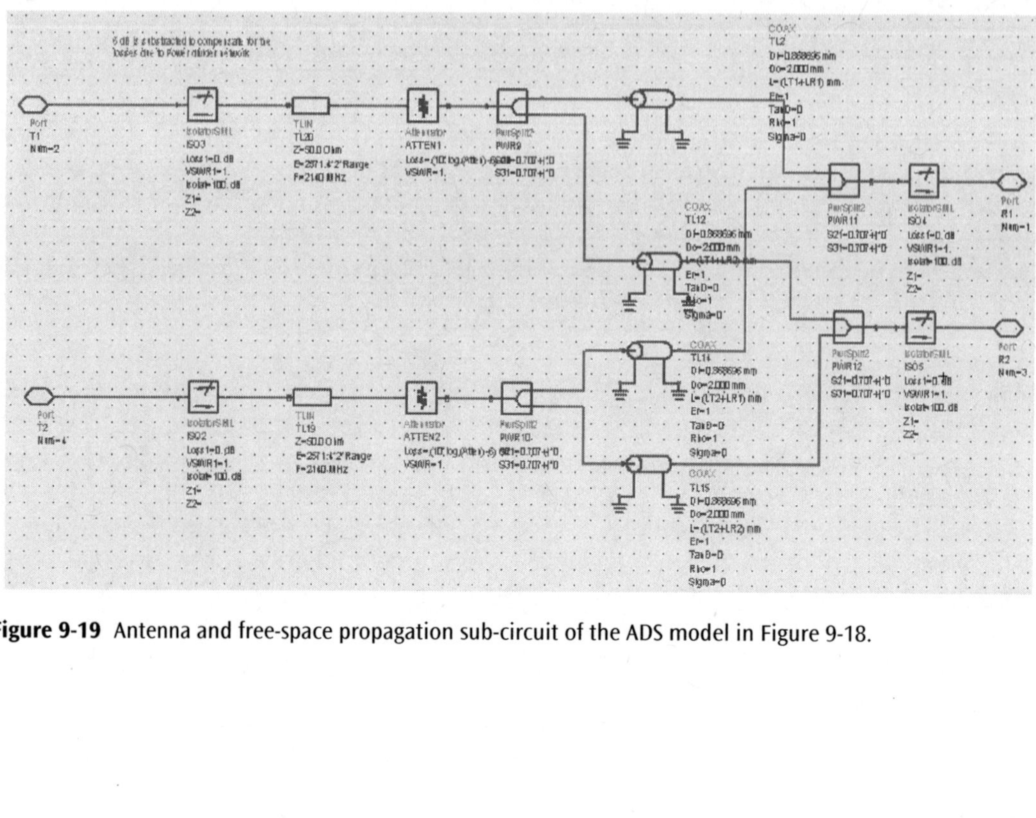

Figure 9-19 Antenna and free-space propagation sub-circuit of the ADS model in Figure 9-18.

by using an ideal 50-Ω microstrip delay line and a phase rotation of $(360/\lambda_0) \times 2R$ for the electrical delay parameter. More subtle is the important difference in phase between neighboring elements of the array caused by spacing and the radiation angle, $\phi_n = nd\sin(\theta)$, where n is the element index. This is modeled using four coaxial lines interconnected by power splitters and combiners as shown in Figure 9-19.

The far-left signal-conditioning and processing block in Figure 9-18 is for noise generation and channel switching. Additive white Gaussian noise (AWGN) is generated separately for each channel and injected into each using a timed summer. Because the synchronicity between this block and the remainder of the circuit is critical, a time step is defined in this block that is equal to the subcircuit time domain simulation time step (typically 67 ps). This top-level time step then controls the numeric DSP simulation and allows for the all-important time-frequency domain co-simulation. This capability is what separates ADS from all other known RF system tools, and is critical to accurately simulating the dynamic behavior of the RNC radar.

Some key results from the ADS simulations of a two-channel S-band RNC radar are plotted in Figures 9-20, and 9-21. Figure 9-20 shows the dynamics of target acquisition via I and Q cross-correlation between the two adjacent channels. Once the target is present, both the I and Q cross-corelator output signals increase rapidly and monotonically with time, approaching a steady-state value determined by the duty cycle. The target angle is determined unambiguously by the relation $\theta = \tan^{-1}$ (Q/I). Since neither cross-correlation is frequency selective, they both increase in strength (and signal-to-noise ratio) through the integrated power from Figure 9-20, not just one longitudinal mode. This means that the acquisition time and angular accuracy are both enhanced by the retrodirective gain — an important issue in justifying the retrodirective architecture versus more traditional noise radars.

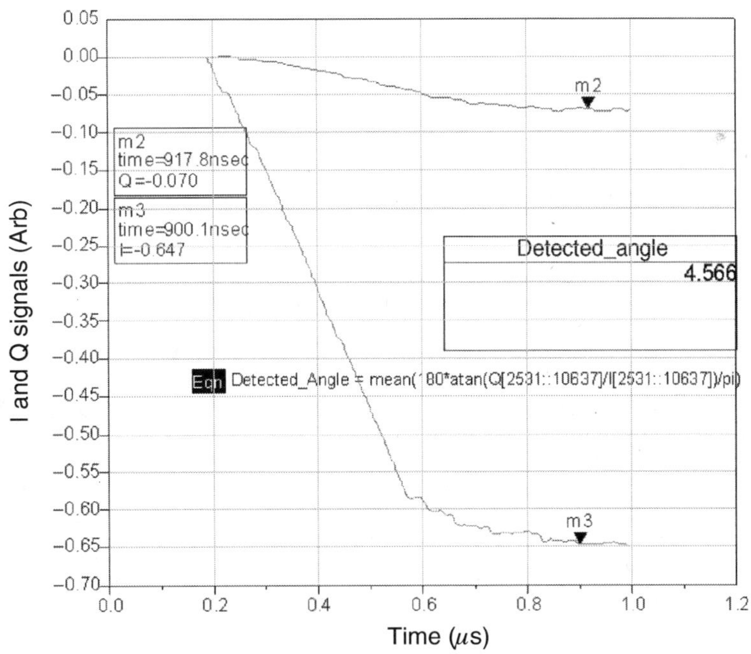

Figure 9-20 ADS simulation of target acquisition (I and Q detection) in the time domain.

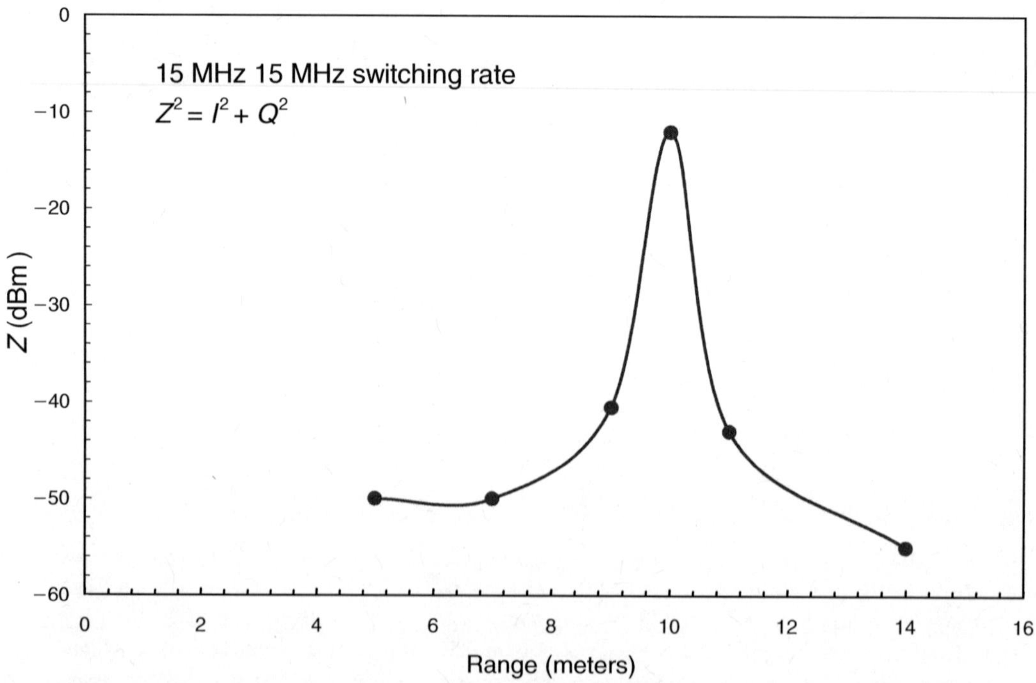

Figure 9-21 ADS-generated plot of two-channel radar output vs range for a fixed radar switching frequency and a target of fixed RCS.

Figure 9-21 displays a plot of total corelator output $[Z = (I^2 + Q^2)^{1/2}]$ versus range for a target of fixed RCS and for the radar operating with a fixed switching frequency of 15 MHz. This investigates the effect of "mode locking" on target range resolution. There is a dramatic peak in the output signal strength at the range 10 m corresponding to $c/2f_s$, as expected. What the ADS simulations show that cannot be predicted analytically is the degree of range discrimination provided by this effect. Figure 9-21 shows clearly that a movement of the target to just 2 m away from this particular range reduces the output signal by about 40 dB on the high side and 35 dB on the low side.

9.7 CONCLUSION

This chapter presented a common-aperture radar-communications system which demonstrated the feasibility of simultaneous operation of radar and communications using a shared aperture. The range resolution was demonstrated at 10 m to be 63 cm and the probability of detection was 80% in the worst-case scenario. The S-band retrodirective noise correlating radar system has been demonstrated with comparable range and targets. It should be emphasized that simulations indicate that much greater range and minimum detectable RCS are possible from both systems.

References

[1] E. D. Adler, M. C. Calcatera, J.-F. Luy, W. D. Palmer, and D. S. Purdy, "Special issue on multifunctional RF systems—guest editorial," *IEEE Trans. Microw. Theory Tech.*, Vol. 53, No. 3, March 2005, pp. 1005–1008.

[2] S. Lindenmeier, K. Boehm, and J. F. Luy, "A wireless data link for mobile applications," *IEEE Microw. Wirel. Compon. Lett.*, Vol. 13, No. 8, Aug. 2003, pp. 326–328.

[3] G. Boeck, D. Plenkowski, R. Circa, M. Otte, B. Heyne, P. Rykaczewski, R. Wittmann, and R. Kakerow, "RF front-end technology for reconfigurable mobile systems," *Proc. 2003 SBMO/IEEE MTT-S Int. Microw. Optoelectron. Conf. (IMOC 2003)*, Vol. 2, 20–23 Sept. 2003, pp. 863–868.

[4] B. Schauwecker, K. M. Strohm, W. Simon, and J.-F. Luy. "RF-MEMS components for broadband applications," *Dig. of Papers 2003 Topical Meeting on Silicon Monolithic Integrated Circuits in RF Systems*, 9–11 April 2003, pp. 166–169.

[5] M. Streifinger, T. Muller, J.-F. Luy, and E. A. Biebl, "A software-radio front-end for microwave applications," *Dig. of Papers 2003 Topical Meeting on Silicon Monolithic Integrated Circuits in RF Systems*, 9–11 April 2003, pp. 53–56.

[6] Defense Advanced Research Projects Agency, "Intelligent RF Front-Ends (IRFFE)." US Department of Defense, 24 Aug. 2007, Available: http://www.darpa.mil/MTO/programs/irffe/index.html

[7] Department of the Navy, "Advanced Multifunction RF Concept (AMRFC)." US Department of Defense,24 Aug. 2007, Available: http://radar-www.nrl.navy.mil/navbar/Research%20Programs/AMRFC.html

[8] P. K. Hughes and J. Y. Choe, "Overview of advanced multifunction RF system (AMRFS)," *Proc. 2000 IEEE Int. Conf. on Phased Array Systems and Technology*, 21–25 May 2000, pp. 21–24.

[9] Department of the Air Force, "The Sensorcraft Challenge." US Department of Defense, 24 Aug. 2007, Available: http://www.afrlhorizons.com/Briefs/Jun04/SN0312.html

[10] P. F. McManamon, E. A. Watson, and M. T. Eismann, "Suggestions for low cost multifunction sensing," *Proc. IEEE Aerospace Conf.*, 1998., Vol. 1, 21–28 March 1998, pp. 283–307.

[11] M. Conn, F. Koenig, G. Goldman, and E. Adler, "Waveform generation and signal processing for a multifunction radar system," *Proc. 2004 IEEE Radar Conf.*, 26–29 April 2004, pp. 161–165.

[12] S. Weiss, E. Adler, O. Kilic, E. Viverros, and R. A. Dahlstrom, "Multifunctional Ka-band electronic scanning antenna (ESA)," *2003 IEEE Int. Symp. on Phased Array Systems and Technology*, 14–17 Oct. 2003, pp. 371–376.

[13] Joint Tactical Radio System (JTRS) Operational Requirements Document (ORD), Version 3.2, JROC Approved, JROCM 087-03, 09 April 2003, pp. E1–E6.

[14] Software Defined Radio Forum, 24 Aug. 2007, Available: http://www.sdrforum.org/

[15] K. Gard, H. Gutierrez, and M. B. Steer, "Characterization of spectral regrowth in microwave amplifiers based on the nonlinear transformation of a complex gaussian process," *IEEE Trans. Microw. Theory Tech.*, Vol. 47, July 1999, pp. 1059–1069.

[16] K. M. Gharaibeh, K. G. Gard, and M. B. Steer, "In-band distortion of multisines," IEEE Trans. Microw. Theory Tech., Aug. 2006, pp. 3227–3236.

[17] K. G. Gard, L. E. Larson, and M. B. Steer, "The impact of RF front-end characteristics on the spectral regrowth of communications signals," *IEEE Trans. Microw. Theory Tech.*, June 2005, pp. 2179–2186.

[18] K. M. Gharaibeh and M. B. Steer, "Modeling distortion in multichannel communication systems," *IEEE Trans. Microw. Theory Tech.*, Vol. 53, No. 5, May 2005, pp. 1682–1692.

[19] H. Gutierrez, K. Gard, and M. B. Steer, "Nonlinear gain compression in microwave amplifiers using generalized power series and transformation of input statistics," *IEEE Trans. Microw. Theory Tech.*, Nov. 2000, pp. 1774–1777.

[20] K. M. Gharaibeh, K. G. Gard, and M. B. Steer, "The applicability of noise power ratio (NPR) in real communication signals," *67 th ARTFG*, June 2006.

[21] K. M. Gharaibeh, K. Gard, and M. B. Steer, "Characterization of in-band front-ends using multi-sine excitation," *Radio and Wireless Symp. 2006*, Jan. 2006.

[22] K. M. Gharaibeh, K. Gard, and M. B. Steer, "Estimation of in-band distortion in digital communication system," *2005 IEEE MTT-S Int. Microw. Symp.*, June 2005.

[23] K. M. Gharaibeh, K. Gard, and M. B. Steer, "Accurate estimation of nonlinear distortion metrics in digital communication system—SNR, EVM and Rho," *Proc. of the 42'th Automatic Radio Frequency Testing Group Meeting*, Dec. 2004, pp. 41–63.

[24] K. Gard and M. B. Steer, "The impact of nonlinear amplifier distortion on the performance of CDMA systems," *IEEE Radio and Wireless Conf. (RAWCON)*, Sept. 2004, pp. 83–86.

[25] K. Gharaibeh and M. B. Steer, "Statistical modeling of cross modulation in multichannel power amplifiers using a new behavioral modeling technique," *2003 IEEE MTT-S Int. Microw. Symp.*, June 2003.

[26] M. Li, L. Hoover, K. G. Gard, and M. B. Steer, "Behavioral modeling of quadrature modulators for characterization of nonlinear distortion," *2006 IEEE MTT-S Int. Microw. Symp.*, June 2006.

[27] M. Li, K. M. Gharaibeh, K. Gard, and M. B. Steer, "Accurate multisine representation of digital communication signals for characterization of nonlinear circuit," *Radio and Wireless Symp. 2006*, Jan. 2006.

[28] K. Gharaibeh, K. Gard, H. Gutierrez, and M. B. Steer, "The importance of nonlinear order in modeling intermodulation distortion and spectral regrowth," *IEEE Radio and Wireless Conf. (RAWCON)*, Aug. 2002, pp. 161–164.

[29] W. R. Curtice and M. Ettenberg, "N-FET, a new software tool for large-signal GaAs FET circuit design," *RCA Review*, Vol. 46, Sept. 1985, pp. 321–340.

[30] W. R. Curtice, "GaAs MESFET modeling and nonlinear CAD," *IEEE Trans. Microw. Theory Tech.*, Vol. 36, Feb. 1988, pp. 220–230.

[31] G. W. Rhyne, *Computer-aided analysis of nonlinear microwave analog circuits*, Ph.D. Dissertation, CCSP-TR-23, Department of Electrical and Computer Engineering, North Carolina State University, Raleigh, NC, 1988.

[32] C. R. Chang *Computer-aided analysis of nonlinear microwave analog circuits using frequency domain spectral balance*, Ph.D. Dissertation, Department of Electrical and Computer Engineering, North Carolina State University, Raleigh, NC, 1990.

[33] IEEE Standard Definition of Terms for Nonlinear, Active and Nonreciprocal Waveguide Components, IEEE Std. 457–1982, 1982.

[34] H. T. Friis, "Noise figure of radio receivers," *Proc. IRE*, Vol. 32, No. 6, July 1944, pp. 419–422.

[35] W. W. Mumford and E. H. Schelbe, *Noise Performance Factors in Communication Systems*, Horizon House, 1968.

[36] R. C. Sagers, "Intercept point and undesired responses," *IEEE Trans. Veh. Technol.*, Vol. 32, No. 1, Feb. 1983, pp. 121–133.

[37] A. M. Victor and M. B. Steer, "Transceiver cascade system analysis and design via a contribution method," *Int. J. on RF and Microwave Computer Aided Engineering*, July 2006, pp. 338–345.

[38] S. A. Maas, "Third-order intermodulation distortion in cascaded stages," *IEEE Microw. Guid. Wave Lett.*, Vol. 5, No. 6, June 1995, pp. 189–191.

[39] R. Pettai, *Noise in Receiving Systems*, John Wiley & Sons, 1984.

[40] J. F. Sevic and M. B. Steer, "On the significance of envelope peak-to-average ratio for estimating the spectral regrowth of an RF/microwave power amplifier," *IEEE Trans. Microw. Theory Tech.*, Vol. 48, 2000, pp. 1068–1071.

[41] N. Blachman, "Band-pass nonlinearities," *IEEE Trans. Inf. Theory* , Vol. 10, 1964, pp. 162–164.

[42] G. W. Rhyne and M. B. Steer, "Generalized power series analysis of intermodulation distortion in a mesfet amplifier: simulation and experiment," *IEEE Trans. Microw. Theory Tech.*, Vol. 35, 1987, pp. 1248–1255.

[43] N. Blachman, "Detectors, bandpass nonlinearities, and their optimization: Inversion of the Chebyshev transform," *IEEE Trans. Inf. Theory*, Vol. 17, 1971, pp. 398–404.

[44] A. Kaye, D. George, and M. Eric, "Analysis and compensation of bandpass nonlinearities for communications," *IEEE Trans. on Commun.*, Vol. 20, 1972, pp. 965–972.

[45] O. Shimbo, "Effects of intermodulation, AM-PM conversion, and additive noise in multicarrier TWT systems," *Proc. of the IEEE*, Vol. 59, 1971, pp. 230–238.

[46] J. G. Proakis and D. G. Manolakis, *Digital Signal Processing: Principles, Algorithms, and Applications*, 2 ed. New York: Macmillan, 1992.

[47] N. Wiener, "Generalized harmonic analysis," *Acta Math.*, Vol. 55, 1930, pp. 117–258.

[48] N. M. Blackman, "Band-pass nonlinearities," *IEEE Trans. Inf. Theory*, Vol. 10, April 1964, pp. 162–164.

[49] N. M. Blackman, "Detectors, band-pass nonlinearities and their optimization: inversion of the chebishev transform," *IEEE Trans. Inf. Theory*, Vol. 17, July 1971, pp. 398–404.

[50] A. Walker, M. B. Steer, and K. Gard, "A vector intermodulation analyzer applied to behavioral modeling of nonlinear amplifiers with memory effect," *IEEE Trans. Microw. Theory Tech.*, Vol. 54, No. 5, May 2006, pp. 1991–1999.

[51] G. M. Rebeiz and J. B. Muldavin, "RF MEMS switches and switch circuits," *IEEE Microw. Magazine*, Vol. 2, No. 4, Dec. 2001, pp. 59–71.

[52] C. Goldsmith, J. Ehmke, A. Malczewski, B. Pillans, S. Eshelman, Z. Yao, J. Brank, and M. Eberly, "Lifetime characterization of capacitive RF MEMS switches," *2001 IEEE MTT-S Int. Microw. Symp. Dig.*, Vol. 1, May 2001, pp. 227–230.

[53] J. B. Muldavin and G. M. Rebeiz, "High-isolation CPW MEMS shunt switches–Part1: Modeling," *IEEE Trans. Micrvw. Theory Tech.*, Vol. 48, No. 6, June 2000, pp. 1045–1052.

[54] T. Itoh, G. Haddad, and J. F. Harvey (Editors), *RF Technologies for Low Power Wireless Communications*, Wiley Interscience, Chapter 10, pp. 349–382.

[55] D. Peroulis, S. P. Pacheco, K. Sarabandi, and L.P.B. Katehi, "Electromechanical considerations in developing low-voltage RF MEMS switches," *IEEE Trans. Microw. Theory Tech.*, Vol. 51, No. 1, Jan. 2003, pp. 259–270.

[56] S. P. Pacheco, L. P. B. Katehi, and C. T. Nguyen, "Design of low actuation voltage RF MEMS switch," *2002 IEEE MTT-S Int. Microw. Symp. Dig.*, Vol. 1, June 2000, pp. 165–168.

[57] E. K. Chan, E. C. Kan, and R. W. Dutton, "Non-linear dynamic modeling of micromachined microwave switches," *1997 IEEE MTT-S Int. Microw. Symp. Dig.*, Vol. 3, June 1997, pp. 1511–1514.

[58] R. K. Gupta and S. D. Senturia, "Pull-in time dynamics as a measure of absolute pressure," *IEEE Tenth Int. Workshop on MEMS*, June 1997, pp. 290–294.

[59] G. K. Fedder, *Simulation of microelectromechanical systems*, Ph.D. Dissertation, University of California at Berkeley, Berkeley, CA, 1994.

[60] S. P. Timoshenko and J. N. Goodier, *Theory of Elasticity*, 3 ed., New York: McGraw-Hill, 1970.

[61] CADRE Analytic, "CADRE light v. 1.3 finite element structural analysis software," 1999, Samammish, WA.

[62] M. T. A. Saif and N. C. MacDonald, "Planarity of large MEMS," *IEEE J. Microelectromech. Syst.*, Vol. 5, No. 2, June 1996, pp. 79–97.

[63] J. A. Thornton and D. W. Hoffman, "Stress related effects in thin films," *Thin Solid Films*, Vol. 171, No. 1, April 1989, pp. 5–31.

[64] S. M. Hu, "Stress related problems in silicon technology," *J. Appl. Phys.*, Vol. 70, No. 6, Sept. 1991, pp. R53–80.

[65] W. Fang and J. A. Wickert, "Determining mean and gradient residual stresses in thin films using micromachined cantilevers," *J. Micromech. Microeng.*, Vol. 6, No. 3, Sept. 1996, pp. 301–309.

[66] L. Lin, A. P. Pisano, and R. T. Howe, "A micro strain gauge with mechanical amplifier," *IEEE J. Microelectromech. Syst.*, Vol. 6, No. 4, Dec. 1997, pp. 313–321.

[67] Berkeley Sensors and Actuators Center, "SUGAR simulation software," http://bsac.eecs.berkeley.edu/cfm/, Berkeley, CA.

[68] D. Peroulis, S. Pacheco, K. Sarabandi, and L. P. B. Katehi, "Alleviating the adverse stress effects of residual stress in RF MEMS switches," *Proc. of the European Microw. Conf.*, Vol. 1, Sept. 2001, pp. 173–176.

[69] D. Peroulis, S. Pacheco, K. Sarabandi, and L. P. B. Katehi, "RF MEMS devices for high isolation switching and tunable filtering," *IEEE MTT-S Int. Microw. Symp. Dig.*, Vol. 2, June 2000, pp. 1217–1220.

[70] R. J. Roark and W. Young, *Formulas for Stress and Strain*, 6 ed., New York: McGraw-Hill, 1989.

[71] S. Duffy, C. Bozler, S. Rabe, J. Knecht, L. Travis, P. Wyatt, C. Keast, and M. Gouker, "MEMS microswitches for reconfigurable microwave circuitry," *IEEE Microw. Wirel. Compon. Lett.*, Vol. 11, No. 3, March 2001, pp. 106–108.

[72] G. L. Tan, *High-performance RF MEMS circuits and phase shifters*, Ph.D. Dissertation, University of Michigan, Ann Arbor, MI, 2002.

[73] J. B. Muldavin and G. M. Rebeiz, "Nonlinear electro-mechanical modeling of MEMS switches," *IEEE MTT-S Int. Microw. Symp. Dig.*, Vol. 3, June 2001, pp. 2119–2122.

[74] F. Shi, P. Ramesh, and S. Mukherjee, "Dynamic analysis of micro-electro-mechanical systems," *Int. J. Numer. Methods Eng.*, Vol. 39, 1996, pp. 4119–4139.

[75] J. B. Muldavin, *Design and analysis of aeries and shunt MEMS switches*, Ph.D. Dissertation, University of Michigan, Ann Arbor, MI, 2001.

[76] T. Veijola, H. Kuisma, J. Lahdenpera, and T. Ryhanen, "Equivalent circuit model of the squeezed gas film in a silicon accelerometer," *Sensors and Actuators A Physical*, Vol. 48, No. 3, May 1995, pp. 239–248.

[77] J. J. Blech, "On isothermal squeeze films," *J. Lubr. Technol.*, Vol. 105, Oct. 1983, pp. 615–620.

[78] D. Peroulis, S. Pacheco, K. Sarabandi, and L.P.B. Katehi, "Tunable lumped components with applications to reconfigurable MEMS filters," *IEEE MTT-S Int. Microw. Symp. Dig.*, Vol. 1, May 2001, pp. 341–344.

[79] C. L. Goldsmith, Z. Yao, S. Eshelman, and D. Denniston, "Performance of low-loss RF MEMS capacitive switches," *IEEE Microw. Guid. Wave Lett.*, Vol. 8, No. 8, Aug. 1998, pp. 269–271.

[80] Ansoft Corporation, "Ansoft Maxwell 2D Student Version, v. 3.1," Pittsburgh, PA.

[81] D. Peroulis, S. P. Pacheco, and L. P. B. Katehi, "RF MEMS switches with enhanced power-handling capabilities," *IEEE Trans. Microw. Theory Tech.*, Vol. 52, No. 1, Jan. 2004, pp. 59–68.

[82] R. M. Reano, D. Peroulis, J. F. Whitaker, and L. P. B. Katehi, "Electro/thermal measurements of RF MEMS capacitive switches," *IEEE MTT-S Int. Microw. Symp. Dig.*, Vol. 3, June 2003, pp. 1923–1926.

[83] J. Rizk, E. Chaiban, and G. M. Rebeiz, "Steady state thermal analysis and high-power reliability considerations of RF MEMS capacitive switches," *IEEE MTT-S Int. Microw. Symp. Dig.*, Vol. 1, June 2002, pp. 239–242.

[84] G. M. Rebeiz, *RF MEMS Theory, Design, and Technology*, New York: John Wiley & Sons, 2003.

[85] R. E. Mihailovich, M. Kim, J. B. Hacker, E. A. Sovero, J. Studer, J. A. Higgins, and J. F. DeNatale, "MEM relay for reconfigurable RF circuits," *IEEE Microw. Wirel. Compon. Lett.*, Vol. 11, No. 2, Feb. 2001, pp. 53–55.

[86] P. M. Zavracky, N. E. McGruer, R. H. Morrison, and D. Potter, "Microswitches and microrelays with a view toward microwave applications," *Int. J. RF Microw. Comput.-Aided Des.*, Vol. 9, No. 4, July 1999, pp. 338–347.

[87] J. B. Muldavin and G. M. Rebeiz, "All-metal high-isolation series and series/shunt MEMS switches," *IEEE Microw. Wirel. Compon. Lett.*, Vol. 11, No. 9, Sept. 2001, pp. 373–375.

[88] D. Peroulis and L. P. B. Katehi, "DC-contact RF MEMS switches for wireless communication systems," *Int. J. RF and Microw. Comput.-Aided Eng.*, Vol. 14, July 2004, pp. 345–355.

[89] Y. Loke, G. H. McKinnon, and M. J. Brett, "Fabrication and characterization of silicon micromachined threshold accelerometers," *Sensors and Actuators A Phys.*, Vol. A29, No. 3, Dec. 1991, pp. 235–240.

[90] R. B. Marks, "A multiline method of network analyzer calibration," *IEEE Trans. Microw. Theory Tech.*, Vol. 39, No. 7, July 1991, pp. 1205–1215.

[91] R. B. Marks and D. F. Williams, "Program multical," Tech. Rep., NIST, 1995, Rev. 1.

[92] Sonnet Software, "Sonnet high frequency electromagnetic software, v. 9.52," North Syracuse, NY 13212.

[93] F. Ericson, S. Greek, and J. Söderkvist, "High-sensitivity surface sicromachined structures for internal stress and stress gradient evaluation," *J. Micromech. Microeng.*, Vol. 7, No. 1, March 1997, pp. 30–36.

[94] S. Hannoe and H. Hosaka, "Characteristics of electrical contacts used for microrelays-development of measurement apparatus and resistance measurement of ultra low force contacts," *Proc. of the Int. Symp. on Microsyst., Intell. Mater. and Robots*, Sept. 1995, pp. 173–176.

[95] S. Majumder, N. E. McGruer, G. G. Adams, A. Zavracky, P. M. Zavracky, R. H. Morrison, and J. Krim, "Study of contacts in an electrostatically actuated microswitch," *Proc. of the Forty-fourth IEEE Holm Conf. on Electrical Contacts*, Oct. 1998, pp. 127–132.

[96] S. Timoshenko, "Analysis of bi-metal thermostats," *J. Opt. Soc. Am.*, Vol. 11, 1925, pp. 233–255.

[97] W.-H. Chu, M. Mehregany, and R. L. Mullen, "Analysis of tip deflection and force of a bimetallic cantilever microactuator," *J. Micromech. Microeng.*, Vol. 3, No. 1, March 1993, pp. 4–7.

[98] D. C. Miller, *Micromachined, flip-chip assembled, actuatable, normally-closed contacts for use in high density electronics packaging*, M.S. Thesis, University of Colorado, Boulder, CO, 2000.

[99] P. Krulevitch and G. Johnson, "Curvature of a cantilever beam subjected to an equi-biaxial bending moment," *Proc. of the Microelectromechanical Structures for Materials Research*, April 1998, pp. 67–72.

[100] J. Soderkvist, "Similarities between piezoelectric, thermal and other internal means of exciting vibrations," *J. Micromech. Microeng.*, Vol. 3, No. 1, March 1993, pp. 24–31.

[101] C. L. Goldsmith, A. Malczewski, Z. J. Yao, S. Chen, J. Ehmke, and D. H. Hinzel, "RF MEMS variable capacitors for tunable filters," *Int. J. RF Microw. Comput.-Aided Eng.*, Vol. 9, No. 4, July 1999, pp. 362–374.

[102] J. Zou, C. Liu, and J. E. Schutt-Aine, "Development of a wide-tuning-range two-parallel-plate tunable capacitor for integrated wireless communication systems," *Int. J. RF Microw. Comput.-Aided Eng.*, Vol. 11, No. 5, Sept. 2001, pp. 322–329.

[103] L. Dussopt and G. M. Rebeiz, "High-Q millimeter-wave MEMS varactors: Extended tuning range and discrete-position designs," *IEEE MTT-S Int. Microw. Symp. Dig.*, Vol. 2, June 2002, pp. 1205–1208.

[104] J. J. Yao, S.-T. Park, and J. DeNatale, "High tuning-ratio MEMS-based tunable capacitors for RF communications applications," *Solid-State Sensor and Actuator Workshop Technical Dig.*, June 1998, pp. 124–127.

[105] N. S. Barker and G. M. Rebeiz, "Optimization of distributed MEMS transmission-line phase shifters–U-band and W-band design," *IEEE Trans. Microw. Theory Tech.*, Vol. 48, No. 11, Nov. 2000, pp. 1957–1966.

[106] A. Dec and K. Suyama, "Micromachined electro-mechanically tunable capacitors and their applications to RF ICs," *IEEE Trans. Microw. Theory Tech.*, Vol. 46, No. 12, Dec. 1998, pp. 2587–2596.

[107] D. Peroulis and L. P. B. Katehi, "Electrostatically-tunable analog RF MEMS varactors with measured capacitance range of 300%," *IEEE MTT-S Int. Microw. Symp. Dig.*, Vol. 3, June 2003, pp. 1793–1796.

[108] D. Peroulis and L. P. B. Katehi, "Highly reliable analog MEMS varactors," *Proc. of 2003 IEEE MTT-S Int. Microw. Symp.*, June 2003, Vol. 2, June 2003, pp. 869–872.

[109] D. Peroulis and L. P. B. Katehi, "A Novel device for in situ experimental characterization and reliability analysis of DC-contact RF MEMS switches," *Proc. of the 12th Int. Conf. on Solid State Sensors, Actuators and Microsystems, Transducers 2003*, Vol. 1, June 2003, pp. 867–870.

[110] M. R. Douglass, "Lifetime estimates and unique failure mechanisms of the digital micromirror device (DMD)," *Proc. of the 36th Annual Int. Reliability Physics Symp.*, March 1998, pp. 9–16.

[111] D. J. Vickers-Kirby, R. L. Kubena, F. P. Stratton, R. J. Joyce, D. T. Chang, and J. Kim, "Anelastic creep phenomena in thin metal plated cantilevers for MEMS," *Mater. Sci. Microelectromech. Syst. (MEMS) Devices III*, Vol. 657, 2001, pp. EE2.5.1–EE2.5.6.

[112] F. A. McClintock and A. S. Argon, *Mechanical Behavior of Materials*, Chapter 12, pp. 432–434 and Chapter 19, pp. 640-644, Fairfax, VA: TechBooks (reprinted by arrangement with Addison-Wesley Publishing Company), 1965.

[113] Y. Lu, L. P. B. Katehi, and D. Peroulis, "High-power MEMS varactors and impedance tuners for millimeter-wave applications," *IEEE Trans. Microw. Theory Tech.*, Vol. 53, No. 11, Nov. 2005, pp. 3672–3678.

[114] Sonnet Software, North Syracuse, NY 13212. http://www.sonnetusa.com/

[115] Advanced Design Tool (ADS), Agilent technologies, Version 2003A.

[116] H. A. Atwater, "Circuit design of the loaded-line phase shifter," *IEEE Trans. Microw. Theory Tech.*, Vol. 33, No. 7, July 1985, pp. 626-634.

[117] J. B. Muldavin, and G. M. Rebeiz, "High-isolation CPW MEMS shunt switches–Part 2: Design," *IEEE Trans. Microw. Theory Tech.*, Vol. 48, No. 6, June 2000, pp. 1053–1056.

[118] D. Peroulis, *RF MEMS Devices for Multifunctional Integrated Circuits and Antennas*, Ph. D. Dissertation, The University of Michigan, Ann Arbor, MI, 2003.

[119] A. R. Brown, *High Q integrated micromachined components for a 28 GHz front-end transceiver*, Ph. D. Dissertation, The University of Michigan, Ann Arbor, MI, 1999.

[120] K. J. Herrick, *W-band Three Dimensional Integrated Circuits Utilizing Silicon Micromachining*, The University of Michigan, Ann Arbor, MI, 2000.

[121] D. M. Pozar, "*Microw. Engineering, 2nd Edition*," New York: John Wiley & Sons, , 1998.

[122] E.-C. Park, Y.-S. Choi, J.-B. Yoon, S. Hong, and E. Yoon, "Fully integrated low phase-noise VCOs with on-chip mems inductors," *IEEE Trans. Microw. Theory Tech.*, Vol. 51, No. 1, Jan. 2003, pp. 289–296.

[123] H. Lakdawala, X. Zhu, H. Luo, S. Santhanam, L. R. Carley, and G. K. Fedder, "Micromachined high-Q inductors in a 0.18-μm copper interconnect low-k dielectric CMOS process," *IEEE J. of Solid-State Circuits*, Vol. 37, No. 3, March 2002, pp. 394–403.

[124] R. Ramachandran and A.-V. H. Pham, "Development of RF/microwave on-chip inductors using an organic micromachining process," *IEEE Trans. on Advanced Packaging*, Vol.25, No. 2, May 2002, pp. 244–247.

[125] B. C. Wadell, "*Transmission Line Design Handbook*," Norwood, MA: Artech House, 1991.

[126] C. L. Goldsmith, A. Malczewski, J. Papapolymerou, K. L. Lange, and J. Kleber, "Reconfigurable double-stub tuners using MEMS switches for Intelligent RF front-ends," *IEEE Trans. Microw. Theory Tech.*, Vol.51, No. 1, Jan. 2003, pp. 271–278.

[127] K. Kang, J.-H. Park, Y.-K. Kim, H.-T. Kim, S. Jung, and Y. Kwon, "Low-Loss analog and digital micromachined impedance tuners at the Ka-Band," *IEEE Trans. Microw. Theory Tech.*, Vol. 49, No. 12, Dec. 2001, pp. 2394–2400.

[128] J. Tuovinen, T. Vaha-Heikkila, J. Varis, and G. M. Rebeiz, "Reconfigurable 6-20 GHz RF MEMS impedance tuner," *IEEE MTT-S Int. Microw. Symp. Dig.*, Vol. 2, June 2004, pp. 729–732.

[129] T. Vaha-Heikkila and G. M. Rebeiz, "A 20-50 GHz reconfigurable matching network for power amplifier applications," *IEEE MTT-S Int. Microw. Symp. Dig.*, Vol. 2, June 2004, pp. 717–720.

[130] J. Tuovinen, T. Vaha-Heikkila, J. Varis, and G. M. Rebeiz, "A V-Band single-stub RF MEMS impedance tuner," *Proc. of the European Microw. Conf.*, Oct. 2004, pp. 1301–1304.

[131] R. E. Collin, "*Foundations For Microwave Engineering*," New York: McGraw-Hill, 1997.

[132] Y. Lu, L. P. B. Katehi, and D. Peroulis, "High-power MEMS varactors and impedance tuners for millimeter-wave applications," *IEEE Trans. Microw. Theory Tech.*, Vol. 53, No. 11, Nov. 2005, pp. 3672–3678.

[133] N. O. Sokal and A. D. Sokal, "Class-E—a new class of high-efficiency tuned single-ended wwitching power amplifers," *IEEE J. Solid-State Circuits*, Vol. 10, No. 6, June 1975, pp. 168–176.

[134] F. H. Raab, P. Asbeck, S. Cripps, P. B. Kenington, Z. Popovic, N. Pothecary, J. Sevic, and N. O. Sokal, "Power Amplifiers and transmitters for RF and microwave," *IEEE Trans. Microw. Theory Tech.*, Vol.50, No. 3, March 2002, pp. 814–826.

[135] S. C. Cripps, "*RF Power Amplifiers for Wireless Communications*," Norwood, MA: Artech House, 1999.

[136] T. Mader, E. W. Bryerton, M. Markovic, M. Forman, and Z. Popovic, "Switched-mode high-efficiency microwave power amplifier in a free-space power-combiner array," *IEEE Trans. Microw. Theory Tech.*, Vol. 46, No. 10, Oct. 1998, pp. 1391–1397.

[137] A. Mediano, P. Molina, and J. Navarro, "Class E RF/microwave power amplifier: linear 'equivalent' of transistors nonlinear output capacitance, normalized design and maximum operating frequency vs. output capacitance," *2000 IEEE MTT-S Int. Microw. Symp. Dig.*, Vol. 2, June 2000, pp. 783–786.

[138] Raytheon Company, Andover, MA, 100 μm gate width, 0.25 μm gate length, 12 finger PHEMT, unpackaged discrete device.

[139] F. H. Rabb, "Suboptimum operation of Class-E power amplifier," Feb. 1989, *Proc. of RF Technology Expo.*, pp. 85–98.

[140] R. Mallavarpu, D. Teeter, and M. Snow, "The importance of gate charge formulation in large-signal phemt modeling," *Proc. of Gallium Arsenide Integrated Circuit (GaAs IC) Symp.*, 1998, pp. 87–90.

[141] S. Pajic and Z. Popovic, "An efficient x-band 16-element spatial combinet of switched-mode power amplifiers," *IEEE Trans. Microw. Theory Tech.*, Vol. 51, Number 7, July 2003, pp. 1863–1870.

[142] T. Quach, P. M. Watson, W. Okamura, E. N. Kaneshiro, A. Gutierrez-Aitken, T. R. Block, J. W. Eldredge, T. J. Jenkins, L. T. Kehias, A. K. Oki, D. Sawdai, R. J. Welch, and R. D. Worley, "Ultrahigh-efficiency power amplifier for space radar applications," *IEEE J. Solid-State Circuits*, Vol.37, No. 9, Sept. 2002, pp. 1126–1134.

[143] R. Tayrani, "A monolithic X-band class-E power amplifier," *Proc. of Gallium Arsenide Integrated Circuit (GaAs IC) Symp.*, 2001, pp. 205–208.

[144] D. Peroulis, K. Sarabandi, and L. P. B. Katehi, "Design of reconfigurable slot antennas," *IEEE Trans. Antennas Propag.*, Vol. 53, No. 2, Feb. 2005, pp. 645–654.

[145] D. L. Sengupta, "Resonant frequency of a tunable rectangular patch antenna," *Electron. Lett.*, Vol. 20, No. 15, July 1984, pp. 614–615.

[146] D. L. Sengupta, "Transmission line model analysis of rectangular patch antennas," *Electromagn.*, Vol. 4, No. 4, 1984, pp. 355–376.

[147] K. Guney, "Resonant frequency of a tunable rectangular microstrip patch antenna," *Microw. Opt. Technol. Lett.*, Vol. 7, No. 12, Aug. 1994, pp. 581–585.

[148] N. Fayyaz, S. Safavi-Naeini, E. Shin, and N. Hodjat, "A novel electronically tunable rectangular patch antenna with one octave bandwidth," *Proc. of the IEEE Canadian Conf. on Electrical and Computer Engineering*, Vol. 1, May 1998, pp. 25–28.

[149] S. H. Al-Charchafchi and M. Frances, "Electronically tunable microstrip patch antennas," *IEEE Antennas and Propag. Soc. Int. Symp.*, Vol. 1, June 1998, pp. 304–307.

[150] K. A. Jose, V. K. Varadan, and V. V. Varadan, "Experimental investigations on electronically tunable microstrip antennas," Feb. *Microw. and Opt. Technol. Lett.*, Vol. 20, No. 3, 1999, pp. 166–169.

[151] P. J. Rainville and F. J. Harackewiez, "Magnetic tuning of a microstrip patch antenna fabricated on a ferrite film," *IEEE Microw. Guid. Wave Lett.*, Vol. 2, No. 12, Dec. 1992, pp. 483–485.

[152] R. K. Mishra, S. S. Pattnaik, and N. Das, "Tuning of microstrip antenna on ferrite substrate," I*EEE Trans. Antennas Propag.*, Vol. 41, No. 2, Feb. 1993, pp. 230–233.

[153] S. Kawasaki and T. Itoh, "A slot antenna with electronically tunable length," *IEEE Antennas Propag. Soc. Int. Symp.*, Vol.1, June 1991, pp. 130–133.

[154] M. A. Forman and Z. B. Popovic, "A tunable second-resonance cross-slot antenna," *IEEE Antennas Propag. Soc. Int. Symp.*, Vol.1, June 1997, pp. 18–21.

[155] D. J. Roscoe, L. Shafai, A. Ittipiboon, M. Cuhaci, and R. Douville, "Tunable dipole antennas," *IEEE Antennas Propag. Soc. Int. Symp.*, Vol. 2, June 1972, pp. 672–675.

[156] Agilent Technologies, Linear Models for Active Diode Surface Mount Packages, Application Note 1124.

[157] Agilent Technologies, Applications of PIN Diodes, Application Note 922.

[158] Rogers Microwave Products, http://www.rogerscorporation.com.

[159] C. A. Balanis, "*Advanced Engineering Electromagnetics*," Chapter 8, New York: John Wiley & Sons, 1989.

[160] W. L. Stutzman and G. A. Thiele, "*Antenna Theory and Design*," 2 Ed., Chap. 9, New York: John Wiley & Sons, 1998.

[161] S. Roberts, "Dielectric and piezoelectric properties of barium titanate," *Phys. Rev.*, Vol. 71, June 1947, pp. 890–895.

[162] L. Davis and L. G. Rubin, "Some dielectric properties of barium–strontium titanate ceramics at 3000 megacycles" *J. Appl. Phys.*, Vol. 24, Sept 1953, pp. 1194–1197.

[163] W. J. Gemulla and R. D. Hall, "Ferroelectrics at microwave frequencies," *Microw. J.*, Vol. 3, June 1960, pp. 47–51.

[164] M. DiDomenico, D. A. Johnson and R. H. Pantell, "Ferroelectric harmonic generator and the large–signal microwave characteristics of a ferroelectric ceramic," *J. Appl. Phys.*, Vol. 33, May 1962, pp. 1697–1706.

[165] M. DiDomenico and R. H. Pantell, "An X-band ferroelectric phase–shifter," *IEEE Trans. Microw. Theory Tech.*, Vol. 10, May 1962, pp. 179–185.

[166] M. Cohn and A. F. Eikenburgl, "Ferroelectric phase shifters for VHF and UHF," *IEEE Trans. Microw. Theory Tech.*, Vol. 10, Nov. 1962, pp. 536–548.

[167] J. W. Amoss, M. R. Donaldson, L. J. Donaldson, A. L. Stanford, and J. E. Pippin, "A ferroelectric Microwave switch," *IEEE Trans. Microw. Theory Tech.*, Vol.13, Nov. 1965, pp. 789–793.

[168] G. F. Derbenwick and A. F. Isaacson, "Ferroelectric memory: on the brink of a breakthrough," *IEEE Circuits Devices Mag.*, Jan 2001, pp. 20–30.

[169] F. Jona and G. Shirane, *Ferroelectric Crystals*, New York: Dover, 1993.

[170] A. von Hippel, *Dielectrics and Waves*, Boston: Artech House. See also A. von Hippel, "Ferroelectricity, domain structure, and phase transitions in Barium Titanate," *Rev. Mod. Phys.*, Vol. 22, July 1950, pp. 221–245.

[171] O. G. Vendik, I. G. Mironenko, and L. T. Ter-Martirosyan, "Superconductors spur application of ferroelectric films," *Microw. RF* , Vol. 33, 1994, pp. 67–70.

[172] D. E. Kotecki, J. D. Baniecki, H. Shen, R. B. Laibowitz, K. L. Saenger, J. J. Lian, T. M. Shaw, S. D. Athavale, C. Cabral, Jr. , P. R. Duncombe, M.Gutsche, G. Kunkel, Y.-J. Park, Y.-Y. Wang, and R. Wise, "(Ba,Sr)TiO_3 dielectrics for future stacked-capacitor DRAM," *IBM J. Res. Dev.*, Vol. 43 No. 3, May 1999, pp. 367–382.

[173] G. D. Wilk, R. M. Wallace, and J. M. Anthony, "High-κ gate dielectrics: Current status and materials properties considerations," *J. Appl. Phys.*, Vol.89, May 2001, pp. 5243–5275.

[174] C. Basceri, S. K. Streiffer, A. I. Kingon, and R. Waser, "The dielectric response as a function of temperature and film thickness of fiber textured (Ba,Sr)TiO3 thin films grown by chemical vapor deposition," *J. Appl. Phys.*, Vol. 82, No. 5 , 1997, 2497–2504. see also C. Basceri, *Electrical and Dielectric Properties of (Ba,Sr)TiO3 Thin-Film Capacitors for Ultra-High Density Dynamic Random Access Memory, Ph. D. Dissertation*, North Carolina State University.

[175] K. R. Carroll, J. M. Pond, D. B. Chrisey, J. S. Horwitz, R. E. Leuchtner, and K. S. Grabowski, "Microwave measurement of the dielectric constant of $Sr_{0.5}Ba_{0.5}TiO_3$ ferroelectric thin films," *Appl. Phys. Lett.*, Vol. 62, April 1993, pp. 1845–1847; see erratum in *Appl. Phys. Lett.*. Vol. 63, 1993, pp. 1291.

[176] A. T. Findikoglu, Q. X. Jia, D. W. Reagor, and X. D. Wu, "Electrical characteristics of coplanar waveguide devices incorporating nonlinear dielectric thin films of $SrTiO_3$ and $Sr_{0.5}Ba_{0.5}TiO_3$," *Microw. Opt. Tech. Lett.*, Vol 9, No. 6 , 1995, pp. 306.

[177] A. T. Findikoglu, Q. X. Jia, I. H. Campbell, X. D. Wu, D. Reagor, C. B. Mombourquette, and D. McMurry, "Electrically tunable coplanar transmission line resonators using $YBa_2Cu_3O_{7-x}$ /$SrTiO_3$ bilayers," *Appl. Phys. Lett.*, Vol. 66, June 1995, pp. 3674–3676.

[178] A. T. Findikoglu, Q. X. Jia, X. D. Wu, G. J. Chen, T. Venkatesan, and D. W. Reagor, "Tunable and adaptive bandpass filter using a nonlinear dielectric thin film of $SrTiO_3$," *Appl. Phys. Lett.* Vol. 68, March 1996, pp. 1651–1653.

[179] O. G. Vendik, E. F. Carlsson, P. K. Petrov, R. A. Chakalov, S. S. Gevorgian, and Z. G. Ivanov, "HTS/ferroelectric CPW structures for voltage tuneable phase shifters," *27th Europea Microw. Conf. and Exhibition*, Sept. 8–12, 1997, pp. 196–202.

[180] G. Subramanyam, F. Van Keuls, and F. A. Miranda, "A K-Band tunable microstrip bandpass filter using a thin-film conductor/ferroelectric/dielectric multilayer configuration," *IEEE Microw. Guid. Wave Lett.*, Vol. 8, Feb. 1998, pp. 78–80.

[181] V. N. Keis, A. B. Kozyrev, M. L. Khazov, J. Sok, and J. S. Lee, "20 GHz tunable filter based on ferroelectric (Ba,Sr)TiO_3 films," *Electron. Lett*, Vol. 34, No. 11, May 1998, pp. 1107–1109.

[182] F. A. Miranda, G. Subramanyam, F. W. Van Keuls, and R.R. Romanofsky, "A K-band (HTS, gold)/ferroelectric thin films diplexer for a discriminator-locked tunable oscillator," *IEEE Trans. Appl. Supercond.*, Vol. 9, June 1999, pp. 3581–3584.

[183] T. M. Shaw, S. Trolier-McKinstry, and P. C. McIntyre, "The properties of ferroelectric films at small dimensions," *Annu. Rev. Mater. Sci.* 2000. 30:263–98.

[184] S. K. Streiffer, C. Basceri, C. B. Parker, S. E. Lash, and A. I. Kingon, "Ferroelectricity in thin films: The dielectric response of fiber-textured (BaxSr1-x)Ti1+y03+z thin films grown by chemical vapor deposition, *J. Appl. Phys.*, Vol. 86, No. 8, 15 Oct. 1999, pp. 4565–4575.

[185] T. R. Taylor, P. J. Hansen, B. Acikel, N. Pervez, R. A. York, S. K. Streiffer, and J. S. Speck, "Impact of thermal strain on the dielectric constant of sputtered barium strontium titanate thin films," *Appl. Phys. Lett.* Vol. 80, 2002, pp. 1978.

[186] T. Taylor, *Stoichiometry and thermal mismatch effects on sputtered barium strontium titanate thin films*, Ph.D. Dissertation, University of California Santa Barbara, 2003.

[187] S. Nagata *et al.*, "A GaAs MMIC chip set for mobile communications using on-chip ferroelectric capacitors," *Int. Solid-State Circuits Conf., ISSCC 1993*, pp. 172–175. See also K. Miyatsuji *et al.*, "A GaAs high-power RF single-pole double-throw switch IC for digital mobile communication system," *Int. Solid-State Circuits Conf., ISSCC 1994*, pp. 34–35.

[188] M. Nishitsuji *et al.*, "Advanced GaAs-MMIC process technology using high-dielectric constant thin film capacitors by low-temperature rf sputtering method," *GaAs IC Sympsium* Oct 1993, pp. 329–332. See also M. Nishitsuji et al.,"New GaAs MMIC process technology using low-temperature-deposited SrTiO3 thin-film capacitors," *Electron. Lett.*, Vol. 30, June 1994, pp. 1045–1046, and also T. Nakatsuka et al., "A Highly miniaturized receiver front-end hybrid IC using on-chip high-dielectric constant capacitors for mobile communications equipment," *IEEE Microw. Millim.-wave Monolithic Circuits Symp.*, 1995, pp. 85–88.

[189] D. Ueda, "Implementation of GaAs monolithic microwave integrated circuits with on-chip BST capacitors," *J. Electroceram.*, Vol. 3, June 1999, pp. 105–113.

[190] P. Penfield and R. P Rafuse, *Varactor Applications*, Cambridge, Massachusetts: MIT Press, 1962.

[191] A. K. Tagantsev, V. O. Sherman, K. F. Astafiev, J. Venkatesh, and N. Setter, "Ferroelectric materials for microwave tunable applications," *J. Electroceram.*, Vol. 11, Sept 2003, pp. 5–66.

[192] D. Dimos and C. H. Mueller, "Perovskite thin films for high-frequency capacitor applications," *Annual Review of Mater. Sci.*, Vol. 28, Aug. 1998, pp. 397–419.

[193] A. I. Kingon and S. Srinivasan "Lead zirconate titanate thin films directly on copper electrodes for ferroelectric, dielectric and piezoelectric applications," *Nature Materials*, Vol. 4, 2005, pp. 233–237.

[194] S. W. Liu, Y. Lin, J. Weaver, W. Donner, X. Chen, C. L. Chen, J. C. Jiang, and E. I. Meletis, A. Bhalla, "High-dielectric-tunability of ferroelectric (Pb,Sr)TiO3 thin films on (001) LaAlO3," *Appl. Phys. Lett.*,, Vol. 85, Oct. 2004, pp. 3202–3204.

[195] Q-Y Shao, A-D Li, Y-D Xia, D. Wu, Z-G Liu, and N-B Ming, "Strontium-modified lead zirconate titanate thin films for electrically tunable device applications," *J. Appl. Phys.*, Vol. 100, 2006, pp. 036102.

[196] W. Ren, S. Trolier-McKinstry, C. A. Randall, and T. R. Shrout, "Bismuth zinc niobate pyrochlore dielectric thin films for capacitive applications," *J. Appl. Phys*,Vol. 89, 2001, pp. 767–774.

[197] J. Lu and S. Stemmer, "Low-loss, tunable bismuth zinc niobate films deposited by rf magnetron sputtering," *Appl. Phys. Lett.*, Vol. 83, 2003, pp. 2411.

[198] J. C. Nino, M. T. Lanagan, and C. A. Randall, "Dielectric relaxation in Bi_2O_3-ZnO-Nb_2O_5 cubic pyrochlore," *J. Appl. Phys.*, Vol. 89, 2001, pp. 4512–4516.

[199] J. Park, J. W. Lu, S. Stemmer, and R. A. York, "Microwave dielectric properties of tunable capacitors employing bismuth zinc niobate thin films," *J. Appl. Phys.*, Vo. 97, 2005, pp. 084110.

[200] A. K. Tagantsev, J. W. Lu, and S. Stemmer, "The temperature dependence of the dielectric tunability of pyrochlore bismuth zinc niobate thin films," *Appl. Phys. Lett.*, Vol. 86, 2005, pp. 032901.

[201] J. W. Lu, S. Schmidt, D. S. Boesch, N. Pervez, R. A. York, and S. Stemmer, "Low-loss tunable capacitors fabricated directly on gold bottom electrodes," *Appl. Phys Lett.*, Vol. 88, 2006, pp. 112905.

[202] J. Park, J. W. Lu, D. S. Boesch, S. Stemmer, and R. A. York, "Distributed phase shifter with pyrochlore bismuth zinc niobate thin films," *IEEE Microw. Wirel. Compon. Lett.*, Vol. 16, May 2006, pp. 264–266.

[203] P. Hansen, D. Hennings, and H. Schreinemacher, "Dielectric properties of acceptor-doped (Ba, Ca) (Ti, Zr)O3 ceramics," *J. Electroceramics*, Vol. 2 No.2, 1998, pp. 85–94.

[204] W. Hofman, S. Hoffmann, and R. Waser, "Dopant influence on dielectric losses, leakage behaviour, and resistance degradation of $SrTiO_3$ thin films," *Thin Solid Films*, Vol. 305, 1997, pp. 66–73.

[205] M. Copel, J. D. Baniecki, P. R. Duncombe, D. Kotecki, R. Laibowitz, D. A. Neumayer, and T. M. Shaw, "Compensation doping of $Ba_{0.7}Sr_{0.3}TiO_3$ thin films," *Appl. Phys. Lett.*, Vol. 73, No. 13, 28 Sept. 1998, pp. 1832–1834.

[206] S. S. Kim and C. Park, "Leakage current behaviors of acceptor- and donor-doped $(Ba_{0.5}S_{r0.5})TiO_3$ thin films," *Appl. Phys. Lett.*, Vol. 75, 1999, pp. 2554.

[207] P. C. Joshi and M. W. Cole, "Mg-doped $Ba_{0.6}Sr_{0.4}TiO_3$ thin films for tunable microwave applications," *Appl. Phys. Lett.*, Vol. 77, July 2000, pp. 289–291.

[208] K. H. Ahn, S. Baik, and S. S. Kim, "Significant suppression of leakage current in $(Ba,Sr)TiO_3$ thin films by Ni or Mn doping," *J. Appl. Phys.*, Vol. 92, 2002, pp. 2651.

[209] R-V. Wanga, P. C. McIntyre, J. D. Baniecki, K. Nomura, T. Shioga, K. Kurihara, and M. Ishii, "Effect of Y doping and composition-dependent elastic strain on the electrical properties of $(Ba,Sr)TiO_3$ thin films deposited at 520°C," *Appl. Phys. Lett.*, Vol. 87, 2005, pp. 192906.

[210] D. S. Jeong, C. S. Hwang, J. D. Baniecki, T. Shioga, K. Kurihara, N. Kamehara, and M. Ishii, "Dielectric constant dispersion of yttrium-doped (Ba,Sr)TiO3 films in the high-frequency (10 kHz–67 GHz) domain" *Appl. Phys. Lett.*, Vol. 87, 2005, pp. 232903.

[211] J. Wang, S. Bharadwaja and S. Trolier-McKinstry, "DC reliability of doped dc reliability of doped $Ba_{0.7}Sr_{0.3}TiO_3$ films films prepared by chemical solution deposition," *Int. Symp. on Integrated Ferroelectrics* (ISAF), July 2006.

[212] K. F. Astafiev, V. O. Sherman, A. K. Tagantsev, N. Setter, "Can the addition of a dielectric improve the figure of merit of a tunable material?" *J. Eur. Ceramic Soc.*, Vol. 23, 2003, pp. 2381–2386.

[213] O. Auciello, S. Saha, D.Y. Kaufman, S. K. Streiffer, W. Fan, B. Kabius, J. Im, and P. Baumann, "Science and technology of high dielectric constant thin films and materials integration for application to high frequency devices," *J. Electroceram,*, Vol. 12, 2004, pp. 119–131.

[214] Agilent Technologies, *Impedance Measurement Handbook*, 2nd ed., 2000. Available:http://cp.literature.agilent.com/litweb/pdf/5950-3000.pdf

[215] A. West, D. C. Sinclair, and N. Hirose, "Characterization of electrical materials, especially ferroelectrics, by impedance spectroscopy," *J. Electroceram*, Vol.1 No.1, 1997, pp. 65–71.

[216] Baki Acikel, *High performance barium strontium titanate varactor technology for low cost circuit applications*, Ph.D. Dissertation, University of California, Santa Barbara, 2002.

[217] D. Chase, L.-Y. Chen, and R. A. York "Modeling the capacitive nonlinearity in thin-film bst capacitors," *IEEE Trans. Microw. Theory Tech.*, Vol. 53, No. 10, Oct 2005, pp. 3215–3220.

[218] O. G. Vendik, and S. P. Zubko, "Modeling the dielectric response of incipient ferroelectrics," *J. Appl. Phys.*, Vol. 82, Nov 1987, pp. 4475–4483.

[219] C. Zhou and D. M. Newns, "Intrinsic dead layer effect and the performance of ferroelectric thin film capacitors," *J. Applied Physics*, Vol. 82, No. 6, 15 Sept. 1997, pp. 3081–3088.

[220] C. T. Black and J. J. Welser, "Electric-field penetration into metals: consequences for high-dielectric-constant capacitors," *IEEE Trans. Electron. Devices*, Vol. 46, No. 4, April 1999, pp. 776–780.

[221] H. Rohdin, N. Moll, A. M. Bratkovsky, and C.-Y. Su, "Dispersion and tunneling analysis of the interfacial gate resistance in schottky barriers," *Phys. Rev. B*, Vol. 59, May 1999, pp. 13102–13113.

[222] N. K. Pervez, P. J. Hansen, and R. A. York "High tunability barium strontium titanate thin films for RF circuit applications," *Appl. Phys. Lett.*. Vol. 85, 2004, pp. 4451.

[223] M. S. Tsai, S. C. Sun, and T. Y. Tseng, "Effect of oxygen to argon ratio on properties of $(Ba,Sr)TiO_3$ thin films prepared by radio-frequency magnetron sputtering," *J. Appl. Phys.*, Vol. 82, Oct. 1997, pp. 3482–3487.

[224] J. Im, O. Auciello, P. K. Baumann, S. K. Streiffer, D. Y. Kaufman, and A. R. Krauss, "Composition-control of magnetron-sputter-deposited $(Ba_xSr_{1-x})Ti_{1+y}O_{3+z}$ thin films for voltage tunable devices," *Appl. Phys. Lett.*, Vol. 76, 31 Jan. 2000, pp. 625–627.

[225] T. R. Taylor, P. J. Hansen, B. Acikel, N. Pervez, R. A. York, and J. S. Speck, "The influence of stoichiometry on the dielectric properties of sputtered strontium titanate thin films," *J. Appl. Phys.*, Vol. 94, No. 5, Sept 2003, pp. 3390–3396.

[226] Y. Imanaka, T. Shioga, and J. D. Baniecki, "Decoupling capacitor with low inductance for high-frequency digital applications," *Fujitsu Sci. Tech. J.*, Vol. 38, No. 1, June 2002, pp. 22–30.

[227] J. Scarpulla, D. C. Eng, S. R. Olson, C.-S. Wu, "A TDDB model of Si_3N_4-based capacitors in GaAs MMICs," *IEEE Int. Reliability Physics Symp. Proc.*, March 1999, pp. 128–137; see also J. Scarpulla, K. Kho, and S. Olson, "Process monitoring for nitride dielectric defect density," *Int. Conf. on Compound Semiconductor Manufacturing Technology* (GaAs Mantech), 1999.

[228] R. Waser, T. Baiatu, and K.-H. Hardtl, "DC electrical degradation of perovskite-type titanates: I. ceramics," *J. Amer. Ceram. Soc.*, Vol. 73, No. 6, June 1990, p. 1645.

[229] T. Horikawa, T. Kawahara, M. Yamamuka, and K. Ono, "Degradation in $(Ba,Sr)TiO_3$ thin films under DC and dynamic stress conditions," *35th annual IEEE Int. Reliability Physics Symp. Proc.*, 1997, pp. 82–89.

[230] H. Reisinger, H. Wendt, G. Beitel, and E. Fritsch, "Dielectric breakdown, reliability and defect density of $(Ba_{0.7}Sr_{0.3})TiO_3$ (BST)," *1998 Symp. on VLSI Technology*, June 1998, pp. 58–59.

[231] C. Basceri *et al.*, "An important failure mechanism in MOCVD (Ba,Sr)TiO3 thin films–resistance degradation," *MRS Proc.: Ferroelectric thin films VI*, Boston, MA, 30 Nov.–4 Dec. 1997, pp. 9–14.

[232] S.-C. Huang, H.-M. Chen, S. C. Wu, and J. Y.-M. Lee, "Time dependent dielectric breakdown of paraelectric barium-strontium-titanate thin film capacitors for memory device applications," *J. Appl. Phys.*, Vol. 84, No. 9, 1 Nov. 1998, pp. 5155–5157.

[233] S. Zafar, "Resistance degradation in barium strontium titanate thin films," *J. Applied Physics*, Vol. 86, Oct. 1999, pp. 3890–3895.

[234] C. B. Parker, *Size effects and reliability of $(Ba,Sr)TiO_3$ thin films*, PhD. Dissertation, North Carolina State University, 2002; see also C.B. Parker, Sung-Jin Kim, and A.I. Kingon, "Investigation of Failure Modes Under AC and DC Stress and Implications for Device Reliability in LS-MOCVD (Ba,Sr)TiO 3 Thin Films," *ISAF Conf.*, Hawaii, July 2000.

[235] R. Gerson, T. C. Marshall, "Dielectric breakdown in porous ceramics," *J. Appl. Phys.*, Vol. 30, Nov. 1959, pp. 1650–1653.

[236] J. D. Baniecki *et al.*, "Hydrogen induced tunnel emission in $Pt/(Ba_xSr_{1-x})Ti_{1+y}O_{3+z}/Pt$ thin film capacitors," *J. Appl. Phys.*, Vol. 89, No. 5, 1 March 2001.

[237] J. D. Baniecki, T. Shioga, K. Kurihara, and N. Kamehara,"Investigation of the importance of interface and bulk limited transport mechanisms on the leakage current of high dielectric constant thin film capacitors," *J. Appl. Phys.*, Vol. 94, No. 10, Nov 2003, pp. 6741–6748.

[238] J. D. Baniecki, T. Shioga, K. Kurihara, and N. Kamehara, "A study of current transport in $(Ba_xSr_{1-x})Ti_{1+y}O_{3+z}$ thin-film capacitors containing a voltage-dependent interface state charge distribution," *J. Appl. Phys.*, Vol. 97, 2005, pp. 114101.

[239] A. K. Jonscher, *Dielectric Relaxation in Solids*, London: Chelsea Dielectrics Press, 1983. See also A. K. Jonscher, "Dielectric relaxation in solids," *J. Phys. D: Appl. Phys.*, Vol.32, 1999, pp. R57–R70.

[240] T. Horikawa, T. Makita, T. Kuroiwa, and N. Mikami, "Dielectric relaxation of $(Ba,Sr)TiO_3$ thin-films," *Japanese J. Appl. Phys.*, Vol. 34, Sept. 1995, pp. 5478–5482.

[241] J. D. Baniecki, R. B. Laibowitz, T. M. Shaw, P. R. Duncombe, D. A. Neumayer, D. E. Kotecki, H. Shen, and Q. Y. Ma, "Dielectric relaxation of $Ba_{0.7}Sr_{0.3}Ti0_3$ thin films from 1 mHz to 20 GHz," *Appl. Phys. Lett.*, Vol. 72, No. 4, 26 Jan. 1998.

[242] N. K. Pervez, J. Park, J. Lu, S. Stemmer, and R. A. York "High frequency loss modeling using universal relaxation," *Int. Symp. on Integr. Ferroelectr. (ISIF 2005)*, Shanghai China, April 2005.

[243] H. Fröhlich, *Theory of Dielectrics, 2nd ed.*, Oxford: Clarendon Press, 1958.

[244] N. Pervez, *Investigation of loss mechanisms inthin-film barium strontium titanate capacitors*, Ph.D. Dissertation, University of California, Santa Barbara, 2006.

[245] Y. Watanabe, M. Okano, and A. Masuda, "Surface conduction on insulating $BaTiO_3$ crystal suggesting an intrinsic surface electron layer," *Phys. Rev. Lett.*, Vol. 86, Jan 2001, pp. 332–335.

[246] R. A. York, A. Nagra, E. Erker, T. Taylor, P. Periaswamy, J. Speck, S. Streiffer, O. Auciello, "Microwave integrated circuits using thin-film BST," *Proc. 12th Intl Symp. Appl. Ferroelectr. (ISAF)*, Vol 1, July 2000, pp. 195–200.

[247] A. Kassam, I. Koutsaroff, L. McNeil, J. Obeng, P. Woo, and M. Zelner, "An effective interlayer dielectric and passivation scheme using reactively-sputtered Al_2O_3 for $(Ba,Sr)TiO_3$ capacitors," *J. Integr. Ferroelectr.*, Vol. 47, 2002, pp. 259–264.

[248] R. A. York, A. S. Nagra, P. Periaswamy, O. Auciello, S. K. Streiffer, and J. Im, "Synthesis and characterization of $(Ba_xSr_{1-x})Ti_{1+y}O_{3+z}$ thin films and integration into microwave varactors and phase shifters," *J. Integr. Ferroelectr.*, Vol. 34, April 2000, pp. 177–188.

[249] R. A. York, United States Patent 6,683,341, "Voltage-variable capacitor with increased current conducting perimeter," issued 27 Jan. 2004.

[250] Amit Nagra, *Varactor based Technologies for the Tuning and Control of Microwave Circuits and Antennas*, Ph.D. Dissertation, University of California, Santa Barbara, 1999.

[251] K. Ikuta, Y. Umeda, and Y. Ishii, "Measurement of high-frequency dielectric characteristics in the mm-wave band for dielectric thin-films on semiconductor substrates," *Jpn. J. Appl. Phys.*, Vol. 34, 1995, pp. L1211–L1213. See also K. Ikuta, Y. Umeda, and Y. Ishii, "Upper-bound frequency for measuring mm-wave-band dielectric characteristics of thin films on semiconductor substrates," *Jpn. J. Appl. Phys.*, Vol. 37, 1998, pp. 210–214.

[252] Y. Iwazaki, K. Ohta, and T. Suzuki, "Elimination of parasitic effects due to measurement conditions of $SrTiO_3$ thin films up to 40 GHz," *J. Eur. Ceramic Soc.*, Vol. 26, 2006, pp. 1841—1844. See also T. Suzuki, K. Morito, and Y. Iwazaki, "The Latest advances in high-dielectric thin-film capacitor technology for GHz-RF devices," *J. Integr. Ferroelectr.*, Vol. 76, 2005, pp. 47–57.

[253] J. Zhang, A. Tombak, J.-P. Maria, B. Boyette, G. T. Stauf, A. I. Kingon, , and A. Mortazawi, "Microwave characterization of thin film BST material using a simple measurement technique," *2002 IEEE MTT-S Int. Microw. Symp. Dig.*, Vol. 2, June 2002, pp. 1201–1204.

[254] G. Kwan, "Sensitivity analysis of one-port characterized devices in vector network analyzer calibrations: theory and computational analysis," NCSL Int. 2002 Symp., Available: http://metrologyforum.tm.agilent.com/ncsli2002_kwan.shtml.

[255] P. K. Petrov, N. McN. Alford, and S. Gevorgyan, "Tech. for microwave measurements of ferroelectric thin films and their associated error and limitations" *Meas. Sci. Technol.* Vol. 16, 2005, pp. 583–589.

[256] L. F. Tiemeijer and R. J. Havens, "A calibrated Lumped-element de-embedding technique for on-wafer RF characterization of high-quality inductors and high-speed transistors," *IEEE Trans. Electron Devices*, pp. 822–829, March 2003.

[257] X. Zhu, D.-Y. Chen, Z. Jin, J. D. Phillips, A. Mortazawi, "Characterization of thin film BST tunable capacitors using a simple two port measurement technique," *2005 IEEE MTT-S Int. Microwave Symp. Dig.*, June 2005, pp. 611–614.

[258] W. T. Beyene and J. E. Schutt-Aine, "Accurate frequency-domain modeling and efficient circuit simulation of high-speed packagign interconnects," *IEEE Trans. Microwave Theory Tech.*, Vol. 45, Oct. 1997, pp. 1941–1947.

[259] W. H. Press, S. A. Teukolsky, W. T. Vetterling, and B. P. Flannery, *Numerical Recipes in C++*, Cambridge Press, 2002.

[260] M. P. McNeal, S.-J. Jang, and R. E. Newnham, "The effect of grain and particle size on the microwave properties of barium titanate $(BaTiO_3)$," *J. Applied Physics*, Vol. 83, Nov March 1998, pp. 3288–3297.

[261] C. Hubert, J. Levy, A. C. Carter, W. Chang, S. W. Kiechoefer, J. S. Horwitz, and D. B. Chrisey: *Appl. Phys. Lett.* 71 (1997) 3353. see also C. Hubert, J. Levy, E. J. Cukauskas, S. W. Kiechoefer, "Mesoscopic Microwave Dispersion in Ferroelectric Thin Films," *Phys. Rev. Lett.*, Vol. 85, No. 9, Aug 2000, pp. 1998–2001.

[262] T.-G. Kim, J. Oh, Y. Kim, T. Moon, K. S. Hong, and B. Park, "Crystallinity dependence of microwave dielectric properties in $(Ba,Sr)TiO_3$ thin films," *Jpn. J. Appl. Phys.* Vol. 42, 2003, pp. 1315–1319.

[263] A. Vorobiev, P. Rundqvist, K. Khamchane, and S. Gevorgian, "Microwave loss mechanisms in $Ba_{0.25}Sr_{0.75}TiO_3$ thin film varactors," *J. Appl. Phys.* Vol. 96, Oct. 2004, pp. 4642–4649.

[264] J. C. Booth, I. Takeuchi, and K.-S. Chang, "Microwave-frequency loss and dispersion in ferroelectric $Ba_{0.3}Sr_{0.7}TiO_3$ thin films," *Appl. Phys. Lett.* 87, 802908 (2005).

[265] A. Vorobiev, P. Rundqvist, K. Khamchane, and S. Gevorgian, "Silicon substrate integrated high Q-factor parallel-plate ferroelectric varactors for microwave/millimeterwave applications," *Appl. Phys. Lett.* 83, Oct. 2003, p. 3144.

[266] W. P. Mason, "Electrostrictive effect in barium titanate ceramics," *Phys. Rev.*, Vol. 74, Nov 1948, pp. 1134–1147.

[267] K. Morito, Y. Iwazaki, T. Suzuki, and M. Fujimoto, "Electric field induced piezoelectric resonance in the micrometer to millimeter waveband in a thin film $SrTiO_3$ capacitor," *J. Applied Phys.*, Vol. 94, Oct 2003, pp. 5199–5205.

[268] S. Tappe, U. Böttger, and R. Waser, "Electrostrictive resonances in $Ba_{0.7}Sr_{0.3}TiO_3$ thin films at microwave frequencies," *Applied Physics Lett.*, Vol. 85, 26 July 2005, pp. 624–626.

[269] R. Krimholtz, D. A. Leedhom, and G. L. Matthaei, "New equivalent circuit for elementary piezoelectric transducers," *IEE Electron Lett.*, Vol. 6, 1970, pp. 398–399.

[270] J. F. Rosenbaum, *Bulk Acoustic Wave Theory and Devices*, Artech House: Boston, 1988.

[271] United States Patent, 6,674,321 "Circuit configuration for DC-biased capacitors," issued Jan 4, 2004.

[272] Y.-K. Yoon, D. Kim, M. G. Allen, J. S. Kenney, "A reduced intermodulation distortion tunable ferroelectric capacitor: architecture and demonstration," *2004 IEEE MTT-S Int. Microwave Symp.*, June 2003, Vol. 3, pp. 1989–1992.

[273] J.-S. Fu, X. A. Zhu, D.-Y. Chen, J. D. Phillips, and A. Mortazawi, "A linearity improvement technique for thin-film barium strontium titanate capacitors," *2006 IEEE MTT-S Int. Microwave Symp. Dig.*, June 2006.

[274] H. Katta, H. Kurioka, Y. Yashima, "Tunable Power amplifier using thin-film BST capacitors" *2006 IEEE MTT-S Int. Microwave Symp. Dig.*, June 2006.

[275] L. Yan, L. B. Kong, L. F. Chen, K. B. Chong, C. Y. Tan, and C. K. Ong, "$Ba_{0.5}Sr_{0.5}TiO_3$–$Bi_{1.5}Zn_{1.}0Nb_{1.5}O_7$ composite thin films with promising microwave dielectric properties for microwave device applications" *Applied Physics Letters*, Vol. 85, Oct. 2004, pp. 3522–3524.

[276] D. C. Collier, "Ferroelectric phase shifters for phased array radar applications," *8th Int. Symp. on Applications of Ferroelectrics. ISAF 1992*, Aug 1992 pp. 199–201.

[277] R. W. Babbitt, T. E. Koscica, W. E. Drach, "Planar microwave electro-optic phase-shifters," *Microwave Journal*, Vol. 35, June 1992, pp. 63–79.

[278] V. K. Varadan, D. K. Ghodgaonkar, V. V. Varadan, J. F. Kelly, and P. Glikerdas, "Ceramic phase shifters for electrically steerable antenna systems," *Microwave Journal*, Vol. 35(1), 1992, pp. 116.

[279] R. W. Babbitt, T. E. Koscica, W. E. Drach, and D. Didomenico, *Integrated Ferroelectrics*, Vol. 8, 1995, pp. 65.

[280] V. K. Varadan, K. A. Jose, V. V. Varadan, R. Hughes, and J. F. Kelly, "A novel microwave planar phase shifter," *Microwave Journal*, April 1995, pp. 244–254.

[281] F. D. Flaviis and N. G. Alexopoulos, "Planar microwave integrated phase-shifter design with high purity ferroelectric material," *IEEE Trans. Microwave Theory Tech.*, Vol. 45, No. 6, 1997, pp. 963.

[282] A. S. Nagra and R. A. York, "Distributed analog phase shifters with low insertion loss," *IEEE Trans. Microwave Theory Tech.*, Vol. 47, No. 9, Sept. 1999, pp. 1705–1711.

[283] A. S. Nagra, T. R. Taylor, P. Padmini, J. S. Speck and R.A. York, "First demonstration of a periodically loaded line phase shifter using BST capacitors," *MRS Symp.: Materials Issues for Tunable RF and Microwave Devices*, Vol. 603, pp. 37–42.

[284] E. G. Erker, A. S. Nagra, Y. Liu, P. Periaswamy, T. R. Taylor, J. Speck and R. A. York, "Monolithic Ka-band phase shifter using voltage tunable $BaSrTiO_3$ parallel plate capacitors," *IEEE Microwave Guided Wave Lett.*, Vol. 10, Jan. 2000, pp. 7–10.

[285] Y. Liu, A. S. Nagra, E. G. Erker, P. Periaswamy, T. R. Taylor, J. S. Speck and R. A. York, "$BaSrTiO_3$ interdigitated capacitors for distributed phase shifter applications," *IEEE Microwave Guided Wave Letters*, Vol.10, No.11, Nov. 2000, pp. 448–450.

[286] B. Acikel, Y. Liu, A. S. Nagra, T.R. Taylor, P. J. Hansen, J. S. Speck, R. A. York, "Phase shifters using (Ba,Sr)TiO3 thin films on sapphire and glass substrate," *2001 IEEE MTT-S Int. Microwave Symp. Dig.*, May 2001, pp. 1191–4; see also Y. Liu, B. Acikel, P.J. Hansen, T.R. Taylor, A.S. Nagra, J. S. Speck, R. A. York, "Distributed phase shifters using $(Ba,Sr)TiO_3$ thin films on sapphire and glass substrates," *J. Integrated Ferroelectrics*, Vol. 39, No. 1–4, pp. 313–20.

[287] B. Acikel, T. R. Taylor, P. J. Hansen, J. S. Speck, R. A. York, "A new X-Band 180° high performance phase shifter using $(Ba,Sr)TiO_3$ thin films," *2002 IEEE MTT-S Int. Microwave Symp. Dig.*, Seattle, June 2002, pp. 1467–1469.

[288] B. Acikel, T. R. Taylor, P. J. Hansen, J. S. Speck, and R. A. York, "A new high performance phase shifter using $BaSrTiO_3$ thin films," *IEEE Microw. Wirel. Compon. Lett.*, Vol.12, No.7, July 2002, pp. 237–239.

[289] G. Velu, K. Blary, L. Burgnies, J. C. Carru, E. Delos, and A. Marteau, D. Lippens, "A 310°/3.6-dB K-band phase shifter using paraelectric BST thin films," *IEEE Microw. Wirel. Compon. Lett.*, Vol. 16, Feb. 2006 pp. 87–89.

[290] S. Shi, J. Purden, J. Jin, and R. A. York, "A 24 GHz wafer scale electronically scanned antenna using bst phase shifters for collision avoidance systems," *IEEE Antennas & Propagation Symp. Dig.*, June 2005.

[291] Q. Meng, X. Zhang, F. Li, J. Huang, X. Zhu, D. Zheng, B. Cheng, Q. Luo, C. Gu, Y. He, "An impedance matched phase shifter using $BaSrTiO_3$ thin film," *IEEE Microw. Wirel. Compon. Lett.*, Vol. 16, June 2006 pp. 345–347.

[292] F. Ellinger, H. Jäckel, and W. Bächtold, "Ultra compact, low loss, varactor tuned phase shifter MMIC at C-band" *IEEE Microw. Wirel. Compon. Letts.*, Vol. 11, March 2001, pp. 10–106.

[293] F. Ellinger, H. Jäckel, and W. Bächtold, "Varactor-loaded transmission-line phase shifter at C-band using lumped elements," *IEEE Trans. Microwave Theory Tech.*, Vol. 51, April 2003, pp. 1135–1140.

[294] J. Serraiocco, B. Acikel, P. Hansen, T. Taylor, H. Xu, J. S. Speck, and R. A. York, "Tunable passive integrated circuits using bst thin films," *Proc. Int. Symp. on Integrated Ferroelectrics*, Nara, Japan, June 2002.

[295] J. Serraiocco, P. J. Hansen, T. R. Taylor, J. S. Speck, and R. A. York, "Compact distributed phase shifters at X-Band using BST," *J. Integrated Ferroelectrics*, Vol. 56, Iss. 1, 2003, pp. 1087–1095.

[296] T. O'Sullivan, R.A. York, B. Noren, and P. Asbeck, "Adaptive duplexer implemented using feedforward technique with a BST phase shifter," *2004 IEEE MTT-S Int. Microwave Symp. Dig.*, Vol. 3, June 2004, pp. 1299–1302.

[297] T. O'Sullivan, R. A. York, B. Noren, P. M. Asbeck, "Adaptive duplexer implemented using single-path and multipath feedforward techniques with BST phase shifters," *IEEE Trans. Microwave Theory Tech.*, Vol. 53, Jan. 2005, pp. 106–114.

[298] S. K. Koul and B. Bhat, *Microwave and Millimeter-wave Phase Shifters, Volume II: Semiconductor and Delay Line Phase Shifters*, Artech House: Norwood, MA, 1991.

[299] R. V. Garner, "360° varactor linear phase modulator," *IEEE Trans. Microwave Theory Tech.*, Vol. 17, No. 3, March 1969, pp. 137–147.

[300] D. M. Krafscik, S. A. Tmhoff, D. E. Dawson, and A. L. Conti, "A dual-varactor, analog phase shifter operating from 6 to 18 GHz," *IEEE 1988 Microwave and Millimeter-Wave Monolithic Circuits Symp.*, pp. 83–88.

[301] J. Lange, "Interdigitated stripline quadrature coupler," *IEEE Trans. Microwave Theory Tech.*, Vol. 17, Dec. 1969, pp. 1150–1151.

[302] J. Serraicco, *Compact phase-shifter deign using barium strontium titanate thin-film varactors*, M.S. Thesis, University of California Santa Barbara, 2003.

[303] D. Kim, Y. Choi, M. G. Allen, J. S. Kenney and D. Kiesling, "A wide-band reflection-type phase shifter at S-band using BST coated substrate," *IEEE Trans. Microwave Theory and Tech.*, Vol. 50, No. 12, Dec. 2002, pp. 2903–2909; also D. Kim, Y. Choi, Allen, M.G., Kenney, J.S., Kiesling, D., "A wide bandwidth monolithic BST reflection-type phase shifter using a coplanar waveguide Lange coupler" *2002 IEEE MTT-S Int. Microwave Symp. Dig.*, Volume: 3, 2–7 June 2002, pp. 1471–1474.

[304] T. Hirota, A. Minakawa, and M. Muraguchi, "*Reduced-size branch-line and rat-race hybrids for uniplanar MMIC's*," *IEEE Trans. Microwave Theory Tech.*, Vol. 38, No. 3, March 1990, pp. 270–275.

[305] R. W. Vogel, "Analysis and design of lumped and lumped-distributed element directional couplers for MIC and MMIC applications," *IEEE Trans. Microwave Theory Tech.*, Vol. 40, No. 2, Feb. 1992, pp. 253–262.

[306] F. Ellinger, R. Vogt, and W. Bachtold, "Compact reflective-type phase-shifter MMIC for C-Band using a lumped-element coupler," *IEEE Trans. Microwave Theory Tech.*, Vol. 49, No. 5, May 2001, pp. 913–917.

[307] J. Serraiocco, P. J. Hansen, T. R. Taylor, J. S. Speck, and R. A. York "Compact ferroelectric reflection phase shifters at X-Band," *IEEE Int. Microwave Symp. Dig.*, June 2003, pp. 1993–1996.

[308] N. E. Hodges and M. H. Yam, "A precise analog phase-shifter for SHF SATCOM phased-arrays," *1992 IEEE GaAs IC Symp. Dig.*, pp. 29–32.

[309] H. Hayashi, and M. Muraguchi, "A MMIC active phase-shifter using a variable resonant circuit," *IEEE Trans. Microwave Theory Tech.*, Vol. 47, Oct 1999, pp. 2021–2026.

[310] D. Viveiros, D. Consonni, and A. K. Jastrzebski, "A tunable all-pass MIC active phase-shifter," *IEEE Trans. Microwave Theory Tech.*, Vol. 50, Aug 2002, pp. 1885–1889.

[311] D. Kim, Y. Choi, M. Ahn, M. G. Allen, J. S. Kenney and P. Marry, "2.4 GHz continuously variable ferroelectric phase shifters using all-pass networks," *IEEE Microw. Wirel. Compon. Lett.*, Vol. 13, Oct. 2003 pp. 434–436.

[312] D. Kim; S.-S. Je, J. S. Kenney, and P. Marry, "Design of ferroelectric phase shifters for minimum performance variation over temperature," *2004 IEEE MTT-S Int. Microwave Symp.*, Vol. 1, June 2004 pp. 257–260.

[313] A. T. Findikoglu, Q. X. Jia, X. D. Wu, G. J. Chen, T. Venkatesan, and D. W. Reagor, "Tunable and adaptive bandpass filter using a nonlinear dielectric thin film of $SrTiO_3$," *Appl. Phys. Lett.* Vol. 68, March 1996, pp. 1651–1653.

[314] G. Subramanyam, F. Van Keuls, and F. A. Miranda, "A K-band tunable microstrip bandpass filter using a thin-film conductor/ferroelectric/dielectric multilayer configuration," *IEEE Microwave Guided-Wave Lett*, Vol. 8, Feb. 1998, pp. 78–80.

[315] F. A. Miranda, G. Subramanyam, F. W. Van Keuls, and R. R. Romanofsky, "A K-band (HTS, gold)/ ferroelectric thin films diplexer for a discriminator-locked tunable oscillator," *IEEE Trans. Applied Superconductivity*, Vol. 9, June 1999, pp. 3581–3584.

[316] F. A. Miranda, G. Subramanyam, F. W. Van Keuls, R. R. Romanofsky, J. D. Warner, C. H. Mueller, "Design and development of ferroelectric tunable microwave components for Ku- and K-Band satellite communication systems," *IEEE Trans. Microw. Theory Tech.*, Vol. 48, July 2000, pp. 1181–1189.

[317] P. G. Clem, J. Sigman, and C. D. Nordquist, "Integrated (Ba,Sr) TiO3 (BST) films on copper and alumina for monolithic microwave integrated circuit (MMIC) applications," Sixteenth *2007 IEEE Int. Symp. on Applications of Ferroelectrics* (ISAF 2007), May 2007, pp. 205–206.

[318] J. Papapolymerou, C. Lugo, Z. Zhao, S. Wang, and A. Hunt, "A miniature low-loss slow-wave tunable ferroelectric bandpass filter from 11–14 GHz," *2006 IEEE MTT-S Int. Microwave Symp. Dig.*, June 2006, pp. 556–559.

[319] D. Kuylenstierna, A. Vorobiev, S. Gevorgian, "40 GHz lumped element tunable bandpass filters with transmission zeros based on thin $Ba_{0.25}Sr_{0.75}TiO_3$ (BST) film varactors," *Silicon Monolithic Integrated Circuits in RF Systems*, Jan. 2006.

[320] M. K. Roy and J. Richter, "Tunable ferroelectric filters for software defined tactical radios," *15th IEEE Int. Symp. on the Application of Ferroelectrics* (ISAF 2006), July 2006; see also M. K. Roy, C. Kalmar, R. R. Neurgaonkar, J. R. Oliver, D. Dewing, "A highly tunable radio frequency filter using bulk ferroelectric materials," *14 th Int. Symp. on Appl. Ferroelect., ISAF 2004*, Aug. 2004, pp. 25–28.

[321] V. L.-Y. Chen, R. Forse, D. Chase, R. A. York, "Analog tunable matching network using integrated thin-film BST capacitors," *2004 IEEE MTT-S Int. Microw. Symp. Dig.*, Vol. 1, 2004, pp. 261–264.

[322] S. Summerfelt, "(Ba,Sr)TiO_3 thin films for DRAM," Chapter 1 of *ThinFilm Ferroelect. Mater. Devices*, R. Ramesh, ed., Kluwer: Boston, 1997.

[323] N. Mikami, "(Ba,Sr)TiO_3 films and process integration for DRAM capacitors," Chapter 2 of *ThinFilm Ferroelectric Materials and Devices*, R. Ramesh, ed., Kluwer: Boston, 1997.

[324] K. Eisenbeiser, J. M. Finder, Z. Yu, J. Ramdani, J. A. Curless, J. A. Hallmark, R. Droopad, W. J. Ooms, L. Salem, S. Bradshaw, and C. D. Overgaard, "Field effect transistors with $SrTiO_3$ gate dielectric on Si," *Applied Phys. Lett*, Vol. 76, No.10, 6 March 2000, pp. 1324–1326.

[325] H. Xu, N. K. Pervez, P. J. Hansen, L.Shen, S. Keller, J. S. Speck, U. Mishra, R. A. York, "Integration of $Ba_xSr_{1-x}TiO_3$ thin-films with AlGaN/GaN HEMT Circuits," *IEEE Electron Device Letters*, Vol. 25, Feb 2004, pp. 49–51.

[326] P. J. Hansen, L. Shen, Y. Wu, A. Stonas, Y. Terao, S. Heikman, D. Buttari, T. R. Taylor, S. P. DenBaars, U. K. Mishra, R. A. York, and J. S. Speck, "AlGaN/GaN Metal-oxide-semiconductor heterostructure field effect transistors using barium strontium titanate," *J. Vacuum Science and Technology B*, Vol. 22, No. 5, Sept/Oct 2004, pp. 2479–2485.

[327] H. Xu, C. Sanabria, N. K. Pervez, S. Keller, U. K. Mishra, and R. A. York, "Low phase-noise 5 GHz AlGaN/GaN HEMT oscillator integrated with $Ba_xSr_{1-x}TiO_3$ thin films," *2004 IEEE Int. Microw. Symp. Dig.*, Vol. 3, June 2004, pp. 1509–1512.

[328] Y. Liu, T. R. Taylor, J. Speck, and R. A. York, "High-isolation BST-MEMS switches," *IEEE Int. Microw. Symp. Dig.*, June 2002, pp. 227–230.

[329] G. Wang, T. Polley, A. Hunt, and J. Papapolymerou, "A high performance tunable RF MEMS switch using barium strontium titanate (BST) dielectrics for reconfigurable antennas and phased arrays," *IEEE Antennas Wirel. Propag. Lett.*, Vol. 4, 2005, pp. 217–220.

[330] A. Victor, J. Nath, D. Ghosh, B. Boyette, J.-P. Maria, M. B. Steer:, A. I. Kingon and G. T. Stauf, "A voltage controlled oscillator using barium strontium titanate (BST) thin film varactor," *IEEE Radio Wirel. Conf.*, Sept. 2004, pp. 91–94; see also A. Victor, J. Nath, D. Ghosh, B. Boyette, J-P. Maria, M. B. Steer, A. I. Kingon, G. T. Stauf, "Noise characteristics of an oscillator with a barium strontium titanate (BST) varactor," *IEE Proc. Microw. Antennas Prop.*, Vol. 153, Feb 2006, pp. 96–102.

[331] G. Subramanyam, F. Ahamed and R. Biggers, "A Si MMIC compatible ferroelectric varactor shunt switch for microwave applications," *IEEE Microw. Wirel. Compon. Lett.*Vol. 15, Nov. 2005, pp. 739–741.

[332] J. Mitola, "The software radio architecture," *IEEE Commun. Mag.*, May 1995, pp. 26–38.

[333] C. Bergstrom, S. Chuprun, and D. Torrieri, "Adaptive spectrum exploitation using emerging software defined radios," *Proc. IEEE RAWCON*, 1999.

[334] G. Matthei, L. Young, and E. M. T. Jones, *Microwave Filters, Impedance Matching Networks, and Coupling Structures*, McGraw-Hill, 1964.

[335] L. P. B. Katehi, G. M. Rebeiz, and C. T.-C. Nguyen, "MEMS and micromachined components for low power, high frequency communication systems," *IEEE MTT-S Int. Microwave Symp. Dig.*, 1998, pp. 331–333.

[336] J. Papapolymerou, K. L. Lange, C. L. Goldsmith, A. Malczewski, and J. Kleber, "Reconfigurable double-stub tuners using MEMS switches for intelligent RF front-ends," *IEEE Trans. on Microwave Theory & Tech.*, Vol. 51, No. 1, Pt. 2, Jan. 2003, pp. 271–278.

[337] D. Qiao, R. Molfino, S. M. Lardizabal, B. Pillans, P. M. Asbeck, and G. Jerinic, "An intelligently controlled RF power amplifier with a reconfigurable MEMS-varactor tuner," *IEEE Trans. on Microwave Theory & Tech.*, Vol.53, No.3, Pt.2, March 2005, pp. 1089–95.

[338] S. Cripps, *RF Power Amplifiers for Wireless Communications*, Artech House, 1999.

[339] W. H. Doherty, "A new high efficiency power amplifier for modulated waves," *Proc. IRE* Vol. 24, No. 9, Sept. 1936, pp. 1163–1182.

[340] F. H. Raab, "Efficiency of Doherty RF power-amplifier systems," *IEEE Trans. Broadcast.* Vol. 33, 1987, pp. 77–83.

[341] M. Iwamoto, *et al.*, "An extended doherty amplifier with high efficiency over a wide power range," *IEEE Trans. Microw. Theory Tech.*, Vol. 49, Dec. 2001, pp. 2472–2479.

[342] Y. Zhao, *et al.*, "Doherty amplifier with DSP control to improve performance in CDMA operation," *2003 IEEE MTT-S Int. Microwave Symp. Dig.*, 200, pp. 687–6903.

[343] D. Qiao, Y. Zhao, T. Hung, D. Kimball, M. Li, P. Asbeck, D. Choi and D. Kelly, "Antenna impedance mismatch measurement and correction for adaptive cdma transceivers," *2005 IEEE MTT-S Int. Microwave Symp.*, June 2005, pp. 783–786.

[344] G. A. Garcia, R. E. Reedy, and M. L. Burgener, "High-quality CMOS in thin (100 nm) silicon on sapphire," *IEEE Electron Device Lett.*, Vol. 9, No. 1, 1988, pp. 32–34.

[345] P. B. Kenington, *High Linearity RF Amplifier Design*, Artech House, 2000.

[346] C. J. Clark, C. P. Silva, A. A. Moulthrop and M. S. Muhal, "Power-amplifier characterization using a two-tone measurement technique," *IEEE Trans. Microw. Theory Tech.*, Vol 50, No. 6, Jun. 2002, pp. 1590–1602.

[347] S. W. Chen, W. Panton, and R. Gilmore, "Effects of nonlinear distortion on CDMA communications systems," *IEEE Trans. Microw. Theory Tech*., Vol. 44, No. 12, Dec. 1996, pp. 2743–2750.

[348] H. Ku, M. D. McKinle and J. S. Kenney, "Extraction of accurate behavioral models for power amplifiers with memory effects using two tone measurements," *2002 IEEE MTT-S Int. Microwave Symp. Dig.*, Jun. 2002, pp. 139–142.

[349] K. Gard, M. B. Steer, and L. E. Larson, "Generalized autocorrelations of spectral regrowth from bandpass nonlinear circuits," *2001 IEEE MTT-S Int. Microw. Symp. Dig.*, Jun. 2001, pp. 9–12.

[350] G. T. Zhou and J. S. Kenney, "Predicting spectral regrowth of nonlinear power amplifiers," *IEEE Trans. Commun.*., Vol 50, No.5, May 2002, pp. 718–722.

[351] E. G. Jeckeln, F. M. Ghannouchi and M. A. Sawan, "A new adaptive predistortion technique using software-defined radio and DSP technologies suitable for base station 3G power amplifiers," *IEEE Trans. Microw. Theory Tech.*, Vol 9, No. 9, Jun. 2002, pp. 2139–2148.

[352] M. Y. Li, I. Galton, L. E. Larson and P. M. Asbeck, "Correlation technique for estimation of amplifier nonlinearity," *2004 IEEE Radio Wireless Conf. (RAWCON) Dig.*, Sep. 2004, pp. 179–182.

[353] A. M. Saleh, "Frequency-independent and frequency-dependent nonlinear models of TWT amplifiers," *IEEE Trans. Commun.*, Vol. 29, No.11, Nov. 1981, pp. 1715–1720.

[354] S. P. Stapleton, G. S. Kandolà and J. K. Cavers, " Simulation and analysis of an adaptive predistorter utilizing a complex spectral convolution," *IEEE Trans. MTT.*, Vol. 41, No. 4, Nov. 1992, pp. 387–393.

[355] J. Rode, J. Hinrichs, and P. Asbeck, "Transmitter architecture using digital generation of RF signals," *Proc. 2003 IEEE Radio Wireless Conf. (RAWCON)*, 2003, pp. 245–248.

[356] J. Keyzer, *et al.*, "Digital generation of rf signals for wireless communications with band-pass delta-sigma modulation," *2001 IEEE MTT-S Int. Microwave Symp. Dig.*, Vol. 3, pp. 2127–2130.

[357] I. Galton, "Delta–sigma data conversion in wireless transceivers," *2002 IEEE MTT-S Int. Microwave Symp. Dig.*, Vol.50, pp. 302–315.

[358] P. Wagh, P. Midya, P. Rakers, and J. Caldwell, T. Schooler, "An all-digital universal RF transmitter" *IEEE Custom Integr. Circuit Conf.*, 3–6 Oct. 2004, pp. 549–552.

[359] R. B. Staszewski, D. Leipold, K. Muhammad and P. T. Balsara, "Digitally controlled oscillator (DCO)-based architecture for RF frequency synthesis in a deep-submicrometer CMOS process," *IEEE Trans. Circuits Systems II*, Vol. 50, 2003, pp. 815.

[360] T. O'Sullivan, R. York, B. Noren, and P. Asbeck, "Adaptive duplexer implemented using feedforward technique with a BST phase shifter," *2004 IEEE MTT-S Int. Microw. Symp. Dig.*, Vol. 2, 2004, pp. 1299–1302.

[361] S. B. Cohn, "Parallel-coupled transmission-line resonator filters," *IRE Trans. Microwave Theory Tech.*, Vol. 6, Apr. 1958, pp. 223–231.

[362] B. J. Minnis, "Printed circuit coupled-line filters for bandwidths up to and greater than an octave," *IEEE Trans. Microw. Theory Tech.*, Vol. 29, Mar. 1981, pp. 215–222.

[363] E. G. Cristal, and S. Frankel, "Hairpin-line and hybrid hairpin-line/half-wave parallel-coupled-line filters," *IEEE Trans. Microw. Theory Tech.*, Vol. 20, Nov. 1972, pp. 719–728.

[364] U. H. Gysel, "New theory and design for hairpin-line filters," *IEEE Trans. Microw. Theory Tech.*, Vol. 22, May 1974, pp. 523–531.

[365] G. I. Zysman and A. Kent Johnson, "Coupled transmission line networks in an inhomogeneous dielectric medium," *IEEE Trans. Microw. Theory Tech.*, Vol. No. 17, Oct. 1969, pp. 753–759.

[366] I. Hunter, *Theory and Design of Microwave Filters*, IEE, 2001.

[367] M. Horton and R. Wenzel, "General theory and design of optimum quarter-wave TEM filters," *IEEE Trans. Microw. Theory Tech.*, Vol. 13, May 1965, pp. 316–327.

[368] J. A. G. Malherbe, *Microwave Transmission Line Filters*, Artech House, 1979.

[369] I. Bahl, "Capacitively compensated high performance parallel coupled microstrip filters," *1989 IEEE MTT-S Int. Microwave Symp. Dig.*, June 1989, pp. 679–682.

[370] S. M. Wang, C. H. Chi, M. Y. Hsieh, C. Y. Chang, "Miniaturized spurious passband suppression microstrip filter using meandered parallel coupled lines," *IEEE Trans. Microw. Theory Tech.*, Vol. 53, Feb. 2005, pp. 747–753.

[371] M. C. Velazquez-Ahumada, J. Martel, and F. Medina, "Parallel coupled microstrip filters with ground-plane aperture for spurious band suppression and enhanced coupling," *IEEE Trans. Microw. Theory Tech.*, Vol. 52, Mar. 2004, pp. 1082–1086.

[372] M. C. Velazquez-Ahumada, J. Martel, and F. Medina, "Parallel coupled microstrip filters with floating ground-plane conductor for spurious-band suppression," *IEEE Trans. Microw. Theory Tech.*, Vol. 53, May 2005, pp. 1823–1828.

[373] L. Zhu, H. Bu, and K. Wu, "Broadband and compact multi-pole microstrip bandpass filters using ground plane aperture technique," *IEE Proc. Microw. Antennas Propag.*, Vol. 149, Feb. 2002, pp. 71–77.

[374] J. T. Kuo, M. Jiang, and H. J. Chang, "Design of parallel-coupled microstrip filters with suppression of spurious resonances using substrate suspension," *IEEE Trans. Microw. Theory Tech.*, Vol. 52, Jan. 2004, pp. 83–89.

[375] J. T. Kuo and M. Jiang, "Enhanced microstrip filter design with a uniform dielectric overlay for suppressing the second harmonic response," *IEEE Microw. Wirel. Compon. Lett.*, Vol. 14, Sept. 2004, pp. 419–421.

[376] T. Lopetegi, M. A. G. Laso, J. Hernandez, M. Bacaicoa, D. Benito, M. J. Garde, M. Sorolla, and M. Guglielmi, "New microstrip "wiggly-line" filters with spurious passpand suppression," *IEEE Trans. Microw. Theory Tech.*, Vol. 49, Sept. 2001, pp. 1593–1598.

[377] T. Lopetegi, M. A. G. Laso, F. Falcone, F. Martin, J. Bonache, J. Garcia, L. Perz-Ceivas, M. Sorolla, and M. Guglielmi, "Microstrip "wiggly-line" filters with multispurious rejection," *IEEE Microw. Wirel. Compon. Lett.*, Vol. 14, Nov. 2004, pp. 531–533.

[378] B. S. Kim, J. W. Lee, and M. S. Song, "An implementation of harmonic-suppression microstrip filters with periodic grooves," *IEEE Microw. Wirel. Compon. Lett.*, Vol. 14, Sept. 2004, pp. 413–415.

[379] J. T. Kuo, W. H. Hsu, and W. T. Huang, "Parallel coupled microstrip filters with suppression of harmonic response," *IEEE Microw. Wirel. Compon. Lett.*, Vol. 12, Oct. 2002, pp. 383–385.

[380] J. T. Kuo, W. H. Hsu, and W. T. Huang, "Tapped wiggly-coupled technique applied to microstrip bandpass filters for multi-octave spurious suppression," *Electron. Lett.*, Vol. 40, Jan. 2004, pp. 46–47.

[381] W. M. Fathelbab and M. B. Steer, "Parallel-coupled line filters with enhanced stopband performance," *IEEE Trans. Microw. Theory Tech.*, Dec. 2005, No. 12, pp. 3774–3781.

[382] J. Garcia-Garcia, F. Martin, F. Falcone, J. Bonache, I. Gil, T. Lopetegi, M. Laso, M. Sorolla, and R. Marques, "Spurious passband suppression in microstrip coupled line band pass filters by means of split ring resonators," *IEEE Microw. Wirel. Compon. Lett.*, Vol. 14, Sept. 2004, pp. 416–418.

[383] M. Makimoto and S. Yamashita, "Bandpass filters using parallel coupled striplines stepped impedance resonators," *IEEE Trans. Microw. Theory Tech.*, Vol. 28, Dec. 1980, pp. 1413–1417.

[384] R. Mongia, I. Bahl and P. Bhartia, *RF and Microwave Coupled-line Circuits*, Artech House, 1999, pp. 411–435.

[385] I. D. Robertson and S. Lucyszyn, *RFIC and MMIC Design and Technology*, IEE Circuits, Devices and Systems Series 13, 2001.

[386] D. Kuylenstierna and P. Linner, "Design of broad-band lumped-element baluns with inherent impedance transformation," *IEEE Trans. Microw. Theory Tech.*, Vol. 52, No. 12, Dec. 2004, pp. 2739–2745.

[387] B. J. Minnis and M. Healy, "New broadband balun structures for monolithic microwave integrated circuits," 1991 *IEEE MTT-S Int. Microw. Symp. Dig.*, June 1991, pp. 425–428.

[388] W. M. Fathelbab and M. B. Steer, "New classes of miniaturized planar marchand baluns," *IEEE Trans. Microw. Theory Tech.*, Vol. 53, No. 4, April 2005, pp. 1211–1220.

[389] C. L. Goldsmith, A. Kikel, and N. L. Wilkens, "Synthesis of Marchand baluns using multilayer microstrip structures," *Int. J. Microw. Millim.-Wave Compu.-Aided Eng.*, Vol. 2, No. 3, 1992, pp. 179–188.

[390] N. Marchand, "Transmission line conversion transformers," *Electron.*, Dec. 1944, pp. 142–145.

[391] W. M. Fathelbab and M. B. Steer, "New classes of miniaturized planar Marchand baluns," *IEEE Trans. Microw. Theory Tech.*, Vol. 53, No. 4, April 2005, pp. 1211–1220.

[392] J. Cloete "Graphs of circuit elements for the Marchand balun," *Microw. J.*, Vol. 24, May 1981, pp. 125–128.

[393] J. A. G. Malherbe, "Realization of elliptic function bandstop filters by means of resonated prototype," *IEEE Trans. Microw. Theory Tech.*, Vol. 25, Aug. 1977, p. 717.

[394] H. J. Orchard and G. C. Temes, "Filter design using transformed variable," *IEEE Trans. Circuit Theory*, Vol. 15, Dec 1968, pp. 385–408.

[395] R. J. Wenzel, "Synthesis of combline and capacitively loaded interdigital bandpass filters of arbitrary bandwidth," *IEEE Trans. Microw. Theory Tech.*, Vol. 19, Aug. 1971, pp. 678–686.

[396] Infineon Technologies, Silicon tuning diodes, BB833.

[397] D. M. Pozar, *Microwave Engineering*, John Wiley & Sons, 1998.

[398] Mini-Circuits HELA-10B, 50 MHz to 1 GHz amplifier.

[399] H. Ozaki and J. Ishii, "Synthesis of a class of strip-line filters," *IRE Trans. Circuit Theory*, Vol. 5, June 1958, pp. 104–109.

[400] G. Matthaei, L. Young and E. M. T. Jones, *Microwave Filters Impedance-Matching Networks, and Coupling Structures*, Artech House, 1980.

[401] R. Levy and L. F. Lind, "Synthesis of symmetrical branch-guide directional couplers," *IEEE Trans. Microw. Theory Tech.*, Vol. No. 16, Feb. 1968, pp. 80–89.

[402] B. J. Minnis, "Decade bandwidth bias Ts for MIC applications up to 50 GHz," *IEEE Trans. Microw. Theory Tech.*, Vol. No. 35, June 1987, pp. 597–600.

[403] R. J. Wenzel, "Application of exact synthesis methods to multi-channel filter design," *IEEE Trans. Microw. Theory Tech.*, Vol. 13, No. 1, Jan. 1965, pp. 5–15.

[404] R. Levy, "Synthesis of general asymmetric singly- and doubly-terminated cross-coupled filters," *IEEE Trans. Microw. Theory Tech.*, Vol. 42, No. 12, Dec. 1994, pp. 2468–2471.

[405] B. J. Minnis, *Designing Microwave Circuits by Exact Synthesis*, Artech House, 1996.

[406] H. J. Carlin and P. P. Civalleri, *Wideband Circuit Design. CRC Press LCC*, 1998.

[407] E. Cristal, "Tapped-line coupled transmission lines with applications to interdigital and Combline filters," *IEEE Trans. Microw. Theory Tech.*, Vol. 23, No. 12, Dec. 1975, pp. 1007–1012.

[408] P. Cheong, S. W. Fok, and K. W. Tam, "Miniaturized parallel coupled-line bandpass filter with spurious-response suppression," *IEEE Trans. Microw. Theory Tech.*, Vol. 53, May 2005, pp. 1810–1816.

[409] J. A. G. Malherbe, *Microwave Transmission Line Filters*, Artech House, 1979, pp. 71–108.

[410] R. Levy, R. V. Snyder, and S. Shin, "Bandstop filters with extended upper passbands," *IEEE Trans. Microw. Theory Tech.*, Vol. 54, No. 6, June 2006, pp. 2503–2515.

[411] H. C. Bell, "L-Resonator bandstop filters," *IEEE Trans. Microw. Theory Tech.*, Vol. 44, No. 12, Dec. 1996, pp. 2669–2672.

[412] P. L. Field, I. C. Hunter, and J. G. Gardiner, "Asymmetric bandpass filter using a novel microstrip circuit," *IEEE Microw. Guid. Wave Lett.*, Vol. 2, June 1992, pp. 247–249.

[413] P. L. Field, I. C. Hunter, and J. G. Gardiner, "Asymmetric bandpass filter using a ceramic structure," *IEEE Microw. Guid. Wave Lett.*, Vol. 2, Sept. 1992, pp. 361–363.

[414] R. J. Cameron, A. R. Harish, and C. J. Radcliffe, "Synthesis of advanced microwave filters without diagonal cross-couplings," *2002 IEEE MTT-S Int. Microw. Symp. Dig.*, pp. 1437–1440.

[415] R. Hershtig, R. Levy, and K. Zaki, "Synthesis and design of cascaded trisection (CT) dielectric resonator filters," *27th Eur. Microw. Conf. Proc*, 1997, pp. 784–791.

[416] R. J. Cameron, "General prototype network synthesis methods for microwave filters," *European Space Agency (ESA) J.*, Vol. 6, 1982, pp. 193–206.

[417] I. C. Hunter and J. D. Rhodes, "Electronically tunable microwave bandpass filters," *IEEE Trans. Microw. Theory Tech.*, Vol. 30, Sep. 1982, pp. 1354–1360.

[418] J. Nath, D. Ghosh, J.-P. Maria, A. I. Kingon, W. Fathelbab, P. D. Franzon, and M. B. Steer, "An electronically-tunable microstrip bandpass filter using thin-film barium strontium titanate (BST) varactors," *IEEE Trans. Microw. Theory Tech.*, Vol. 53, No.9, Sep. 2005, pp. 2707–2712.

[419] J. G. Hong and M. K. Lancaster, *Microstrip Filters for RF/Microwave Applications*, John Wiley & Sons, 2001.

[420] R. Levy and P. Petre, "Design of CT and CQ filters using approximation and optimization," *IEEE Trans. Microw. Theory Techn.*, Vol. 49, Dec. 2001, pp. 2350–2356.

[421] R. Levy and J. D. Rhodes, "A comb-line elliptic filter," *IEEE Trans. Microw. Theory Techn.*, Vol. 19, Jan. 1971, pp. 26–29.

[422] A. I. Zverev, *Handbook of Filter Synthesis*, John Wiley & Sons, 1967.

[423] T. C. Edwards and M. B. Steer, *Foundations of Interconnect and Microstrip Design*, John Wiley & Sons, 3 Ed., 2000, pp. 350–352.

[424] T. Samoilova, A. Ivanov, A. Kozyrev, O. Soldatenkov, M. Zelner, T. A. Bernacko, and A. Cervin-Lawry, "Power handling capability of ferroelectric film varactors and tunable microwave devices," *2006 IEEE MTT-S Int. Microw. Symp. Dig.*, June 2006, pp. 1273–1276.

[425] J. Nath, D. Ghosh, J. Maria, M. B. Steer, A. I. Kingon and G. T. Stauf, "Microwave properties of BST thin film interdigital capacitor on low cost alumina substrate," *Proc. 34 th Eur. Microw. Conf.*, Oct. 2004, pp. 1497–1500.

[426] D. Ghosh, B. J. Laughlin, J. Nath, A. I. Kingon, M. B. Steer, and J. P. Maria, "Tunable high quality factors interdigitated (BaSr)TiO_3 capacitors fabricated on low cost substrates with copper metallization," *Thin Solid Films*, Vol. 496, No. 2, Feb. 2006, pp. 669–673.

[427] J. D. Kraus, "The corner-reflector antenna," *Proc. IRE*, Vol. 28, 513–519, 1940.

[428] L. C. Van Atta, "Electromagnetic Reflector," U. S. Patent No. 2,908,002; Oct. 6, 1959.

[429] E. Sharp, M. Diab, "Van Atta reflector array," *IEEE Trans. Antennas Propag.*, Vol. 8, No. 4, Jul. 1960, pp. 436–438.

[430] C. Y. Pon, "RDA using the heterodyne technique," *IEEE Trans. Antennas Propag.*, Vol. 12, Mar. 1964, pp. 176–180.

[431] Y. Chang, H. R. Fetterman, I. Newberg, and S. K. Panaretos, "Microwave phase conjugation using antenna arrays," *IEEE Trans. Microw. Theory Tech.*, Vol. 46, No. 11, Nov. 1998, pp. 1910–1919.

[432] R. Y. Miyamoto, K. M. K. H. Leong, S. Jeon, Y. Wang, Y. Qian, and T. Itoh, "Digital wireless sensor server using an adaptive smart-antenna/retrodirective array," *IEEE Trans. Vehicular Technol.*, Vol. 52, No. 5, Sept. 2003, pp. 1181–1188.

[433] K. M. K. H. Leong, R. Y. Miyamoto, S.-S. Jeon, Y. Wang, and T. Itoh, "A frequency autonomous retrodirective array transponder," *2002 IEEE MTT-S Int. Microw. Symp. Dig.*, June 2002, pp. 1349–1352.

[434] Y. Qian, W. R. Deal, N. Kaneda, and T. Itoh, "Microstrip-fed quasi-Yagi antenna with broadband characteristics," *Electron. Lett.*, Volume 34, No. 23, Nov.1998, pp. 2194–2196.

[435] K. M. K. H. Leong, Y. Wang, and T. Itoh, "A full duplex capable retrodirective array system for high-speed beamtracking and pointing applications," *IEEE Trans. Microw. Theory Tech.*, Vol. 52, No. 5, May 2004, pp. 1479–1489.

[436] D. S. Goshi, K. M. K. H. Leong, and T. Itoh, "A secure high-speed retrodirective communication link," *IEEE Trans. Microw. Theory Tech.*, Vol. 53, No. 11, Nov. 2005, pp. 3548–3556.

[437] D. S. Goshi, K. M. K. H. Leong, and T. Itoh, "A sparsely designed retrodirective transponder," *IEEE Antennas Wirel. Propag. Lett.*, Vol. 5, No. 1, Dec. 2006, pp. 339–342.

[438] D.S. Goshi, K. M. K. H. Leong, and T. Itoh, "A scheme for hardware reduction in wireless retrodirective transponders," *2006 IEEE MTT-S Int. Microwave Symp. Dig.*, Jun. 2006.

[439] D. S. Goshi, K. M. K. H. Leong, and T. Itoh, "A sparse retrodirective transponder with a time shared phase-conjugator", *IEEE Trans. Antennas Propag.*, Vol. 55, No. 8, Aug. 2007, pp. 2367–2372.

[440] Austeng, A., Holm, S., "The impact of 'non-half-wavelength' element spacing on sparse array optimization," *5 th Nordic Signal Processing Symp.*, NORSIG-2002 2002.

[441] M. G. Bray, D. H. Werner, D. W. Boeringer and D. W. Machuga, "Optimization of thinned aperiodic linear phased arrays using genetic algorithms to reduce grating lobes during scanning," *IEEE Trans. Antennas Propag.*, Vol. 50, No. 12, Dec. 2002, pp. 1732–1742.

[442] D. S. Goshi, Y. Wang, and T. Itoh, "A compact digital beamforming SMILE array for mobile communications," *IEEE Trans. Microw. Theory Tech.*, Vol. 52, No. 12, Dec. 2004, pp. 2732–2738.

[443] P. K. Hughes and J. Y. Choe, "Overview of advanced multifunction RF systems (AMRFS)," in *Proc. IEEE Int. Conf. on Phased Array Systems and Technology*, May 2000, pp. 21–24.

[444] P. Antonki, R. Bonneau, R. Brown, S. Ertan, V. Vannicola, D. Weiner, and M. Wicks, "Bistatic radar denial/embedded communications via waveform diversity," in *Proc. IEEE Radar Conf.*, May 2001, pp. 41–45.

[445] W. Gugler, A. Springer, and R Weigel, "A robust SAW-based chirp–Pi/4 DQPSK system for indoor applications," *IEEE Int. Conf. Commun.*, Vol. 2, June 2000, pp. 773–777, 18–22.

[446] C. Cook and M. Bernfeld, *Radar Signal — An Introduction to Theory and Application.* New York: Academic, 1967.

[447] M. R. Winkler, "Chirp signals for communications," in *IEEE WESCON Conv. Rec.*, Paper 14.2, 1962.

[448] G. W. Barnes, D. Hirst, and D. J. James, "Chirp modulation system in aeronautical satellites," in *AGARD Conf. Proc. 87 on Avionics in Spacecraft*, North Atlantic Treat Organization, 1971, pp. 30.1–30.10.

[449] G. W. Judd and V. H. Estrick, "Applications of surface acoustic wave (SAW) filters-an overview," in *Proc. Soc. Photo-Opt. Instrum. Eng.*, Vol. 239, Jul. 1980, pp. 220–235.

[450] W. Hirt and S. Pasupathy, "Continuous phase chirp (CPC) signals for binary data communication—Part I: Coherent detection and Part II: Non-coherent detection," in *IEEE Trans. Commun.*, Vol. 29, pp. 836–856.

[451] G. F. Gott and J. P. Newsome, "H.F. data transmission using chirp signals," *Proc. Inst. Elec. Eng.*, Vol. 118, Sept. 1971, pp. 1162–1166.

[452] A. J. Berni and W. D. Gregg, "On the utility of chirp modulation for digital signaling," *IEEE Trans. Commun.*, Vol. Com–21, June 1973, pp. 748–751.

[453] A. Springer, M. Huemer, L. Reindl, C. C. W. Ruppel, A. Pohl, F. Seifert, W. Gugler, and R. Weigel, "A robust ultra-broad band wireless communication system using SAW chirped delay lines," *IEEE Trans. Microwave Theory Tech.*, Vol. 46, Dec. 1998, pp. 2213–2218.

[454] A. Springer, W. Gugler, M. Huemer, R. Keller, and R. Weigel, "A wireless spread spectrum communication system using SAW chirped delay lines," *IEEE Trans. Microwave Theory and Tech.*, Vol. 49, Apr. 2001, pp. 754–760.

[455] K. Siwiak, D. McKeown, *Ultra-Wideband Radio Technology*, John Wiley & Sons., 2004.

[456] J. R. Andrews, "UWB signal sources, antennas and propagation," *IEEE Tropical Conf. on Wirel. Communication Technology*, Oct. 2003, pp. 439–440, 15–17.

[457] J. D. Taylor, *Ultra-Wideband Radar Technology*, CRC, Boca Raton, Fl, 2000.

[458] A. Yarovoy, "Ultra-wideband systems," *33rd European Microwave Conf.*, Vol. 2, Oct. 2003, pp. 597–600, 7–9.

[459] K. Siwiak, "The potential of ultrawideband communications," Twelfth Int. Conf. on Antennas and Propagation, Vol. 1, March 2003, pp. 225–228.

[460] M. Roberton and E. R. Brown, "Integrated Radar and Communication based on Chirped Spread-Spectrum Tech.," *2003 IEEE MTT-S Int. Microwave Symp. Dig.*, Vol. 1, June 2003, pp. 611–614.

[461] G. N. Saddik and E. R. Brown, "Towards a Multifunctional LFM-Waveform Communication/Radar System: Single-Crystal AlN-on-SiC SAW Filters for X-Band," *Government Microcircuit Applications & Critical Tech. Conf.*, Las Vegas, April, 2005, pp. 399–402.

[462] D. K. Barton, C. E. Cook, and P. H. Hamilton, *Radar Evaluation Handbook.* Massachusetts: Artech House, Inc. 1991.

[463] B. Razavi, *Design of ICs for Optical Communications*, McGraw-Hill, 2003.

[464] L. V. Blake, *Radar Range-Performance Analysis*, (Artech House, Norwood, MA, 1986), Table 3-1, pp. 129.

[465] *Principles of Optics* 5 th Ed., M. Born and E. Wolf, New York: Pergamon, 1975.

[466] L. C. Van Atta, "Electromagnetic reflector," U.S. Patent No. 2908002, 1959.

[467] R. Y. Miyamoto, Y. Qian, and T. Itoh, "A retrodirective array using balanced quasi-optical FET mixers with conversion gain," *1999 IEEE MTT-S Int. Microwave Symp. Dig.*, June 1999, pp. 655–658.

[468] B. Y. Toh, V. F. Fusco and N. B. Buchanan, "Assessment of performance limitations of retrodirective arrays," *IEEE Trans. on Antennas and Propagation*, Vol. 50, No. 10, Oct. 2002, pp. 1425–1432.

[469] L. D. DiDomenico and G. M. Rebeiz, “Digital communications using self-phased arrays,” *IEEE Trans. Microw. Theory Tech.*, Vol.49, No.4, Part 1, April 2001, pp. 677–84.

[470] M. Dawood and R. M. Narayanan, “Receiver operating characteristics for the coherent UWB random noise radar,” *IEEE Trans. on Aerospace and Electronic Systems*, Vol.37, No. 2, April 2001, pp. 586–94.

[471] K. A. Lukin, “Developments of noise radar technology in LNDES IRE NASU,” *Proc. of the First Int. Conf. on Noise Radar Technology*, 18-22 Sept. 2002, Yalta, Ukraine, pp. 90–96.

[472] E. R. Brown, A. C. Cotler, A. Umali and S. Gupta, “First demonstration of a retrodirective noise-correlating radar in S Band,” *2004 IEEE MTT-S Int. Microwave Symp. Dig.*, June 2004, pp. 751–754.

Contributors

Elliott Brown is Professor of Electrical and Computer Engineering at the University of California. Santa Barbara. He is teaching courses in solid-state engineering, RF sensors, and THz science, and is conducting research in the THz field in several areas including ultra-low-noise rectifiers, photomixing sources, the THz phenomenology of biomaterials, and THz remote sensor and imager design and simulation. Other areas of research include multifunctional RF electronics and systems, biomedical ultrasonic imaging in and around hard tissue (in collaboration with the UCLA Dental and Medical Schools), and electronic and photonic transport in nanostructures. Prior to UCSB Dr. Brown was a Professor of Electrical Engineering at the University of California, Los Angeles, and prior to that was a Program Manager at DARPA in the Electronics Technology Office, Arlington, VA. He earned a Ph.D. in Applied Physics from the California Institute of Technology in 1985, and did his post-doctoral work at Lincoln Laboratory, Massachusetts Institute of Technology. Dr. Brown is a Fellow of the IEEE (since 2000) and a Member of the American Physical Society. In 1998 he received an Award for Outstanding Achievement from the U.S. Office of the Secretary of Defense.

Austin Cotler received his Bachelor of Science degree in electrical engineering from UCLA in 2000. Austin is currently employed as an electrical engineer at Rockwell Collins at the commercial systems facility in Tustin, California. Prior to this, Austin spent several years working as a staff engineer in Professor Elliot R. Brown's research group at both UCLA and UCSB. His research included the design and demonstration of a Retro-Directive Noise Correlating Radar in S-band, the simulation and fabrication of vivaldi antennas, thermal management of LDMOS and GaN power amplifiers, and variable matching networks.

Kevin Gard received the B.S. and M.S. degrees in electrical engineering from North Carolina State University, Raleigh, in 1994 and 1995, respectively, and the Ph.D. degree in electrical engineering from the University of California at San Diego, La Jolla, in 2003. He is currently the William J. Pratt Assistant Professor with the Electrical and Computer Engineering Department, North Carolina State University. From 1996 to 2003, he was with Qualcomm Inc., San Diego, CA, where he was a Staff Engineer and Manager responsible for the design and development of RF integrated circuits (RFICs) for code-division multiple-access (CDMA) wireless products. He has designed SiGe BiCMOS, Si BiCMOS, and GaAs metal–semiconductor field-effect transistor (MESFET) integrated circuits for cellular and personal communication systems (PCSs) CDMA, wideband code-division multiple-access (WCDMA), and AMPS transmitter applications. His research interests are in the areas of integrated circuit design for wireless applications and analysis and modeling of nonlinear microwave circuits with digitally modulated signals. Dr. Gard is a member of the IEEE Microwave Theory and Techniques and Solid-State Circuits Societies, Sigma Xi, Eta Kappa Nu, and Tau Beta Pi.

Darren Goshi received the B.S. degree in electrical engineering from the University of Hawaii at Manoa in 2002 and the M.S. and Ph.D. degrees in electrical engineering from the University of California, Los Angeles in 2004 and 2007. He is currently a postdoctoral scholar at UCLA working on retrodirective array development and genetic algorithms for phased array radar imaging systems.

His research interests include active microwave circuits, smart antennas, phased arrays and radar systems.

Wael Fathelbab received the Bachelor of Engineering with first class honors from the University of Bradford in 1995. In 1995 he was awarded the prestigious 'Overseas Research Students' award from the U.K. Department of Higher Education with Filtronic Comtek (U.K.) Ltd. (now Powerwave Technologies Inc.) industrial sponsorship to pursue the degree of Doctor of Philosophy (Ph.D.) at the University of Bradford. In 1999 he was awarded his Ph.D. degree which was focused on the synthesis of passive microwave filters with finite Q circuit elements. From 1999 to 2001, he was a Radio Frequency (RF) engineer at Filtronic Comtek (U.K.) Ltd., where he was involved in the design and development of different classes of filters and multiplexers for various cellular base-station applications. He was subsequently involved in the architecture of novel RF front-end transceivers for the U.K. and European markets when he joined the Mobile Handset Division at NEC Technologies (U.K.) Ltd. Between the period of 2002 to 2006, he was appointed as Research Fellow with the Department of Electrical and Computer Engineering, North Carolina State University, U.S.A. Dr. Fathelbab is currently an Assistant Professor at the Department of Electrical and Computer Engineering, South Dakota School of Mines and Technology, U.S.A. His research interests are focused on the synthesis of novel reconfigurable/tunable microwave coupling structures and the design of ultra-broadband differential and meta-material microwave circuits. He has been a senior member of the IEEE since 2005; is a reviewer of the IEEE Transactions on Microwave Theory and Techniques (TMTT) since 2003; is a reviewer of the IET Microwaves, Antennas & Propagation since 2005; and is presently an Associate Editor of the International Journal of Electronics.

Tatsuo Itoh received the Ph.D. Degree in Electrical Engineering from the University of Illinois, Urbana in 1969. He is currently Northrop Grumman Professor of Microwave and Millimeter Wave Electronics. Dr. Itoh is a Fellow of the IEEE, served as the Editor-in-Chief of IEEE Microwave Theory and Techniques (MTT) for 1983–1985, was President of the MTT Society in 1990 and was a founding Editor-in-Chief of IEEE Microwave and Guided Wave Letters from 1991 through 1994. He was elected as an Honorary Life Member of MTT Society in 1994. He received a number of awards including IEEE Third Millennium Medal and IEEE MTT Distinguished Educator Award. He is a member of the National Academy of Engineering. He was a Distinguished Microwave Lecturer for IEEE MTT Society on Microwave Applications of Metamaterial Structures for 2004–06. He has over 1000 publications and generated 68 Ph.D's in the area of microwave and millimeter-waves, computational electromagnetics, active and integrated antennas, microwave photonics, photonic bandgap structures, retrodirective transponders and negative index left handed metamaterials.

Linda Katehi is the Provost and Vice Chancellor for Academic Affairs at the University of Illinois at Urbana Campaign and Professor of Electrical and Computer Engineering. She holds a joint appointment with the Program of Gender and Women Studies at the University of Illinois. As a faculty member, Professor Katehi has focused her research on the development and characterization of three-dimensional integration and packaging of high-frequency circuits with particular emphasis on MEMS devices, high-Q passives and embedded filters. She pioneered the development of on-wafer packaging for high-density, high-frequency monolithic Si-based circuit and antenna architectures that lead to low-cost, high-performance integrated circuits for radar, satellite, and wireless applications. Her work in this area has led to numerous national and international technical awards and to distinctions as an Educator. Professor Katehi holds 14 U.S. patents and has authored more than 500 papers published in refereed journals and symposia proceedings. Professor Katehi is a member of the National Academy of Engineering, a Fellow of AAAS and a Fellow of IEEE. She has been a member of many scientific committees including the Nominations Committee for the National Medal of Science and the Committee for the National Medal of Technology, the Board of AAAS, the Kauffman National Panel for Entrepreneurship, the NSF Advisory Committee to the Engineering, the NRC Telecommunications Board, the NRC Army Research Lab Advisory Committee on Sensors and Electronics Division, the NSF Advisory Committee to CISE, the NASA Aeronautics Technical Advisory Committee (ARAC), and the DoD Advisory Group on Electron Devices. Professor Katehi earned her diploma degree from the National Technical University of Athens, Greece, in 1977 from the School of Mechanical and Electrical Engineering. Following her undergraduate studies, she worked for two years as a senior engineer in the Naval Research Lab and joined UCLA as a graduate student in the fall of 1979, completing an MSEE in December of 1981 and a PhD in Electrical Engineering in 1984. From 1984 to 2002 she was a faculty member of the Electrical Engineering and Computer Science department of the University of Michigan in Ann Arbor, MI, where she served as the Associate Dean for Academic Affairs from 1998 to 2002. From 2002 until 2006 she served as the Dean of Engineering and as faculty member of the Electrical and Computer Engineering at Purdue University.

Kevin Leong received the B.S. degree in electrical engineering from the University of Hawaii at Manoa in 1999 and the M.S. degree and Ph.D. in electrical engineering from the University of California at Los Angeles in 2001 and 2004, respectively. Dr. Leong was the first place winner of the best-student paper contest at the 2001 European Microwave Conference and was also awarded the Microwave Prize at the Asia Pacific Microwave Conference in 2006. From 2004 to 2007 he was a postdoctoral researcher at UCLA. He is currently working at Northrop Grumman Space Technology. His research interests include planar antennas, millimeter-wave circuits, and mobile communication systems.

W. Devereux Palmer W. Devereux Palmer received the B.A. degree in physics in 1980, and the M.S. and Ph.D. degrees in electrical engineering in 1988 and 1991 from Duke University in Durham, North Carolina. His field of graduate study was electromagnetic theory, and design, construction, and testing of microwave circuits and systems for practical applications. From 1991 to 2001 he served on the technical staff at MCNC Research and Development Institute where he worked on a number of technologies including silicon vacuum microelectronics for microwave power amplifiers, polymeric MEMS structures, high-Q HTS filters, wide-bandgap semiconductors for power electronics applications, radio and optical communications systems, and optical and electronic packaging. In 2000, he became the Director of the MCNC-RDI Optical and Electronic Packaging group, where he managed programs in development of lead-free flip-chip bumping processes, bumping and assembly of high-density tiled detector arrays for particle accelerators, and packaging for OC-768 optical components. He holds the position of Adjunct Professor in the Department of Electrical and Computer Engineering at Duke University where he collaborates on research activities with other faculty members and graduate students, serves on graduate examination committees, and occasionally teaches introductory electromagnetics. Since his assignment to the US Army Research Office (Research Triangle Park, NC) in 2001, he manages extramural basic research programs in computational electromagnetics, microwave and millimeter-wave circuit integration, compact and multifunctional antenna design, low-power communications systems, and power electronics. Dr. Palmer is a Professional Engineer registered in North Carolina and a member of the IEEE, USNC/URSI Commission C and Commission D, the American Vacuum Society, the Materials Research Society, the Acoustical Society of America, and Sigma Xi. He serves as the 2006-2008 Chair of the USNC/URSI Commission C. Within the IEEE, he participates in the Antennas and Propagation; Components, Packaging, and Manufacturing Technology; Electron Devices; Microwave Theory and Techniques; Power Electronics; and Professional Communications societies, served on the Vacuum Devices Technical Committee from 1997 to 2003, and served as Guest Editor for the T-ED Special Issue on Vacuum Electronics (January 2001) and the T-MTT Special Issue on Multifunctional RF Systems (March 2005). He currently serves as Vice-Chair for the Eastern North Carolina Section in Region 3, and is a founding member and past Chair of the ACME (AP/CPMT/MTT/ED) local chapter.

Dimitrios Peroulis received the Diploma degree in Electrical and Computer Engineering from the National Technical University of Athens, Athens, Greece, in 1998, and the M.S.E. and Ph.D. degrees in electrical engineering from The University of Michigan, Ann Arbor, Michigan, in 1999 and 2003, respectively. Since August 2003, he has been an Assistant Professor at the School of Electrical and Computer Engineering, Purdue University, West Lafayette, Indiana. His current research work is focused on micro-electromechanical systems (MEMS) for multifunctional communications systems and sensors. Dr. Peroulis has received several teaching awards from Purdue University including the Teaching for Tomorrow Award in 2006. He has also been the recipient of three Student Paper Awards at IEEE MTT-S International Microwave Symposium in 2001 and 2002, and the IEEE Antennas and Propagation Society Symposium in 2001.

George N. Saddik received his Bachelor of Science degree in electrical engineering from the California State Polytechnic University, Pomona, in 1998. He earned the master degree in electrical engineering from the University of California Santa Barbara (UCSB) in 2006 and is currently working toward his Ph.D. degrees at UCSB. He was employed by Watkins-Johnson Company in Palo Alto, California, USA, as a Product Engineer working on microwave transceivers, and later as an Application Engineer working on RF and microwave semiconductors for commercial applications. His research interests include design and fabrication of solidly mounted resonator filters in AlN and BST, their integration with GaN-based HEMT technology, and their application to communication systems.

Rahul S. Singh received his M.S. and Ph.D. in electrical engineering from UCLA in 1999 and 2005, respectively, and received his B.Sc. in electrical engineering and mathematics at Southern Methodist University in 1997. He is currently a post-doctoral researcher at the University of California, Santa Barbara working in Professor Brown's research group. He is currently conducting research in RF systems and integration, biomedical ultrasound including both hard tissues (dental) and soft tissues, and terahertz imaging, including biomedical applications. His research interests include piezoelectric materials (perovskites and nitrides), transducer design, analog and digital signal processing, radar applications, and system research.

Michael Steer received his B.E. and Ph.D. in Electrical Engineering from the University of Queensland, Brisbane, Australia, in 1976 and 1983 respectively. Currently he is Lampe Professor of Electrical and Computer Engineering at North Carolina State University. Professor Steer is a Fellow of the Institute of Electrical and Electronic Engineers cited for contributions to the computer aided engineering of non-linear microwave and millimeter-wave circuits. He is active in the Microwave Theory and Techniques Society (MTT-S) . In 1997 he was Secretary of the Society and from 1998 to 2000 was an Elected Member of its Administrative Committee. He was Editor-In-Chief of the IEEE Transactions on Microwave Theory and Techniques from 2003 to 2006. In

1999 and 2000 he was Professor and Director of the Institute of Microwaves and Photonics at the University of Leeds where he held the Chair in Microwave and Millimeterwave Electronics. He has authored more than 400 publications on topics related to nonlinear RF effects; RF behavioral modeling; RF circuit simulation; microwave and millimeter-wave systems; high-speed digital design; and RF/microwave design methodology. He is an expert on circuit-field interactions. He is coauthor of the book *Foundations of Interconnect and Microstrip Design*, John Wiley, 2000. He is a 1987 Presidential Young Investigator (USA) and in 1994, and again in 1996, he was awarded the *Bronze Medallion* by U.S. Army Research for "Outstanding Scientific Accomplishment." He received the Alcoa Foundation Distinguished Research Award from North Carolina State University in 2003.

Alan Victor received his Bachelor of Science in Electrical Engineering degree from the University of Florida and Master of Science in Engineering degree from Florida Atlantic University in 1971 and 1981 respectively. He is currently a senior scientist with Harris Stratex Networks, formerly Harris Microwave, in research Triangle Park, North Carolina. At Harris Stratex Networks he is involved with millimeter wave transceiver design. Prior to joining Harris, Mr. Victor was with Motorola Communications, IBM Microelectronics, and co-founded a wireless LAN manufacturing company providing data communication products to the auto ID industry. He is currently pursuing a Ph.D. at North Carolina State University. His main interests are the design of power oscillators and the application of ferroelectric materials. He is a senior member of the IEEE and Eta Kappa Nu.

Robert York is professor of electrical engineering at UC Santa Barbara. A native of New Hampshire, Bob received his B.S. degree in electrical engineering from the University of New Hampshire in 1987, and the M.S. and Ph.D. degrees in electrical engineering at Cornell University in 1989 and 1991, respectively. He joined the faculty at UCSB in 1991, and has led an active research group in RF and wireless electronics. In 2002 Dr. York founded AgileRF Inc. in Goleta, a privately-held company that develops and produces adaptive antennas and electronics for the wireless industry. Dr. York received an IEEE Outstanding Young Engineer Award in 2004, and an Office of Naval Research Young Investigator award in 1996. He has been voted the Outstanding Faculty Member in electrical engineering by the graduating senior class, most recently in 2007.

Index

C

D

E

F

G

M

P

Q

R

S

T

U

V

W

X

Y

Z

Glossary

A	area, 168
A	effective area, 75
A	switch area, 39
α	isolation factor, 146
α_1	inverse of the zero-bias permittivity, 168
α_3	nonlinearity, 168
ACPR	Adjacent Channel Power Ratio, 214
ADC	Analog-to-Digital Converter, 2
AM	amplitude modulation, 361
APDF	Amplitude Probability Density Function, 28
AWGN	additive white Gaussian noise, 401
B	bandwidth, 22
B	noise equivalent RF bandwidth, 404
BAW	Bulk-Acoustic-Wave device, 188
BER	Bit Error Rate, 1, 365, 392
BPF	ideal bandpass filter, 12
BPFs	BandPass Filters, 2
BST	barium strontium titanate, 159
BTO	barium-titanate, 159
BVDSS	drain-source breakdown voltage, 220
C	capacitance, 101
C	constant, 406
C_{v0}	geometric average of the maximum and minimum varactor capacitance, 198
C_b	bulk film capacitance, 171
C_f	fringing capacitance, 179
C_i	fixed capacitance, 171
$C_{\max}$	maximum capacitance, 167
$C_{\min}$	minimum capacitance, 167
c_d	capacitance density, 181
CDR	Data Recovery Circuit, 393
CMRR	Common Mode Rejection Ratio, 259
CPS	coplanar-stripline, 196
CT	Cascaded Trisection, a filter element, 337
δ	deflection, 83
$\pi/4$-DQPSK	$\pi/4$-differential quadrature phase-shift keying, 389
Δf	fraction bandwidth, 389
$\Delta \nu$	RF instantaneous bandwidth, 405
ΔT	temperature difference, 82
D	flux density, 168

d	capacitor thickness, 168
d_0	nominal material thickness, 171
DBF	digital beamforming receiver, 356
DFT	Discrete Fourier Transform, 31
DOA	Direction of Arrival, 374
DR	Dynamic Range, 14
DS – CDMA	direct sequence-CDMA, 224
DSP	Digital Signal Processing, 2
DUT	Device-Under-Test, 182
DVM	digital voltmeter, 400
E	Young's modulus, 75
ε_∞	high-frequency asymptote, 175
F	noise figure, 404
$F(\theta s \phi)$	element beam pattern, 404
f	frequency, 101
f_0	center frequency, 240
f_r	commensurate frequency, 243
f_c	center frequency, 394
f_0	reference frequency, 175
f_R	resonant frequency, 147
f_{COMMS}	communication waveform, 394
f_{RADAR}	radar wavefrom, 394
f_{BRAGG}	Bragg frequency, 195
FFLB	Fixed-Fixed Lower Beam, 90
FFT	Fast Fourier Transform, 31
FOM	Figure-Of-Merit, 195
FSK	Frequency Shift Keying, 28
FSLB	Folded-Suspension Lower Beam, 90
G_e	element gain, 404
G_{LINK}	free-space link gain, 404
g_0	the gap between switch and actuation electrode, 39
g_E	electronic gain, 409
g_{nm}	free-space coupling gain between the n th Tx and m th Rx elements, 408
Γ	reflection coefficient, 184
G	conductance, 175
G_p	periphery dependent conductance, 181
G_{DC}	leakage term, 175
G_R	gain of receive antenna, 392
G_T	gain of transmit antenna, 392
GaN	gallium nitride, 205
h	hour, 91
h_{COMMS}	impulse response of communications matched filter, 394
I	beam moment of inertia, 75

IFF	Identification Friend or Foe, 353
IIP3	input-referred third-order intercept point, 18
IL	Insertion Loss, 317
ISI	intersymbol interference, 391
J_c	critical current density, 190
K	inverter, 261
$K(S)$	characteristic polynomial, 276
K_z	equivalent spring constant, 39
k	modulation rate, 394
K	characteristic function (of a filter), 317
κ	Permittivity or dielectric constant, 159
KNO	potassium niobate, 163
KTO	potassium tantalate, 164
L	movable beam length, 75
L	length of BST parallel-plate varactor, 166
LFM	Linear-Frequency-Modulated, 388
$L_{\mathrm{FREESPACE}}$	free space link loss, 392
LHCP	Left-Hand Circularly Polarized, 391
LNA	ow Noise Amplifier, 26
LRM	Line-Reflect-Match calibration technique, 183
LSFR	linear feedback shift register, 233
MARRS	A Multi-functional, Adaptive Radio Radar and Sensor systems, xiii
MARRS	Multi-functional, Adaptive Radio Radar and Sensor systems, xiii
MIM	Metal-Insulator-Metal capacitor, 151
m_{MIN}	minimum number of passes required for detection, 404
MOCVD	Metal-Organic Chemical Vapor Deposition, 161
MURI	Multidisciplinary University Research Initiative funded by the U.S. Department of Defense, xvi
F	noise factor, 21
N_e	number of elements, 404
N_i	input noise power, 21
N_o	output noise power, 21
N_{Tn}	noise transmitted by the n th transmitter element, 409
n	capacitance nonlinearity exponent, 175
NCSU	North Carolina State University, 352
NF	noise figure, 20
Ni	nickel, 39
NRZ	Non-Return to Zero, 393
ω_r	self-resonant frequency, 189
OIP3	output-referred third-order intercept point, 18
P	axial tensile load, 50

P	device periphery, 181
P_d	probability of detection, 387
P_T	power transmitted, 392
P_{fa}	probability of false alarm, 387
PAE	Power Added Efficiency, 137, 214
PAR	Peak-to-Average power Ratio, 28
PBR	PassBand Ripple, 318
PCB	Printed Circuit Board, 342
PCL	Parallel Coupled Line, 238
PD	Polynomial Predistorter, 226
PDE	partial differential equation, 180
PECVD	Plasma Enhanced Chemical Vapor Deposition, 39
PLL	Phase-Locked Loop, 358
PN	pseudorandom noise, 224
PRF	Pulse Repetition Frequency, 392
Pt	platinum, 169
PTO	lead titanate, 163
PZT	lead zirconium titanate, 163
Q_0	zero-bias asymptotic Q, 187
Q_f	film Q, 177
QAM	Quadrature Amplitude Modulation, 28
R_0	range to the target, 408
R_{nt}	range factors between the n th Tx element and the target, 408
R_{tm}	range factors between the target and the m th Rx element, 408
r_r	sheet resistance of the top electrode, 178
R_{MATERIAL}	contact resistance term, 178
R_{MAX}	maximum range, 392
r	radius, 190
r_c	normalized contact resistance term for a BST varactor, 178
r_b	sheet resistance of the bottom electrode, 178
RCS	Radar Cross Section, 355, 387, 392
RDA	Retrodirective Array, 353
$R_{\text{ELECTRODE}}$	BST varactor series resistance, 178
RFIC	Radio Frequency Integrated Circuit, 242
RFID	Radio Frequency Identification, 353
RHCP	Right-Hand Circularly Polarized, 391
RM method	retrodirective matrix, 410
RNC	retrodirective noise correlating, 398
R_s	series resistance, 186
Rx	receiver, 399
σ	radar cross section, 392
σ_0	constant mean stress, 46
σ_1	gradient stress, 46
σ_{bistatic}	bistatic radar cross section, 355
σ'	effective conductivity, 181

$\sigma_{\text{monostatic}}$	monostatic radar cross section, 355
S	Richards variable, 240
SAW	Surface Acoustic Wave, 388
SFDR	Spurious Free Dynamic Range, 2
SIR	Signal-to-Interference Ratio, 1
SNR_{MIN}	minimum signal to noise ratio required for a communication or radar system to detect a signal, 17
SNR	signal to noise ratio, 17
SNR_{MIN}	minimally acceptable SNR, 404
SNR_{OUT}	output signal-to-noise ratio, 406
SNR_{POST}	post-detection signal-to-noise ratio, 405
SNR_{PRE}	pre-detection signal-to-noise ratio, 405
SOL	Short-Open-Load calibration technique, 183
SOM	Self-Oscillating Mixer, 355
SOS	Silicon-on-Sapphire switch, 219
SPST	Single-Pole, Single-Throw, 393
SSPA	Solid-State Power Amplifier, 400
STO	strontium titanate, 159
SVD	Singular-Value Decomposition, 185
T_N	receiver noise temperature, 405
T_p	pulse width, 394
t_{ACQ}	acquisition time, 405
t_{BW}	time-bandwidth product, 394
t_d	detection time, 404
t_g	total electronic (antenna to antenna) group delay, 405
τ	integration time, 405
T	transmission coefficient, 317
t_c	top interconnect thickness, 190
τ	tunability factor, 167
T_c	Curie temperature, 159
$\tan\delta$	loss tangent, 174
TPR	Transducer Power Ratio, 315
TSD	Through-Short-Delay calibration technique, 183
TWT	Traveling Wave Tube, 62
Tx	transmitter, 398
UE	Unit Element, 240
UWB	Ultra Wideband, 389
V_2	voltage at which BST varactor capacitance is half of its zero-bias value, 169
V_q	fitting parameter, 187
VSA	Vector Signal Analyzer, 226
VSWR	Voltage Standing Wave Ratio, 217
W	transverse concentrated load, 50
w	half the gap width between the Tx and Rx array, 408
W	width of BST parallel-plate varactor, 166

x	loading factor, 194
XOR	exclusive or logic operation, 233
y_{COMMS}	output communications matched filter signal, 394
y_{radar}	output radar matched filter signal, 394
Z_d	equivalent impedance of the diode, 147